STRATEGIES AND TACTICS
IN ORGANIC SYNTHESIS
Volume 6

STRATEGIES AND TACTICS IN ORGANIC SYNTHESIS
Volume 6

Edited by

Michael Harmata
University of Missouri – Columbia

With a foreword by

Professor Paul A. Wender
Stanford University

2005

ELSEVIER

Amsterdam • Boston • Heidelberg • London • New York • Oxford • Paris
San Diego • San Francisco • Singapore • Sydney • Tokyo

ELSEVIER B.V.
Radarweg 29
P.O. Box 211, 1000 AE
Amsterdam, The Netherlands

ELSEVIER Inc.
525 B Street
Suite 1900, San Diego
CA 92101-4495, USA

ELSEVIER Ltd
The Boulevard
Langford Lane, Kidlington,
Oxford OX5 1GB, UK

**ELSEVIER Ltd
84 Theobalds Road
London WC1X 8RR
UK**

First edition 2005

Transferred to digital print 2007

Library of Congress Cataloging in Publication Data
A catalog record is available from the Library of Congress.

British Library Cataloguing in Publication Data
A catalogue record is available from the British Library.

ISBN: 0-12-450289-X (Paperback)
ISBN: 0-12-450288-1 (Hardbound)

∞ The paper used in this publication meets the requirements of ANSI/NISO Z39.48-1992 (Permanence of Paper).
Printed and bound by CPI Antony Rowe, Eastbourne.

CONTENTS

14. **DIASTEREOSELECTIVE
INTRAMOLECULAR 4+3 CYCLOADDITION
AND AN ENANTIOSELECTIVE TOTAL
SYNTHESIS OF (+)-DACTYLOL**   **437**
Paitoon Rashatasakhon and Michael Harmata

CONTRIBUTORS

MERRITT B. ANDRUS, *Department of Chemistry and Biochemistry, Brigham Young University, Provo, Utah 84602*

PAUL R. BLAKEMORE, *Department of Chemistry, Oregon State University, Corvallis, Oregon 97331*

F. CHARLES CAMPBELL, *The Department of Surgery, Queen's University of Belfast, Grosvenor Road, Belfast, Northern Ireland BT12 6BJ, United Kingdom*

MARCO A. CIUFOLINI, *Laboratoire de Synthèse et Méthodologie Organiques (LSMO), CNRS UMR 5622, Université Claude Bernard Lyon 1 and Ecole Supérieure de Chimie, Physique, Electronique de Lyon 43, Bd. du II Novembre 1918, 69622 Villeurbanne, France*

SCOTT E. DENMARK, *Department of Chemistry, University of Illinois, at Urbana-Champaign, Urbana, Illinois 61801*

MATHIAS M. DOMOSTOJ, *The UCL Centre for Chemical Genomics, The Christopher Ingold Laboratories, The Department of Chemistry, University College London, 20 Gordon Street, London WC1H0AJ, United Kingdom*

MOHAMED EL-TANANI, *The Department of Surgery, Queen's University of Belfast, Grosvenor Road, Belfast, Northern Ireland BT12 6BJ, United Kingdom*

ELLEN FEUSTER, *Department of Chemistry, University of Missouri, Columbia, Missouri 6521J and Department of Chemistry, Pennsylvania State University, University Park, Pennsylvania 16802*

KRISTEN N. FLEMING, *Department of Chemistry and Biochemistry and The Walther Cancer Research Center, 251 Nieuwland Science Hall, University of Notre Dame, Notre Dame, IN 46556*

KELLY M. GEORGE, *Department of Chemistry, University of Pennsylvania, Philadelphia, Pennsylvania 19104*

TIMOTHY GLASS, *Department of Chemistry, University of Missouri, Columbia, Missouri 6521J and Department of Chemistry, Pennsylvania State University, University Park, Pennsylvania 16802*

KARL J. HALE, *The UCL Centre for Chemical Genomics, The Christopher Ingold Laboratories, The Department of Chemistry, University College London, 20 Gordon Street, London WC1H0AJ, United Kingdom*

JOHN HANLEY, *Department of Chemistry, University of Missouri, Columbia, Missouri 6521J and Department of Chemistry, Pennsylvania State University, University Park, Pennsylvania 16802*

MICHAEL HARMATA, *Department of Chemistry, University of Missouri-Columbia, Columbia, Missouri 65211*

BRIAN R. HEARN, *Department of Chemistry and Biochemistry and The Walther Cancer Research Center, 251 Nieuwland Science Hall, University of Notre Dame, Notre Dame, IN 46556*

ERIK J. HICKEN, *Department of Chemistry and Biochemistry, Brigham Young University, Provo, Utah 84602*

LYLE ISAACS, *Department of Chemistry and Biochemistry, University of Maryland, College Park, Maryland 20742*

JASON LAGONA, *Department of Chemistry and Biochemistry, University of Maryland, College Park, Maryland 20742*

CHARLENE K. MASON, *The Department of Surgery, Queen's University of Belfast, Grosvenor Road, Belfast, Northern Ireland BT12 6BJ, United Kingdom*

ERIK L. MEREDITH, *Department of Chemistry and Biochemistry, Brigham Young University, Provo, Utah 84602*

STUART J. MICKEL, *Chemical and Analytical Development, Novartis Pharma AG, CH4002 Basel, Switzerland*

GARY A. MOLANDER, *Department of Chemistry, University of Pennsylvania, Philadelphia, Pennsylvania 19104*

RICARDO MORAN, *Department of Chemistry, University of Missouri, Columbia, Missouri 6521J and Department of Chemistry, Pennsylvania State University, University Park, Pennsylvania 16802*

CHISATO MUKAI, *Division of Pharmaceutical Sciences, Graduate School of Natural Science and Technology, Kanazawa University, Kakuma-machi, Kanazawa 920-1192, Japan*

TODD D. NELSON, *Merck Research Laboratories Inc., Department of Process Research, Merck & Co., Wayne, Pennsylvania 19087*

JOSEPH RAKER, *Department of Chemistry, University of Missouri, Columbia, Missouri 6521J and Department of Chemistry, Pennsylvania State University, University Park, Pennsylvania 16802*

PAITOON RASHATASAKHON, *Department of Chemistry, University of Missouri-Columbia, Columbia, Missouri 65211*

TAREK SAMMAKIA, *Department of Chemistry, University of Colorado, Boulder, Colorado 80309*

KRISTEN SECOR, *Department of Chemistry, University of Missouri, Columbia, Missouri 6521J and Department of Chemistry, Pennsylvania State University, University Park, Pennsylvania 16802*

ERIC L. STANGELAND, *Department of Chemistry, University of Colorado, Boulder, Colorado 80309*

RICHARD E. TAYLOR, *Department of Chemistry and Biochemistry and The Walther Cancer Research Center, 251 Nieuwland Science Hall, University of Notre Dame, Notre Dame, IN 46556*

MARK C. WHITCOMB, *Department of Chemistry, University of Colorado, Boulder, Colorado 80309*

JAMES D. WHITE, *Department of Chemistry, Oregon State University, Corvallis, Oregon 97331*

SHYH-MING YANG, *Department of Chemistry, University of Illinois, at Urbana-Champaign, Urbana, Illinois 61801*

Foreword

Synthesis is evolving in many dramatic ways. Since its genesis in the seminal work of Wöhler, whose one-step synthesis of urea marked a transition from the 19th century belief that natural products could not be made in the laboratory, the field has successfully addressed a dazzling array of natural and non-natural compounds, profoundly impacting chemistry, medicine, biology, materials, and most disciplines whose future is inexorably coupled to understanding processes at the molecular level. As we enter the 21st century with approximately one total synthesis being reported per day, it is clear that we can make many types of molecules. The challenges now are increasingly related to which molecules merit attention and whether those molecules could be made in a practical fashion. These challenges and the current state of the field are brilliantly addressed by Mike Harmata and the superb group of contributors to this outstanding volume of *Strategies and Tactics in Organic Synthesis*.

This volume offers the *unique* perspective on the evolution of synthesis that only those involved directly in the science can give: the vision, ideas, plans, insights, setbacks, troubleshooting, discoveries and, with hard work and creativity, the successes that attend the process. From the dramatic start of the first chapter – "The story you are about to read began to unfold in 1987" – to the always hoped for ending – "...the total synthesis of (+)-dactylol was successfully executed", this collection of organic syntheses is rich with information, insights, and inspiration that only those involved with the work can present. The student of the field is exposed to the many factors that motivate an interest in synthesis from targets of profound medical potential or value to materials and molecules of functional interest on to opportunities to advance methods, reactions, strategies, and our fundamental understanding of structure and mechanism.

The range of synthetic problems in this volume is also impressive; including alkaloids, heterocycles, macrolides, terpenoids, polyketides, designed targets, sensors and many other systems that defy easy description. Infused in these analyses is the preeminent importance of step economy as it is becoming increasingly clear that the length of a synthesis influences if not determines most other economies and measures. This is clearly addressed by several contributors and perhaps

most poignantly in a contribution from the process sector, another bonus
of this volume, where in one accounting an estimated 33,000 kilograms
of solvent are used to produce 60 grams of final product. One often
overlooks this measure of atom loss and its direct connection to synthesis
length.

Strategies and Tactics in Organic Synthesis originated over two decades
ago from the desire to capture the creative process of synthesis and the
excitement of the field from the unique, behind-the-scenes perspectives
of those innovators involved in its advancement. This richness of insight
and information is generally neither available in the primary literature
due to space constraints nor in secondary sources due mostly to the lack
of input from those who actually designed and executed the work. Mike
Harmata and his contributing authors use the powerful impact of this
format to advantage in chronicling their beautiful and varied
contributions to synthesis. The reader will be richly rewarded with
exciting science, analyses, and even art. This is at once a great scholarly
analysis of this exciting field and an inspiration for its future.

Paul A. Wender
Stanford University, CA
March 2005

Preface

Proofreading and editing are time-consuming and it is not easy to do things both quickly and well at the same time. I hope I have done better than in the past to avoid typographical errors.

I want to thank all the contributors for their efforts and patience. It was a pleasure to be exposed to so much great science and art.

Thanks to the folks at Elsevier for their support of this series.

I am at the end of my contractual obligations with respect to this series. It is not clear at this writing that more volumes will appear. I hope they do and I stand ready to edit them, but market forces will determine whether any more of this series appear. To be not at all subtle: Buy the books: for yourself, friends and loved ones!

Finally and quite importantly, I have an active research program and without it, there would not likely be time to invest in "synergistic activities" like book editing. My thanks go out to the National Institutes of Health, the National Science Foundation and the Petroleum Research Fund for their continuing support of our work.

Michael Harmata

Dedication

This volume is dedicated to my wonderful wife, Judy L. Snyder.

Chapter 1

THE TOTAL SYNTHESIS OF LUZOPEPTINS

Marco A. Ciufolini
Laboratoire de Synthèse et Méthodologie Organiques (LSMO), CNRS UMR 5622
Université Claude Bernard Lyon 1 and
Ecole Supérieure de Chimie, Physique, Electronique de Lyon
43, Bd. du 11 Novembre 1918, 69622 Villeurbanne, France

I. Introduction

The story you are about to read began to unfold in 1987. At that time, the AIDS crisis was rampant. The viral nature of the disease had been established only 5 years earlier, and predictions regarding the spread of the virus were dire. None of the extraordinary drugs available today to combat the devastating effects of the AIDS virus (HIV) existed: AZT (3'-deoxy-3'-azidothymidine, **1**) was essentially the only treatment available. AZT is a nucleoside analogue that functions as an inhibitor of reverse transcriptase (RT), a unique RNA-templated DNA polymerase found only in retroviruses such as HIV. RT is essential for retroviral infection and replication; therefore, it is a major target in AIDS therapy. However, AZT and congeners tend to inhibit also other cellular DNA polymerases. This causes a variety of unfortunate side effects.

In the midst of the societal and scientific maelstrom created by the AIDS epidemic, a noteworthy paper by Inouye and collaborators appeared in the January 1987 issue of the *Journal of Antibiotics*.[1] Thus, three natural products known as luzopeptins A, **2**, B, **3**, and C, **4**, were found to possess significant inhibitory activity against RT from the avian myeloblastosis virus (AMV).

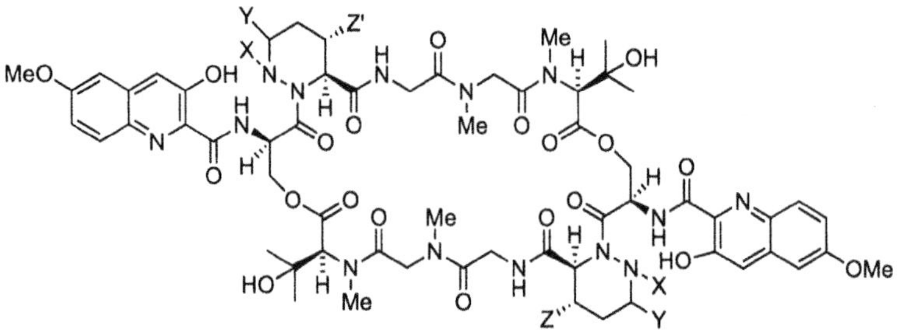

2	Luzopeptin A :	X, Y = π bond; Z = Z' = OAc
3	Luzopeptin B :	X, Y = π bond; Z = H, Z' = OAc
4	Luzopeptin C :	X, Y = π bond; Z = Z' = H
5	Luzopeptin E2:	X = Y = Z = H

SCHEME 1

This noteworthy observation raised a number of intriguing questions. Like its deadly cousin HIV, AMV is a retrovirus, and RT's from various retroviruses are very similar, if not essentially identical.

Could luzopeptins also be inhibitors of HIV-RT?

Second, virtually all inhibitors of RT (or of other DNA polymerases) known at that time were nucleoside analogues such as AZT. Not so luzopeptins. The mechanism of action of nucleoside analogues is well understood.

But what is the mechanism of RT inhibition by a peptide agent?[2]

Third, and interestingly, anti-RT activity was especially pronounced in luzopeptin C, **4**, which happens to be the least cytotoxic member of the family.

Could luzopeptin C inhibit retroviral replication at non-cytotoxic doses?

Answers to the above were not late coming: in a remarkable follow-up paper, Inouye confirmed the activity of luzopeptin C against HIV-RT, and disclosed that, indeed, **4** suppresses HIV replication in infected MT-4 cells at non-cytopathic concentrations.[3] Luzopeptin C must thus be selective for RT *vis-à-vis* human DNA polymerases!

These results had major ramifications in the AIDS field, but they also left a number of other fundamental questions unanswered: what is the mechanism of action of luzopeptins? What are the minimal structural requisites for anti-HIV activity? Could **2-4** serve as anti-HIV drugs, and if so, how could they be modified to further enhance potency and selectivity? Since their isolation, luzopeptins have attracted considerable interest in the biomedical[4,5] arena, but their rarity and chemical sensitivity have precluded any sort of chemical modification that may be required for a thorough medicinal chemistry study. A viable synthesis was needed to address these issues. Thus commenced an epic that spanned nearly ten years and two continents: the total synthesis of luzopeptins.

II. Background

Luzopeptins are cytotoxic peptides produced by *Actinomadura luzonensis*, and were discovered at Bristol-Meyers Japan in the early 1980's during research on new potential anticancer agents.[6] The luzopeptin complex consists of three major (luzopeptins A, B and C, **2-4**) and three minor (luzopeptins D, E and F) components. The structure of **2-4**[7] was confirmed by X-ray crystallography.[8] The structures of luzopeptins D and F remain unknown.[9] The E series of antibiotics is a mixture of compounds, among which only luzopeptin E2 (**5**) has a secure structure (Scheme 1).[10] Together with quinoxapeptins,[11] sandramycin,[12] and related natural products,[13] luzopeptins form what may be referred to as the *peptin* family of natural products. Peptins display a dimeric peptide scaffold of C_2-like symmetry that supports a pair of heteroaroyl segments

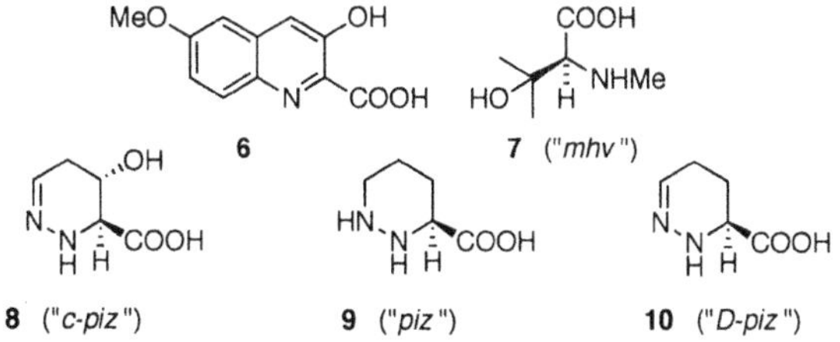

8 ("*c-piz*")   9 ("*piz*")   10 ("*D-piz*")

SCHEME 2

(quinaldic acid **6** in the case of **2-5**). These appear to be intimately involved in the expression of bioactivity. In addition, **2-4** incorporate two unusual amino acids previously unknown in nature: (L)-*N*-methyl-3-hydroxyvaline ("*mhv*", **7**), and (3*S*,4*S*)-4-hydroxy-2,3,4,5-tetrahydro-pyridazine-3-carboxylic acid (**8**), and differ only for the degree of acetylation of component **8**. Piperazic acid ("*piz*", **9**) replaces **8** in luzopeptin E2 (Scheme 2). For the sake of convenience, we shall refer to **8** as *c-piz* (= piperazic acid-like subunit of luzopeptin C), and to its desoxy analog, **10**, which ultimately played a major role in the effort toward **5**, as *Δ-piz* (dehydropiperazic acid).

The synthesis of luzopeptins is fraught with extreme difficulties.[14] Indeed, these molecules have remained elusive goals for nearly 20 years,[15] until in 1999 Boger announced the first total synthesis of **2-4**[16] and of quinoxapeptins,[17,18] and in 2000 we described the culmination of much background work[19] in the form of the total synthesis of luzopeptin E2,[20] followed shortly thereafter by that of luzopeptin C.[21]

But what is it that makes a synthesis of luzopeptins so difficult?

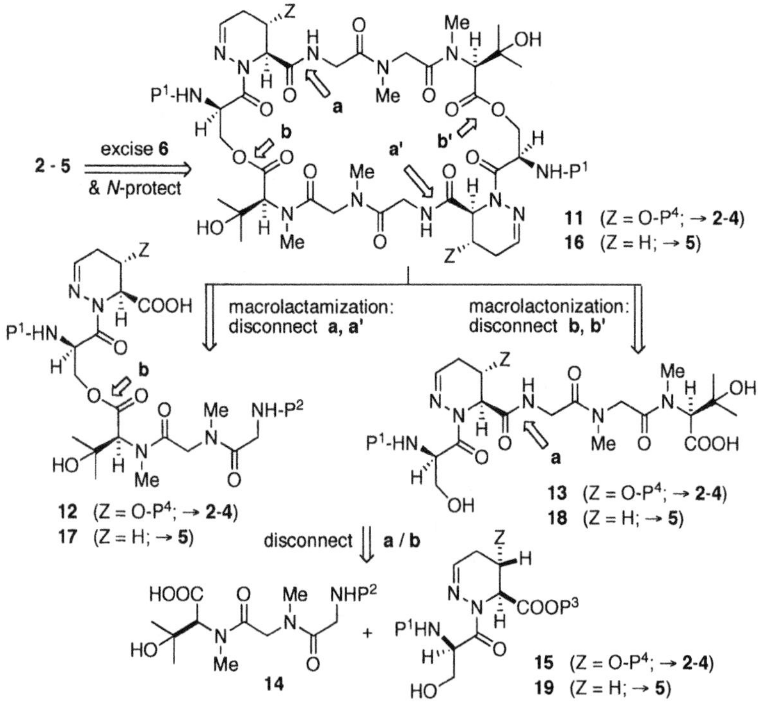

SCHEME 3

Guidelines for the synthesis of cyclopeptides[22] have it that pendant groups should be connected to the cyclic scaffold after ring closure. Moreover, peptide bond formation is generally preferred to ester bond formation as a means to accomplish macrocyclization of depsipeptide frameworks, and it is also well established that primary amino groups react more efficiently than secondary ones in peptide bond forming reactions; i.e., secondary amides are more readily formed than tertiary ones.[23] Retrosynthetic dissection of the only pair of secondary amide bonds in luzopeptin A-C precursor 11 leads to pentadepsipeptide 12 (Scheme 3).

And therein lies the first potential source of problems.

At some point, a linear dimer of 12 must undergo cyclization. However, the efficiency of cyclization of linear peptides is known to be sequence-dependent and sensitive even to minor structural changes in the substrate, including the nature of protecting groups.[24] At the onset of our journey, the literature contained no reports of cyclization of linear depsipeptides incorporating unsaturated piperazic acids; furthermore, the chemistry of those unusual amino acids was not well developed. Without the benefit of precedent, what might happen due to the presence of *two c-piz* subunits in the macrocyclization substrate was anybody's guess.

It is perhaps to minimize the probability of unpleasant surprises at a

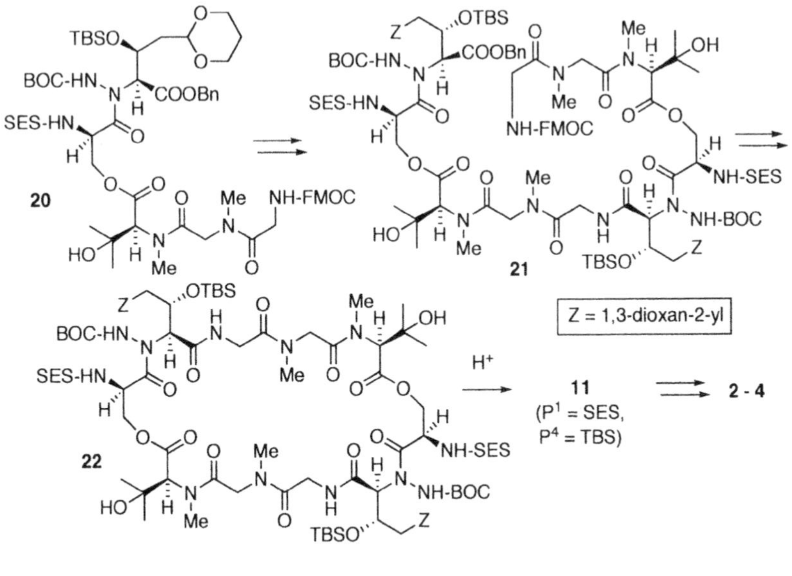

SCHEME 4

late stage of the synthesis that Boger avoided macrocycle formation with *c-piz*-containing substrates, favoring instead a strategically safer cyclization of preformed intermediate **21**. Subsequent acid treatment of the resultant **22** installed the *c-piz* segments (Scheme 4).

By contrast, we focused on the cyclization of *piz*-containing intermediates right from the beginning. We felt that such a strategy would require ancillary investigations on the chemistry of piperazic acids: it seemed likely that this could lead to interesting new findings. At the same time, it did not seem prudent to exclude alternative cyclization strategies at this early stage of our work, including *macrolactonization*, due to the uncertain behavior of our projected depsipeptide substrate. Fortunately, release of the depsi bonds in macrolactonization substrate **11** (Scheme 3) produces intermediate **13**, whereas further disconnection of either **12** or **13** leads to an identical pair of precursors: dipeptide **14** and tripeptide **15**. The same pair of building blocks would thus enable exploration of both cyclization sequences. On the other hand, a "hybrid" cyclization technique involving sequential formation of ester and amide bonds (e.g, **a**, **b'**, **a'**, **b**, cf. **11**, Scheme 3) would be considerably less convergent, and was never seriously considered.

Regardless of which macrocyclization method would ultimately emerge as the winner, the COOH group of the *mhv* residue in **14** had to be activated at some point, in order to effect ester bond formation. But the α-amino group of *mhv* in **14** is part of a *tertiary amide function*: two forces thus create significant potential for major annoyances at this juncture.

First, derivatives of α-amino acids in which the amino group is part of an *amide* are considerably more prone to α-epimerization/racemization upon COOH activation than their *carbamate* congeners. Loss of stereochemical integrity is attributed to cyclization of a generic activated species, **23**, to oxazolone **24**, which exists in prototropic equilibrium with the aromatic tautomer **25** (Scheme 5). An amide being more nucleophilic than a carbamate, cyclization to an intermediate of type **24** is easier. Second, the tendency to cyclize is even more pronounced if the α-amido

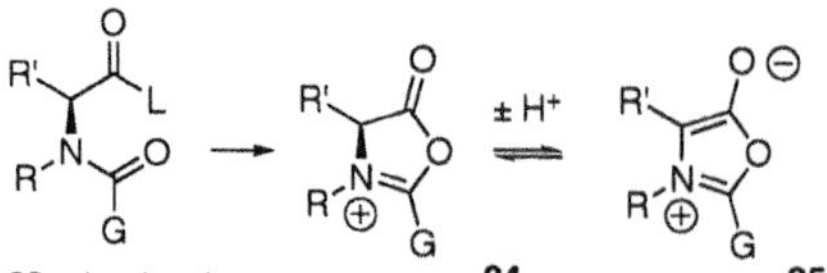

nature of R, G	ease of cylization
R = H	less facile
R = alkyl	more facile
G = OAlkyl	less facile
G = Alkyl	more facile

23   L = leaving grp.   **24**   **25**

SCHEME 5

group is tertiary, as it is in **14**. Conclusion: stereochemical difficulties were likely to materialize upon activation of **14**

The above notwithstanding, it ultimately transpired that the battle for luzopeptins would actually be won or lost at the stage of fragment **15**, the synthesis of which turned out to be a major challenge. Unexpectedly, the acylation of what may be termed the "inner" nitrogen atom of an unsaturated *piz* structure, such as **8** or **10**, proved to be impossible. Circumventing the obstacles presented by this seemingly trivial operation required a major dose of experimentation and, in the end, the development of new serinylation technology.

A bane we had to confront repeatedly during our synthetic journey was the proclivity of derivatives of serine and of **8** to suffer β-elimination of the OH functionality, free or protected, during steps requiring the use of basic reagents. Because of this, our final sprint toward the natural products centered initially on luzopeptin E2, **5**. The Δ-*piz* component present in precursors to **5** is not liable to that troublesome side reaction, facilitating the search for solutions to the other problems described above. Our synthetic plan for **5** parallels that outlined for **2-4**. It was already known that the imino linkage of Δ-*piz* may be reduced with acidic NaBH$_3$CN.[25,19b] One may thus envision that **5** could be obtained from **16**, via **18** and **19**.

Right from the start, we were conscious to have settled on the riskier approach to the target molecules, but we felt that there was much to be learned from the chosen path. The obstacles that we were likely to encounter are the lifeblood of academic science: they reveal gaps in our knowledge, unravel weaknesses of current synthetic technology, and provide incentives to devise innovative solutions. Of course, that also meant that mayhem could break out at any stage of our research.

We just hoped we would be resourceful (lucky?) enough to parry any blows that luzopeptins may hurl our way.

III. Synthesis of Quinaldic Acid and *gly-sar-mhv* Tripeptide Fragments

The preparation of quinaldic acid **6** presented no significant difficulties. The compound was made from quinoline **29**, which in turn is readily available through the Tennant quinoline synthesis. This noteworthy reaction provides a straightforward avenue to 3-hydroxyquinolines such as **29** via NaOH treatment of the product of *C*-acylation of acetylacetone with 2-nitrobenzoyl chlorides.[26] It is likely that such an intermediates proceeds to the ultimate quinoline *via* a Smiles-type rearrangement of the corresponding dianion (cf. **27** → **28**) A variant of this transformation developed by us involves reaction of a 2-nitrobenzoyl chloride with acetylacetone (1 equiv.) and NaH (5 equiv.). The yield of **29** from **26** is thus 60% (Scheme 6). Protection of the phenol, haloform reaction,[27] and final deprotection afforded **6**.

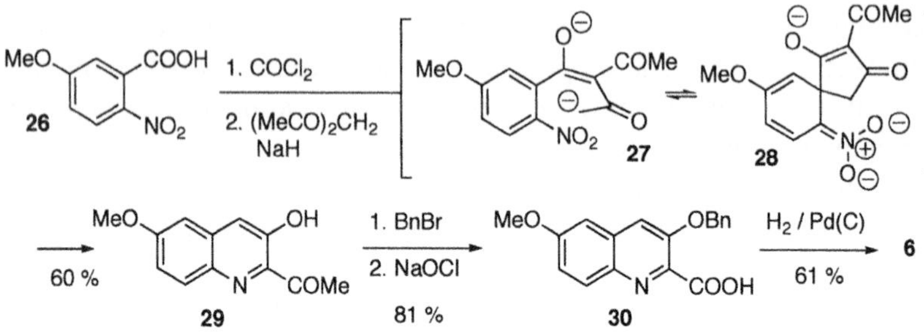

SCHEME 6

A particularly efficient avenue to tripeptide **14** involves coupling of a preformed glycine-sarcosine dipeptide to an ester of *mhv*. This minimizes the number of chemical operations to be effected on synthetic intermediates. A Rapoport "serine inversion"[28] strategy seemed to be well-suited for the creation of *mhv*.[19a] Accordingly, reaction of methyl (D)-serinate derivative **31** with MeMgBr, followed by treatment with NaH / MeI led to **33**. Vigorous base hydrolysis and *N*-BOC protection afforded **34**. Direct oxidation of small quantities of the corresponding alcohol **35** to acid **37** was possible, in moderate yield, by a Garner-type oxidation (basic KMnO$_4$).[29] Unfortunately, this technique soon proved to be unsuitable for large-scale work. A more efficient avenue[30] proceeds

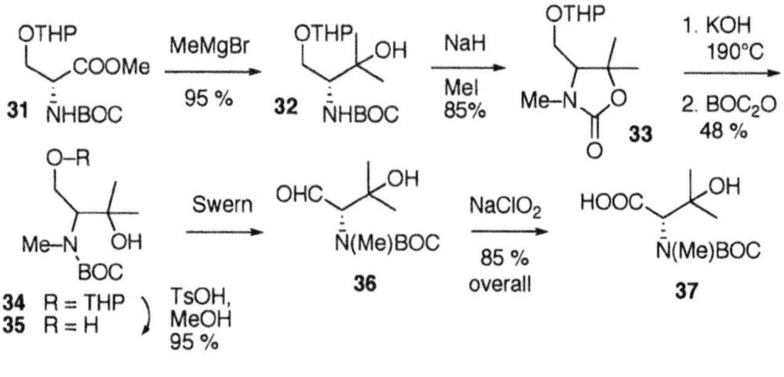

SCHEME 7

through Swern oxidation[31] of **35** and treatment of the resultant, sensitive aldehyde[32] **36** with $NaClO_2$[33] (Scheme 7). It is worthy of note that no chromatography is required in this sequence, which, however, worked well only with BOC-protected substrates. Thus, *N*-CBZ, *N*-FMOC and *N*-TROC analogues tended to undergo β-elimination of the OH group upon oxidation to **36**, probably due to enolization. Curiously, the Sharpless catalytic RuO_4 procedure,[34] which has often produced excellent results in our hands, was unsatisfactory for the direct oxidation of **35** to **37**.[35]

The coupling of an N-protected glycine with a sarcosine ester ran into annoying technical difficulties originating from the poor solubility of many derivatives of glycine and sarcosine in common organic solvents. It was surmised that the use of a glycine equivalent lacking H-bonding donor sites, that is, lacking a secondary amide group, could alleviate these problems, and an especially good solution emerged as follows. Reaction of free sarcosine with chloroacetyl chloride in a biphasic medium composed of aqueous Na_2CO_3 and CH_2Cl_2 afforded acid **38** in 89 % yield (Scheme 8). This material was amenable to silica gel chromatography (100% EtOAc). Subsequent reaction of **38** with NaN_3 afforded *N*-azidoacetyl sarcosine **39** (85% yield), wherein the azido group acts as a surrogate of a primary amino function. The same chemistry did not perform adequately with the methyl ester of sarcosine, because attempted LiOH saponification of the final intermediate, the methyl ester analog of **39**, was plagued by an unexpectedly facile cleavage of the azidoacetamide unit, leading to a mixture of the desired **39**, plus azidoacetic acid and free **37**.

An additional benefit of using an azide in lieu of a protected NH_2 group became evident during the subsequent exploration of the chemistry of *mhv* and of its derivatives. Initial experiments aiming to couple **39** with an *mhv* fragment were carried out with methyl ester **40**. We presumed that the free acid form of tripeptide **14** could ultimately be retrieved by saponification under mild basic conditions. To our dismay, we discovered that esters of *mhv* were stable in acidic environments, but quite sensitive to basic agents. To illustrate, treatment of **40** with TFA

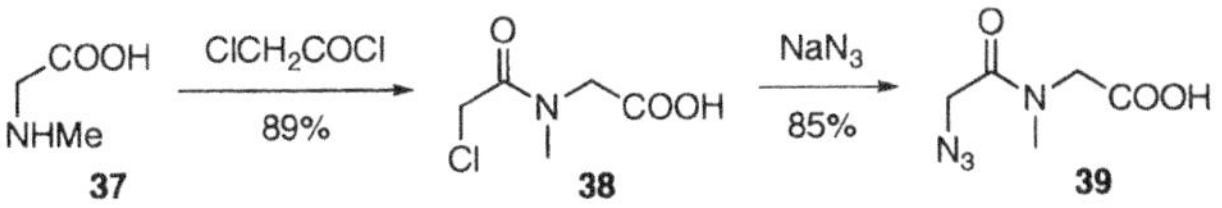

SCHEME 8

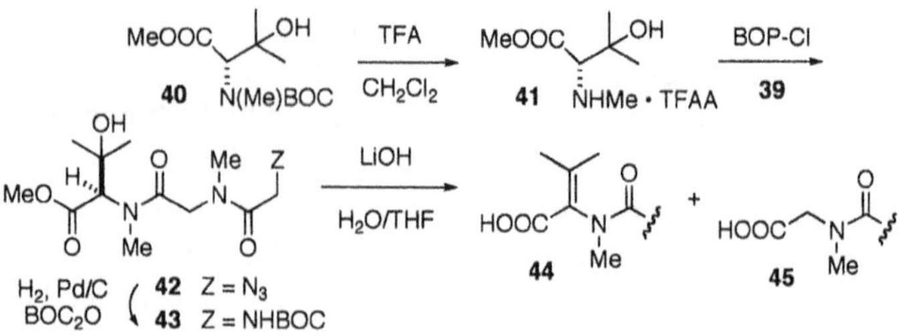

SCHEME 9

yielded **41**, which, without purification, was coupled with **39** (Scheme 9). Not unexpectedly, formation of the tertiary amide was less than straightforward. A good condensing agent for this step proved to be *bis*(2-oxo-3-oxazolidinyl)phosphonic chloride (BOP-Cl). Standard treatment of the resultant **42** with aqueous LiOH / THF resulted in a complex mixture that appeared to contain products of dehydration (cf. **44**, Scheme 9) as well as of retro-aldol loss of acetone (cf. **45**) from the *mhv* unit. Substances lacking the azidoacetyl unit were also detected. A maneuver aiming to suppress at least the loss of the terminal glycine equivalent involved hydrogenolysis of the azide and *in situ* N-protection (Pd/C in the presence of BOC_2O), resulting in formation of peptide **43**. Attempted basic hydrolysis of the methyl ester again inflicted extensive injury to the *mhv* subunit. This, and a number of other observations, indicated that all manipulations involving esters of **7**, *and in particular deprotection steps*, had to avoid basic reagents. The azido unit nicely fulfills this requirement, since it may be converted to an NH_2 group under neutral conditions by catalytic or chemical reduction. An allyl ester represented an ideal form of COOH protection, inasmuch as cleavage may be effected under the catalytic influence of Pd(0). A suitably

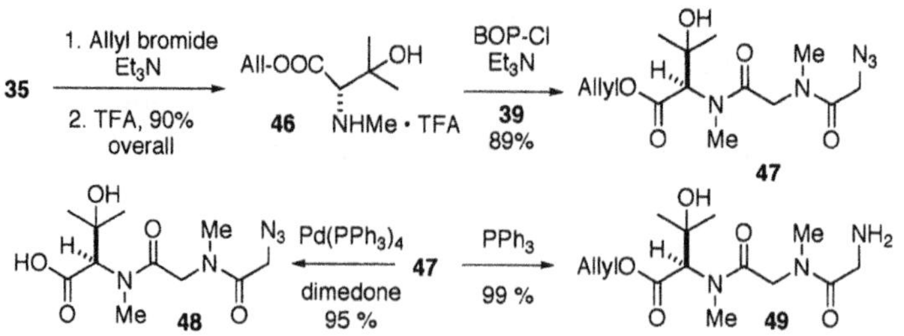

SCHEME 10

blocked version of **14** is thus **47** (Scheme 10). Selective release of COOH and amino termini in **47** was straightforward.[36] Deprotection of the allyl ester was best accomplished using catalytic Pd(PPh$_3$)$_4$ and dimedone as the acceptor. The sensitivity of **47** to basic agents caused the formation of mixtures of products when amines such as pyrrolidine or morpholine were utilized in place of dimedone. Free acid **48** was amenable to a quick, filtration-type chromatography to remove gross amounts of contaminants. Staudinger reduction of the azide,[37,38] gave the very polar amine **49**, which was more difficult to purify because it retained Ph$_3$P=O rather strongly.

Attempts to shorten the sequence leading to **47** were not particularly successful. For instance, introduction of the *N*-methyl substituent in **33** by reduction of the BOC group in **32** would save two steps. But reduction of a BOC group can be difficult, and indeed, treatment of **32** with LiAlH$_4$ in refluxing THF accomplished no more than about 50% conversion after 32 hours. Unreacted **32** was separable from **33** — at the price of an unpleasant large-scale chromatography during process work.

The use of a more readily reducible *N*-blocking group on **31**, e.g., formyl or CBZ, created new problems, because now the corresponding serine methyl esters gave mixtures of products in the Grignard addition. Furthermore, CBZ-protected materials did not perform well in the oxidation sequence (Scheme7), as indicated earlier.

A second possibility that might have eliminated some protection-deprotection steps involved direct formation of tripeptide acids through oxidation of intermediates **50** and **51**. However, the Sharpless catalytic RuO$_4$ oxidation of either was problematic. Attempted Garner oxidation induced rapid destruction of the azido substrate, whereas only diketopiperazine **52** was identifiable in mixtures arising from **51**.

Other options likewise led to little or no improvement. All in all, the procedure of Scheme 10 represents the best avenue to **47**: up to 20 g of

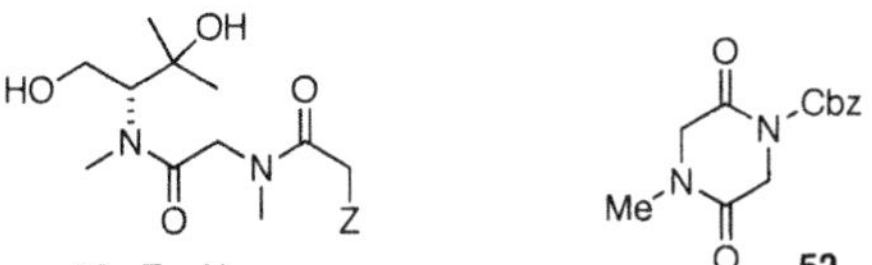

50 Z = N$_3$
51 Z = NHCBZ

SCHEME 11

final product have been obtained in a single run of this sequence.

IV. Synthesis and Chemistry of the Piperazic Acid Fragment

Hughes and Clardy described the first synthesis of **8** in 1989.[39] This important work contains no indication of the chemical properties of the new amino acid, except for a passing comment regarding its instability. It seemed to us that some knowledge of the chemistry of **8** was essential to the success of the luzopeptin effort. In that context, the issue of

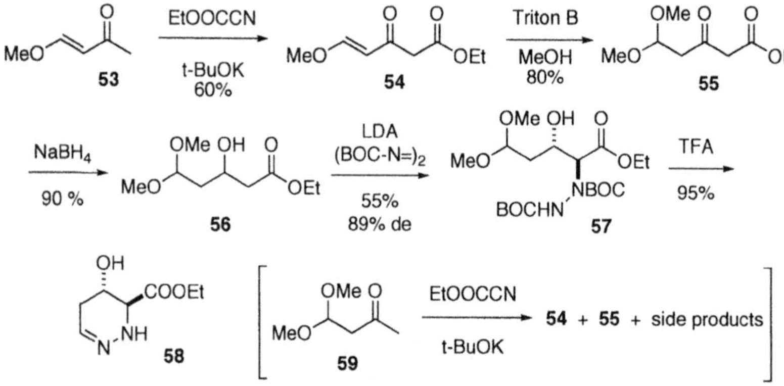

SCHEME 12

enantiopurity was secondary. Therefore, we initially targeted a rapid synthesis of the racemate. The work of Guanti[40] suggested that a Gennari-Evans-Vederas ("GEV") hydrazination[41] of the dianion of β-hydroxy ester **56** would lead to **57**, which should cyclize to *c-piz* ethyl ester **58** under acidic conditions (Scheme 12). The hydroxyester was obtained by selective reduction of β-ketoester **54**, which was prepared[42] by condensation of commercial enone **53**[43] with the Mander reagent,[44] followed by conjugate addition of MeOH (methanolic Triton B).[45] A more direct route to **55** from commercial acetal **59** was less satisfactory. Elimination of MeOH during the condensation step led to a mixture of **55** and **54**, necessitating a subsequent treatment with methanolic Triton B to effect complete conversion to **55**. Furthermore, accumulation of liberated KOMe seemed to have an adverse effect and promoted formation of side products.

The dianion of **56** reacted with di-*tert*-butyl azodicarboxylate ("DBAD") to furnish an 18:1 mixture (500 MHz ^{1}H NMR) of *anti* (major) and *syn* (minor) diastereomers, which were separated by column chromatography and recrystallization. Exposure of **57** to TFA in CH$_2$Cl$_2$ induced rapid and quantitative conversion to **58**, a delicate substance that was difficult to purify, but that, fortunately, emerged in high purity when pure **57**, mp 90-91 °C, was used in this step. An exploration of the reactivity of *c-piz*, and in particular of its acylation chemistry, was now possible.

To our surprise, compound **58** refused to undergo *N*-acylation, decomposing more or less rapidly during all such attempts. Exposure to mild acylating agents, such as *N*-acylimidazoles or 4-nitrophenyl esters in the presence of HOBt, induced slow degradation and produced none of the desired products. Complete destruction of the substrate ensued upon

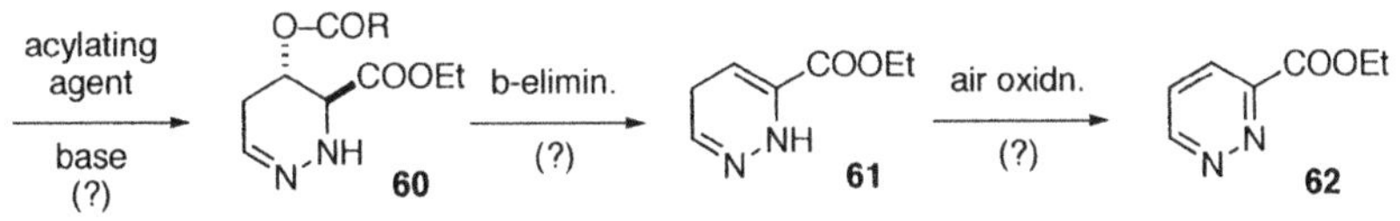

SCHEME 13

treatment with acid chlorides, with symmetrical or mixed anhydrides, or with a carboxylic acid plus any of the common condensing agents (DCC, BOP-Cl, etc.).

Various experiments revealed that the NH group in **58** suffers from an innate lack of nucleophilicity. This, however, was only *one* of the reasons why **58** had failed to undergo acylation. A relatively facile β-elimination of the oxygenated functionality in the presence of the bases commonly employed to promote acylation (triethylamine, etc.) is another. Scrutiny of the complex mixtures resulting from attempted acylation often indicated the presence of fully aromatic 3-carbethoxypyridazine (**62**), which may have formed as shown in Scheme 13. The use of milder bases, such as *N*-methylmorpholine, collidine or lutidine, or of other *O*-protecting groups, such as silyl ethers, was no cure for these woes. Compound **58** was also sensitive to aqueous bases, sustaining serious damage even under mild saponification conditions (LiOH, THF). The substance was more tolerant of strong protonic acids, such as TFA in CH$_2$Cl$_2$, showing little or no evidence of decomposition after up to 2 hours' exposure.

A modestly successful acylation protocol was devised as outlined in Scheme 14. Pioneering studies by Hassall had shown that *piz* derivatives of the type **64** may be *N*-acylated, but only with acid chlorides.[46]

Accordingly, NaBH$_3$CN reduction of **58** furnished a presumed hydrazine, which was intercepted *in situ* with BOC$_2$O to yield **63** in 35% yield. Interestingly, prolonged contact with BOC$_2$O induced formation of the *O*-BOC derivative of **63**, signaling that the OH group is more nucleophilic than the abnormally unreactive "internal" nitrogen atom. In complete accord with the Hassall report, reaction of **63** with Ac$_2$O provided largely **64** (90%), plus a small amount of diacetyl derivative **65**. Conversion of **64** to **65** ensued only upon treatment with acetyl chloride in the presence of N-methylmorpholine. Fortunately, the ester groups in **65** underwent hydrolysis (aq. LiOH) without incident, signaling that the base sensitivity of **58** and congeners was partly due to the presence of additional unsaturation. Likewise, the resulting acid **66** condensed normally with methyl glycinate under the influence of DCC. No epimerization was

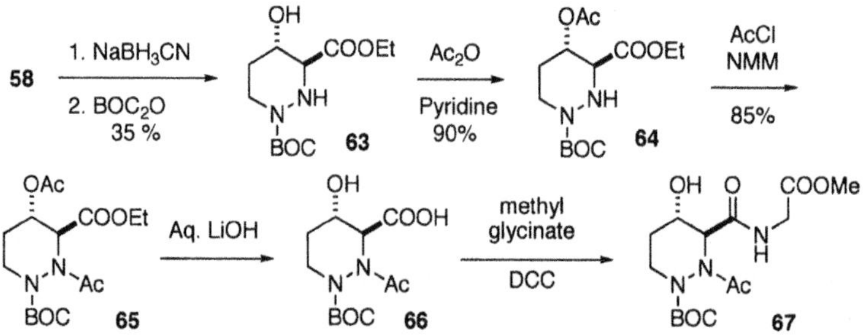

SCHEME 14

apparent, even though the α-amino group was now part of an amide. It is likely that the strong preference for the axial orientation of the carboxy group in **66** and related systems[47] disfavors formation of an intermediate of the type **24** (cf. Scheme 5). It was already known that oxidation of intermediates such as **67** back to Δ-*piz*-type structures is possible (*t*BuOCl),[48] if inefficient. Whereas the solution we had devised might have been serviceable for the total synthesis, it was already clear that better acylation technology was needed.

This preliminary phase of the luzopeptin program had nonetheless validated our approach to **58**, which we now wanted in nonracemic form. This required an enantioenriched variant of **56**, which would be available from an appropriate β-ketoester either by a Noyori asymmetric hydrogenation[49] or through baker's yeast reduction.[50] We ultimately opted for a technically simpler yeast reduction on the basis of the excellent results reported with substrate **68**,[51] and we were not to be disappointed. The resultant **69** was then processed to **72** to as shown in Scheme 15. It is worthy of note that debenzylation of **69** had to be run in cyclohexane, because the use of polar solvents (MeOH, EtOAc) promoted cyclization of alcohol **70** to the corresponding valerolactone. A test run of the GEV reaction on **72** led to enantiopure *c-piz* ester **74** in satisfactory yield.

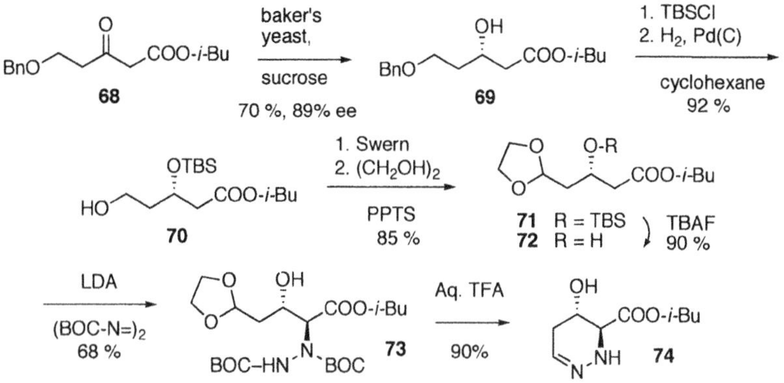

SCHEME 15

Parallel studies directed toward luzopeptin E2 focused on the synthesis of Δ-*piz* (Scheme 16). Accordingly, Schreiber ozonolysis[52] of cyclopentene furnished **75**, which was elaborated to Evans oxazolidinone **76**. Reaction of the latter with DBAD gave **77** in 64% yield and in at least 98 % de (one diastereomer in ^{1}H NMR). Release of the chiral auxiliary and acid treatment produced **78**, which like its hydroxylated congener **58** was entirely resistant to *N*-acylation, but at least it was considerably more stable.

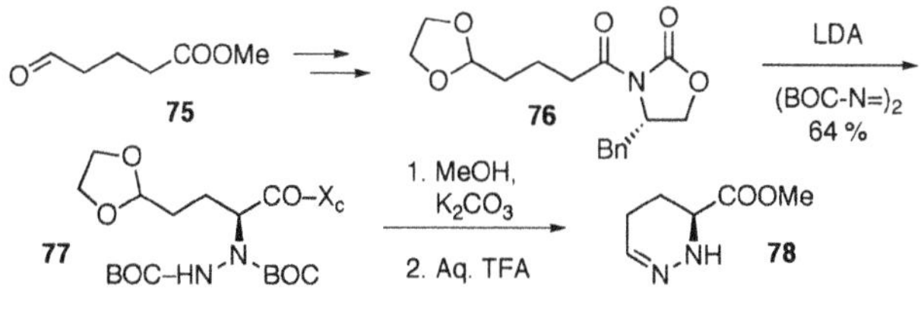

SCHEME 16

The work of Hassall and the positive results obtained with **64** made us wonder whether a Δ-*piz* precursor such as **79**, which is an acyclic variant of **64**, might also be *N*-acylated. If so, we could obtain the requisite *N*-acyl *Δ-piz* intermediates simply through acid-promoted cyclization of **80** (Scheme 17). Various routes to terminally monoprotected hydrazines of the type **79** were thus explored, and a particularly good process was devised on the basis of the hypothesis delineated in Scheme 17. Suppose

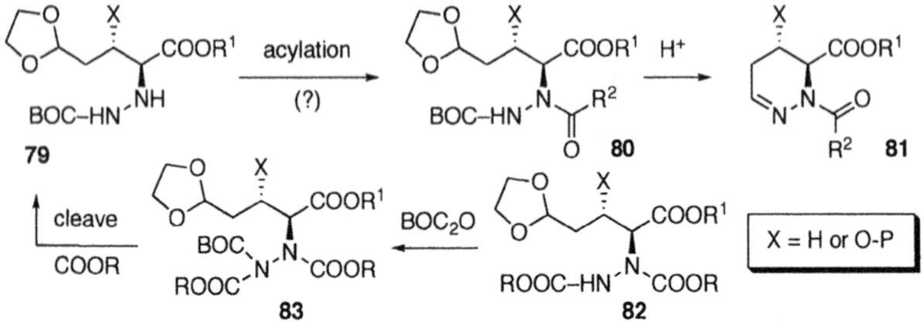

SCHEME 17

that the COOR segments in a generic GEV compound **82** were orthogonal to a BOC group. One could then install a BOC unit on the terminal N atom and release the COOR group to reach **79**. This may be possible if R were benzyl, which is cleavable by hydrogenolysis.

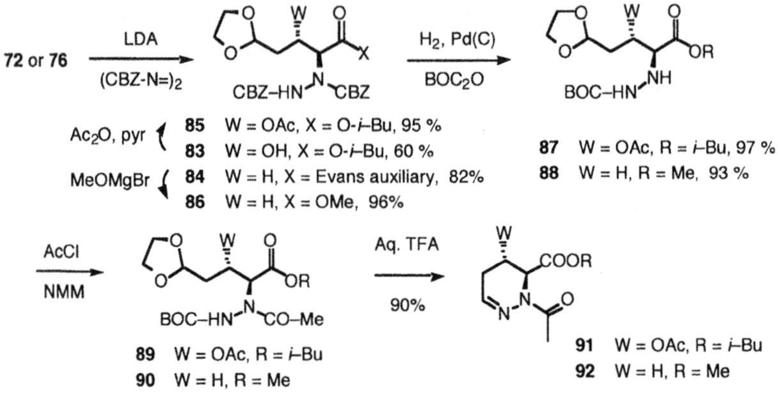

SCHEME 18

Successive refinements of more circuitous procedures ultimately led to a delightfully simple solution: compounds **85** and **86** (Scheme 18) were rapidly and cleanly converted to the desired **87** / **88** upon hydrogenolysis in MeOH in the presence of BOC_2O.[53] Borrowing terminology introduced by Speckamp,[54] we describe this transformation as a "trans-protection." To this date, we have not ascertained whether a free hydrazine is involved in this reaction, which would imply that terminal N-BOC protection is faster than hydrogenolysis of the N-N bond, or whether the pathway to **87 / 88** proceeds through selective release of the terminal CBZ, reaction of the emerging carbazate with BOC$_2$O, and slower cleavage of the inner CBZ group. Regardless, acetylation as described earlier for **64** and TFA treatment produced — finally! — acylated derivatives **91** and **92** of our dehydropiperazic acids. Interestingly, the room temperature ^{1}H and ^{13}C NMR spectra of **91 - 92** were sharp and showed the presence of a single rotamer; yet, *two* rotamers of structurally related tertiary amides, e.g., N-acetyl derivatives of proline and of pipecolinic acid, are normally discernible under the same conditions. The significance of this observation escaped us completely until much later.

But why is acylation of piperazic acids and their precursors so difficult?

We have researched this matter fairly extensively, and we think we know the answer.[19h] Briefly, two major electronic effects conspire to diminish the nucleophilicity of the NH center in **74 / 78** as well as **87 / 88**. An additional conformational effect present in **74 / 78** all but obliterates nucleophilic reactivity.

Whatever the "real" reasons for this peculiar behavior, interesting papers centering on the synthesis of *piz/Δ-piz* containing natural products had appeared in the chemical literature by this time (1995), notably from the laboratories of U. Schmidt,[55] but also of C. Shin and K. Hale.[56] All these chemists had encountered serious difficulties during the acylation of unsaturated piperazic acids and their precursors. In particular, Schmidt had discovered that only N-Cbz-valinyl chloride is sufficiently reactive to acylate a terminally protected hydrazine very similar to **88**. Furthermore, Olson and Rebert had determined earlier[57] that *serinylation* of piperazic acids and of its precursors by any of the common methods fails. All we could do is to confirm this conclusion: even cyclic N-carboxy anhydrides (NCA's)[58] are impotent *vis-a-vis* **87–88**.

The hurdles that our colleagues (and ourselves!) had encountered, as well as the results of acetylation experiments outlined in Scheme 13, indicated that our only hope to obtain a serinyl derivative of *Δ-piz* or *c-piz* rested on a reaction employing a serinyl chloride,[59] or possibly a fluoride.[60]

Serinyl chloride **93** (Scheme 19) is a known compound,[61] so we chose this agent for an initial foray into the acylation chemistry of **87-88**. Try as we might, we uniformly failed to effect the desired serinylation with this acid chloride, despite numerous attempts involving a variety of techniques. Acylations with amino acid chlorides often work best in biphasic aqueous-organic media (Schotten-Baumann conditions, cf. **38**, Scheme 8).[62] In our case, the use of biphasic systems (CH$_2$Cl$_2$ or THF with aq. NaHCO$_3$ or MgO as the bases) simply caused hydrolysis of the chloride back to the acid. We then turned to acylation in homogeneous media, cognizant that such reactions could be problematic.[60] Indeed, β-elimination of AcOH from **93** and polymerization of the resulting dehydroalanine competed effectively with the coupling reaction, regardless of solvent (CH$_2$Cl$_2$, THF, CHCl$_3$, EtOAc), base (Et$_2$N, NMM, 2,6-lutidine and 2,4,6-collidine), or temperature (–40 to 25 °C) employed. In many cases, immediate darkening of the solution was observed, and although much of the starting hydrazine could be recovered, no substance related to serine could be discerned in the NMR spectra of crude reaction mixtures. Identical results were obtained with the then-unknown chlorides **94 - 97**, which were prepared by a procedure analogous to that used for **93**. Replacement of an acetoxy group with a less nucleofugal substituent, in an effort to repress β-elimination, was of no use. Attempted use of the powerful acylation promoter, AgCN,[63] led to unmitigated disasters. Not only did the acid chlorides still decompose, but even the substrate now sustained major damage, perhaps due to some oxidative process induced by Ag(I) ion. High expectations were placed on **96** and **97**, on the basis of Carpino's discovery that *N*-Fmoc amino acid chlorides are especially well behaved acylation reagents.[64] But reality mercilessly crushed such hopes.

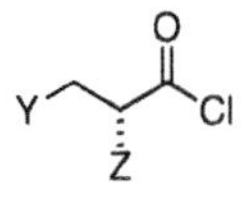

93	Y = OAc, Z = NPhtaloyl
94	Y = OTIPS, Z = NPhtaloyl
95	Y = OAc, Z = NCBZ
96	Y = OAc, Z = NFMOC
97	Y = OBn, Z = NFMOC

SCHEME 19

We had lost our skirmishes against serinyl chlorides because these agents had consistently undergone β-elimination faster than they would condense with poorly nucleophilic **87-88**. Ergo, a powerful inhibition mechanism had to be harnessed to prevent β-elimination. An attractive possibility was to marshal a stereoelectronic effect thoroughly studied by Baldwin[65] against the nucleofugal inclinations of the oxygenated functionality. If the N and O atoms of the serine were connected to form a 5-membered ring, β-elimination would become tantamount to a strongly disfavored, reverse "5-endo-trig" process. This may subdue any proclivity of the serinyl chloride to eliminate the β-oxygenated functionality and permit acylation. We tested this hypothesis with some serine-derived *N*-acyl-oxazolidines, which are deblocked by gentle acidic hydrolysis given their acetal-like nature, and oxazolidinones ("oxazolones"), which are cleavable under mild basic conditions after introduction of an *N*-BOC group.[66]

Acetal-type intermediates were not amenable to conversion to acid chlorides, doubtless because of the acid-sensitive nature of the cycle, but the sturdier oxazolones were. A literature method for the synthesis of **98**[67] was improved by bubbling phosgene into an aqueous solution of D-serine, NaOH and Na_2CO_3. However, oxazolone **98** itself was unsuitable for conversion to an acid chloride: reaction with $(COCl)_2$ produced a mixture of **101** or **102**, while **102** was the end product of attempted chlorination with $SOCl_2$ (Scheme 20). This signaled that the NH group had to be blocked. Introduction of an *N*-BOC at this stage seemed unwise, because the HCl liberated during the subsequent conversion to an acid chloride could release this protection. In the end, a simple acetyl group proved to be quite satisfactory. The resulting **100** was

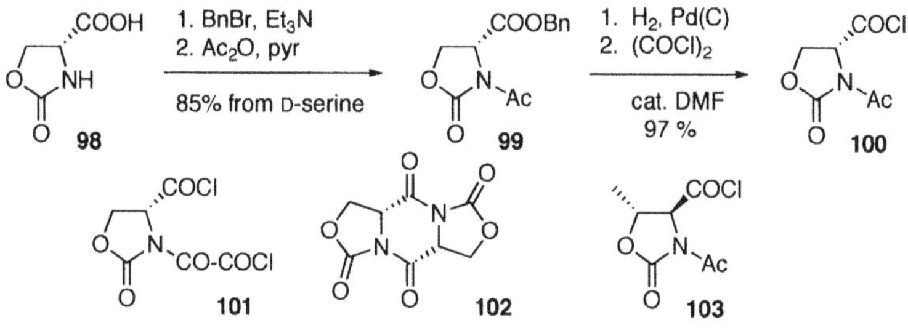

SCHEME 20

obtained as a thick oil not amenable to further purification. Fortunately, it was free from organic contaminants (^1H, ^{13}C NMR), and it displayed satisfactory shelf life (months) at –20°C. An analogous acid chloride **103** derived from L-threonine was prepared in a like manner and found to possess identical properties.

As the first chlorides of their kind, their ability to acylate L-proline and L-pipecolinic acid derivatives, which are more difficult to acylate than ordinary amino acids, was investigated in detail. Uniformly good to excellent yields of the expected products were obtained in each case.[19c]

Room temperature NMR spectra of these substances confirmed the expected presence of two amide rotamers in a ratio comprised between 4:1 and 1.5:1, but only one species was discernible when the spectra were recorded at 70°C. The ease of acylation of the pipecolinic system was noteworthy, given its notoriously poor reactivity. All such reactions were best carried out in the presence of *sym*-collidine at 0° C. Stronger bases such as Et₃N undermined the stereochemical integrity of **100**, possibly through ketene formation. Even so, no evidence of β-elimination could be garnered upon scrutiny of crude reaction mixtures.

We were relieved to find that reaction of **87 - 88** with **100** delivered the long-sought **104-105** in good yield (Scheme 21). Acid-catalyzed cyclization led nearly quantitatively to **106** and **107**. Once again, NMR spectra of **104-105** were broad at room temperature, necessitating spectroscopic observation at 50°C or higher for full structural characterization. However, the room temperature NMR spectra of **106-107** were sharp and betrayed the presence of a single amide rotamer, in contrast to the case of proline, pipecolinic acid, or **104-105**.

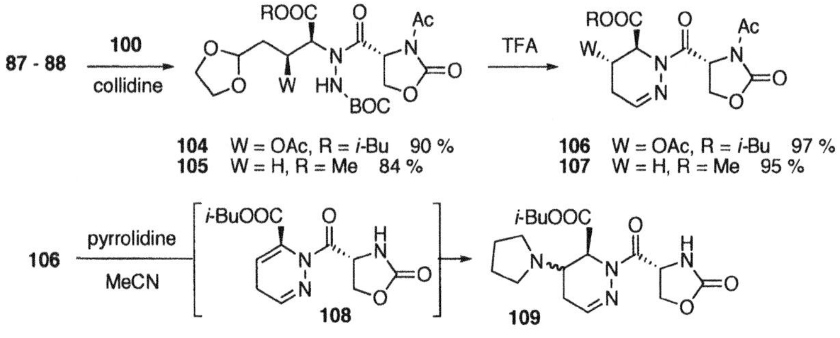

SCHEME 21

Like most other aspects of the luzopeptin effort, deblocking of hard-won dipeptides **106** and **107** turned out to be a delicate proposition. In derivatives of proline and pipecolinic acid, the serine segment had been extricated from its protective webbing by *N*-deacetylation with pyrrolidine in acetonitrile and Kunieda cleavage of the oxazolone. However, reaction of **106** with pyrrolidine furnished **109** (30%) as the sole identifiable compound. Evidently, pyrrolidine was basic enough to promote β-elimination of acetate and subsequent Michael-type addition to intermediate **108**, a significant fraction of which may have been lost to polymerization. Other amines, such as ethylamine, diethylamine, piperidine and morpholine, or aqueous bases such as LiOH, K_2CO_3 and KCN, fared no better with either **106** or its precursor **104**. Available evidence suggests that β-elimination of acetate was again at the root of such ills.

This string of disappointments was broken when highly nucleophilic, but feebly basic, hydrazine was brought to bear on **106** (Scheme 22). One equivalent of this reagent cleanly *N*-deacetylated the oxazolone in just 30 minutes. The resulting **110** was accompanied by a small amount of alcohol **111** (2-3 %). Treatment of **110** with BOC_2O afforded imide **113** in 95% yield. In a like manner, **107** was elaborated to **114**. Our elation, however, was short-lived. Exposure of **113** to methanolic Cs_2CO_3 led to a mixture of products, among which we identified pyridazine **117**. A similar outcome obtained upon reaction with aqueous LiOH. Either set of conditions was too mild to possibly cleave an amide. Again, we imputed the result to an initial β-elimination to intermediate **115**, followed by air-oxidation to **116** and solvolysis.

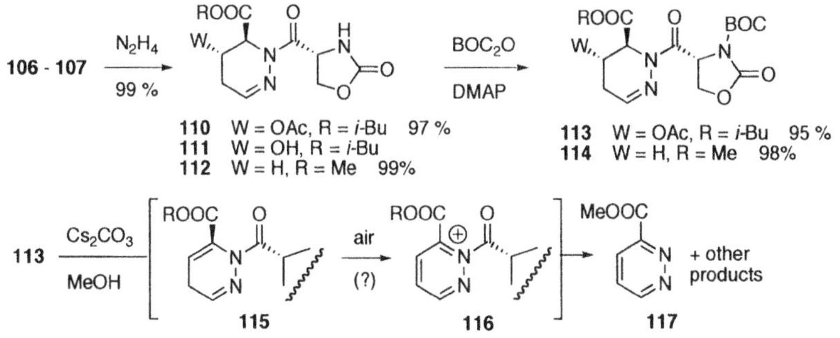

106 - 107 $\xrightarrow[\text{99 \%}]{N_2H_4}$

110   W = OAc, R = *i*-Bu   97 %
111   W = OH, R = *i*-Bu
112   W = H, R = Me   99%

$\xrightarrow[\text{DMAP}]{BOC_2O}$

113   W = OAc, R = *i*-Bu   95 %
114   W = H, R = Me   98%

113 $\xrightarrow[\text{MeOH}]{Cs_2CO_3}$ 115 $\xrightarrow[\text{(?)}]{\text{air}}$ 116   117 + other products

SCHEME 22

Analogy with important work by a Merck group[68] induced us to examine oxazolone cleavage in the considerably less readily enolizable amide **118**, which indeed underwent the Kunieda reaction in excellent yield (Scheme 23). This provided additional evidence that β-elimination was the source of the difficulties adumbrated above. Furthermore, deblocking of **114**, which is not subject to β-elimination, proceeded smoothly and produced **120** in 84% yield. In the end, it transpired that protection of the OH group as a TBS ether was all it was needed to contain – if not suppress – that pernicious side reaction. Slightly more than 2 equivalents of hydrazine removed both acetyl groups from **106** in 4 hours. Manipulation of **111** as outlined in Scheme 23 culminated with an efficient release of the oxazolone and formation of the desired **123**. In no case did the basic conditions of this transformation cause any harm to the β-hydroxyamide unit of the serine sector. Evidently, the amide nature of the serine carbonyl renders enolization less facile, thus disfavoring β-elimination.

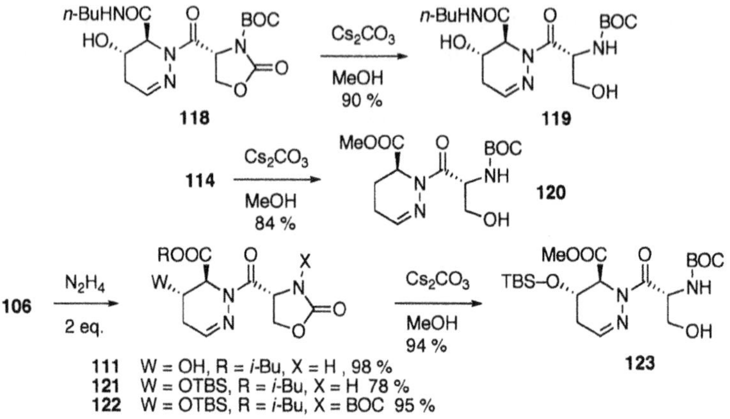

SCHEME 23

A brief digression is in order at this point. The NMR spectra of **113**, **114**, **119**, **120**, and **123** once again showed the presence of a single amide rotamer. One of my collaborators on the luzopeptin problem, Ning Xi, suggested that we should look more closely at the conformational properties of acylated piperazic acids. The apparent presence as a single rotamer at room temperature implied that either rotation about the N–CO bond was extremely fast, or that, for some unknown reasons, that amide linkage strongly favored one conformation. Correctly, he assessed that either conclusion would be interesting and worth reporting.

To our complete surprise, variable temperature ^{1}H NMR spectra of representative N-acyl piperazic acids showed no evidence of conformational motion from −89 to +120°C. This implied that the molecules in question possess an unusual degree of conformational

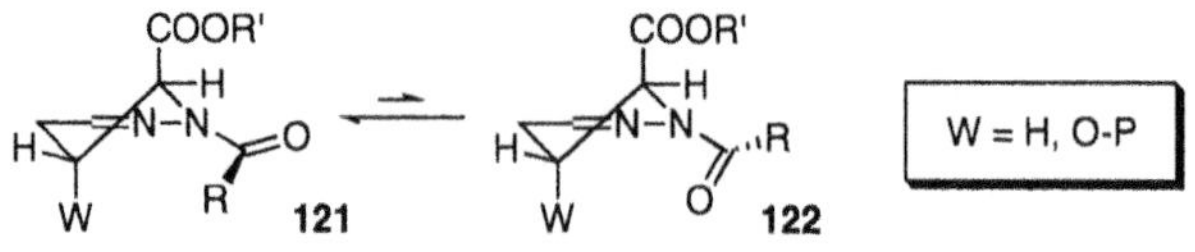

SCHEME 24

rigidity.[19f] Further spectroscopic and crystallographic studies ascertained that the preferred conformation is the one shown as **121** in Scheme 24.

It did not escape us that conformer **121** nicely maps onto the corresponding section of the luzopeptin macrocycle. Furthermore, molecular models as well as MM+ simulations indicated that *the conformational rigidity of this piz-containing segment might predispose pentapeptides such as **12** / **17** or **13** / **18** to spontaneous cyclodimerization upon activation of the carboxy terminus.*

An exercise that had started almost as intellectual overkill had produced an idea for a superb endgame.

V. Initial Cyclization Experiments

The greater stability of **120** relative to **123** induced us to center our efforts on luzopeptin E2. It seemed logical to examine first a macrolactonization strategy involving cyclodimerization of **125**, because we estimated that the coupling of acid **124** with amine **49** would not be subject to the stereochemical difficulties alluded to in the introduction, and that no base treatment would be necessary subsequent to the union of **124** and **49** through peptide bond formation. Recall that peptides incorporating *mhv* are base-sensitive: this property was likely to be carried over into **125**.

The desired condensation was best achieved with O-benzotriazol-1-yl-N,N,N',N'-tetramethyluronium hexafluorophosphate (HBTU) in the absence of any base (55%; Scheme 25). To our puzzlement and dismay,

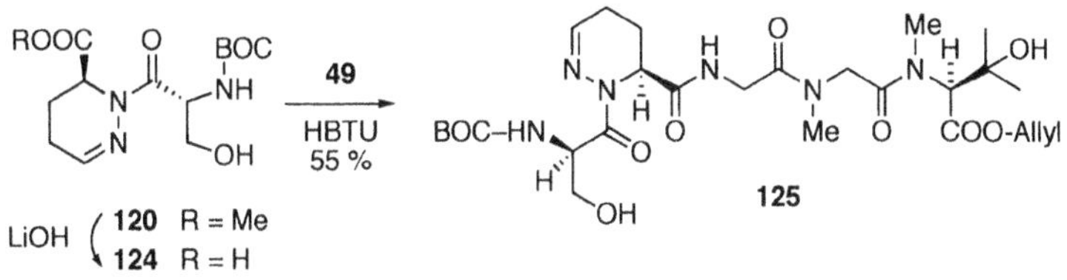

SCHEME 25

contact of **125**, a thick oil, with sources of Na^+ resulted in formation of a hard, crystalline mass soluble in water, moderately soluble in coordinating organic media such as DMSO or pyridine, but insoluble in common solvents.

The new substance turned out to be mono-sodiated **125**.

Remarkably, electron impact mass spectra of the new material displayed signals of sodiated daughter ions arising through loss of organic fragments from the parent ion, signifying that break-up of molecular appendages was more facile than unraveling of the complex! A number of ^{23}Na NMR measurements ascertained that only the 1:1 metal-peptide complex forms, even in the presence of excess Na$^+$, and allowed us to follow the sodiation of **125** by a simple titration experiment. Such a great affinity of a peptide for sodium ion appears to be unprecedented in the literature, and it may have important implications in the area of biological ion transport, ion channels, etc., but in the present context it sealed the fate of the macrolactonization approach, given the ubiquitous nature of Na$^+$ and the insolubility of the resulting complex.

A macrolactamization avenue to **5** required pentapeptide **128**, which would result upon deprotection of a suitable forerunner, **127**. It was not prudent to obtain **127** as a methyl ester (R = Me), i.e., directly from **120**. The basic conditions required for deblocking seemed inconsistent with the anticipated base sensitivity of *mhv*-containing substrates, and the depsi bond present in the molecule might also interfere. What was needed here was an ester cleavable under reductive, neutral, or acidic conditions.

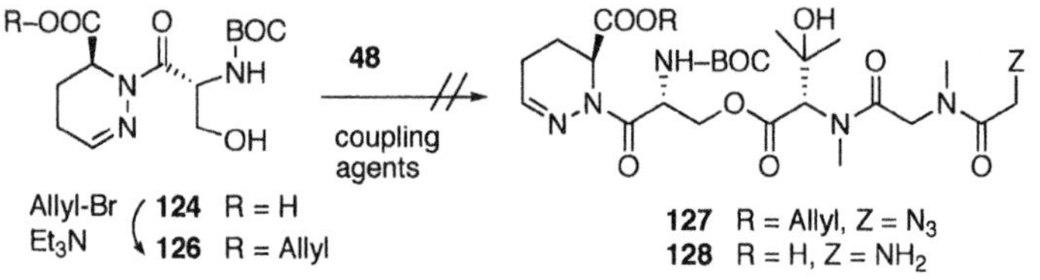

Allyl-Br ⎧ **124** R = H
Et$_3$N   ⎩ **126** R = Allyl

127 R = Allyl, Z = N$_3$
128 R = H, Z = NH$_2$

SCHEME 26

Benzyl and bromophenacyl esters proved to be unsuitable for subsequent transformations, prompting a switch to allyl ester **126** (Scheme 26). Unexpectedly, attempted condensation with **48** using various coupling agents afforded intractable mixtures, due to rapid cleavage of the allyl ester in **126** and consequent formation of various self– and cross-coupling products. The culprit was **48**, because other simple acids reacted normally with **126**. Recall that **48** had only been subject to a quick filtration over silica gel before use. Evidently, traces of palladium inadvertently carried into the coupling step were responsible for the untimely liberation of the carboxy group.

But how to remove contaminating Pd from **48**, a polar substance not amenable to extensive purification?

It was by now mid-1996, a time of great excitement in the area of solid phase synthesis. A flurry of papers touted ever more diverse application the new methodology, and a large selection of resins was becoming commercially available. Inspired by all this ferment, we surmised that if amine **49** were to be anchored to a resin prior to C-deallylation, all traces of palladium and other reaction debris may be completely removed simply by repeated washings. Subsequent coupling with **126**, deallylation and formation of an active ester could all be conducted on solid support. Release from the resin would then produce an activated peptide that might cyclodimerize spontaneously.

A dichlorotrityl resin was used to explore this hypothesis. Accordingly, **49** was anchored on resin **129** (Scheme 29).[69] Subsequent treatment of **130** with dimedone / Pd(PPh$_3$)$_4$ deblocked the C-terminus, as

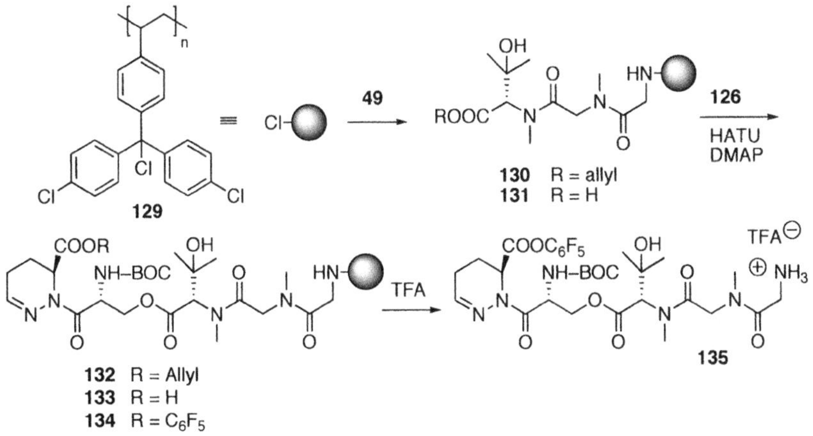

SCHEME 27

apparent also from the presence of mono- and bis-allyl derivatives of dimedone in the liquid phase. The resin was now in form **131**. Esterification with dipeptide **126** was problematic, but it did occur upon activation of the acid with 2-(7-aza-1H-benzotriazole-1-yl)-1,1,3,3-tetramethyluronium hexafluorophosphate (HATU) in the presence of a large excess of DMAP. The pentapeptide was deallylated, converted to a pentafluorophenyl ester (HATU, C$_6$F$_5$OH), and released from the support with 3% TFA in CH$_2$Cl$_2$. Such a weak concentration of TFA effectively excised the N-dichlorotrityl linkage, but readily ensured survival of the BOC groups. Treatment of the resultant trifluoroacetate salt **135** with Hünig's base triggered formation of the macrocyclic dimer **136** in 8%

overall yield, as well as of cyclic monomer **137** (3% overall). Luzopeptin E2 "mockup" **139** was obtained in a like fashion when the sequence was conducted with dipeptide **138** in lieu of **126**.

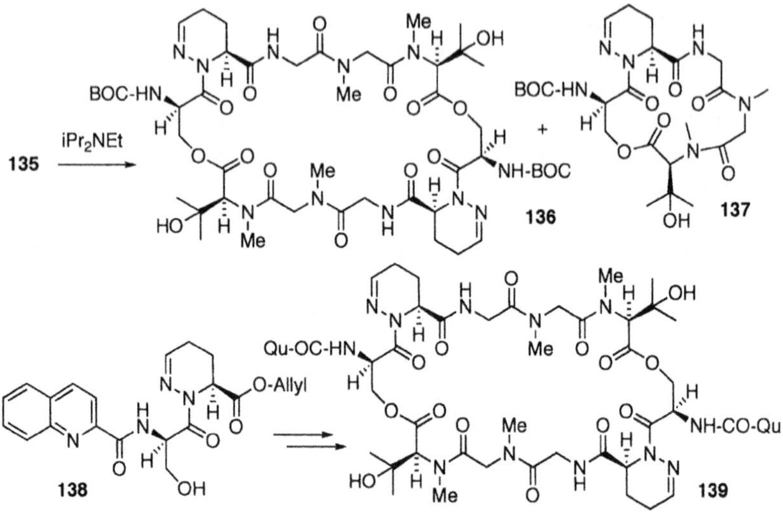

SCHEME 28

Unfortunately, the success of this phase of the research provided little cause for celebration. The protocol just developed was wasteful, because it required a large excess of valuable synthetic intermediates at each and every step. Recovered unreacted materials were all but unusable, because they were badly contaminated with debris of coupling reagents and promoters, while their sensitive nature made them difficult to purify. The low yield of macrocycle was untenable in light of the subsequent steps we still had to execute in order to advance **137** to luzopeptin E2. Finally, overall yields seemed to depend on the batch of resin that was employed. A bad batch would obliterate quantities of precious synthetic peptides for greatly diminished — or null — overall yields. In sum, a major battle was over, and we had won, to be sure.

Now we had to win the war.

VI. The Total Synthesis

The final stages of the luzopeptin campaign took place on French soil. Our foray into supported synthesis had reaffirmed the necessity of a good solution-phase approach to the macrocycle. This entailed avoiding premature release of the allyl group in **126**. It is well established that progressively bulkier analogs of allyl esters are increasingly more resistant to Pd catalyzed deprotection.[70] Indeed, experiment revealed that both the crotyl and the β,β-dimethylallyl ester analogs of **126** were fully competent in subsequent manipulations, but while the crotyl unit was rapidly released upon standard treatment with dimedone / Pd(PPh$_3$)$_4$, the dimethylallyl ester resisted cleavage. The synthesis was thus completed with crotyl ester **140**.

Compound **140** may be obtained directly from **114** by running the oxazolone cleavage reaction in crotyl alcohol. Unfortunately, the transesterification step is inefficient, providing a separable mixture of **140** and of the corresponding methyl ester in variable ratios. Attempts to drive transesterification to completion by lengthening the reaction time resulted in degradation of the substrate to a variable extent. It was best to prepare **140** as described earlier for **126** (Scheme 29). The synthesis of pentapeptide **141** was uneventful. By this time, the Boger synthesis of luzopeptins had appeared in print and we simply duplicated the solution-

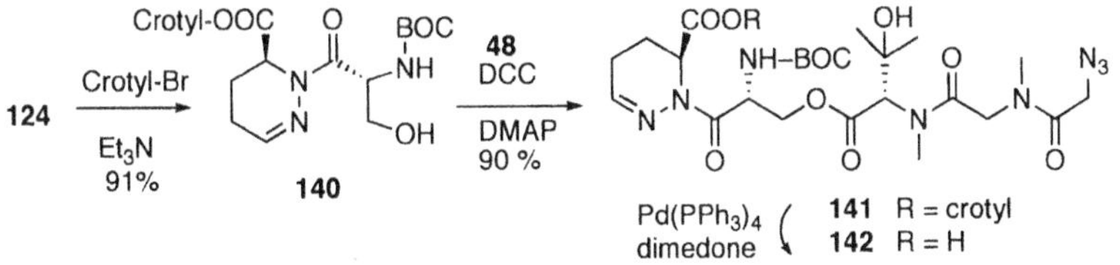

SCHEME 29

phase procedure utilized by the Scripps researchers to effect an analogous esterification. Thus, coupling of **140** with **48** in the presence of excess DCC and excess DMAP furnished **141** in 90 % yield with little or no epimerization of the *mhv* segment. This was also in agreement with the results of our solid-phase experiments (cf. **131**→**132**, Scheme 27).

Compound **141** produced broad spectra at room temperature, due to slow rotation of about the N-CO bond of the two *N*-methyl tertiary amides. Spectroscopic studies were therefore carried out at 50°C, at which temperature sharp, well-resolved spectra were obtained. Cleavage of the crotyl ester and preparative TLC removed most contaminants from **142**, which, however, strongly retained Ph$_3$P=O (from the Pd catalyst). It was difficult to completely eliminate this contaminant, but at this juncture we were not overly concerned about its presence. In fact, a Russian group had found that certain heterocyclic amine *N*-oxides act as nucleophilic catalysts in phosphate ester formation, and thus they are effective adjuvants in oligonucleotide synthesis.[71] One may surmise that $O=PPh_3$ could function in a like fashion. In the absence of evidence to the contrary, we estimated that $O=PPh_3$ would either not interfere with the coupling reaction, or, at best, assist it.

In line with results of solid-phase experiments (Scheme 28), the first solution-phase cyclodimerization attempts were conducted with 142 by conversion to a pentafluorophenyl ester and subsequent reduction of the azide. This strategy proved to be unsatisfactory, leading to a 1:1 mixture of cyclodimer 136 and cyclomonomer 137, each in a low 5 % chromatographed yield from 141 (Scheme 30). The poor efficiency of this step might have been due, at least in part, to the sensitivity of the pentafluorophenyl ester to the water required in the Staudinger reaction.

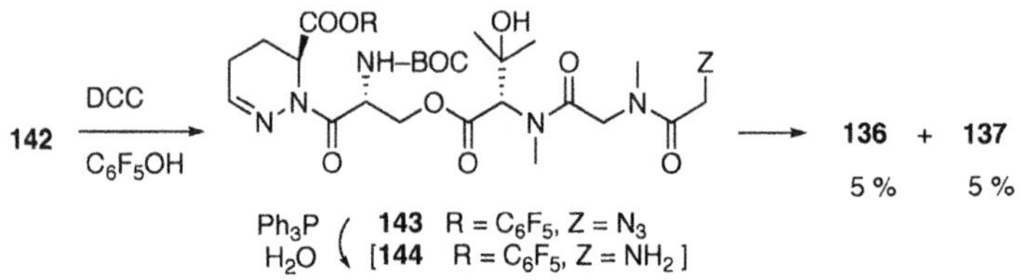

SCHEME 30

We then turned to an alternative approach involving carboxy activation of amino acid 145, readily obtained by Staudinger reaction of 142. Again, it was extremely difficult to separate triphenylphosphine oxide from 145, but fortunately, the crude material was quite suitable for use in the cyclization step.

Activation of the carboxy group in 145 resulted in significantly improved yields of 136. The results of key experiments are summarized in Table 1. A concentration of 30-40 mg / mL (ca. 50-70 μM) seemed optimal, and all reactions presented in the table were carried out at this concentration. Addition of a weak organic base such as lutidine to the reaction medium increased the proportion cyclic monomer (entry b). Reactions conducted in the presence of a stronger mineral base such as cesium carbonate, under otherwise identical conditions, or upon activation of 145 with HATU[77] (entry e) or with BOP-Cl[68] (entry f), furnished complex mixtures containing little or no cyclic product. Best results were obtained by using four equivalents each of 1-(3-dimethylaminopropyl)-3-ethylcarbodiimide hydrochloride and 1-hydroxy-7-azabenzotriazole[72] (entry d). Cyclodimer 136 and cyclic monomer 137 were obtained, respectively, in 25 % and 13 % chromatographed overall yield from **141**. The balance of the crude material consisted of starting materials (about 10%), polymeric peptides and degradation products. It is noteworthy that no uncyclized dimer of **145** was ever detected in the crude reaction mixtures (mass spectrometry). Contrary to the case of linear peptides such as **141** and congeners, cyclic peptides **136**, **137** and **139** produced sharp room temperature NMR spectra, a clear reflection of their conformational rigidity.

TABLE 1
Results of Representative Macrocyclization Experiments

entry	conditions	% yield 136	% yield 137
a	C_6F_5OH, DCC, EtOAc	15	1
b	C_6F_5OH, DCC, EtOAc, lutidine	17	14
c	C_6F_5OH, DCC, EtOAc, Cs_2CO_3, DMF	–	–
d	EDCI, HOAt, CH_2Cl_2	25	13
e	HATU, DIEA, DMF	–	–
f	BOP-Cl, Et_3N, CH_2Cl_2	–	–

Substance **146**, which may be termed "desoxy luzopeptin C", was obtained from **136** in 88% yield upon BOC cleavage and introduction of chromophore **7** by reaction with excess EDCI/HOBt in DMF[16,17] (Scheme 31). It is tempting to speculate that this yet undescribed material might be one of the uncharacterized luzopeptins. Only a reduction separated **146** from luzopeptin E2.

On the basis of our exploratory work,[19b] we attempted a $NaBH_3CN$ reduction under acidic conditions. This led in high yield to a product that was very soluble in water, but essentially insoluble in common organic media. Available evidence suggests that what we had was a complex of **5** with boron, and probably with other inorganic cations. All attempts to retrieve free luzopeptin E2 from this complex failed. Reduction had to avoid exposure of the molecule to inorganic ions.

Saturation of imines may be effected by hydrogenation over Pd or Pt catalysts. Attempts in this sense with **146** led to inseparable mixtures of products arising not only from reduction of the imines, but also from a variable extent of hydrogenation of the quinolines. Efforts to limit reductive damage to the quinolines were fruitless. Fortunately, hydrogenation of **136** over PtO_2 yielded **147** in 99 % yield. Release of the BOC groups and coupling with quinaldic acid **7** resulted in selective acylation of the serine amino groups and furnished a product (50% overall), whose ^1H and ^{13}C spectra were superimposable with those of authentic material.

We had accomplished the first total synthesis of luzopeptin E2.[20]

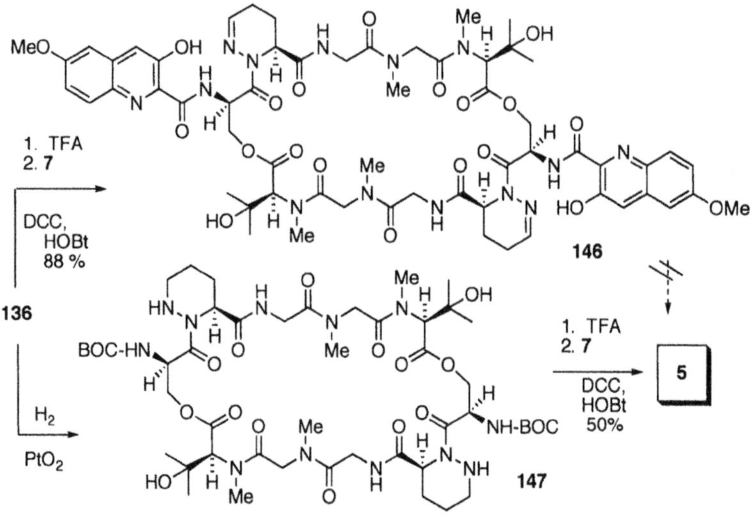

SCHEME 31

Processing of compound **123** as detailed earlier for **124** led to **148**, and thence to **151** (Scheme 32). The cyclodimerization protocol devised for luzopeptin E2 worked equally well in the case of **151**, providing macrocycle **153** in 26 % overall yield from **123**. As expected, cyclic monomer **152** was also obtained in 10% yield. Proton and ^{13}C NMR spectra of these cyclic substances, and of all subsequent intermediates, again betray a considerable degree of conformational rigidity. The *bis*-

TBS ether derivative of luzopeptin C was subsequently reached in 55 % yield.

Luzopeptin C had in store a final agonal act before yielding to total synthesis: attempted deprotection of **154** with TBAF resulted in complete obliteration of the substrate. This was arguably a consequence of the sensitivity of the molecule to basic agents such as TBAF. Deprotection with the acidic pyridine-HF complex performed adequately and carried us through the finish line.[21]

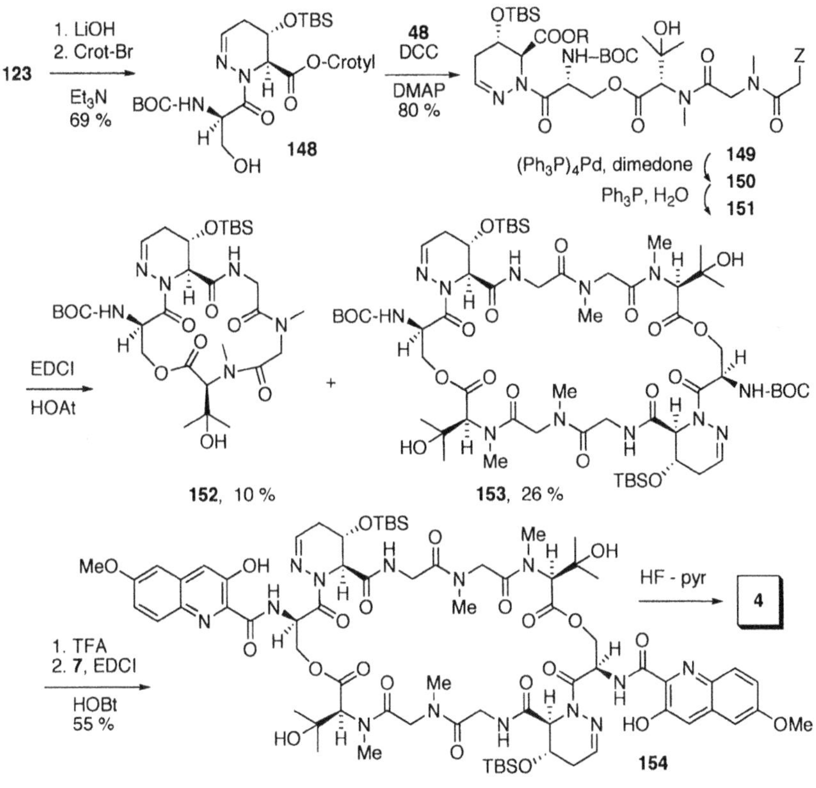

SCHEME 32

VII. Epilogue

"It is difficult today to get excited about peptide chemistry."

This bit of personal wisdom came from a speaker at a conference I was attending some time ago, shortly after completing the total synthesis of luzopeptins. For reasons that, I hope, are obvious by now, my immediate reaction was an emotional:

Oh, really!?

Perhaps the colleague was referring to ordinary peptides composed of proteinogenic amino acids, and not to substances as complex and sensitive as luzopeptins. Even so, our experience with the preparation of a number of "common" peptides, not to mention the deceptively simple tripeptide **47**, led us to a thorough appreciation of a statement that is found in the introduction to M. Bodanszky's landmark book, *"Peptide Chemistry,"*[23] and that goes as follows:

> *The synthesis of a peptide with a well defined*
> *sequence of amino acid residues is a fairly*
> *complex process which appears simple only*
> *for those who have never been involved in it.*

The effort recounted here amply vindicated the veracity of the above truism. Virtually every step of the synthesis imposed a choice among multiple options, none of which had ever been explored before. True to the wisdom of Yogi Berra,[73] we did take — *we had to take* — most of the forks we came upon on the road to luzopeptins, only to find that many of them led to stalemates, quagmires, or disasters. Yet, each paved the way to a hand-to-hand fight, and winning each confrontation required determination, ingenuity, and a large dose of work.

If that is not exciting, I do not know what *is*.

The pursuit of our synthetic objective identified new research opportunities and engendered useful methodology that has already found practical applications. For instance, the conformational properties of piperazic acids may have important ramifications in the area of peptide secondary structure, and our acylation technique has been employed by an industrial laboratory for the preparation of a drug intermediate unavailable by other methods.

This, I submit, is compelling evidence that Synthesis is as central as ever in delivering new research leads and techniques to the scientific community.

Acknowledgements

This chapter recounts mostly what worked. One could write a *book* about what didn't, and why. One of my collaborators, Ning Xi, actually did.[30] It is thanks to the work of Ning that Delphine Valognes, the first doctoral student to graduate from my group in France, was able to raise the emblem of victory on the luzopeptin battlefield. Singlehandedly, she optimized the chemistry required to prepare the fragments and developed all the procedures necessary for their assembly into luzopeptins. Dr. Shankar Swaminathan, Dr. Philippe Belmont, and Audrey Athlan also deserve credit for their contributions to the luzopeptin effort and to accessory investigations. All these young investigators contributed not only experimental skill, but also and especially, a great deal of their own insight to the luzopeptin project. To all of them I express my heartfelt gratitude and I wish Godspeed as their promising careers unfold.

References and Footnotes

1.   Inouye, Y., Take, Y., Nakamura, S., Nakashima, H., Yamamoto, N., Kawaguchi, H., *J. Antibiot.* **1987**, *40*, 100.

2.   Interestingly, anti-RT activity appears to be limited to luzopeptins and actinomycin D. Quinoxaline antibiotics such as triostin A and echinomycin are inactive, despite their structural similarity to luzopeptins: see ref. 1-2 as well as (a) Wang, A. H.-J., Ughetto, G., Quigley, G. J., Hakoshima, T., van der Marel, G. A., van Boom, J. H., Rich, A., *Science*, **1984**, *225*, 1115. (b) van Dyke, M. M., Dervan, P. B., *Science*, **1984**, *225*, 1122.

3.   Take, Y., Inouye, Y., Nakamura, S., Allaudeen, H. S., Kubo, A., *J. Antibiot.* **1989**, *42*, 107.

4.   Binding studies of luzopeptin A to DNA revealed that the compound is a bifunctional intercalator and that it favors DNA regions containing alternating A and T residues. (a) Fox, K. R., Davies, H., Adams, G. R., Portugal, J., Waring, M. J., *Nucleic Acids Res.* **1988**, *16*, 2489. (b) Fox, K. R., Wooley, C., *Biochem. Pharmacol.* **1990**, *39*, 941. (c) Hung, C.-H., Mong, S., Crook, S. T., *Biochemistry*, **1980**, *19*, 5537. (d) Hung, C.-H., Prestayko, A. W., Crook, S. T., *Biochemistry*, **1982**, *21*, 3704. (e) Hung, C.-H., Crook, S. T., *Cancer Res.*, **1985**, *45*, 3768; (f) Hung, C.-H., Mong, S., Crook, S. T., *Cancer Res.*, **1983**, *43*, 2718. (g) Leroy, J. L., Gao, X., Misra, V., Gueron, M., Patel, D. J., *Biochemistry*, **1992**, *31*, 1407. (h) Zhang, X., Patel, D. J., *Biochemistry*, **1991**, *30*, 4026.

5.   The reported activity of luzopeptin A against P388 leukemia is more than 100 times greater than that of mytomycin C, a clinically used anticancer agent: Ohkuma, H., Sakai, F., Nishiyama, Y., Ohbayashi, M., Imanishi, H., Konishi, M., Miyaki, T., Koshiyama, H., Kawaguchi, H., *J. Antibiot.* **1980**, *33*, 1087. In addition, luzopeptin A is roughly 3-times as potent as luzopeptin B, while luzopeptin C is much less active.

6.   (a) Ohkuma, H., Sakai, F., Nishiyama, Y., Ohbayashi, M., Imanishi, H., Konishi, M., Miyaki, T., Koshiyama, H., Kawaguchi, H., *J. Antibiot.* **1980**, *33*, 1087. (b) Tomita, K., Hoshino, Y., Sasahira, T., Kawaguchi, H., *J. Antibiot.* **1980**, *33*, 1098.

7.   (a) Konishi, M., Ohkuma, H., Sakai, F., Tsuno, T., Koshiyama, H., Naito, T., Kawaguchi, H., *J. Antibiot.* **1981**, *34*, 148. (b) Konishi, M., Ohkuma, H., Sakai, F.,

Tsuno, T., Koshiyama, H., Naito, T., Kawaguchi, H., *J. Am. Chem. Soc.* **1981**, *103*, 1241.

8. Arnold, E., Clardy, J., *J. Am. Chem. Soc.* **1981**, *103*, 1243.

9. Or at least, they are not in the public domain.

10. *Netherlands Pat.*, **1984**, 84 00 237 (*Chem Abs.* **1985**, *102*, 22795q)

11. Lingham, R. B., Hsu, A. H. M., O'Brien, J. A., Sigmund, J. M., Sanchez, M., Gagliardi, M. M., Heimbuch, B. K., Genilloud, O., Martin, I., Diez, M. T., Firsch, C. F., Zink, D. L., Liesch, J. M., Koch, G. E., Gartner, S. E., Garrity, G. M., Tsou, N. N., Salituro, G. M., *J. Antibiot.* **1996**, *49*, 253. Quinoxapeptins inhibit HIV-1 and HIV-2 RT with IC_{50} values (the concentration of compound that elicits 50% inhibition) of 4 and 40 nM, respectively, but they are poor inhibitors of mammalian DNA polymerases (the IC_{50}'s of quinoxapeptin A against DNA polymerase a, b, g and d are equal to 2563, 615, 1798 and 494 nM, respectively). This activity is comparable to that reported for luzopeptins.

12. Matson, J. A., Bush, J. A., *J. Antibiot.* **1989**, *42*, 1763. Matson, J. A., Colson, K. L., Belofsky, G. N., Bleiberg, B. B., *J. Antibiot.* **1993**, *46*, 162.

13. For related structures such as echinomycine, triostine, thiocoraline and BE-22179 see, e.g., Boger, D. L., Ichikawa, S., Tse, W. C., Hendrick, M. P., Jin, Q., *J. Am. Chem. Soc.* **2001**, *123*, 561, and references cited therein.

14. Review: Hale, K. J., in *The Chemical Synthesis of Natural Products*, Hale, K. J., Ed.: CRC Press: Boca Raton, FL, 2000, p. 379 ff.

15. For pioneering work in this area see: Olsen, R. K., Apparao, S., Bhat, K. L., *J. Org. Chem.* **1986**, *51*, 3079.

16. Boger, D. L., Lederboer, M. W., Kume, M., *J. Am. Chem. Soc.* **1999**, *121*, 1098.

17. Boger, D. L., Lederboer, M. W., Kume, M., Jin. Q., *Angew. Chem. Int. Ed. Engl.* **1999**, *38*, 2424.

18. The Scripps team also studied various aspects of the bioactivity of luzopeptins and quinoxapeptins. See: (a) Boger, D. L., Chen, J. H., Saionz, K. W., Jin, Q., *Bioorg. Med. Chem.* **1998**, *6*, 85. (b) Boger, D. L., Saionz, K., W. *Bioorg. Med. Chem.* **1999**, *7*, 315. (c) Boger, D. L., Lederboer, M. W., Kume, M., Searcey, M., Jin, Q., *J. Am. Chem. Soc.* **1999**, *121*, 11375. (d) Boger, D. L., Ichikawa, S., Tse, W. C., Hendrick, M. P., Jin, Q., *J. Am. Chem. Soc.* **2001**, *123*, 561.

19. (a) Ciufolini, M. A., Swaminathan, S., *Tetrahedron Lett.* **1989**, *30*, 3027. (b) Ciufolini, M. A., Xi, N., *J. Chem. Soc., Chem. Commun.* **1994**, 1867. (c) Xi, N., Ciufolini, M. A., *Tetrahedron Lett.* **1995**, *36*, 6595. (d) Ciufolini, M. A., Xi, N., *J. Org. Chem.* **1997**, *62*, 2320. (e) Ciufolini, M. A., Shimizu, T., Swaminathan, S., Xi, N., *Tetrahedron Lett.* **1997**, *38*, 4947. (f) Xi, N., Alemany, L. B., Ciufolini, M. A. *J. Am. Chem. Soc.* **1998**, *120*, 80. (g) Ciufolini, M. A., Hermann, C. Y. W., Dong, Q., Shimizu, T., Swaminathan, S., Xi, N., *Synlett* **1998**, 105. (h) Ciufolini, M. A., Xi, N. *Chem. Soc. Rev.* **1998**, 437. (i) Ciufolini, M. A., Valognes, D., Xi, N., *Tetrahedron Lett.* **1999**, *40*, 3693. (j) Ciufolini, M. A., Valognes, D., Xi, N., *J. Het. Chem.* **1999**, 1409.

20. (a) Ciufolini, M. A., Valognes, D., Xi, N., *Angew. Chem.* **2000**, *112*, 2612. (b) Ciufolini, M. A., Valognes, D., Xi, N., *Angew. Chem., Int. Ed. Engl.* **2000**, *39*, 2493.

21. Valognes, D.; Belmont, P.; Xi, N.; Ciufolini, M. A., *Tetrahedron Lett* **2001**, *42*, 1907.

22. (a) Dory, Y. L.; Mellor, J. M.; McAleer, J. F., *Tetrahedron* **1996**, *52*, 1343. (b) Calvelier, F.; Jacquier, R.; Mercadier, J.-L.; Verducci, J., *Tetrahedron* **1996**, *52*, 6173. (c) Schmidt, U., *Pure. Appl. Chem.* **1986**, *58*, 295.
23. Bodanzsky, M. "*Peptide Chemistry*", Springer-Verlag, Berlin, Germany, 1993.
24. Meng, Q.; Hesse, M. in: "*Tropics in Current Chemistry: Macrocycles*"; Vol. 161, Weber, E.; Vogtle, F., Eds.; Springer-Verlag: Berlin, Germany, 1992; p. 107.
25. Bock, M. G.; Di Pardo, R. M.; Williams, P. D.; Tung, R. D.; Erb, J. M.; Gould, N. P.; Whitter, W. L.; Perlow, D. S.; Lundell, G. F., *et al.*, *Colloq. INSERM 1991*, **208** (vasopressin), 349; *Chem. Abstr.* **1991**, *115*, 115072.
26. Bayne, D. W.; Nicol, A. J.; Tennant G., *J. Chem. Soc., Chem. Commun.*, **1975**, 782.
27. Newman, M. S.; Holmes, H. L., *Org. Syn. Coll. Vol. II*, **1943**, 428.
28. Maurer, P. J.; Takahata, H.; Rapoport, H., *J. Am. Chem. Soc.* **1984**, *106,* 1094.
29. Garner, P., *Tetrahedron Lett.* **1984**, *25*, 5855.
30. Xi, N. *Dissertation*, Rice University, Houston, TX, 1996.
31. Tidwell T. T., *Org. React.*, **1990**, *39*, 297.
32. This oxidation may also be carried out with TPAP/NMO (Ley S. V., Norman J., Griffith W. P., Marsden S. P., *Synthesis*, **1994**, 639), but in diminished yield.
33. (a) Hillis, L. R.; Ronald, R. C., *J. Org. Chem.* **1985**, *50*, 470; (b) Bal, B. S.; Childers, W. E.; Pinnick, H. W., *Tetrahedron Lett.*, **1981**, *37*, 2091.
34. Carlsen P. H. J., Katsuki T., Martin V. S., Sharpless K. B., *J. Org. Chem.*, **1981**, *46*, 3936.
35. However, Boger successfully employed this procedure in a related system: see ref. 16 and 18c.
36. Ciufolini, M. A.; Valognes, D.; Xi, N., *Tetrahedron Lett.* **1999**, *40*, 3693.
37. Scriven, E. F.; Turnbull, K., *Chem. Rev.* **1988**, *88*, 297.
38. Gololobov, Y. G.; Kasukhin, L. F., *Tetrahedron*, **1992**, *48*, 1353.
39. Hughes, P.; Clardy. J., *J. Org. Chem.* **1989**, *54*, 3260.
40. Guanti, G.; Banfi, L.; Narisano, E., *Tetrahedron* **1988**, *44*, 5553.
41. (a) Gennari, C.; Colombo, L.; Bertolini, G., *J. Am. Chem. Soc.* **1986**, *108*, 6394. (b) Evans, D. A.; Britton, T. C.; Dorow, R. L.; Dellaria, J. F., *J. Am. Chem. Soc.* **1986**, *108*, 6395. (c) Trimble, L. A.; Vederas, J. C., *J. Am. Chem. Soc.* **1986**, *108*, 6397.
42. Review on the preparation of β-ketoesters: Benetti, S.; Romagnoli, R.; Risi, C. D.; Spalluto, G.; Zanirato, V., *Chem. Rev.* **1995**, *95*, 1065.
43. In the present case, superior results were obtained when the enolate of **53** was generated with tBuOK (Ganem, B.; Morgan, T. A., *Tetrahedron Lett.* **1980**, *21*, 2773) rather than LDA (Koreeda, M.; Akagi, H., *Tetrahedron Lett.* **1980**, *21*, 1197).
44. Mander, L. N., Sethi, S. P., *Tetrahedron Lett.* **1983**, *24*, 5425.
45. Dasaradhi, L., Fadnavis, N. W., Bhalerao, U. T., *J. Chem. Soc., Chem. Commun.* **1990**, 729.
46. Hassall C. H., Johnson W. H., Thebald J., *J. Chem. Soc., Perkin Trans. 1*, **1979**, 1451.
47. Fraser, R. R., Grindley, T. B., *Tetrahedron Lett.* **1974**, *15*, 4169, as well as refs. 19f and 19h and literature cited therein.
48. Pettibone, D. J., Clineschmidt, B. V., Anderson, P. S., Freidinger, R. M., Lundell, G. F.; Koupal, L. R., Schwartz, C. D., Williamson, J. M., Goetz, M. A., Hensens, O. D., Liesch, J. M., Springer, J. P., *Endocrinology*, **1989**, *125*, 217. Later work

showed that this transformation may also be achieved under Swern conditions: Gentilucci, L., Grijzen, Y., Thijs, L., Zwanenburg, B., *Tetrahedron Lett.* **1995**, *36*, 4665.

49. (a) Noyori, R., Ohkuma, T., Kitamura, M., Takaya, H., Sayo, N., Kumobayashi, H., Akutagawa, S., *J. Am. Chem. Soc.* **1987**, *109*, 5856. (b) Kitamura, M., Tokunaga, M., Ohkuma, T., Noyori, R., *Org. Synth.*, **1992**, *71*, 1.

50. Seebach, D., Sutter, M. A., Weber, R. H., Zuger, M. F., *Org. Synth., Coll. Vol. VII* **1990**, 215.

51. (a) Hirama, M., Nakamine, T., Ito, S., *Chem. Lett.* **1986**, 1381. (b) Brooks, D., Kellogg, R. P., Cooper, C. S., *J. Org. Chem.* **1987**, *52*, 192. Ketoester **68** is readily prepared by alkylation of the dianion (Huckin, S. N., Weiler, L., *Can. J. Chem.* **1974**, *52*, 2157) of isobutyl acetoacetate with benzyloxymethyl chloride (Connor, D. S., Klein, G. W., Taylor, G. N., Bockman, R. K., Medwid, J. B., *Org. Synth., Coll. Vol. VI*, **1988**, 101).

52. Claus, R. E., Schreiber, S. L., *Org. Synth. Coll. Vol. VII* **1990**, 168. Schreiber ozonolysis was decidedly superior to a more classical route proceeding through Michael addition of malonic ester to acrolein (cf. Rebert, N. W. *Dissertation*, Utah State University, Logan, Utah, 1987).

53. For a noteworthy variant of this process see: Maeorg, U., Grehn, L., Ragnarsson, U., *Angew. Chem., Int. Ed. Engl.* **1996**, *35*, 2626.

54. Rutjes, F. P. T. J.: Paz, M. M., Hiemstra, H., Speckamp, W. N., *Tetrahedron Lett* **1991**, *32*, 6621.

55. Schmidt, U., Riedl, B., *Synthesis* **1993**, 809.

56. Ref. 14 presents a detailed review of the work published by these two scientists.

57. Rebert, N. W. *Dissertation*, Utah State University, Logan, Utah, 1987.

58. Saragovi, H. U., Fitzpatrick, D., Raktabutr, A., Nakanishi, H., Kahn, M., Greene, M. I., *Science*, **1991**, *253*, 792.

59. Chemistry of α-aminoacid chlorides: (a) Carpino, L. A., Cohen, B. J., Stephens, K. E., Jr., Sadat-Aalaee, S. Y., Tien, J.-H., Langridge, D. C., *J. Org. Chem.* **1986**, *51*, 3732. (b) Carpino, L. A., Beyermann, M., Wenschuh, H., Bienert, M., *Acc. Chem. Res.*, **1996**, *29*, 268.

60. Kim, H.-O., Gardner, B., Kahn, M., *Tetrahedron Lett.* **1995**, *36*, 6013.

61. (a) Newman, H., *J. Am. Chem. Soc.* **1973**, *95*, 4098. (b) Saitoh, Y., Moriyama, H., Hirota, H., Takahashi, T., Khuong-Huu, Q., *Bull. Chem. Soc. Jpn.* **1981**, *54*, 488.

62. (a) Carpino, L. A., Cohen, B. J., Stephens, K. E., Jr., Sadat-Aalaee, S. Y., Tien, J.-H., Langridge, D. C., *J. Org. Chem.* **1986**, *51*, 3732. (b) Carpino, L. A., Beyermann, M., Wenschuh, H., Bienert, M., *Acc. Chem. Res.*, **1996**, *29*, 268.

63. (a) Durette, P. L., Baker, F., Barker, P. L., Boger, J., Bondy, S. S., Hammond, M. L., Lanza, T. J., Pessolano, A. A., Caldwell, C. G., *Tetrahedron Lett.* **1990**, *31*, 1237. (b) Hale, K. J., Delisser, V. M., Yeh, L.-K., Peak, S. A., Manaviazar, S., Bhatia, G. S., *Tetrahedron Lett.* **1994**, *35*, 7685.

64. (a) Carpino, L. A., Cohen, B. J., Stephens, K. E., Jr., Sadat-Aalaee, S. Y., Tien, J.-H., Langridge, D. C., *J. Org. Chem.* **1986**, *51*, 3732. (b) Carpino, L. A., Beyermann, M., Wenschuh, H., Bienert, M., *Acc. Chem. Res.*, **1996**, *29*, 268.

65. Baldwin, J., *J. Chem. Soc. Chem. Commun.* **1976**, 734.

66. Ishizuka, T., Kunieda, T., *Tetrahedron Lett.* **1987**, *28*, 4185.

67. Kaneko, T., Takeuchi, I., Inui, T., *Bull. Chem. Soc. Jpn.* **1968**, *41*, 974.
68. Askin, D., Verhoeven, T. R., Liu, T. M.-H., Shinkai, I., *J. Org. Chem.* **1991**, *56*, 4929.
69. The yield of each step in this sequence was estimated by releasing the anchored material from a sample of resin and determining its mass.
70. Kunz, H. *Angew. Chem., Int. Ed. Engl.* **1987**, *26*, 294.
71. Efimov, V. A., Chakhmakhcheva, O. G., Ovchinnikov, Yu. A., *Nucl. Acids Res.* **1985**, *13*, 3651.
72. Carpino L. A., *J. Am. Chem. Soc.*, **1993**, *115*, 4397.
73. Baseball Hall of Famer and icon of American popular culture, credited with the dictum "When you come to a fork on the road, take it!" For additional quotes see: http://www.yogi-berra.com/yogiisms.html.

STRATEGIES AND TACTICS IN ORGANIC SYNTHESIS, VOL. 6

Chapter 2

SYNTHESIS OF GELDANAMYCIN USING GLYCOLATE ALDOL REACTIONS

Merritt B. Andrus, Erik J. Hicken, and Erik L. Meredith
Department of Chemistry and Biochemistry
Brigham Young University
Provo, Utah 84602

I. Introduction

In 1970, geldanamycin was isolated as a crystalline antimicrobial compound from culture filtrates of the soil bacterium *Strep. hygroscopicus* (var. geldanus nova) by De Boer and co-workers at the Upjohn Company.[1] It was shown to possess moderate activity against bacteria and fungi (2-100 μg/mL) and potent activity against KB and leukemia L1210 tumor cells (<0.002 μg/mL). Shortly thereafter, Rinehart and co-workers reported the structural assignment of geldanamycin, identifying it as a new member of the ansamycin family of antibiotics.[2] Other members of the ansamycin class, lactams possessing meta-substituted benzenoid or quinone macrocycles, include the rifamycins, streptovaricins, tolypomycins, and maytansines.[3] The closely related benzoquinoid ansamycins, macbecin I and the herbimycins, were reported in 1979 and 1980, and the absolute stereochemistry was assigned by X-ray crystallography.[4] In contrast, the absolute stereochemistry of geldanamycin, assumed to be related to macbecin and herbimycin, remained unestablished until completion of the total synthesis. While macbecin and herbimycin have enjoyed considerable synthetic attention,[5] geldanamycin, the most potent member of the family, has only recently succumbed to total synthesis.[6] This initial lack of attention can be attributed to its ambiguous stereochemistry, more challenging oxygenation pattern, and the unique presence of the additional methoxyl group on the quinone at C17. The impetus to surmount these challenges was the more recent revelation of its ability to bind and inactivate the oncogenic kinase chaperone, heat shock protein 90 (Hsp90).[7] This finding opens up great opportunities for geldanamycin and its analogs as a new class of highly selective, therapeutic agents.

Key to this synthetic effort is the development of two new asymmetric glycolate aldol methods,[8] used to set the *syn* C6-7 and *anti* C11-12 hydroxy-methoxy functionality. Problems associated with the methoxyquinone formation will also be discussed along with a general solution using a 1,4-diprotected precursor.

II. Retrosynthetic Analysis

Geldanamycin is a densely functionalized compound that presents a considerable synthetic challenge. A dealkylative quinone-forming step was selected as the key transformation to finalize the route (Scheme 1). Trimethoxybenzene lactam **1** would be oxidized to give the C17-methoxy quinone of geldanamycin. This approach had been successfully used previously with macbecin and herbimycin, the only difference being the absence of the C17 methoxyl. Ample precedent for this step was found in routes to other quinone natural products and various examples with more simple trimethylbenzene substrates.[9] Opening up the macrocycle gives the

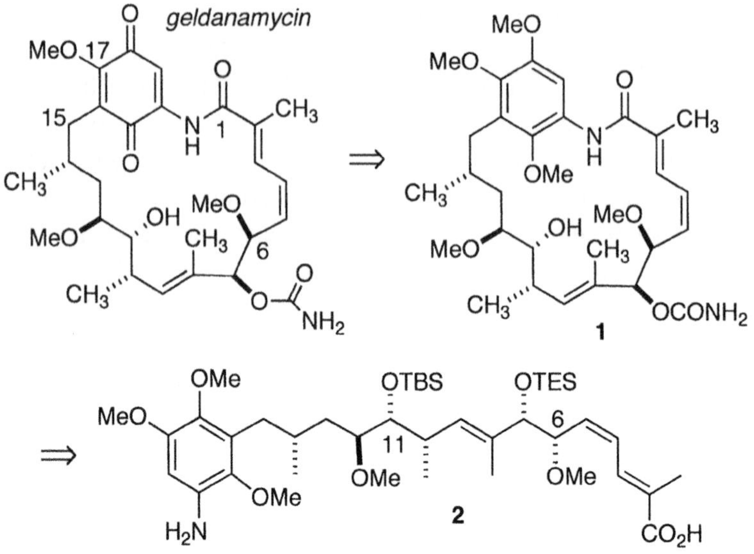

SCHEME 1

seco acid precursor **2**. The *tert*-butyldimethylsilyl (TBS) and triethylsilyl (TES) protecting groups at C11 and C7 were selected to facilitate attachment of the urethane at C7 following macrolactam formation.

Convergent strategies focused in the C8-9 alkene region were considered and investigated for the synthesis of the seco acid to no avail. These included a Horner-Emmons enone formation followed by an S_N2' methyl displacement and Grubbs ring closing (RCM) and intermolecular metathesis approaches. Key aspects of these routes are also discussed below. The successful route involved a less efficient linear approach using an asymmetric hydroboration step as previously employed by Tatsuda for the synthesis of herbimycin A.[5]

The *E,Z*-diene of the seco acid **2** would result from Horner-Emmons and Wittig reactions from aldehyde **3** (Scheme 2). The trimethoxyaniline functionality is carried along as a stable nitrobenzene that would be unmasked at a late stage prior to macrolactam formation. The *syn* methoxy-hydroxy functionality at C6-7 in **3** is installed using the newly developed glycolate aldol reaction with enal **4** and norephedrine glycolate ester auxiliary **5**.[8d] This new method is an application of Masamune's recent work with propionate norephedrine aldol reactions.[10] The C10 methyl is installed using the aforementioned hydroboration reaction,

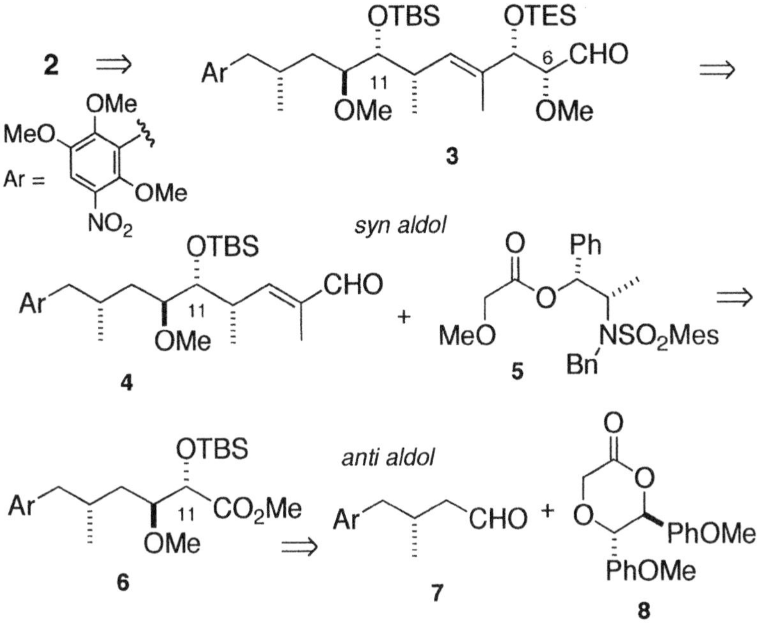

SCHEME 2

which will be discussed below. The precursor to **4**, ester **6**, results from an asymmetric *anti*-selective glycolate aldol reaction from aldehyde **7** and 1,4-dioxan-2-one **8**. This is a novel approach where the boron enolate from **8**, constrained to the *E*-geometry, adds with high *anti*-selectivity via a closed transition state arrangement. This new diphenyl-1,4-dioxan-2-one is readily made from the corresponding stilbene in two steps using a Sharpless AD (asymmetric dihydroxylation) reaction and the products are readily elaborated into intermediates suitable for multistep synthesis.

III. Ansamycin Antitumor Antibiotics

The closely related members of the ansamycin family include the herbimycins and macbecin I (Scheme 3). Structural differences compared to geldanamycin include the absence of a methoxyl at C17 on the quinone portion and the presence of methoxyls at C15 and at C11 (R', R''=OMe). Total syntheses of macbecin I have been reported by Baker and Castro in 1990, Evans and co-workers in 1992, and Panek and Xu in 1995.[5] The total synthesis of herbimycin A was reported in 1992 by Nakata and

Tatsuda. Partial synthetic efforts to these targets have been reported by the Martin, Kallmerten, and Marshall groups.[11]

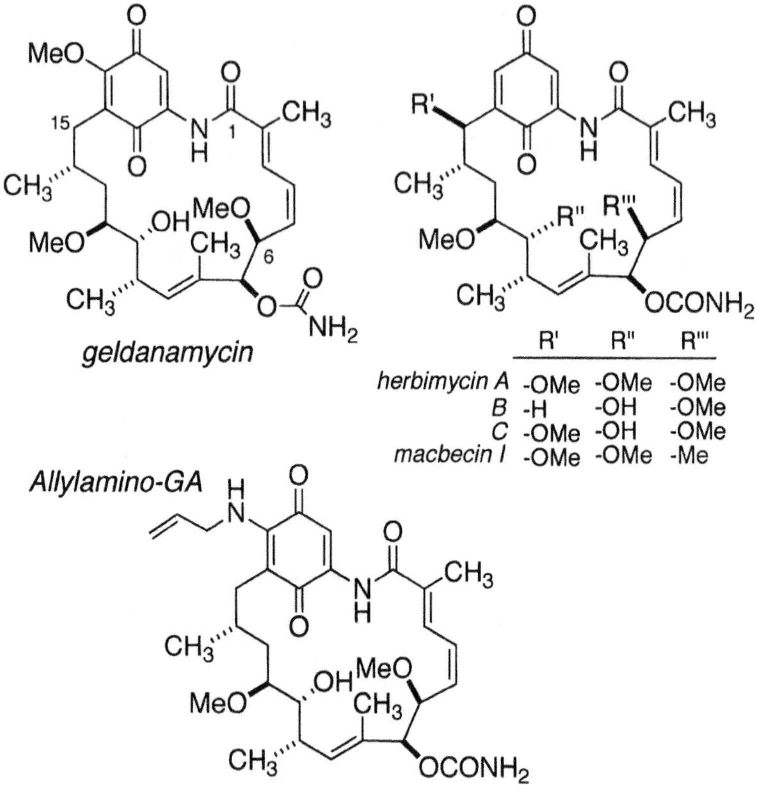

SCHEME 3

Geldanamycin is the most potent member of the class, possessing broad biological activity. It shows a unique profile of action with the NCI (National Cancer Institute) cell-line panel with an average ED_{50} value of 13 nM.[12] Macbecin and herbimycin have similar, but less potent activities (200-400 nM). The C17 methoxyl allows for the introduction of various nucleophiles, hence geldanamycin from the beginning has been a popular template for semi-synthetic analogs. In 1977, Rinehart found that various hydrazone, oxime, phenazine, and phenoxazine derivatives were effective inhibitors of RNA-dependant DNA polymerases found in the Rauscher leukemia virus.[13] Schnur and co-workers at Pfizer reported in 1995 an extensive structure activity study using SKBR-3 human breast cancer cells as a model in an attempt to develop new erbB-2 oncogenic kinase inhibitors.[14] In addition to changes at C17, diene hydrogenation, C7, 9, 11, and N22 derivatives were made and tested. Ring opening, diene alterations, and C7 urethane removal all resulted in loss of activity. Considerable change at C17 was tolerated. Analogs in this case are easily made via conjugate addition of amines or thiols to the quinone followed by loss of the methoxyl. Allylamino-geldanamycin has proven to be a valuable analog in that it maintains comparable activity, and yet possesses attenuated toxicity and improved stability.[15] This compound is currently in phase III clinical trials sponsored by the NCI.

In 1986, Uehara and co-workers reported that herbimycin A and the related ansamycins were able to revert the tyrosine kinase-induced oncogenic phenotype of pp60-v-Src.[16] Initially, it was thought that the ansamycins directly inhibited or induced degradation of the src family of kinases. Whitesell and co-workers dispelled this notion when it was found that levels of phosphotyrosine, src protein, and kinase activity remained unchanged in drug sensitive cells that express src.[17] In addition, it was shown that geldanamycin did not bind to the isolated kinase. In 1994 using an affinity strategy, Neckers and co-workers demonstrated that geldanamycin bound (200 nM) to the Hsp90, a key chaperone protein for v-src.[18] Further studies have shown that geldanamycin dramatically lowers the cellular levels of various kinases including ErB2 (HER2), EGF, Raf-1, and CDK4 kinases, all of which are dependant on Hsp90 for proper folding and stability.[19] X-ray crystal structures for the geldanamycin•Hsp90 complex have been reported by the Pavletich and Pearl groups. These show that geldanamycin adopts a higher energy 'C' shaped conformation as it binds in the ADP site on the chaperone.[20] While this structural work has revealed a great deal concerning the activity of geldanamycin, a key discrepancy persisted concerning the relatively low binding affinity for isolated Hsp90 (1.2 μM) and its potent cellular activity (12 nM). Very recently, this issue was resolved when Burrows and co-workers reported that geldanamycin binds with very high affinity to a chaperone super-complex isolated from transformed cells made up of various proteins including Hsp90, Hsp70, Hsp40, Hop, and p23 (12 nM).[21] These chaperone complexes, typically present at low levels in normal cells, are critical for oncogenic kinase folding and stability in transformed cells. The molecular basis for geldanamycin's potent and selective activity against transformed cells has thus become quite clear.

Various hetero- and homodimeric geldanamycin conjugates have also recently been made and tested in an effort to further improve selectivity and lower toxicity. Estrogen and testosterone hybrids, made by Danishefsky and co-workers, have targeted the estrogen receptor in MCF7 cells and the androgen receptor in a PC3 prostate cell line.[22] This group has also made homo-dimers that show improved selectivity toward HER2 degradation.[23] Brechbiel has produced a Herceptin immunoconjugate[24] and Chiosis and Rosen have investigated a kinase inhibitor-geldanamycin conjugate.[25]

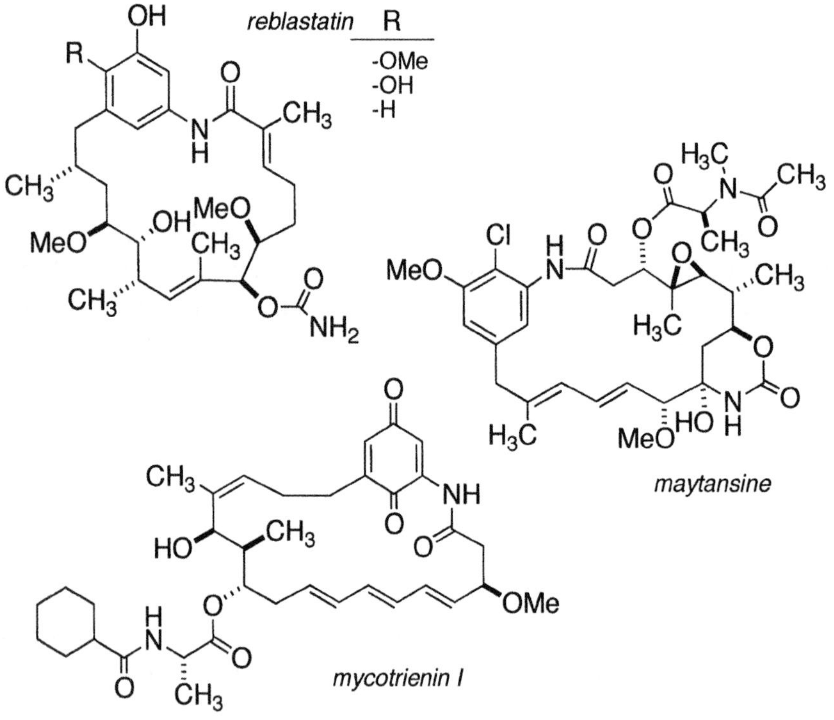

SCHEME 4

Recently, the reblastatins, a new class of directly related benzenoid ansamycins, were reported by a group at Glaxo Wellcome (Scheme 4).[26]

The compounds were shown to possess potent inhibition of oncostatin M signaling activity. The 'ethylene' containing mycotrienins and mycotrienols are another class of related triene ansamycins.[27] Much synthetic effort has been recently focused on the synthesis of these compounds.[28] Historically, the maytansines, a family of chlorobenzenoid ansamycins, have also been very popular as synthetic targets.[29] While a great number of analogs were made and tested in this area, adverse toxicity issues led to termination of their development.

IV.   Synthesis of the C12-C21 Quinone Precursor Portion using Evans Asymmetric Alkylation

The total synthesis of geldanamycin begins with commercially available 1,2,4-trimethoxybenzene **9** (Scheme 5). Following the report of Gilman in 1944,[30] **9** was metallated with *n*-butyllithium and treated with DMF (dimethylformamide) to give the benzaldehyde. Nitric acid in acetic

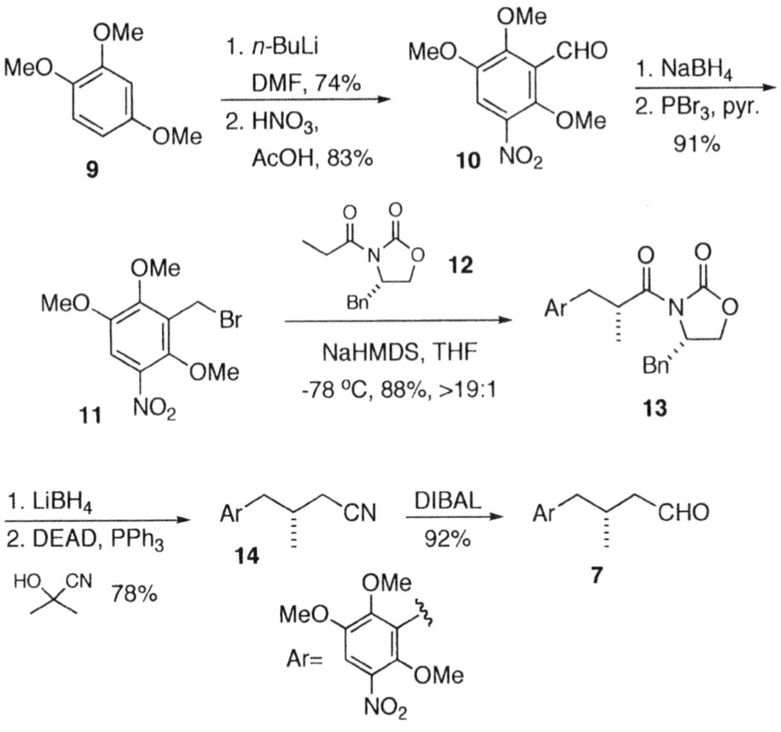

SCHEME 5

acid was then used to generate compound **10** as a light-yellow crystal. Reduction and conversion to the benzyl bromide with phosphorus tribromide gave the crystalline, stable **11** in excellent yield. While at first apprehensive about the di-*ortho*-substituted benzyl bromide, we were please to find that asymmetric Evans alkylation[31] proceeded well with acyl oxazolidinone **12** to give **13** with very high selectivity. The consistent yield of 88% was achieved at 1.1:1 **11:12** ratio even on large scale (10 grams). Reduction with lithium borohydride and homologation under Mitsunobu conditions with acetone cyanohydrin[32] provided compound **14**. Standard treatment with DIBAL (diisobutylaluminum hydride) followed by water then gave the key aldehyde intermediate **7**.

Initially, an alternative approach to C14 methyl incorporation was explored based on asymmetric conjugate addition (Scheme 6). Aldehyde **10** was homologated to aldehyde **15** using a Wittig reaction with the ylide generated from methoxymethyl phosphonium chloride.[33] The unsaturated acyl oxazolidinone **17** was then formed with high *E*-selectivity using phosphonate **16** in the presence of base and lithium chloride. Various

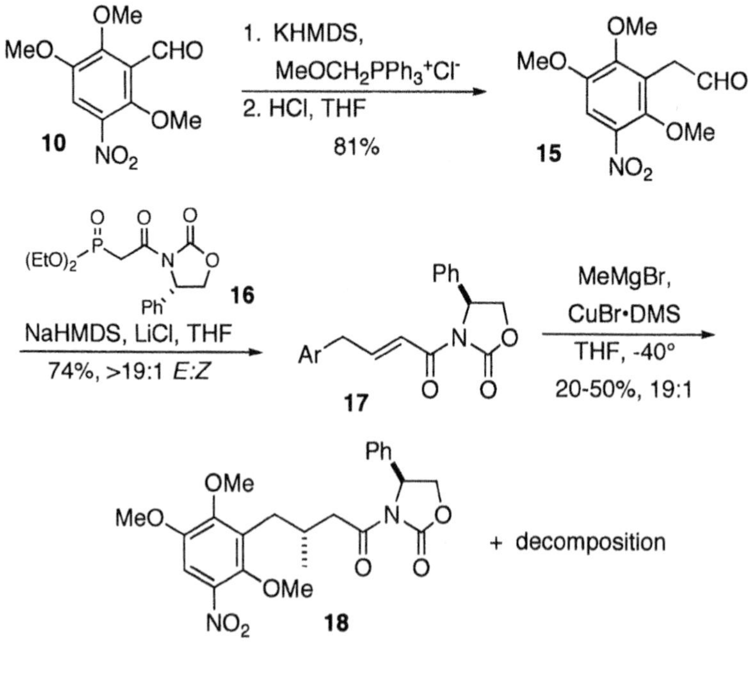

SCHEME 6

known cuprate conditions for methyl addition were then explored.[34] While the selectivity found for the product **18** was in general very high, typically >19:1, the yields for this reaction seldom reached 50%. After various attempts, it was found that considerable material loss was occurring due to methyl addition to the nitrobenzene moiety with methoxy displacement followed by decomposition.

V. Development of Anti-Selective Glycolate Aldol Methodology Based on the 2,3-Diaryl 1,4-Dioxan-5-one Template

Aldol reactions are often used as key steps in the synthesis of various classes of natural products. Of the many variations, the glycolate aldol, or α-hydroxyacetate version, is a useful approach to the generation of differentially protected vicinal diols. While known auxiliaries have been useful for *syn*-glycolate aldols, *anti*-selective versions have been very limited in that the requisite *E*-geometry needed via a closed transition state is not favored.[35] To this end, a new class of cyclic glycolate auxiliaries were developed, constrained to the *E*-geometry to bias for selective *anti*-protected diol formation.[8] Prior to its application to geldanamycin, the development of this new method will be discussed briefly (Scheme 7). The dioxanone synthesis begins with Sharpless asymmetric dihydroxylation using commercial AD-mix-α reacted with *E*-stilbene **19** to give *S,S*-diol **20**.[36] The method is general in that AD-mix-β can be used to give *R,R*-diol leading to the opposite enantiomeric aldol series. The dioxanone **8** is then formed using tin oxide in benzene at reflux followed by addition of *tert*-butyl bromoacetate.[37] When 4,4'-dimethoxystilbene is used, the intermediates are crystalline, more easily purified, and allow for more flexibility upon elaboration. Numerous bases, boron reagents, and metals were screened to promote the aldol reaction with aldehydes leading to product **21**. The optimal conditions employ dicyclohexylboron triflate (2.6 equiv.), added as a standardized 1.0 M hexanes solution, with triethylamine (2.6 equiv.) at -78 °C followed by addition of the aldehyde (1.5 equiv.) with warming to room temperature. The reaction gives very good yields of isolated products with good to excellent selectivity for the *anti*-adduct as determined by ^{1}H NMR. X-ray crystal structure analysis and optical rotation comparison to known products were used to establish the *S,S,S,S*-stereochemistry as shown for **21**. Most classes of aldehydes, alkyl, aryl, branched, and

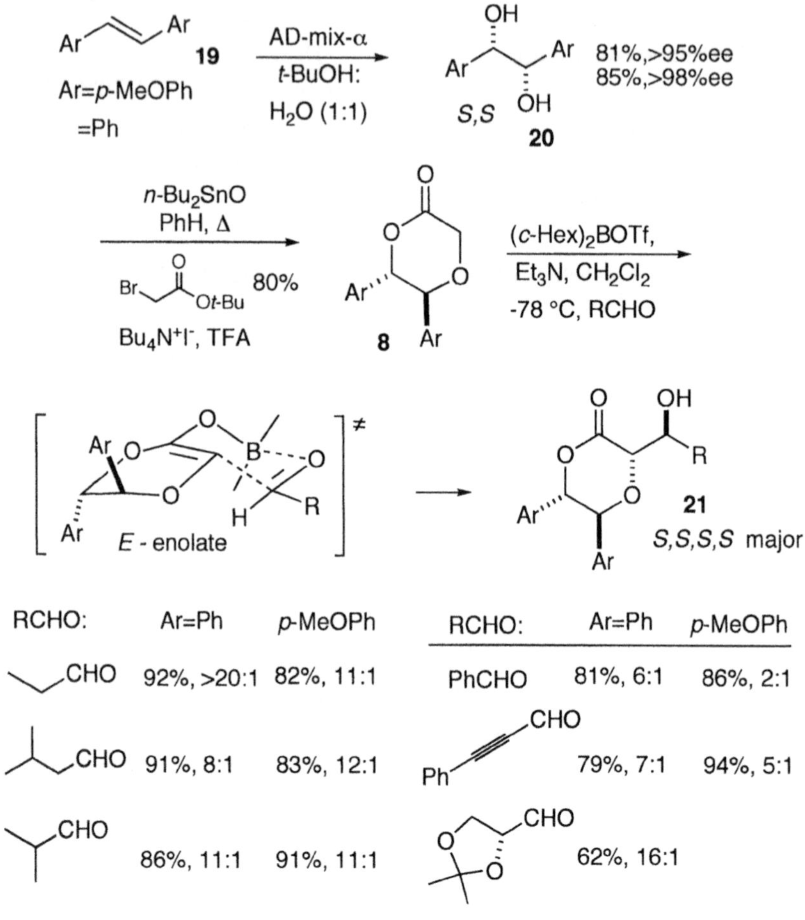

SCHEME 7

unsaturated show high selectivities with **8** (Ar = 4-MeOPh). The origin of the asymmetry of the reaction is due to the proximity of the C5-aryl group on the dioxanone enolate, which directs the aldehyde to the opposite face in the closed, Zimmerman-Traxler type transition state. Interestingly, if the C6 aryl group is omitted on the dioxanone, the selectivity for the aldol reaction is greatly eroded and all four possible diastereomers are obtained (6:4:1:1). This indicates that there is an important synergistic effect between the two aryl groups of the substrate. We have also explored the enantiopure "*cis*" version of the dioxanone **8** and found it to give aldol products with comparable selectivities.

VI.   Installation of the C11-12 Hydroxy Methoxy Functionality

The aldehyde **7** generated above was reacted with dioxanone **8** under the optimized conditions to give aldol adduct **22** (Scheme 8). In this case, the number of equivalents of aldehyde **7** was reduced to 1.2 and the isolated yield was 70% with a selectivity of 15:1. The conditions for

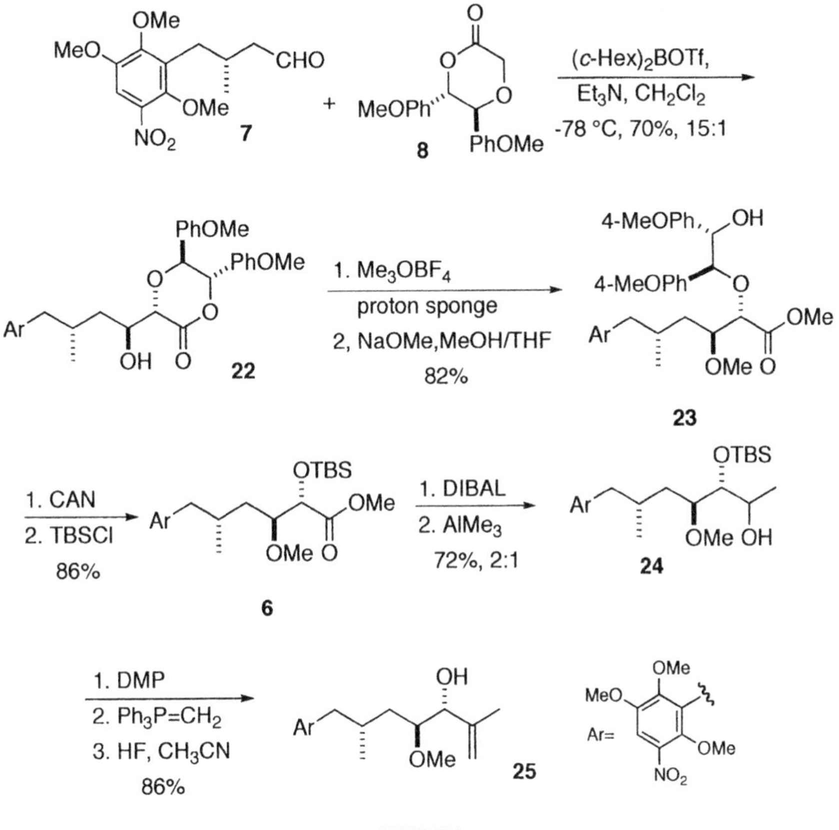

SCHEME 8

auxiliary removal and elaboration of the product are generally applicable. Even though the auxiliary is destroyed at this point, it should be kept in mind that the dioxanone in made in only two steps using a catalytic AD reaction with an inexpensive stilbene. Treatment with trimethyloxonium tetrafluoroborate gives the methyl ether intermediate. Transesterification with catalytic sodium methoxide (0.8 mol%) cleanly produced the methyl ester **23** in excellent overall yield. Cleavage of the activated benzyl ether was achieved using CAN (ceric ammonium nitrate) and the alcohol was protected to give the key intermediate ester **6**.

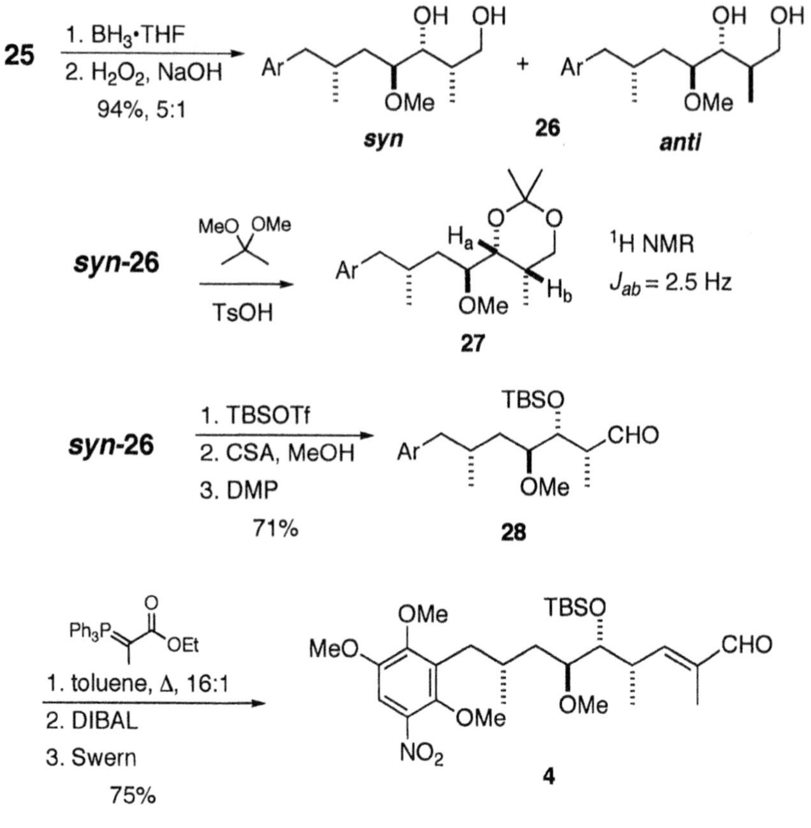

SCHEME 9

The C10 methyl was installed using a sequence of steps, which included an asymmetric hydroboration reaction. This approach was developed by Tatsuda for the synthesis of herbimycin and later followed by Panek for macbecin.[5] Other unsuccessful approaches to this difficult region of the molecule are outlined below. To ready the material for this event, the ester **6** was reduced to the aldehyde and trimethylaluminum was added to generate alcohol **24** (Scheme 8). Oxidation to the ketone, Wittig reaction, and TBS removal with HF provided allyl alcohol **25** in excellent overall yield. Asymmetric hydroboration was then conducted using borane•THF complex at −10 °C (Scheme 9). Oxidative work-up and isolation gave a 5:1 mixture of desired *syn* and *anti* products **26**. All attempts to improve this ratio, including other boranes, protecting groups, and additives, failed. Fortunately the yield was high and the undesired *anti*-isomer **26** was easily removed by chromatography. Evans and Burgess have shown that *syn*-diol products from alkenes of this type can be accessed using rhodium-catalyzed hydroboration.[38] In Panek's approach to macbecin, the terminal, disubstituted alkene gave the undesired *anti*-product under these conditions.[5] While the origin of the asymmetric induction in these cases remains unclear, Houk has proposed a useful model for uncatalyzed hydroboration of allylic alcohols.[39] The conformation with the hydroxyl adopting an 'inside' position over the extended outside position is typically favored, but only slightly, by ~0.1 kcal/mol. Borane then attacks the more nucleophilic face of the alkene, away from the large alkyl chain, leading to the *syn*-diol. In view of these ambiguities, a structure proof on both of the separated diastereomers was performed. *Syn*-**26** was converted to the 6-ring acetonide **27** and a coupling constant of J_{ab} of 2.5 Hz was observed in the ^{1}H NMR. This small value is consistent with an axial-equatorial disposition corresponding to the *syn* diastereomer. The corresponding acetonide from *anti*-**26** gave a J_{ab} of 12 Hz consistent with an axial-axial configuration. Confident in the assignment, the major isomer, *syn*-**26** was converted to the bis-TBS ether, the primary TBS was removed with camphorsulfonic acid, and the primary alcohol was oxidized to give **28**. Stabilized Wittig treatment gave the unsaturated *E*-ester, which was reduced to the allylic alcohol, and converted to the key enal **4**.

VII.   Difficulties Associated with the C10 methyl and the C8-9 Trisubstituted Alkene

Alternative approaches to the hydroboration route were explored in an effort to develop a more selective, convergent means of efficient assembly. Two of these attempts are described in this section and a third route involving a Grubbs RCM (ring closing metathesis) reaction is discussed later. Enone assembly, using an efficient Horner-Emmons coupling, followed by conversion of this functionality to the trisubstituted C8,9 alkene together with methyl installation at C10 would constitute a

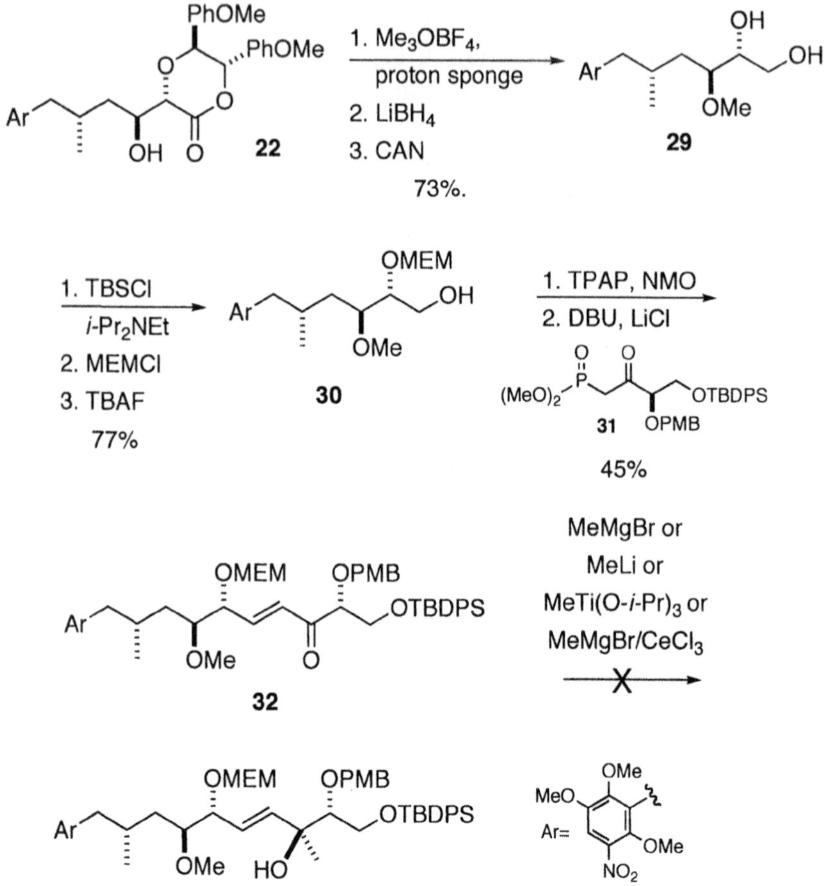

SCHEME 10

significant advance. To that end, aldol adduct **22** was converted to the diol **29**, which readily gave protected primary alcohol **30** (Scheme 10). The corresponding aldehyde was generated and reaction with β-keto phosphonate **31**, derived from glyceraldehyde, generated enone **32** as a model for the new approach.[40] At this point a methyl addition was needed as a precursor to the trisubstituted olefin. Unfortunately, all attempts to add methyl organometallics gave either no reaction or rapid decomposition. The desired tertiary alcohol, needed for S_N2' displacement, was not obtained under these conditions.

A more selective alternative to hydroboration was explored involving a Sharpless AE (asymmetric epoxidation) reaction (Scheme 11).[41] Aldol adduct **22** was readily converted to the unsaturated ester intermediate **34**, which was used to access allylic alcohol **35**. Exposure of this material to the Sharpless reagents provided epoxy alcohol **36** with high selectivity, >19:1, in 60% isolated yield. All attempts to open this epoxide at the internal position with trimethylaluminum to give the needed 1,2-diol product met with failure.[42] At higher temperatures and using smaller protecting groups in place of the TBS ether also lead to decomposition, not diol **37**.

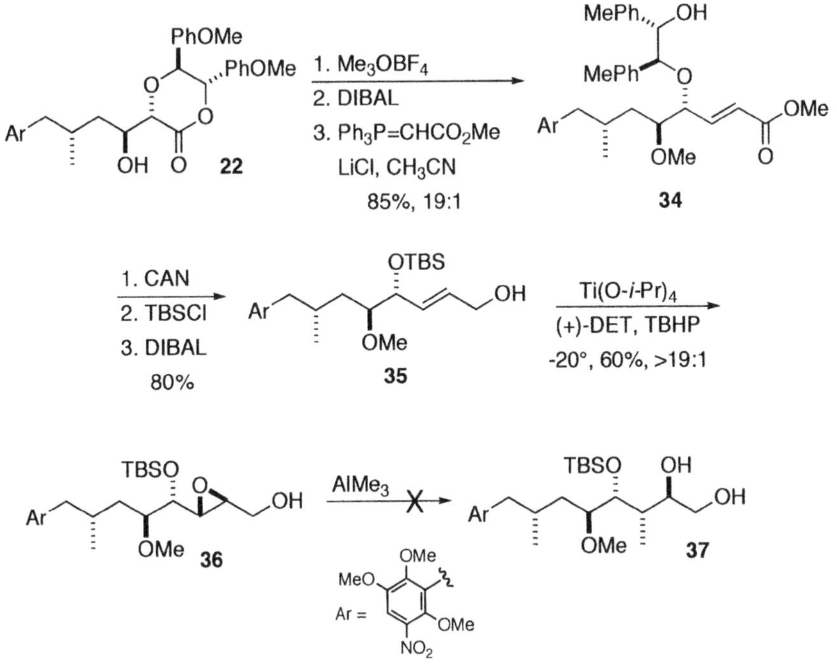

SCHEME 11

VIII. Syn Glycolate Aldol Methodology Based on the Masamune Norephedrine Auxiliary

The new auxiliary developed for the *syn* glycolate aldol, needed for the C6,7 functionality, was based on Masamune's norephedrine propionate ester methodology.[10] While Evans' oxazolidinone auxiliary has been used for various successful glycolate applications,[35] no

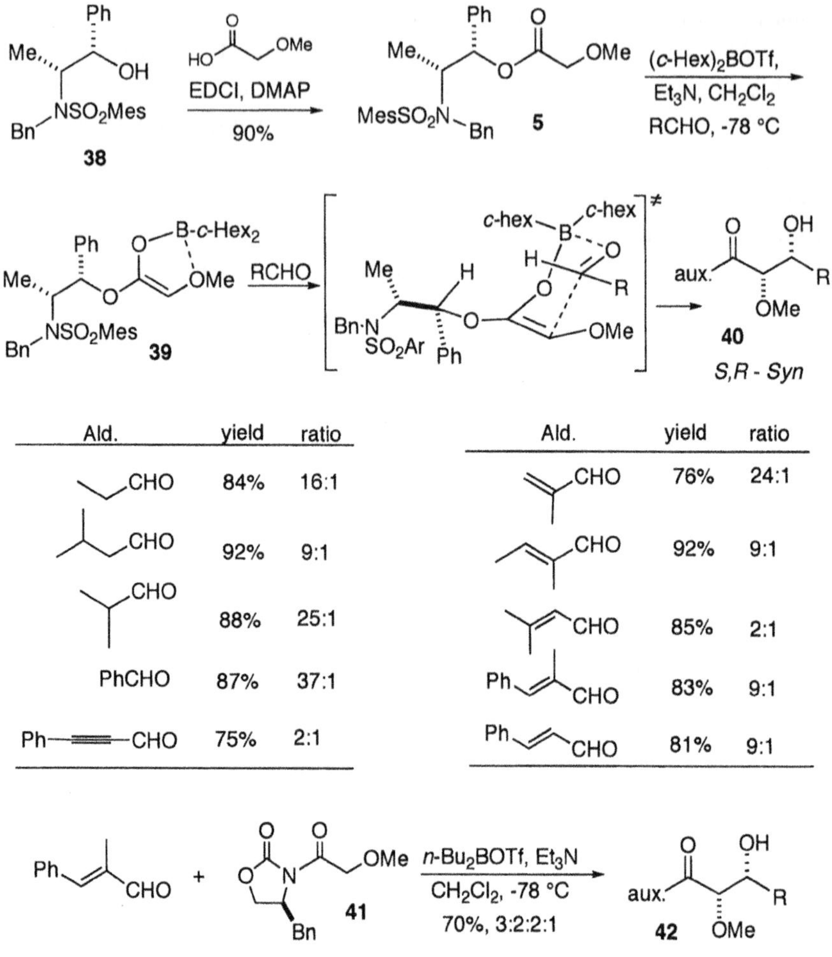

Ald.	yield	ratio
CHO	84%	16:1
CHO	92%	9:1
CHO	88%	25:1
PhCHO	87%	37:1
Ph——CHO	75%	2:1

Ald.	yield	ratio
CHO	76%	24:1
CHO	92%	9:1
CHO	85%	2:1
Ph CHO	83%	9:1
Ph CHO	81%	9:1

SCHEME 12

systematic study of this kind had been reported. Masamune reported that with dicyclohexylboron triflate and triethylamine, the propionate *E*-enolate is generated, giving *anti*-1,2-methyl hydroxy products. With di-*n*-butylboron triflate and Hünig's base, the *Z*-enolate gives *syn* products with high selectivity. We initially explored this template as a means to access the *anti*-aldol series, either through closed or open-transition state arrangements, but to no avail. We were please to find however, that this approach works very well for the more common *syn*-glycolate products (Scheme 12).[8d] Masamune's 1*S*,2*R*-N-benzyl mesityl sulfonamide norephedrine **38** was converted to methyl glycolate **5**. Various boron reagents and bases were then screened for aldol reactivity with aldehydes (1.2 equiv.). Unlike the propionate series of Masamune, aldol reactions in this case, regardless of the boron reagent or base, all gave *syn*-products **40** with high selectivity. The Lewis basic ether at the α-position ensures that the enolate geometry is *Z*-**39**, leading to *syn*-product. The *S,R*-stereochemistry was confirmed by direct comparison to known material. The method is again general, being useful for a wide range of aldehyde substrates. The exceptions are alkyl and β-branched enals at only 2:1 selectivity. Fortunately, α-branched enals, as needed for geldanamycin, typified by α-methyl cinnamaldehyde gave good selectivity (9:1) and 83% yield. In contrast, the known Evans glycolate **41** reacted with this substrate with very poor selectivity (3:2:2:1, **42**).

IX.   Construction of the C2-5 Diene and Macrolactamization

This new aldol method was then applied to enal **4** en route to geldanamycin (Scheme 13). Reaction with norephedrine **5** (2 equiv.) gave adduct **43** with >20:1 selectivity in 90% isolated yield. The ester was hydrolyzed with lithium hydroxide and the methyl ester was formed using TMS diazomethane.[43] The alcohol was protected as the TES ether and the aldehyde intermediate **3** was generated with DIBAL. Treatment with the hexafluorophosphonate reagent under Still-Gennari conditions[44] gave *Z*-unsaturated ester **45** with 13:1 selectivity.

Initially, the reduction of ester **45** proved to be problematic (Scheme 14). When THF was used as solvent, a large excess of DIBAL was required (10 equiv.) and the yield for the desired alcohol was still very low, 50%, with the remainder being unreacted **45**. Use of methylene chloride as solvent led to complete reduction along with unwanted TES removal. Finally, it was found that the use of ether with 2.5 equivalents of DIBAL provided the desired alcohol in 91% yield. Oxidation gave the enal intermediate, which was converted to the diene ester **47** with allyl phosphonate **46**.[45] The selectivity was 19:1 (*E,Z:Z,Z*) and the yield was an excellent 93%. Previously, the more common methyl ester corresponding to **47** was employed. Hydrolysis to the desired acid with this substrate was

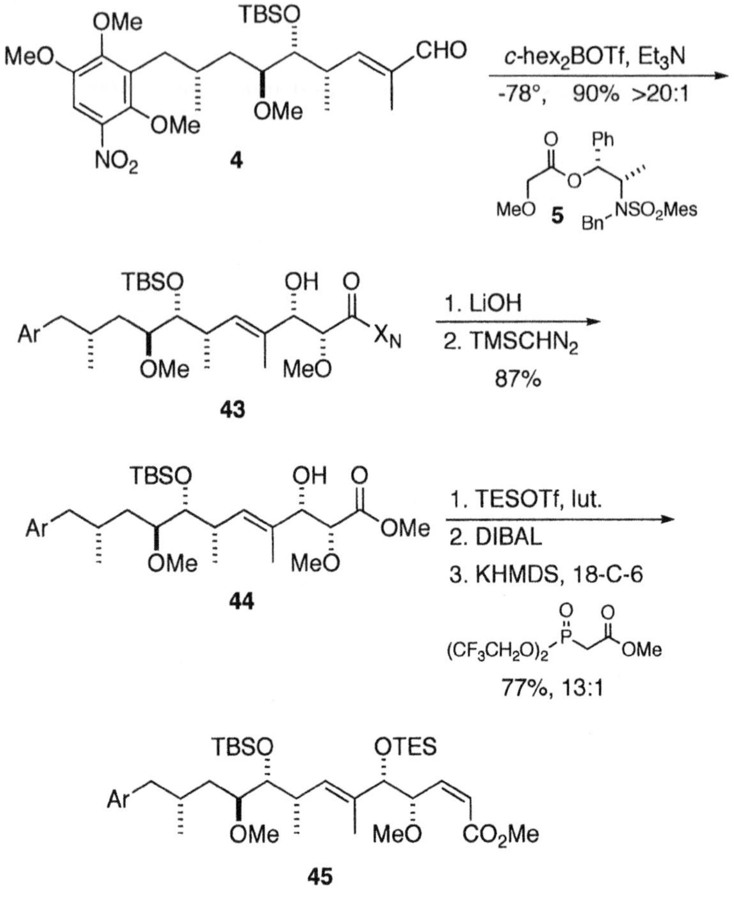

SCHEME 13

very slow and gave multiple products. The allyl ester from **46**, following the lead of Roush,[45] proved to be an ideal solution to this problem. Reduction of the aryl nitro group to the aniline was then performed using the conditions of Lalancette involving $NaBH_2S_3$ formed in situ from sodium borohydride and sulfur yellow.[46] The allyl ester was converted to the carboxylic acid using catalytic palladium (tetrakis)triphenylphosphine and morpholine to give seco acid **2**. Following previous ansamycin synthesis precedent, BOP-Cl (bis(2-oxo-3-oxazolidinyl) phosphinic chloride) was used with Hünig's base in warm toluene under high dilution (0.001 M), to give macrolactam **48** in 76% yield. At this point various

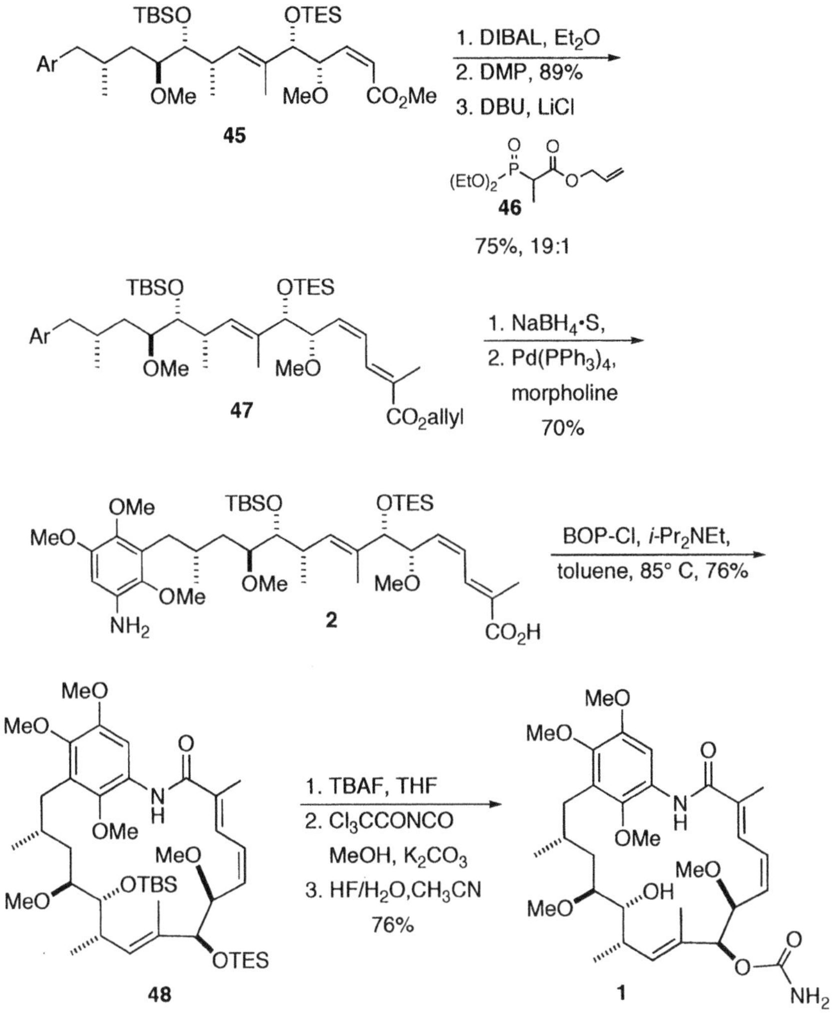

SCHEME 14

ansa ring conformations, including rotational isomers around the amide bond, complicated the NMR analysis. Changes in solvent and temperature gave only minimal improvement in line broadening. The C7 TES ether was removed with TBAF (tetra-*n*-butylammonium fluoride). The new conditions of Kocovsky, involving trichloroacetyl isocyanate,[47] were used to install the urethane in 89% yield. The standard conditions of TBAF or HF•pyridine failed to remove the C11 TBS group even when used in excess. Fortunately, it was found that 10:1 acetonitrile-aqueous HF (48%) gave alcohol 1 in 95% yield.

X.   Unsuccessful RCM-Based Convergent Approach

A highly convergent, RCM-based approach was explored without success (Scheme 15). Its discussion is included to highlight the limitations of this generally efficient method that has been used recently in numerous

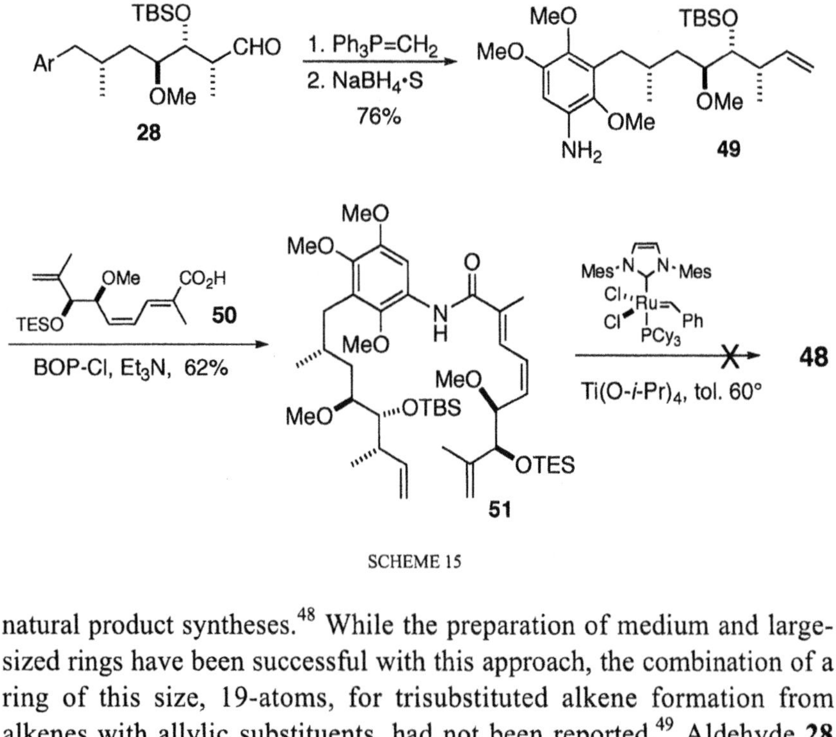

SCHEME 15

natural product syntheses.[48] While the preparation of medium and large-sized rings have been successful with this approach, the combination of a ring of this size, 19-atoms, for trisubstituted alkene formation from alkenes with allylic substituents, had not been reported.[49] Aldehyde 28 was converted to the terminal olefin and the aniline was formed as before. Diene acid 50 was made using a series of analogous steps including glycolate aldol and the allyl phosphonate reagent. Coupling under the BOP-Cl conditions then gave amide 51. The Grubbs imidazolium benzylidene ruthenium catalyst (10 mol%) was used under a variety of

tions, including various solvents, concentrations, modes of
ions, and additives-all without success.[50] Added titanium
opropoxide, used to break up allylic oxygen coordination, in warm
ie, also failed to give the desired macrocycle **48**. RCM reactions
he allylic alcohol, formed by TES removal at C7, and the diol with
'11 TBS removed, were also unsuccessful. A diphenylsiloxane
·ed substrate, formed through the C7, 11 hydroxyls for an 8-ring
also failed in this case.

Unanticipated Problematic Para-Quinone Formation

n the basis of the other ansamycin routes and the various reports of
ssful oxidative demethylation reactions reported by others, it was
ipated that lactam **1** would directly give geldanamycin upon
tion (Scheme 16).[9] Surprisingly, both silver or manganese oxide
·gnated with nitric acid gave the unusual azaquinone product **52** in
and 40% yields, respectively. Other oxidants, including CAN in

acetonitrile, even at −10 °C, gave only decomposition. The structure of **52** was confirmed by conversion to the dihydroquinone **53** using sodium hydrosulfite.[14b] NOE enhancements of 1.54 and 4.1% were observed between the indicated proton signals. Oxidation of the C11 TBS-protected lactam was also attempted and the azaquinone was again produced after only 10 minutes in 70–80% yield. No trace of desired quinone product

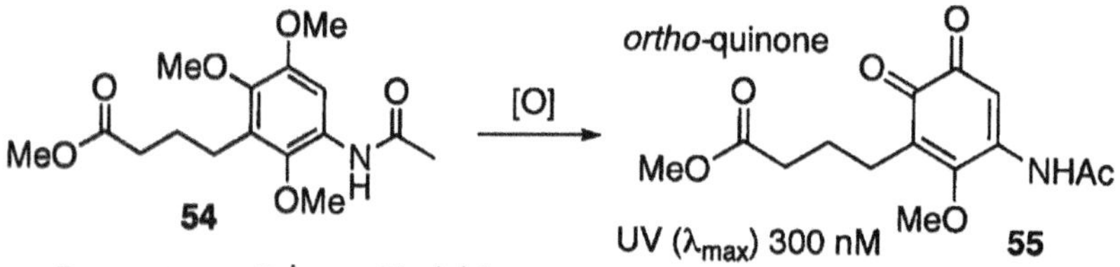

Reagent	quinone % yield
CAN	quant.
AgO/HNO$_3$	quant.
CoF$_3$	50%
MnO$_2$/HNO$_3$	25%
PhI(OAc)$_2$	trace
DDQ	NR
CrO$_3$/AcOH	NR

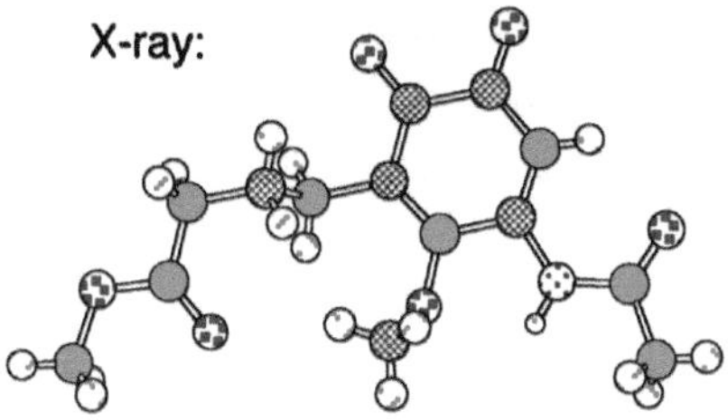

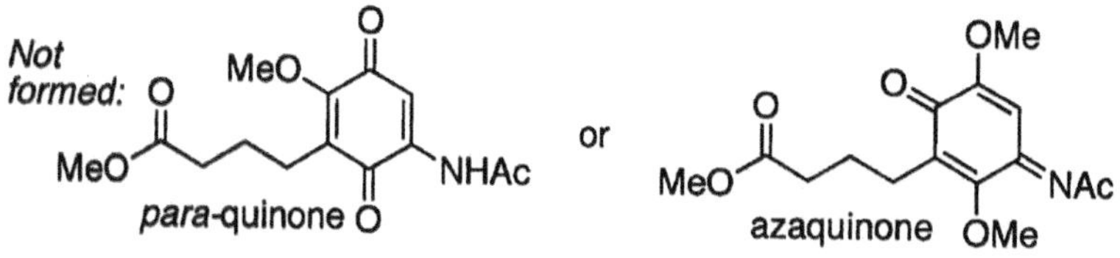

SCHEME 17

was detected by TLC, NMR, or MS in crude or purified fractions.

In anticipation of possible complications, a model trimethoxybenzene amide **54** was made and explored with various oxidants (Scheme 17). CAN and silver oxide rapidly produced a quinone, not azaquinone, in near quantitative yield.[51] Other oxidants as listed were not as successful. Initially, through NMR comparison to related compounds it was assumed that the quinone product obtained was the *para*-quinone. Only after considerable effort was a crystal suitable for X-ray analysis obtained. Surprisingly, this unambiguous result confirmed that the quinone product was in fact the *ortho*-quinone product **55**, not the *p*-quinone. UV data was obtained for **55**, λ_{max} at 300 nM for the $\pi-\pi^*$ (CHCl$_3$, K-band). The corresponding value for geldanamycin is much higher at 311 nM.

The mechanism of dealkylative quinone formation, in accord with the work of Kochi,[52] involves one-electron removal to give a radical cation (Scheme 18). The *para*-disposed methoxyls are best able to stabilize the charge in this case. Addition of water, loss of methanol and a proton, followed by additional loss of an electron then gives an allyl cation **56**. Water can then attack either ortho or para to the carbonyl leading to quinone formation. Preferred attack at the ortho position seems to be due to steric factors and possible destabilization of the allyl cation at the para position. The adjacent amide may force the methyl ether to adopt a conformation that does not optimize lone-pair donation.

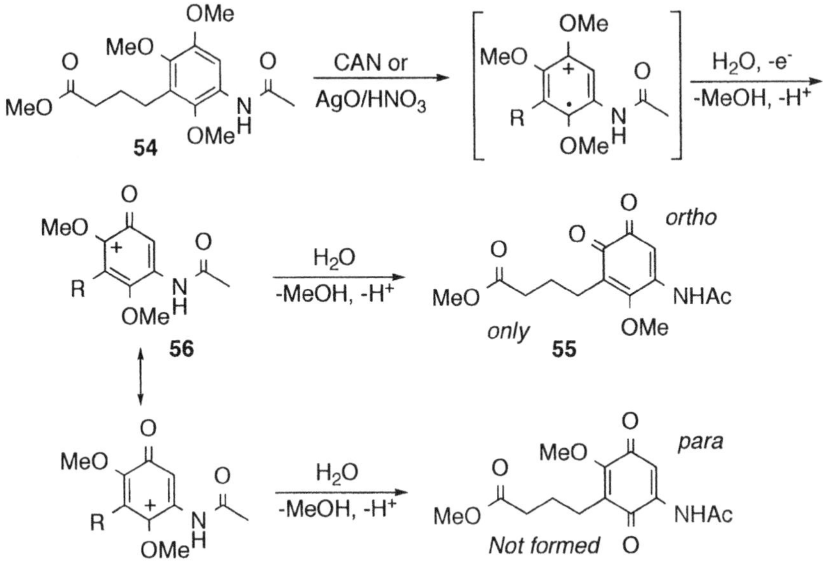

SCHEME 18

The lack of quinone formation in the case of **1**, and other related lactams, under common oxidation conditions, may be rationalized using a conformational argument (Scheme 19). Again, in these examples, no *p*-quinone formation was observed, only azaquinone product **52** was obtained. A stereoelectronic rationale is made difficult by the presence of many low energy conformations that the lactam seems to adopt,[53] as seen by line broadening in the NMR. An important clue can be found by placing lactam **1** in the Hsp90 bound conformation for geldanamycin as shown. The amide nitrogen is found in a twisted *s-cis* conformation where the lone-pair on nitrogen is not fully conjugated with the carbonyl oxygen due to the constraints of the large ring. Instead of the normal planar dihedral angle of 0°, in this case the out of plane twisting shows an angle of 12°. This conformation suggests that the amide nitrogen electronically approximates an aniline-type nitrogen with the potential for strong cation stabilization upon ring oxidation. The less electronegative nitrogen, compared to oxygen, is then more able to donate its lone-pair, leading to azaquinone formation instead of either *p*- or *o*-quinone products. In the case of the acyclic model **54**, the amide is flat and the nitrogen lone pair is not as available due to full amide bond resonance.

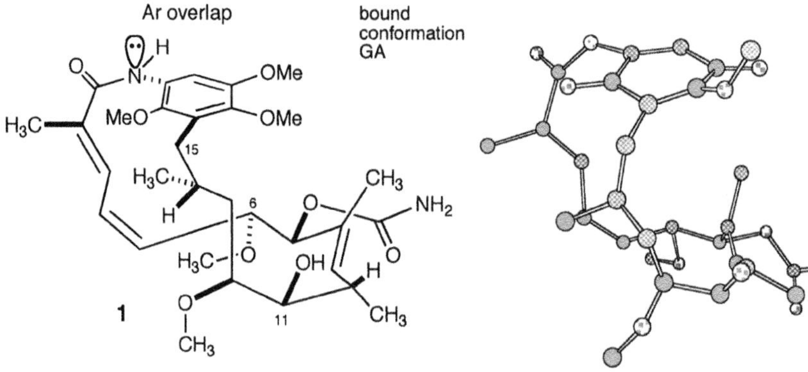

SCHEME 19

To overcome this conformationally induced electronic bias, derivatives were made to further promote nitrogen lone pair delocalization, in an attempt to favor quinone formation (Scheme 20). Following urethane formation with intermediate **48**, treatment with *tert*-butyloxycarbonyl anhydride (BOC$_2$O) gave the *N*-BOC lactam **57**. These conditions also led to BOC protection of the urethane.[54] CAN then gave a quinone type product in low yield from **57**. In this case no aza-quinone was observed.

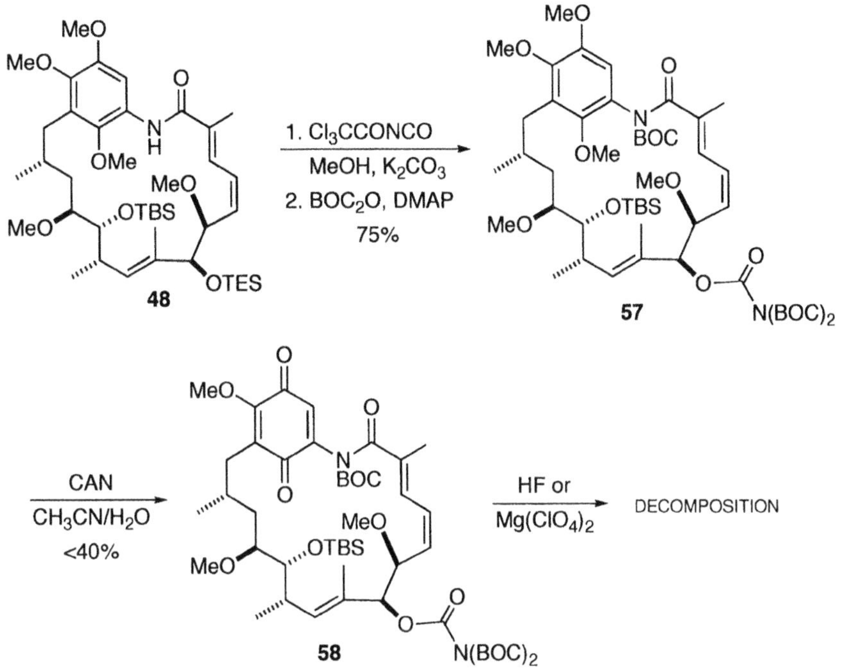

SCHEME 20

Unfortunately, all conditions investigated for BOC removal, dilute HCl,[55] HF, Mg(ClO$_4$)$_2$[56] gave only decomposition, not geldanamycin.

Finally, following a reference by Musgrave,[57] use of nitric acid alone as an oxidant was attempted with **1** (Scheme 21). A glacial acetic acid solution of **1** was treated with 70% nitric acid for 1 min followed by quenching with sodium bicarbonate. Longer reaction times led to extensive decomposition. The more polar, red-orange *ortho*-geldanamycin **59** (UV λ_{max} at 303 nM) was the major product (10:1) over the less polar, yellow-orange natural product geldanamycin, which was shown to be identical in all respects to natural material (NMR, TLC, UV λ_{max} at 311 nM). Multiple runs on 10 mg scale were performed with reproducible results. Success in this case over the other conditions may be attributed to protonation of the amide prior to oxidation, which renders the nitrogen lone pair less available for cation stabilization. As with the model amide **54**, lactam **1** also prefers *ortho*-quinone formation.

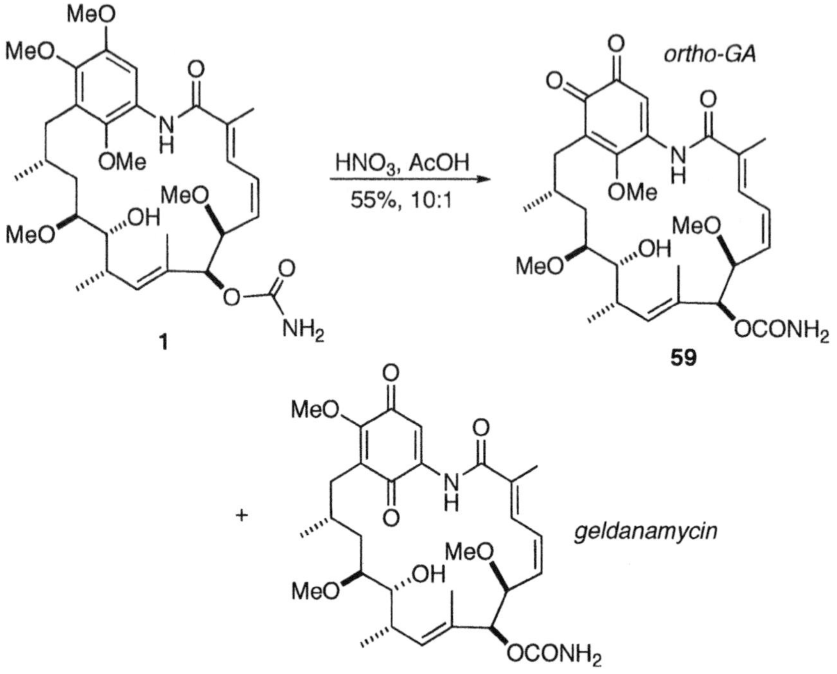

SCHEME 21

XII. Successful Strategy for Para-Quinone Formation Using a 1, 4-Di-Protected Hydroquinone Precursor

A new general approach to selective formation of the methoxy substituted *para*-quinone was developed. Methoxyphenol and 1,4-dihydroquinone routes were both investigated using model systems with unsaturated amides and side chains attached (Scheme 22).[51] The benzene synthesis sequence was modified for the installation of MOM (methoxymethyl) and methyl ethers at the indicated positions on compound **60**. The protected material was converted to the unsaturated amides **61**. TMS iodide, formed in situ from TMS chloride and sodium iodide, was used to cleanly remove the MOM ether giving **62** (R=Me). Surprisingly, the various known oxidants for this transformation proved to be ineffective for the generation of the desired *para*-quinone **63**, including CAN, DDQ (dicyanodichloroquinone),[58] salcomine[59] and Co(salen)[60] oxygen, and iodosobenzene.[61] Importantly, however, the major product obtained in this case, was clearly the desired *para*-quinone **63** (UV λ_{max} at

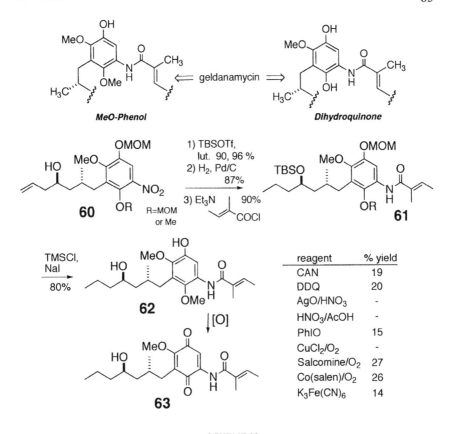

SCHEME 22

310 nM). The salcomine-oxygen combination was best but led to only a 27% isolated yield.

Fortunately, the 1,4-di-MOM model compound **64** proved to be superior (Scheme 23). Removal of both MOM ethers efficiently provided the desired 1,4-dihydroquinone **65** in 79% yield. Use of Rapoport's reported conditions with 10% palladium on carbon in ethyl acetate then gave the previously obtained *p*-quinone product **63** in 98% isolated yield.[62] Surprisingly, these highly efficient conditions have not been used for the synthesis of *p*-quinone containing natural products. We are currently pursuing a second generation of geldanamycin using this methodology.

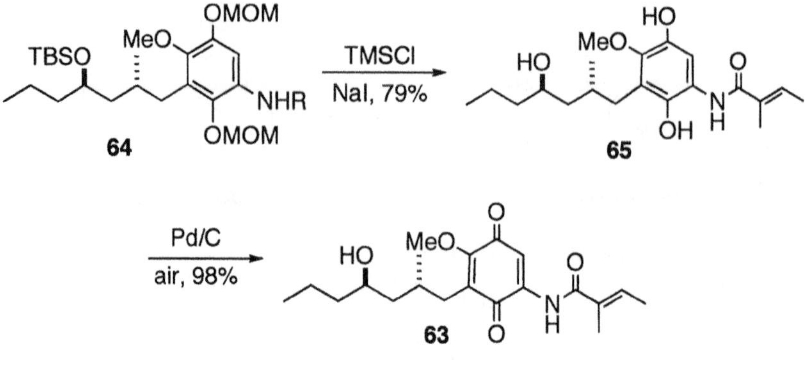

SCHEME 23

XIII. Conclusions

The total synthesis of the important Hsp90 inhibitor geldanamycin has been described. Two new aldol reactions, as general solutions to *anti* and *syn* 1,2 diol formation, were developed for this route. More efficient, convergent approaches, as alternatives to the linear route, were designed

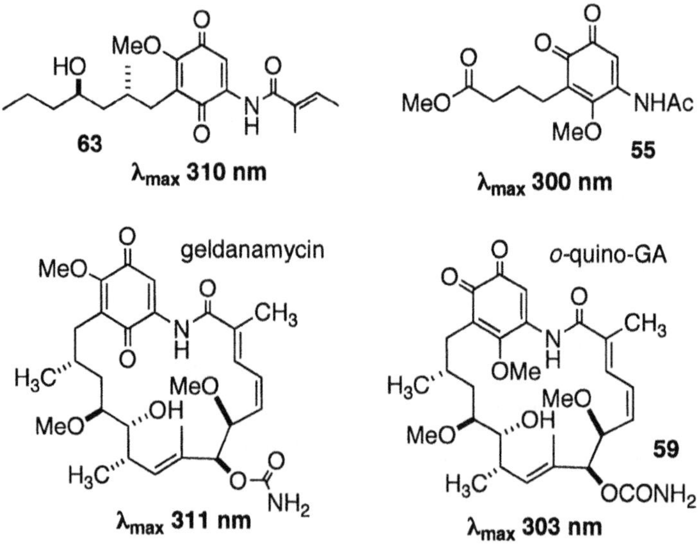

λ_{max} **310 nm**

λ_{max} **300 nm**

geldanamycin

o-quino-GA

λ_{max} **311 nm**

λ_{max} **303 nm**

SCHEME 24

and attempted without success. Important information concerning installation of the *para*-quinone was obtained. UV analysis was used to correlate *ortho* and *para* quinones, both in model systems and with the final ansamycin lactam products (Scheme 24). These findings will prove valuable for the design and synthesis of future compounds based on the geldanamycin template.

Acknowledgments

The successful total synthesis of geldanamycin described herein was possible as a result of the combined efforts of an exceptionally talented and devoted group of co-workers. In addition to Dr. Erik L. Meredith and Mr. Erik J. Hicken, Dr. B. B. V. Soma Sekhar, and Dr. Timothy M. Turner, Mr. Bryon L. Simmons, Mr. Russell R. Glancey, and Mr. Kris G. Mendenhall were all key members of the research team responsible for the design and execution of this synthesis. In addition, we would like to thank Drs John F. Cannon and N. Kent Dalley for X-ray crystallography assistance throughout the project.

References and Footnotes

1. De Boer, C., Meulman, P. A.; Wnuk, R. J., Peterson, D. H., *J. Antibiot.* **1970**, *23*, 442.

2. Rinehart, K. L.; Sasaki, K., Slomp, G.; Grostic, M. F., Olson, E. C., *J. Am. Chem. Soc.* **1970**, *92*, 7591.

3. Rinehart, K. L., *Acc. Chem. Res.* **1972**, *5*, 57.

4. (a) Furusaki, A., Matsumoto, T., Nakagawa, A., Omura, S., *J. Antibiot,* **1980**, *33*, 781. (b) Muroi, M., Haibara, K., Asai, M., Kamiya, K., Kishi, T., *Tetrahedron,* **1981**, *37*, 1123.

5. Herbimycin A: (a) Nakata, M., Osumi, T., Ueno, A., Kimura, T. Tamai, T., Tatsuda, K., *Tetrahedron Lett.* **1991**, *32*, 6015. (b) Carter, K. D.; Panek, J. S. *Org. Lett.* **2004**, *6*, 55-57. Macbecin I: (c) Baker, R., Castro, J. L., *J. Chem. Soc. Perkin Trans.* **1990**, 47. (d) Evans, D. A., Miller, S. J., Ennis, M. D., *J. Org. Chem.* **1993**, *58*, 471. (e) Panek, J. S., Xu, F., Rondon, A. C., *J. Am. Chem. Soc.* **1998**, *120*, 4413.

6. (a) Andrus, M. B., Meredith, E. L., Simmons, B. L., Soma Sekhar, B. B. V., Hicken, E. J., *Org. Lett.* **2002**, *4*, 3549. (b) Andrus, M. B., Meredith, E. L., Hicken, E. J., Simmons, B. L., Glancey, R. R., Ma, W., *J. Org. Chem.* **2003**, *68*, 8162.

7. Schulte, T. W., Neckers, L. M., *Cancer Chemother. Pharmacol.* **1998**, *42*, 273.

8. (a) Andrus, M. B., Soma Sekhar, B. B. V. , Meredith, E. L., Dalley, N. K., *Org. Lett.* **2000**, *2*, 3035. (b) Andrus, M. B., Meredith, E. L. , Soma Sekhar, B. B. V., *Org. Lett.* **2001**, *3*, 259. (c) Andrus, M. B., Mendenhall, K. G., Meredith, E. L. Soma Sekhar, B. B. V., *Tetrahedron Lett.* **2002**, *43*, 1789-1792. (d) Andrus, M. B., Soma Sekhar, B. B. V., Turner, T. M., Meredith, E. L., *Tetrahedron Lett.* **2001**, *42*, 7197.

9.   (a) Cameron, D. W., Feutrill, G. I., Patti, A. F., Perlmutter, P., Sefton, M. A., *Aust. J. Chem.* **1982**, *35*, 1501. (b) Witiak, D. T., Loper, J. T., Anathan, S., Almericao, A. M., Verhoef, V. L., Filppi, J. A., *J. Med. Chem.* **1989**, *32*, 1636. (c) Kozuka, T., *Bull. Chem. Soc. Jpn.* **1982**, *55*, 2415. (d) Cheng, A. C., Castagnoli, N., *J. Med. Chem.* **1989**, *32*, 1636. (e) Michael, J. P, Cirillo, P. F., Denner, L., Hosken, G., Howard, A. S., Tinkler, O. S., *Tetrahedron* **1990**, *46*, 7923. (f) Luly, J. R., Rapoport, H., *J. Org. Chem.* **1981**, *46*, 2745. (g) Kitahara, Y., Nakahara, S., Shimizu, M., Yonezawa, T., Kubo, A., *Heterocycles* **1993**, *36*, 1909.

10.   (a) Abiko, A., Liu, J. –F., Masamune, S., *J. Am. Chem. Soc.* **1997**, *119*, 2586. (b) Yoshimitsu, T., Song, J. J., Wang, G. –Q., Masamune, S., *J. Org. Chem.* **1997**, *62*, 8978. (c) Abiko, A., Liu, J. –F., Buske, D. C., Moriyama, S., Masamune, S., *J. Am. Chem. Soc.* **1999**, *121*, 7168. (d) Liu, J.-F., Abiko, A., Pei, Z., Buske, D. C., Masamune, S., *Tetrahedron Lett.* **1998**, *39*, 1873.

11.   (a) Martin, S. F., Dodge, J. A., Burgess, L. E., Hartmann, M., *J. Org. Chem.* **1992**, *57*, 1070. (b) Martin, S. F., Dodge, J. A., Burgess, L. E., Limberakis, C., Hartmann, M., *Tetrahedron,* **1996**, *52*, 3229. (c) Martin, S. F., Burgess, L. E., Limberakis, C., Hartmann, M., *Tetrahedron,* **1999**, *55*, 3561. (d) Kallmerten, J., Coutts, s. J., *Tetrahedron Lett.* **1990**, *31*, 4305. (e) Marshall, J. A., Sedrani, R., *J. Org. Chem.* **1991**, *56*, 5496.

12.   Whitesell, L., Shifrin, S. D., Schwab, G., Neckers, L. M., *Cancer. Res.* **1992**, *52*, 1721.

13.   Rinehart, K. L., McMillan, M. W., Witty, T. R., Tipton, C. D., Shield, L. S. Li, L. H., Reusser, F., *Bioorg. Chem.* **1977**, *6*, 353.

14.   (a) Schnur, R. C., Corman, M. L., *J. Org. Chem.* **1994**, *59*, 2581. (b) Schnur, R. C., Corman, M. L., Gallaschun, R. J., Cooper, B. A., Dee, M. F., Doty, J. L., Muzzi, M. L., Moyer, J. D., DiOrio, C. I., Barbacci, E. G., Miller, P. E., O'Brien, A. T., Morin, M. J., Foster, B. A., Pollack, V. A., Savage, D. M., Sloan, D. E., Pustilnik, L. R., Moyer, M. P., *J. Med. Chem.* **1995**, *38*, 3806.

15.   Isaacs, J. S., Xu, W., Neckers, L. M., *Cancer Cell,* **2003**, *3*, 213.

16.   Uehara, Y., Hori, M., Takeuchi, T., Umezawa, Y., *Mol. Cell Biol.* **1986**, *6*, 2198.

17.   Whitesell, L., Shifrin, S. D., Schawb, G., Neckers, L. M., *Cancer Res.* **1992**, *52*, 1721.

18.   Whitesell, L., Mimnaugh, E. G., De Costa, B., Myers, C. E., Neckers, L. M., *Proc. Natl. Acad. Sci. USA* **1994**, *91*, 8324.

19.   Lawrence, D. S., Niu, J., *Pharmacol. Ther.* **1998**, *77*, 81.

20.   (a) Stebbins, C. E., Russo, A. A., Schneider, C., Rosen, N., Hartl, F. U., Pavletich, N. P., *Cell,* **1997**, *89*, 239. (b) Roe, S. M., Prodromou, R. O., Ladbury, J. E., Piper, P. W., Pearl, L. H., *J. Med. Chem.* **1999**, *42*, 260.

21.   Kamal, A., Thao, L., Sensintaffar, J., Zhang, L., Boehm, M. F., Fritz, L. C., Burrows, F. J., *Nature* **2004**, *425*, 407.

22.   Kuduk, S. D., Zheng, F. F., Sepp-Lorenzino, L., Rosen, N., Danishefsky, S. J., *Bioorg. Med. Chem. Lett.* **1999**, *9*, 1233.

23.   Kuduk, S. D., Harris, C. R., Zheng, F. F., Sepp-Lorenzino, L., Ouerfelli, Q., Rosen, N., Danishefsky, S. J., *Bioorg. Med. Chem. Lett.* **2000**, *10*, 1303.

24.   Mandler, R., Dadachova, K., Brechbiel, J. K., Waldemann, T. A., Brechbiel, M. W., *Bioorg. Med. Chem. Lett.* **2000**, *10*, 1025.

25.   Chiosis, G., Rosen, N., Sepp-Lorenzino, L., *Bioorg. Med. Chem. Lett.* **2001**, *11*,

909.

26.   Stead, P., Latif, S., Blackaby, A. P., Sidebottom, P. J., Deakin, A., Taylor, N. L., Life, P., Spaull, J., Burrell, F., Jones, R., Lewis, J., Davidson, I., Mander, T., *J. Antibiot.* **2000**, *53*, 657.

27.   Hiramoto, S. , Sugita, M., Ando, C., Sasaki, T., Furihata, K., Seto, H., Otake, N., *J. Antibiot.* **1985**, *38*, 1103.

28.   (a) Smith, A. B., Wood, J. L., Wong, W., Gould, A. E., Rizzo, C. J., Barbosa, J., Komiyama, K., Omura, S., *J. Am. Chem. Soc.* **1996**, *118*, 8308. (b) Masse, C. E., Yang, M., Solomon, J., Panek, J. S., *J. Am. Chem. Soc.* **1998**, *120*, 4123. (c) Smith, A. B., Wan, Z., *J. Org. Chem.* **2000**, *65*, 3738.

29.   (a) Corey, E. J., Weigel, L. O., Chamberlin, A. R., Cho, H., Hua, D. H., *J. Am. Chem. Soc.* **1980**, *102*, 6613. (b) Meyers, A. I., Reider, P. J., Campbell, A. L., *J. Am. Chem. Soc.* **1980**, *102*, 6597. (c) Kitamura, M., Isobe, M., Ichikawa, Y., Goto, T., *J. Org. Chem.* **1984**, *49*, 3517. (d) Benechie, M., Khuong-Huu, F., *J. Org. Chem.* **1996**, *61*, 7133.

30.   Gilman, H., Thirtle, J. R., *J. Am. Chem. Soc.* **1944**, *66*, 858.

31.   Evans, D. A., Ennis, M. D., Mathre, D. J., *J. Am. Chem. Soc.* **1982**, *104*, 1737.

32.   (a) Wilk, B., *Syn. Commun.* **1993**, *23*, 2481. (b) Tsunoda, T., Uemoto, K., Nagino, C., Kawamura, M., Kaku, H., Ito, S., *Tetrahedron Lett.* **1999**, *40*, 7355.

33.   (a) Levine, S. G., *J. Am. Chem. Soc.* **1958**, *80*, 6150. (b) Williams, D. R., Barner, B. A., Nishitani, K., Phillips, J. G., *J. Am. Chem. Soc.* **1982**, *104*, 4708.

34.   (a) Hruby, V. J., Jarosinski, M. A., Li, G., *Tetrahedron Lett.* **1993**, *34*, 2561. (b) Williams, D. R., Kissel, W. S., Li, J. J., *Tetrahedron Lett.* **1998**, *39*, 8593. (c) Wipf, P., Takahashi, H., *J. Chem. Soc. Chem. Commun.* **1996**, 2675. (d) Romo, D. R., Rzasa, R. M., Shea, H. A., *J. Am. Chem. Soc.* **1998**, *120*, 591. (e) Williams, D. R., Kissel, W. S., *J. Am. Chem. Soc.* **1998**, *120*, 11198.

35.   (a) Gennari, C. In *Comprehensive Organic Synthesis*, Trost, B. M., Fleming, I., Eds., Pergamon Press: New York, 1991, Vol. 2, Chapter 2.4. (b) Cameron, J. C., Paterson, I. *Org. Reac.* **1997**, *51*, 1. *Syn* versions: (c) Evans, D. A., Bender, S. L., Morris, J., *J. Am. Chem. Soc.* **1988**, *110*, 2506. (d) Andrus, M. B., Schreiber, S. L., *J. Am. Chem. Soc.* **1993**, *115*, 10420. (e) Evans, D. A., Barrow, J. C., Leighton, J. L., Robichaud, A. J., Sefkow, M., *J. Am. Chem. Soc.* **1994**, 116, 12111. (f) Crimmins, M. T., Choy, A. L., *J. Org. Chem.* **1997**, *62*, 7548. Anti : (g) Evans, D. A., Kaldor, S. W., Jones, T. K., Clardy, J., Stout, T. J., *J. Am. Chem. Soc.* **1990**, *112*, 7001. (h) Evans, D. A., Gage, J. R., Leighton, J. L., Kim, A. S., *J. Org. Chem.* **1992**, *57*, 1961. (i) Evans, D. A., Weber, A. E. *J. Am. Chem. Soc.* **1986**, *108*, 6757-6761. (j) Li, Z., Wu, R., Michalczyk, R., Dunlap, R. B., Odom, J. D., Silks, L. A. III, *J. Am. Chem. Soc.* **2000**, *122*, 386. (k) Kobayashi, S., Kawasuji, T., *Tetrahedron Lett.* **1994**, *35*, 3329. (l) Kobayashi, S., Hayashi, T., *J. Org. Chem.* **1995**, *60*, 1098. (m) Gennari, C., Vulpetti, A., Pain, G. ,*Tetrahedron* **1997**, *53*, 5909. (n) List, B., Lerner, R. A., Barbas, C. F., *J. Am. Chem. Soc.* **2000**, *122*, 2395.

36.   (a) Wang, L., Sharpless, K. B., *J. Am. Chem. Soc.* **1992**, *114*, 7568-7569. (b) Kolb, H.C., VanNieuwenhze, M. S., Sharpless, K. B., *Chem Rev.* **1994**, *94*, 2483.

37.   (a) Burke, S. D., Sametz, G. M., *Org. Lett.* **1999**, *1*, 71. (b) David, S., Thieffry, A., Veyrieres, A., *J. Chem. Soc., Perkin Trans. 1* **1981**, 1796.

38.   (a) Evans, D. A., Fu, G. C., Hoyveda, A. H., *J. Am. Chem. Soc.* **1992**, *114*, 6671. (b) Evans, D. A., Fu, G. C., Anderson, B. A., *J. Am. Chem. Soc.* **1992**, *114*, 6679.

(c) Burgess, K., Cassidy, J., Ohlmeyer, M. J., *J. Org. Chem.* **1991**, *56*, 1020. (d) Burgess, K., van der Donk, W. A., Jarstfer, M. B., Ohlmeyer, M. J., *J. Am. Chem. Soc.* **1991**, *113*, 6139.

39. Houk, K. N., Rondan, N. G., Wu, Y.-D., Metz, J. T. Paddon-Row, M. N., *Tetrahedron* **1984**, *40*, 2257.

40. Blanchette, M. A., Choy, W., Davis, J. T., Essenfeld, A. P., Masamune, S., Roush, W. R., Sakai, T., *Tetrahedron Lett.* **1984**, *25*, 2183.

41. Sharpless, b. K., Hanson, R. M., Klunder, J. M., Yo, S. Y., Masamune, H., Gao, Y., *J. Am. Chem. Soc.* **1987**, *109*, 5765.

42. (a) Suzuki, T., Saimoto, H., Tomioka, H., Oshima, K., Nozaki, H., *Tetrahedron Lett.* **1982**, *23*, 3597. (b) Roush, W. R., Adam, M. A., Peseckis, S. M., *Tetrahedron Lett.* **1983**, *24*, 1377.

43. Hirai, Y., Aida, T., Inoue, S., *J. Am. Chem. Soc.* **1987**, *111*, 3062.

44. Still, W. C., Gennari, C., *Tetrahedron Lett.* **1983**, *24*, 4405.

45. Roush, W. R., Sciotti, R. J., *J. Am. Chem. Soc.* **1998**, *120*, 7411.

46. (a) Lalancette, J. M., Freche, A., Monteaux, R., *Can. J. Chem.* **1968**, *46*, 2754. (b) Lalancette, J. M., Freche, a., Brindle, J. R., Laliberte, M., *Synthesis* **1972**, 526.

47. Kocovsky, P., *Tetrahedron Lett.* **1986**, *27*, 5521.

48. (a) Trnka, T. M., Grubbs, R. H., *Acc. Chem. Res.* **2001**, *34*, 18-29. (b) Furstner, A., *Angew. Chem. Int. Ed.* **2000**, *39*, 3012.

49. (a) Furstner, A., Thiel, O. R., Blanda, G., *Org. Lett.* **2000**, *2*, 3731. (b) Xu, Z., Johannes, C. W., Houri, A. F., La, D. S., Cogan, D. A., Hofilena, G. E., Hoveyda, A. H., *J. Am. Chem. Soc.* **1997**, *119*, 10302.

50. A test substrate, *o*-allylphenyl acrylate, was employed under these conditions with success to verify the activity of the catalyst, >90% isolated yield.

51. Andrus, M. B., Hicken, E. J., Meredith, E. L., Simmons, B. L., Cannon, J. F., *Org. Lett.* **2003**, *5*, 3859.

52. Rathore, R., Bosch, E., Kochi, J. K., *J. Chem. Soc. Perkin Trans. 2* **1994**, 1157.

53. Lee, Y. –S., Marcu, M. G., Neckers, L., *Chem. Biol.* **2004**, *11*, 991.

54. Ragnarrson, U., Grehn, L., *Acc. Chem. Res.* **1998**, *31*, 494.

55. Boger, D. L., McKie, J. A., Nishi, T., Ogiku, T., *J. Am. Chem. Soc.* **1996**, *118*, 2301.

56. Stafford, J. A., Bracken, D. S., Karanewski, D. S., Valvano, N. L., *Tetrahedron Lett.* **1993**, *34*, 7873.

57. Musgrave, O. C., *Chem Rev.* **1969**, 499.

58. (a) Majetich, G., Zhang, Y., *J. Am. Chem. Soc.* **1994**, *116*, 4979. (b) Fukuyama, T., Yang, L., *J. Am. Chem. Soc.* **1987**, *109*, 7881.

59. Boger, D. L., Garbaccio, R. M., *J. Org. Chem.* **1999**, *64*, 8350.

60. (a) Lipshutz, B. H., Mollard, P., Pfeiffer, S. S., Chrisman, W., *J. Am. Chem. Soc.* **2002**, *124*, 14282. With Cu see: (b) Matsumoto, M., Kobayashi, H., Hotta, Y., *Synth. Commun.* **1985**, *15*, 515.

61. (a) Myers, A. G., Kung, D. W., *J. Am. Chem. Soc.* **1999**, *121*, 10828. (b) Layton, M. E., Morales, C. A., Shair, M. D., *J. Am. Chem. Soc.* **2002**, *124*, 773.

62. (a) Luly, J. R., Rapoport, H., *J. Org. Chem.* **1984**, *49*, 1671. For other routes and oxidants involving 1,4-di-MOM substrates see: (b) Roush, W. R., Coffey, D. S., Madar, D. J., *J. Am. Chem. Soc.* **1997**, *119*, 11331. (c) Noland, W. E., Kedrowski, B. L,. *Org. Lett.* **2000**, *2*, 2109.

STRATEGIES AND TACTICS IN ORGANIC SYNTHESIS, VOL. 6

Chapter 3

FROM METHYLENE BRIDGED GLYCOLURIL DIMERS TO CUCURBIT[N]URIL ANALOGS WITH SOME DETOURS ALONG THE WAY

Lyle Isaacs and Jason Lagona
Department of Chemistry and Biochemistry
University of Maryland
College Park, Maryland 20742

I. Introduction

A. STRUCTURES

In 1905, Behrend reported the condensation of glycoluril (1_H) with formaldehyde under strongly acidic conditions to yield an insoluble material now known as Behrend's polymer.[1] To obtain a tractable product, this methylene bridged glycoluril oligomer was recrystallized from hot sulfuric acid yielding a well-defined compound whose co-crystallization with a variety of metal salts was studied extensively

(Scheme 1). In a now classic paper published in 1981, Mock disclosed the structure of Behrend's substance (**CB**[6]) and dubbed it cucurbituril because it is shaped like a pumpkin, which is a prominent member of the *cucurbitaceae* family of plants.[2] In this review, we refer to cucurbituril as **CB**[6] to distinguish it from its homologs (**C B**[n]), subsequently synthesized, which contain a different number of glycoluril units.[3-5]

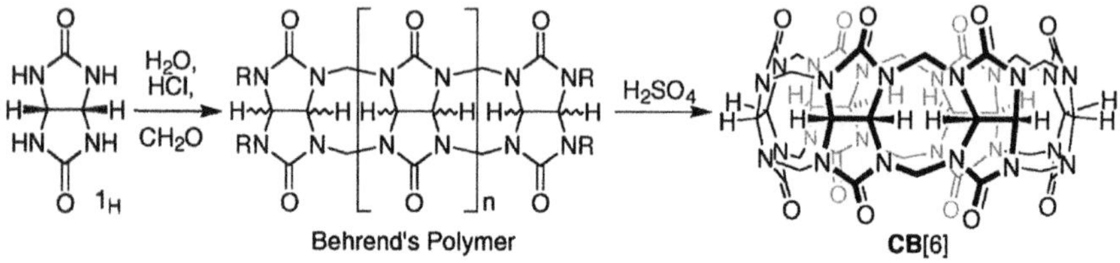

SCHEME 1. Behrend's synthesis of **CB**[6].

B. RECOGNITION PROPERTIES

In a series of elegant papers in the 1980's, Mock and co-workers delineated the recognition properties of **CB**[6].[6-9] **CB**[6] was shown to bind to a wide variety of alkylammonium and alkanediammonium ions in 1:1 $HCO_2H:H_2O$ with dissociation constants (K_d) in the μM range, driven by a combination of the hydrophobic effect and ion-dipole interactions. The binding interactions between **C B**[6] and its guests are highly selective, effectively discriminating between guests on the basis of molecular length, shape, and size (Scheme 2). For example, **CB**[6] binds 305-fold more tightly to $H_3N(CH_2)_6NH_3$ (**2**) than $H_3N(CH_2)_8NH_3$ (**3**). Similarly, **CB**[6] binds tightly to para-substituted **4** but completely rejects the isomeric meta-substituted **5**. Lastly, **C B**[6] binds **6**, which has a volume of 86 $Å^3$, but rejects **7** (89 $Å^3$), whose volume is just slightly larger.[10] The highly selective recognition properties displayed by **CB**[6] were attributed to the relative rigidity of this macrocyclic hexamer. In the late 1980's and 1990's, the groups of Mock,[11-13] Buschmann,[14-16] and Kim[17-20] took advantage of these outstanding recognition properties and used **CB**[6] in advanced applications including the catalysis of a dipolar cycloaddition, the remediation of textile waste streams, molecular switches, and as a bead in the construction of 1-, 2-, and 3-dimensional polyrotaxane networks. The **CB**[n] family has been the subject of a number of recent reviews.[21-23]

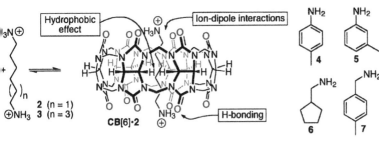

SCHEME 2. Recognition properties of **CB**[6].

ISSUES TO BE ADDRESSED AND CHALLENGES TO BE MET THROUGH
SYNTHETIC CHEMISTRY

Based on the pioneering work of Mock, Buschmann, and Kim it was
that **CB**[6] was a talented host, but a series of perceived deficiencies
limited the range of applications to which **CB**[6] could be applied. These
perceived limitations included: 1) the poor solubility of **CB**[6] in H_2O
which necessitated the use of 1:1 $HCO_2H:H_2O$ for binding studies, 2) the
lack of a homologous series of hosts (e.g. **CB**[n], n = 5, 7, 8) with
different sized cavities similar to α-, β-, and γ-cyclodextrin, 3) the
insolubility of **CB**[6] in organic solvents (e.g. $CHCl_3$, C_6H_6, and CH_3CN),
4) an inability to access tailor-made **CB**[n] derivatives bearing
functional groups along its equator or at its bridging methylene groups.
In recent years, many of these issues have been either partially or fully
addressed.[4,21,23,24]

Shortly after the disclosure of the structure of **CB**[6] by Mock, the
group[25,26] sought to address these questions and attempted the
macrocyclization of diphenylglycoluril **1**Ph with CH_2O in benzene
containing $CH_3C_6H_4SO_3H$ (PTSA) as an acid catalyst (Scheme 3a). This
reaction unexpectedly delivered molecular clip **8** rather than a **CB**[6]
derivative. The Nolte group has subsequently prepared and studied a
range of xylylene walled glycoluril based molecular clips and
demonstrated their unusual recognition and assembly properties.[26]
Compound **8** with its two aromatic and one glycoluril subunit connected
by methylene bridges contains one-half (3/6) of the number of subunits
present in **CB**[6]. In 1992, the Stoddart group reported that
dimethylglycoluril **1**Me undergoes macrocyclization when heated with
formaldehyde in aq. HCl, but delivers the cyclic pentameric macrocycle
CB[5] (Scheme 3b).[27] These studies raised more questions than they
answered. What is the scope of glycoluril derivatives that will undergo

ormation of macrocyclic **CB**[n] derivatives? Why does 1_H give the yclic hexamer **CB**[6] whereas 1_{Me} yields the cyclic pentameric $\text{Me}_{10}\text{CB}[5]$? Given this complexity of the **CB**[6] forming reaction, what actors are responsible for the remarkably high yield obtained (82%)?

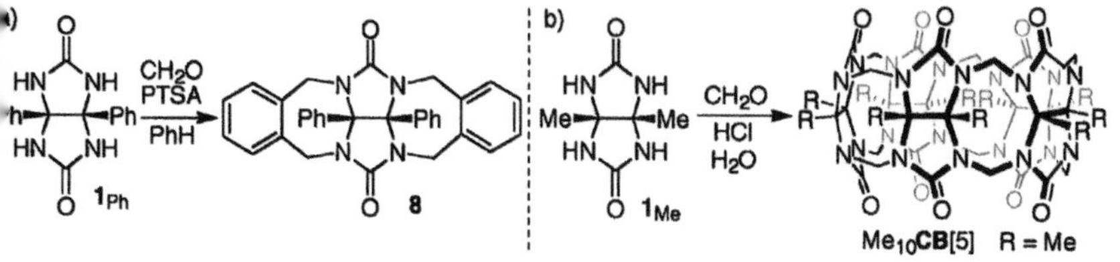

SCHEME 3. Synthesis of a) **8** and b) $\text{Me}_{10}\text{CB}[5]$.

I. Retrosynthetic Analysis of the Cucurbit[n]uril Family

The complexity associated with **CB**[6] synthesis – formation of 6 rings, 12 methylene bridges, with complete control over the relative stereochemistry of 6 glycoluril rings – is daunting. When we started our work in this area in 1998, none of the issues identified above had been addressed. Our retrosynthetic analysis of **CB**[6] identified the methylene bridged glycoluril dimer substructure (shown in bold) as the fundamental building block of the **CB**[n] family (Scheme 4).[28] Within **CB**[6], the methine protons on adjacent glycoluril rings are *cis* to one another. Upon excising the methylene bridged glycoluril dimer substructure from **CB**[6] it becomes clear that two diastereomers are possible, namely **9C** and **9S**. We refer to these compounds as C-shaped and S-shaped because the X-ray structures of **10C** and **11S** (Figure 1) resemble those letters. To restrict our studies to C- and S-shaped dimers – rather than oligomers – we retrosynthetically appended *o*-xylylene groups, which suggested **10C** and **10S** as initial targets. We anticipated that **10C** and **10S** would be accessible by the condensation reaction of glycoluril NH compound **12** and cyclic ether **13**. We hoped that synthetic studies toward the methylene bridged glycoluril dimer substructure would: 1) allow us to shed light on some of the critical mechanistic aspects of **CB**[6] formation, 2) pave the way for the synthesis of cucurbit[n]uril derivatives, analogs, and congeners with enhanced solubility characteristics, and 3) result in acyclic cleft shaped compounds comprising two-thirds (4/6) of the **CB**[6] skeleton. Based on the precedent of Nolte,[26] we anticipated that the self-assembly properties of **10C** and its relatives might even exceed those of their macrocyclic counterparts. Although the current renaissance in

CB[n] chemistry has been based on the work of many groups,[4,21,29-39] this review constitutes a personal account of the work performed at the University of Maryland.

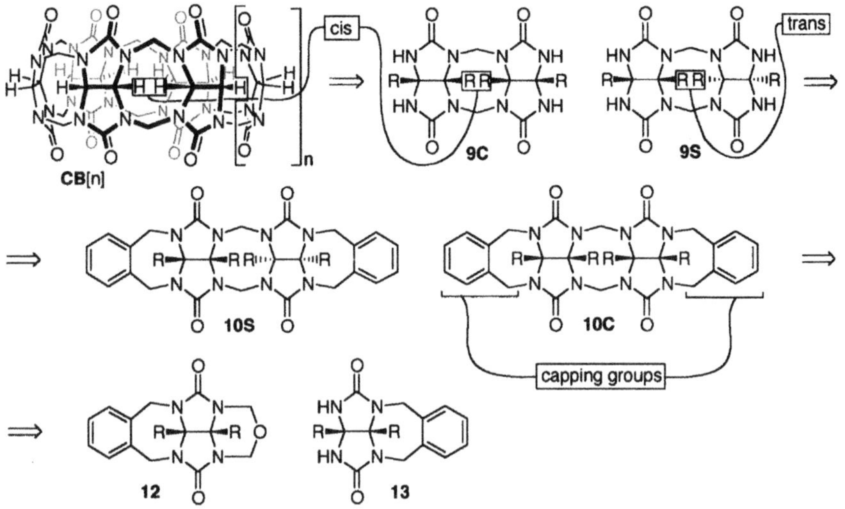

SCHEME 4. Retrosynthetic analysis of CB[6]. R = CO₂Et.

III. A Few Lucky Breaks Lead to C- and S-Shaped Methylene Bridged Glycoluril Dimers

Our first break came on November 25, 1998 about 3 months after Dariusz Witt arrived in College Park as the first postdoctoral fellow in the group.   Darek had prepared **12** and hoped to react it with half an equivalent of 1,4-dimethoxybenzene (**14**) by a double electrophilic aromatic substitution reaction (Scheme 5) to form **15**, which we hoped would behave as an acyclic **CB**[6] congener after deprotection to the water soluble carboxylic acid form.[40]   Although **16** rather than **15** was formed – due to the increased steric demand of the OCH₃ groups when adjacent to the newly formed o-xylylene ring – Darek was also able to isolate an unknown compound in 42% yield. We quickly deduced the structure of the unknown compound C_{2v}-**10C** spectroscopically, and Jim Fettinger corroborated the structure by X-ray crystallography (Figure 1). The C-shape adopted by C_{2v}-**10C**, with the display of all the CO₂Et groups on a single face of the molecule results in a facially amphiphilic topology, which results in well-defined self-assembly properties. When the reaction was performed without added **14**, the C-shaped dimer C_{2v}-**10C** was obtained in 92% yield. In this reaction, we were particularly surprised that the S-shaped diastereomer **10S** was not observed since we did not expect a large diastereoselectivity in the first bond forming reaction between the two equivalents of **12** that establishes the diastereoselectivity of the reaction.

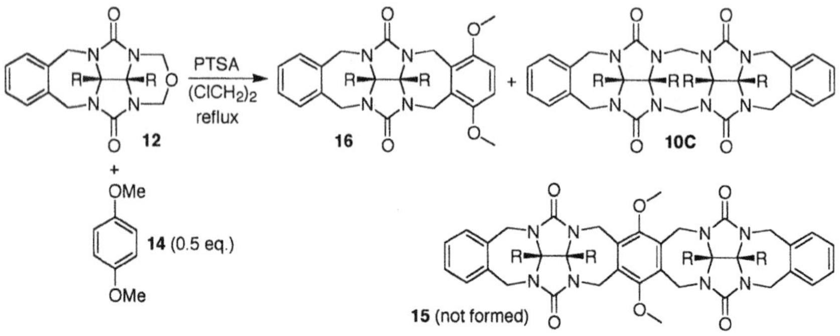

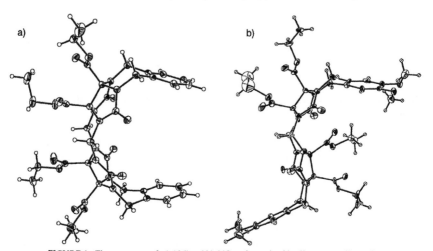

SCHEME 5. The first C-shaped methylene bridged glycoluril dimer. R = CO$_2$Et.

FIGURE 1. The structures of: a) **10C** and b) **11S** as determined by X-ray crystallography.

Our second break came on June 30, 1999 when we were filling in a Table on the scope of the dimerization reaction for our initial communication on the subject.[41] It was a rookie mistake – Jason Lagona set up the dimerization reaction of **17** below reflux and without the azeotropic removal of H$_2$O. Luckily, Jason handled the outcome like a seasoned veteran and isolated **17**, C_{2v}-**11C**, and a new compound. This new compound turned out to be the S-shaped compound (C_{2h}-**11S**) that we had been searching for! The stereochemical assignment was subsequently confirmed by X-ray crystallography (Figure 1b). This finding – along with product resubmission experiments – demonstrated that both the C- and S-shaped compound were the kinetic products of the reaction whereas the C-shaped compound was the thermodynamic product.

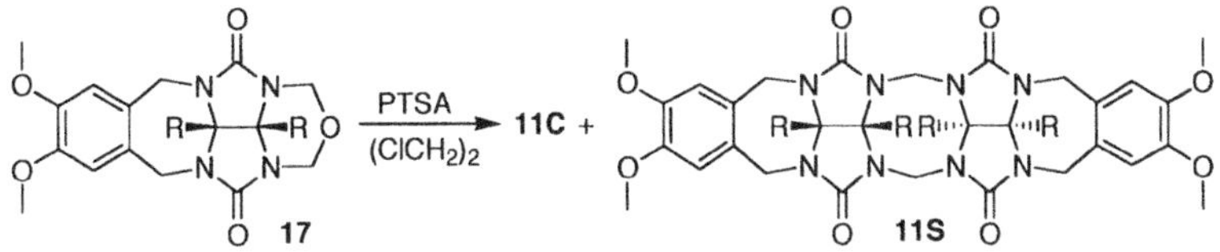

SCHEME 6. The first S-shaped methylene bridged glycoluril dimer. R = CO₂Et.

IV. The Hard Work

To fully explore the scope and limitations of the dimerization reaction we needed to access a range of glycoluril derivatives bearing a variety of substituents on their convex face, on their aromatic rings, and containing either free ureidyl NH compounds or cyclic ether moieties. The following sections outline the synthesis of those compounds.

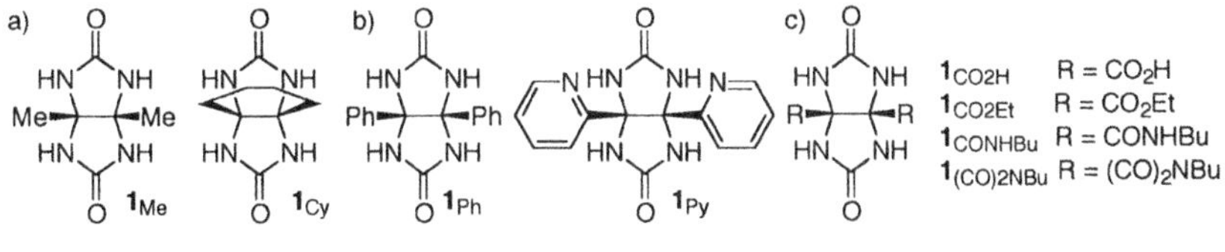

FIGURE 2. a) Alkyl, b) aryl, and c) carboxylic acid derived glycolurils.

A. GLYCOLURILS

The structures of a selection of glycoluril derivatives (1) is shown in Figure 2.[42] To date, the known glycoluril derivatives can be grouped into several broad categories – namely those bearing alkyl or cycloalkyl, aryl, or carboxylic acid derived functional groups their convex face. We have mainly used 1_{CO2Et} in our studies because its ethyl ester groups impart excellent solubility characteristics in $CHCl_3$ solution, and can be transformed into the water soluble carboxylic acids efficiently, and 1_{CO2Et} undergoes smooth methylene bridged glycoluril dimer formation *(vide supra)*.

B. BIS(HALOMETHYL) AROMATICS

Chart 1 shows the structures of the bis(halomethyl) aromatics **18 – 26** that we have used in our studies. These compounds were commercially available or were prepared by the reaction of N-bromosuccinimide with the corresponding *o*-xylene derivatives.

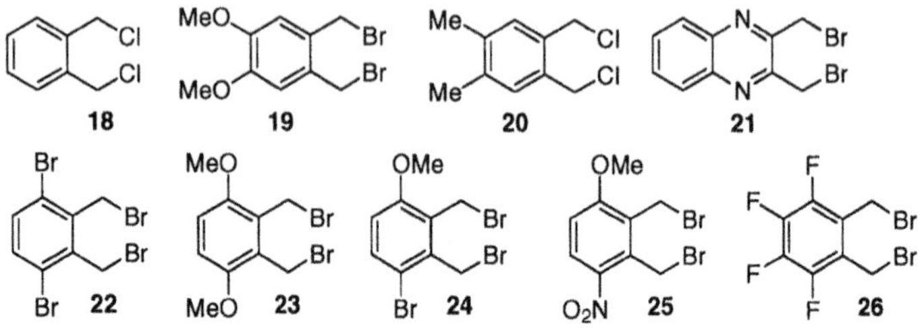

CHART 1. Bis(halomethyl) aromatics.

C. GLYCOLURIL DERIVATIVES BEARING FREE UREIDYL NH GROUPS

For the preparation of a series of glycoluril derivatives bearing a single aromatic sidewall and potentially nucleophilic ureidyl NH groups, we adapted chemistry developed by Nolte and Rebek.[43,44] Deprotonation of glycoluril **1** with *t*-BuOK in DMSO followed by the addition of alkylating

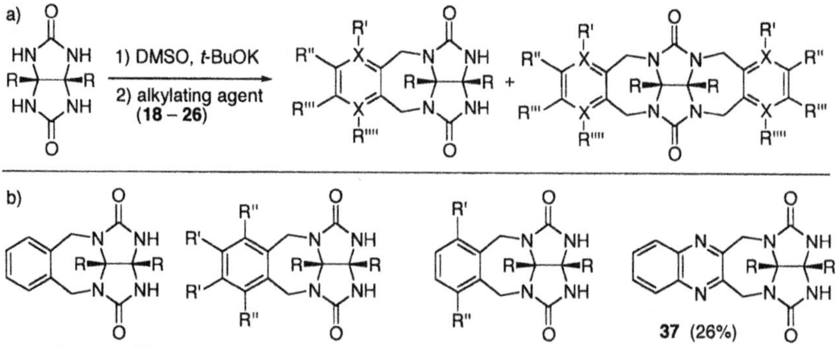

	R	Y (%)
13	CO$_2$Et	84
27	Ph	62
28	2-pyridyl	43
29	(CH$_2$)$_4$	31

	R'	R"	Y (%)
30	OMe	H	45
31	Me	H	68
32	F	F	45

	R'	R"	Y (%)
33	Br	Br	37
34	OMe	OMe	47
(±)-35	OMe	Br	62
(±)-36	OMe	NO$_2$	32

37 (26%)

SCHEME 7. a) Synthesis of glycoluril NH compounds, and b) structures and yields for the specific compounds prepared (**13** and **27 – 37**). R = CO$_2$Et unless stated otherwise.

agent (**18 – 26**) delivered **13** and **27 – 37** invariably contaminated with dialkylated material even when an excess of glycoluril (e.g. 5 eq.) was used (Scheme 7). The relative amount of monoalkylated versus dialkylated product is quite sensitive to the precise experimental conditions (e.g. rate of stirring, reagent quality, and the timing and method of addition of both *t*-BuOK and alkylating agent). Despite some unresolved issues, we were able to access sufficient quantities of **13** and **27 – 37** for the synthetic studies.

D. GLYCOLURIL DERIVATIVES BEARING CYCLIC ETHER FUNCTIONAL GROUPS.

For the transformation of **13**, and **27 – 37** into the corresponding cyclic ether derivatives, we considered the procedure of Nolte that converts **1**$_{Ph}$ into **38**. This involves hydroxymethylation with formaldehyde under basic conditions, followed by cyclic ether formation under strongly acidic conditions (HCl, pH 1, reflux).[45] To circumvent the anticipated issues with ester hydrolysis, we developed a one step procedure that proceeds under anhydrous acidic conditions (TFA, $(CH_2O)_n$) where the esters are stable. In the synthesis of **39** from **1**$_{CO2Et}$, we isolated and crystallographically characterized by-product **40**, which contains an 8-membered dioxygenated ring. Scheme 8 shows the structures of a variety

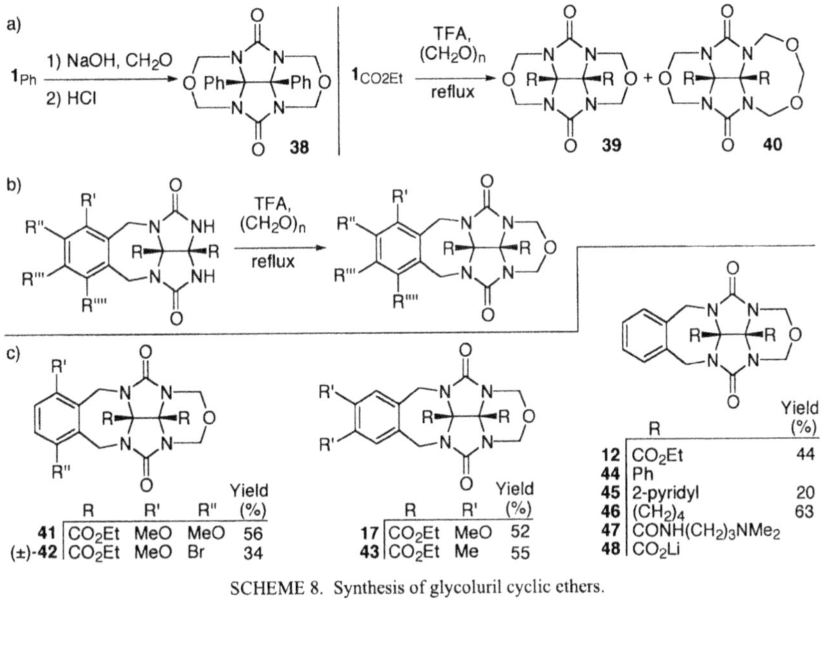

SCHEME 8. Synthesis of glycoluril cyclic ethers.

of cyclic ethers (**12, 17**, and **41 – 48**) obtained by this and related pathways. By-products structurally related to **40** are also formed in the synthesis of **12, 17**, and **41 – 48**, which explains the modest yields obtained. With a variety of cyclic ethers (e.g. **12**) and NH compounds (e.g. **13**) in hand, we set out to study their dimerization reactions.

V. Three Related Synthetic Procedures Lead to S- and C-shaped Methylene Bridged Glycoluril Dimers

As suggested by the retrosynthetic analysis (Scheme 4) described above, there are three pathways to methylene bridged glycoluril dimers – namely – 1) two equivalents of glycoluril cyclic ether (e.g. **12**), 2) two equivalents of glycoluril NH compound (e.g. **13**) with added

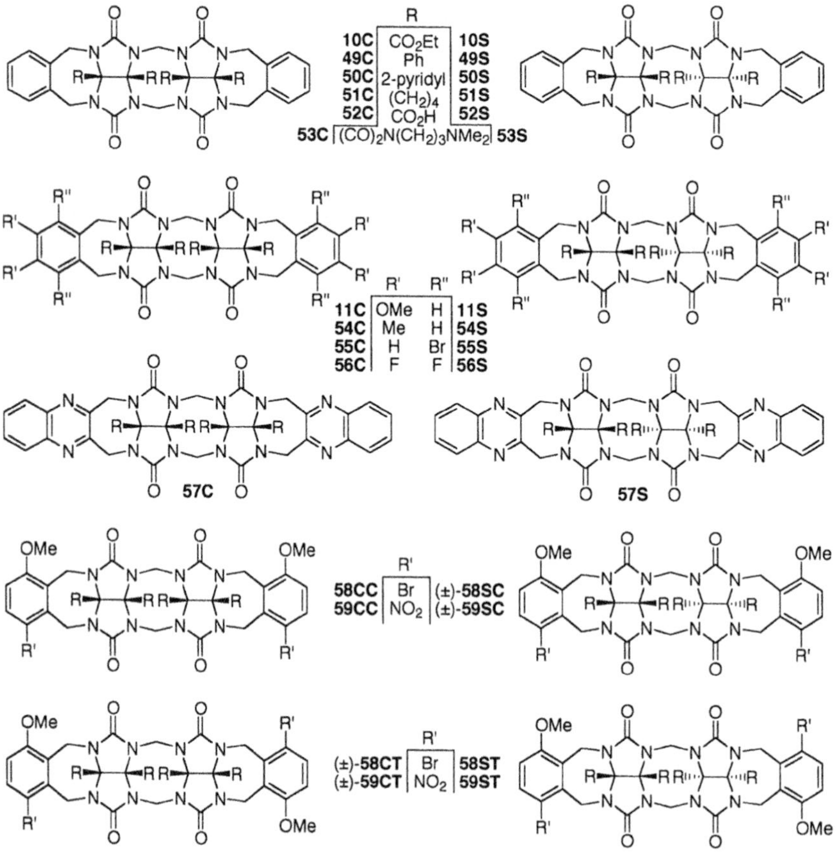

CHART 2. Methylene bridged glycoluril dimers.

paraformaldehyde, and 3) one equivalent each of glycoluril cyclic ether and NH compound. The former reaction occurs with the formal extrusion of formaldehyde whereas the other pathways are condensation reactions. Chart 2 shows the structures of 28 methylene bridged glycoluril dimers (**10, 11**, and **49 – 59**) prepared by these three routes.[28] The separation of the C- and S-shaped diastereomers can be easily accomplished by either column chromatography, since the C-shaped compounds have lower R_f values, or by trituration, since the S-shaped compounds are more soluble in common solvents.

A. HOMODIMERIZATION REACTIONS

Table 1 shows the outcome of a selection of dimerization reactions of glycoluril NH compounds.[28] Entries 1 – 4 show the effect of the substituent on the convex face of the glycoluril derivative on the efficiency of dimerization. For example, **13**, which contains electron withdrawing CO_2Et groups, yields the C-shaped product **10C** in a highly diastereoselective process with no evidence of by-product formation. In sharp contrast, glycoluril derivatives containing alkyl (**29**) or aryl (**27**) groups on their convex face dimerize less efficiently and with the formation of by-products **60 – 63**. A mechanistic rationale for this trend, and implications for the synthesis of **CB**[n] derivatives and congeners is discussed below. Entries 5 – 11 demonstrate that the dimerization reaction is tolerant of a wide range of aromatic ring substituents (e.g. OMe, Me, Br, NO_2, F). Entries 3 and 10 suggest that the presence of aromatic N-atoms disfavor dimerization due to competitive protonation. In all cases, the C-shaped forms are formed in moderate to highly diastereoselective processes. The outcome of the dimerization of (±)-**36** was particularly interesting. All four possible diastereomers – **59CC** (C-shaped, OMe groups _Cis_), (±)-**59CT** (C-shaped, OMe groups _Trans_), (±)-**59SC** (S-shaped, OMe groups _Cis_), and (±)-**59ST** (S-shaped, OMe groups _Trans_) – were formed in nearly equal amounts. Remarkably, all four diastereomers were readily separated by column chromatography and unambiguously identified based on the number, multiplicity, and chemical shifts of ^{1}H and ^{13}C NMR resonances for the newly formed CH_2 bridges. Some of the C-shaped and S-shaped compounds are chiral due to the unsymmetrical distribution of functional groups on their aromatic rings (e.g. (±)-**58CT**, (±)-**59CT**, (±)-**59SC** and (±)-**58SC**). Despite many attempts over the years, we have not been able to separate these enantiomers by HPLC on chiral stationary phases or by derivatization reactions to yield diastereomers. Entries 12 – 18 demonstrate that glycoluril cyclic ethers bearing a range of functional groups on their convex face (e.g. CO_2Et, CO_2Li, $CONH(CH_2)_3NMe_2$, Ph, and $-(CH_2)_4-$) and some functionality on their aromatic rings (e.g. OMe or Me groups) undergo dimerization in comparable yields and slightly higher C-shaped diastereoselectivity than the corresponding glycoluril NH compounds.

TABLE 1
Homodimerization Reactions.

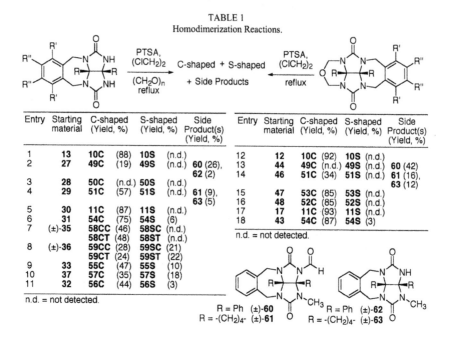

Entry	Starting material	C-shaped (Yield, %)		S-shaped (Yield, %)		Side Product(s) (Yield, %)	Entry	Starting material	C-shaped (Yield, %)		S-shaped (Yield, %)		Side Product(s) (Yield, %)
1	13	10C	(88)	10S	(n.d.)		12	12	10C (92)		10S (n.d.)		
2	27	49C	(19)	49S	(n.d.)	60 (26), 62 (2)	13	44	49C (n.d.)		49S (n.d.)		60 (42)
3	28	50C	(n.d.)	50S	(n.d.)		14	46	51C (34)		51S (n.d.)		61 (16), 63 (12)
4	29	51C	(57)	51S	(n.d.)	61 (9), 63 (5)	15	47	53C (85)		53S (n.d.)		
5	30	11C	(87)	11S	(n.d.)		16	48	52C (85)		52S (n.d.)		
6	31	54C	(75)	54S	(6)		17	17	11C (93)		11S (n.d.)		
7	(±)-35	58CC	(46)	58SC	(n.d.)		18	43	54C (87)		54S (3)		
		58CT	(48)	58ST	(n.d.)		n.d. = not detected.						
8	(±)-36	59CC	(28)	59SC	(21)								
		59CT	(24)	59ST	(22)								
9	33	55C	(47)	55S	(10)								
10	37	57C	(35)	57S	(18)								
11	32	56C	(44)	56S	(3)								

n.d. = not detected.

R = Ph (±)-60
R = -(CH₂)₄- (±)-61

R = Ph (±)-62
R = -(CH₂)₄- (±)-63

B. HETERODIMERIZATION REACTIONS

The previous section demonstrates that both glycoluril NH compounds and glycoluril cyclic ethers form methylene bridged glycoluril dimers in C-shape selective reactions. We hypothesized that an equimolar mixture of NH compound and cyclic ether would undergo heterodimerization. We initially tested the formation of symmetrical homodimers (e.g. **10C** and **10S**) and found similar yields and product distributions as described above for the corresponding homodimerization reactions (Scheme 9). Given that the preparation of the cyclic ethers proceeds by way of the corresponding NH compounds and the generally modest yields obtained in this reaction, the homodimerization of glycoluril NH compounds is our method of choice for the preparation of symmetrical methylene bridged

.uril dimers. We also wondered what would happen if we combined
glycoluril derivatives containing differentially functionalized
atic rings. Would the unsymmetrical heterodimer be formed
ively, or would the position of the CH$_2$ bridges become scrambled
ing in a mixture of homodimer and heterodimer? Scheme 9 also
s the outcome of two such reactions. In both cases, the
dimeric products containing two different aromatic rings were
d selectively (81% and 95%), although a mixture of the C-shaped
S-shaped diastereomers was obtained. This highly selective
dimerization, which provided a useful substrate for self-assembly
s, also suggested that the isomerization between the S-shaped and
ped diastereomers might follow an intramolecular pathway.

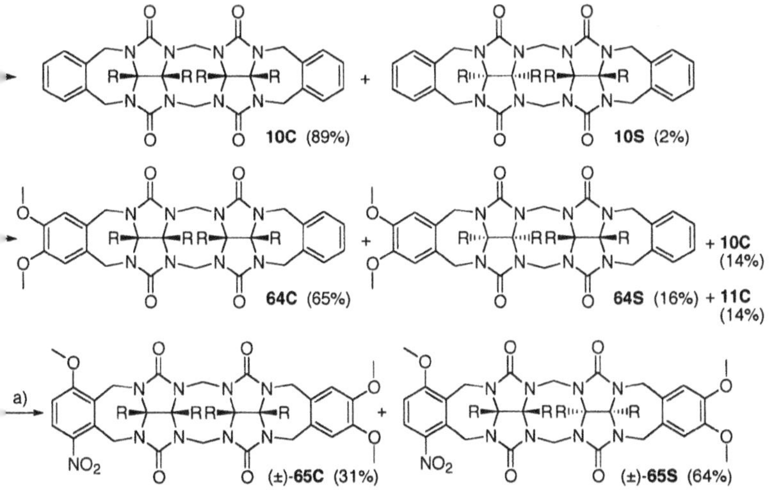

SCHEME 9. Heterodimerization reactions. Conditions: a) ClCH$_2$CH$_2$Cl, PTSA, reflux.

Interlude. **Molecular clips Capable of Enantiomeric Self-Recognition,**
chiral Recognition, and Self-Sorting

e thought that molecular clips based on methylene bridged glycoluril
s – containing two aromatic and two glycoluril rings and
tuting 4/6 of the **CB[6]** skeleton – would possess unusual self-
bly properties. In this section, we demonstrate that this class of
ounds undergo self-association in water and organic solvents and are
le of efficiently distinguishing between themselves, their
iomers, and even other molecules in solution undergoing high
y enantiomeric self-recognition, heterochiral recognition, and self-

orting processes.

Scheme 10 shows the straightforward synthesis of (±)-**66** from (±)-**65C**
ia intermediates (±)-**67** and (±)-**68**. Compound (±)-**66** is chiral but
repared and used in racemic form. When Darek dissolved (±)-**66** in pD
.4 phosphate buffered D$_2$O, he observed a mixture of diastereomeric
ggregates (e.g. (+)-**66**•(−)-**66**, (+)-**66**•(+)-**66**, and (−)-**66**•(−)-**66** with an
nsemble of π-π stacked geometries by ^{1}H NMR. The addition of
en)Pd(ONO$_2$)$_2$, **69**, triggers a high fidelity enantiomeric self-recognition
rocess resulting in the 2:2 aggregate (+)-**66**$_2$•**69**$_2$ and (−)-**66**$_2$•**69**$_2$. This
ggregate features π-π, pyridyl•Pd, and carboxylate•Pd interactions.
Remarkably, the simultaneous application of a second non-covalent
nteraction − in this case metal•ligand interactions − with well-defined
geometrical preferences result in the selection of a single conformation

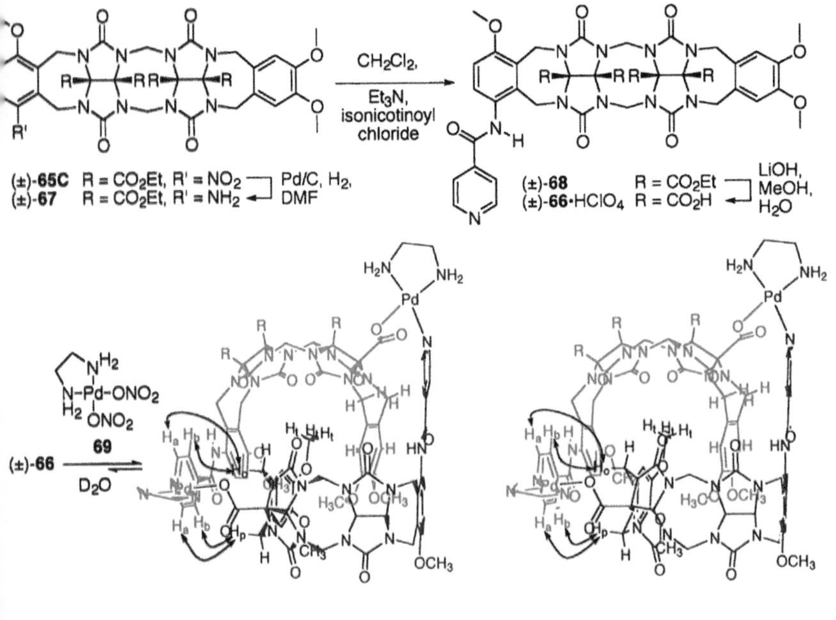

SCHEME 10. a) Synthesis of (±)-**66**, b) a stereoscopic line drawing of one enantiomer of the racemic mixture
((+)-**66**$_2$•**69**$_2$ and ( )-**66**$_2$•**69**$_2$) triggered by the addition of **69**.

from the ensemble of π-π stacked geometries in an enantiomeric self-
recognition process.[46]

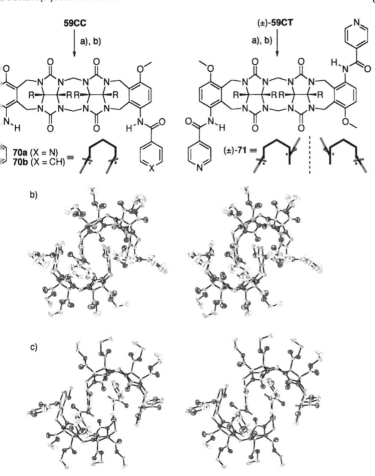

E 11. a) Synthesis of **70 – 71** and schematic illustration of their structures, and cross-eyed stereoviews
-ray crystal structures of b) **70b·70b** and c) **(+)-71·(−)-71**. Conditions: a) Pd/C, H₂, DMF, b) acid
Et₃N, CH₂Cl₂.

.ad the simple-minded idea that if compound (±)-**66**, containing a
pyridyl group, was good, the compounds **70** and (±)-**71** containing
yridyl groups, might display even more interesting self-assembly
rties. Accordingly, Anxin Wu prepared **70a**, **70b** and (±)-**71** from
: and (±)-**59CT** by reduction, acylation, and deprotection reactions
me 11). We studied their self-association properties in the presence
bsence of **69** and did not observe any well-defined aggregates. As
en happens in science, however, we observed something interesting

n the solid state and CDCl$_3$ solution.[47] Furthermore, **70a•70a** was so ghtly dimerized that no dissociation could be observed upon dilution to 0 μM which allows us to place a lower limit on the self-association onstant ($K_s > 9 \times 10^5$ M^{-1}; $-\Delta G > 8.1$ kcal mol^{-1})! Such a large ssociation constant is highly unusual for a system based on the formation f only two hydrogen bonds; clearly π-π interactions and H-bonds act in oncert[48] to drive tight dimerization and the selection of a single well-efined conformation.[47,49] More recently, we have shown that **70a** is imeric across the full range of solvents (C$_6$D$_6$, C$_6$D$_5$CD$_3$, CDCl$_3$, D$_2$Cl$_2$, CH$_3$COCH$_3$, CH$_3$CN, CD$_3$OD, and D$_2$O). Furthermore **70a•70a** s isostructural across this wide range of solvents – that is – it assumes a ommon geometry that is solvent independent.[49] Compound (±)-**71** was lso tightly dimerized in CDCl$_3$ solution. In contrast to the high fidelity nantiomeric self-recognition process observed for (±)-**66**, the imultaneous geometrical requirements of π-π interactions and two H-onds within a sample of (±)-**71** are only satisfied by the heterochiral ggregate (+)-**71•**(–)-**71**.

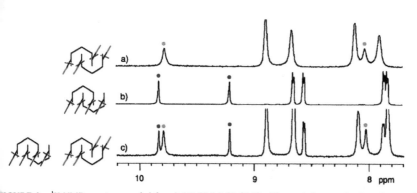

IGURE 3. ^{1}H NMR spectra recorded for: a) (+)-**71•**(–)-**71**, b) **70a•70a**, and c) an equimolar mixture of **70a•70a** and (+)-**71•**(–)-**71**. The green and purple bullets mark the resonances for the amide N-H groups.

Given the high level of diastereoselectivity observed in the omplexation behavior of **70a** and (±)-**71**, we wondered what would aappen if we created mixtures of these methylene bridged glycoluril limers. Would a mixture of **70a** and (±)-**71**, for example, form a mixture of **70a•70a** and (+)-**71•**(–)-**71**, or would heteromeric crossover aggregates e.g. **70a•**(+)-**71**) also be formed? Figure 3 shows the ^{1}H NMR spectra ecorded for an equimolar mixture of **70a** and (±)-**71** in CDCl$_3$.

Remarkably, the ^{1}H NMR spectrum of the mixture is simply equal to the sum of the ^{1}H NMR spectra of its components. This spectroscopic earmark indicates that this mixture of molecules undergoes *self-sorting*. Self-sorting refers to the ability of a molecule or entity to efficiently distinguish between self and non-self even within a complex mixture.[50,51] Self-sorting is commonly observed in Nature (e.g. the immune system) but is still relatively rare in designed supramolecular systems. The molecular clips described here represent a well-defined model system for studies of self-sorting.

VII. Implications for the Mechanism of CB[n] Formation and the Synthesis of Cucurbit[n]uril Derivatives

The preceding sections have discussed the synthesis of methylene bridged glycoluril dimers and demonstrated some of the unusual recognition properties (e.g. enantiomeric self-recognition, heterochiral recognition, and self-sorting) that they possess. This section shows that studies of methylene bridged glycoluril dimers also shed light on some fundamental aspects of the mechanism of CB[n] formation. Scheme 12

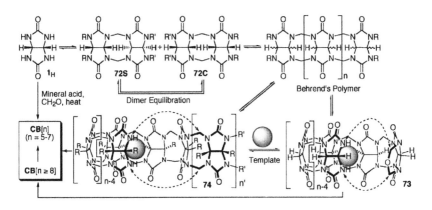

Scheme 12. Proposed mechanism of CB[n] formation.

outlines a mechanistic proposal for CB[n] formation advanced by Day's group[4] and our group.[52] In brief, glycoluril 1_H undergoes dimerization in the presence of formaldehyde under aqueous acidic condition to yield a mixture of the C- and S-shaped dimers **72C** and **72S**. The diastereomeric mixture may undergo S- to C- shaped isomerization in competition with further oligomerization to yield Behrend's polymer. Behrend's polymer must then undergo S- to C-shaped isomerization – potentially aided by suitable templating compounds[4,53] – to yield **73** and **74** which contain a stretch of all C-shaped methylene bridged glycoluril oligomer. Compounds **73** and **74** can then enter the **CB**[n] manifold by direct end-to-end cyclization (**73** to **CB**[n]) or by back-biting (e.g. **74** to **CB**[n]). The studies from our group described below provide strong support for key aspects of this mechanistic proposal.

A. SUBSTITUENT EFFECTS

Why do **1**$_H$ and **1**$_{Me}$ readily cyclize to yield **CB**[6] and Me$_{10}$**CB**[5] exclusively, whereas **1**$_{Ph}$ does not yield a macrocyclic oligomer? The substituent effects described above for the formation of methylene bridged glycoluril dimers **1**$_{CO2Et}$, **1**$_{Ph}$, and **1**$_{Cy}$ (Table 1) and the nature of the side products observed (e.g. **60** – **63**) provide some insight into this question. Scheme 13 proposes a mechanism that accounts for the formation of by-products (±)-**60** and (±)-**61**.[28] In brief, protonation of cyclic ether **75** followed by ring opening yields iminium ion (±)-**76**. Iminium ion (±)-**76** may be in equilibrium with (±)-**77**, which has a resonance form where the positive charge is adjacent to the glycoluril substituent (e.g. CO$_2$Et, (CH$_2$)$_4$, Ph). Intermediate (±)-**77** is formally transformed into aldehydic byproducts (±)-**60** and (±)–**61** by N-acyl

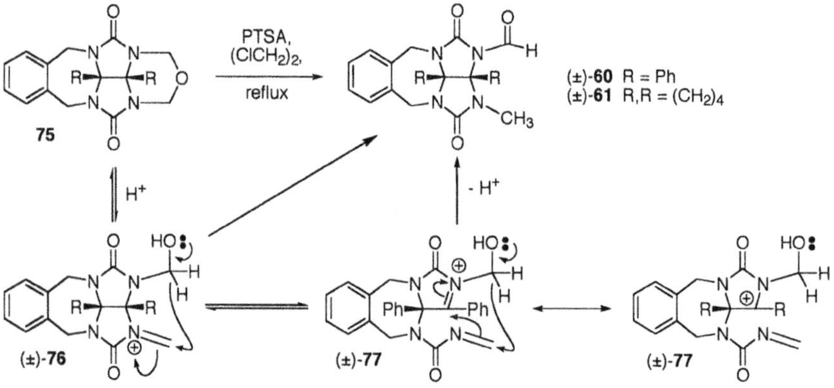

SCHEME 13. Proposed mechanism for the formation of by-product (±)-**60** and (±)-**61**.

iminium ion-imine intramolecular ene reactions. One would expect that cationic intermediate (±)-**77** would be stabilized when R = Ph or R =

$(CH_2)_4$, but destabilized when R = CO_2Et, which enhances the formation of aldehydic by-products in those cases at the expense of the corresponding methylene bridged glycoluril dimers. We have never observed aldehydic by-products in the reactions of CO_2Et substituted glycolurils. These observations suggest that the synthesis of **CB[n]** derivatives is likely to be more successful for substituted glycolurils bearing groups on their convex face that destabilize adjacent positive charge (e.g. electron withdrawing carboxylic acid derivatives) and less successful for those bearing groups that stabilize adjacent positive charge (e.g. alkyl and aryl groups).

B. S- TO C-SHAPED ISOMERIZATION.

How can **CB[6]** be formed in such high yield (e.g. 82%)[54] given that two diastereomeric methylene bridged glycoluril dimers (e.g. **72C** and **72S**) and an even larger number of diastereomeric methylene bridged glycoluril oligomers (e.g. **73** and **74**) can form? To address this question we wanted to firmly establish the presence of an equilibrium between C-shaped and S-shaped methylene bridged glycoluril dimers and to establish a value of ΔG for their interconversion.[52] For this purpose, we isolated diastereomerically pure samples of **10S**, **11S**, and **54S – 57S** and **10C**, **11C**, and **54C – 57C** from dimerization reactions performed to partial conversion. We then separately resubmitted each compound to the reaction conditions and followed its conversion to a mixture of C- and S-shaped diastereomers at equilibrium (Scheme 14). The C-shaped diastereomer predominates at equilibrium by a factor of 9:1 – 99:1. This corresponds to a difference in free energy between the C-shaped and S-shaped forms of 1.55 – 3.25 kcal mol^{-1} in refluxing $ClCH_2CH_2Cl$. Related experiments were performed in 8 different solvents ($CHCl_3$, CCl_4, C_6F_6, THF, $ClCH_2CH_2Cl$, CH_3NO_2, and $MeOCH_2CH_2OMe$) for the equilibrium between **55C** and **55S** to assess the influence of dielectric constant and solvation on the equilibrium between C- and S-shaped diastereomers.[52] Solvent plays only a minor role and the values of -ΔG range from 1.55 to 3.22 kcal mol^{-1}. These experiments represent strong experimental evidence for the intermediacy of **72C** and **72S** in the formation of **CB[n]**, establish the equilibrium between these diastereomers, and provide a value of ΔG for this key equilibration step in the mechanism of **CB[n]** formation.

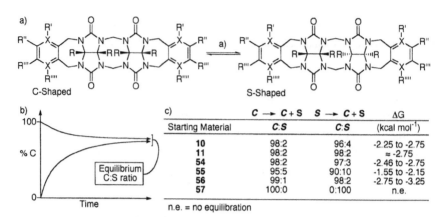

SCHEME 14. a) Equilibrium between S- and C-shaped dimers, b) illustration of the approach to equilibrium, and c) values of the C:S ratios and ΔG for this equilibrium. Conditions: a) PTSA, ClCH₂CH₂Cl, reflux.

VIII. Mechanism of the S- to C-Shaped Interconversion and Implications for the Synthesis of Cucurbit[n]uril Derivatives

The highly heterodimer selective processes (Scheme 9) suggested that the mechanism for the equilibrium between the C- and S-shaped diastereomers might be unusual.[52] Scheme 15 describes three different mechanistic proposals for the interconversion of the C- and S-shaped forms. The three mechanisms are color coded (mechanism 1, blue; mechanism 2, red; mechanism 3, green) with bullets that track the location of imaginary labels during each of the three pathways. Overall, mechanism 1 (blue) represents a pathway characterized by dissociation of the S-shaped dimer into two halves, followed by recombination into the C-shaped diastereomer. Mechanism 2 (red) is an intramolecular pathway characterized by the sequential inversion of configuration at two glycoluril C-atoms. Mechanism 3 is also an intramolecular pathway, but is characterized by a rotation of the two halves of the dimer via a spirocyclic intermediate with retention of configuration.

Mechanisms 1 – 3 have different stereochemical outcomes and can be distinguished on the basis of labelling experiments. Scheme 15 shows how mechanism 1 (blue) leads to a scrambling of the label between two locations, mechanism 2 (red) does not result in the change of position of the labels, and mechanism 3 (green) results in the transposition of one of the labels to the opposite side of the molecule. To realize this labelling experiment in practice we performed separate isomerization reactions of

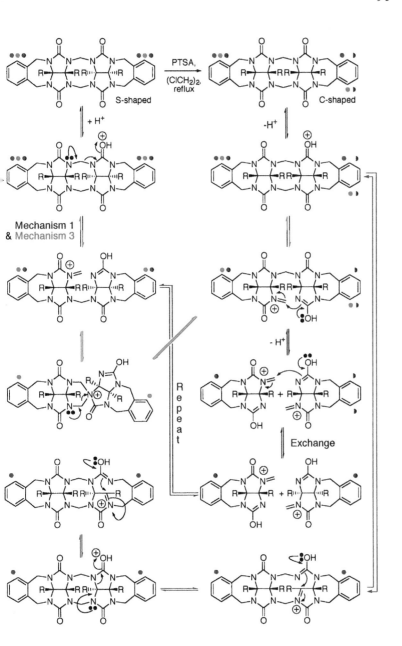

ME 15. Three different mechanisms for the equilibrium between the S- and C-shaped diastereomers.

±)-**59SC** and **59ST**. Under mechanism 1, (±)-**59SC** should yield a nixture of **59CC** and (±)-**59CT**, under mechanism 2 only **59CC**, and ınder mechanism 3 only (±)-**59CT**. Similarly, under mechanism 1, **59ST** hould yield a mixture of **59CC** and (±)-**59CT**, under mechanism 2 only ±)-**59CT**, and under mechanism 3 only **59CC**. In the actual experiment, ±)-**59SC** gave exclusively (±)-**59CT** and **59ST** gave exclusively **59CC** at ¡5% conversion (Scheme 16). These results strongly suggest that the somerization reaction proceeds by mechanism 3 under our standard eaction conditions (ClCH$_2$CH$_2$Cl, PTSA, reflux).

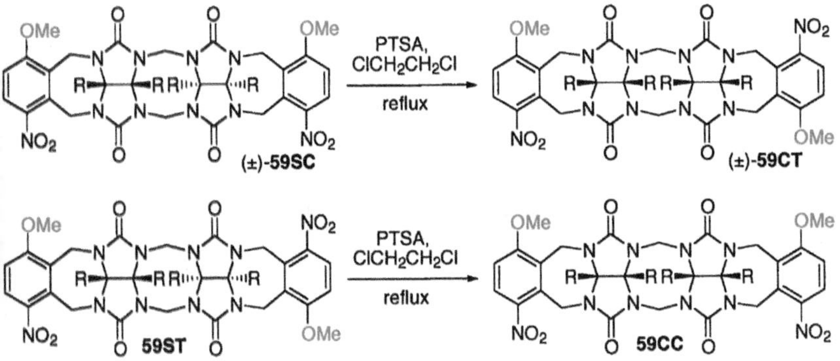

SCHEME 16. Diastereoselective equilibration reactions.

That the isomerization reaction of methylene bridged glycoluril dimers ›roceeds by mechanism 3 has implications for the synthesis of **CB[n]**, ⊃**B[n]** derivatives, and **CB[n]** congeners. For example, in an elegant ›roduct resubmission experiment, Day and co-workers showed that ᴉeating purified **CB[8]** in conc. HCl at 100 ˚C results in the formation of ⊃**B[5]** – **CB[7]**, whereas **CB[5]** – **CB[7]** are stable under these conditions Scheme 12).[4] These results require that two adjacent methylene bridges ᴀre broken upon extrusion of glycoluril, and would by necessity proceed ›y a variation of mechanism 1 under the *aqueous acidic conditions* ᴇmployed. We believe mechanism 1 is not operative in our system ›ecause we use *anhydrous acidic conditions*. Therefore, we suggest that ⊃**B[n]** (n > 8) may display enhanced stability and be best synthesized by ⱳorking under anhydrous acidic conditions. The fact that the S- to C- ̧haped isomerization proceeds intramolecularly suggests that it is the .ength of the methylene bridged glycoluril oligomer (Scheme 12, **73** or **74**) that determines the size of the **CB[n]** produced. The intramolecular ᴉature of the isomerization further suggested that the functionalization ›attern of the intermediate oligomers might be preserved in the **CB[n]** formed. Experimental support for this hypothesis comes from the work of Day and co-workers who performed the heteromeric oligomerization of 1_H and **78** which provided $Me_6CB[3,3]$ in which the substituents alternate (Scheme 17)![24]

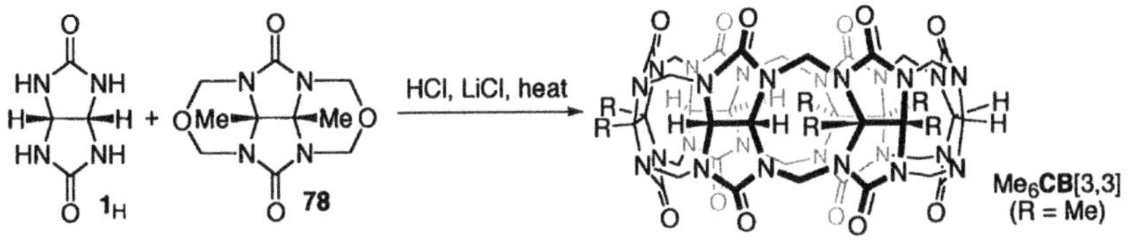

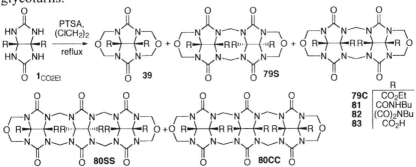

SCHEME 17. Heteromeric reaction of 1_H and **78** to give $Me_6CB[3,3]$.

IX. Methylene Bridged Glycoluril Oligomers

The mechanistic predictions made in the previous section strongly suggest that suitable combinations of nucleophilic and electrophilic glycoluril building blocks would yield control over the size and functionalization pattern of the formed **CB**[n] derivatives. To expand the range of building blocks available to test that prediction, we performed the controlled oligomerization of 1_{CO2Et} to yield **79S**, **79C**, **80CC**, and **80SS** (Scheme 18).[55] The CO_2Et groups of **79C** undergo standard transformations yielding **81 – 83**. Unfortunately, we were unable to isolate new **CB**[n] derivatives from reactions between **79 – 83** and various glycolurils.

SCHEME 18. Controlled oligomerization of 1_{CO2Et}.

X. Phthalhydrazides are Nucleophilic Glycoluril Surrogates

Having failed to obtain **CB**[n] derivatives by heteromeric cyclization reactions between **79C** and **1** we decided to search for substances with enhanced nucleophilicity that might serve as surrogates of glycoluril in their reactions with **79C**. We discovered that phthalhydrazide **84** was a potent nucleophile in methylene bridge forming reactions, presumably due to the α-effect. For example, **84** reacts with **17** to deliver **85** in a rapid reaction without competing homodimerization of **17** to **11C** (Scheme 19).

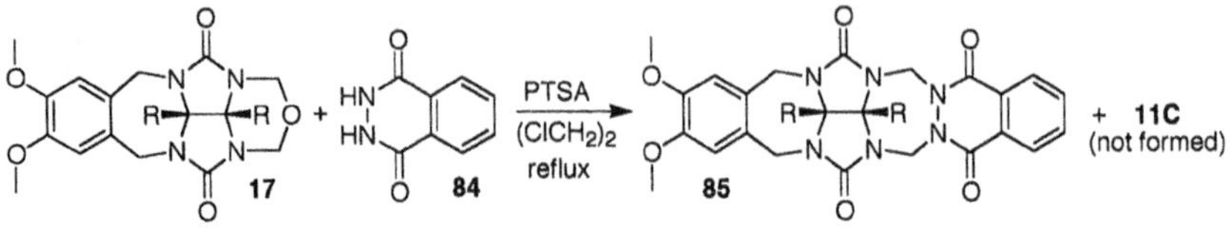

SCHEME 19. Methylene bridge forming reactions with **84**.

XI. Cucurbit[n]uril Analogs

The next obvious step was to prepare bis(phthalhydrazide) **86** that would allow for cyclizations reactions. We obtained **86** quite easily, but were disappointed when it failed to dissolve in common solvents (e.g. $ClCH_2CH_2Cl$, DMSO, CF_3CO_2H) that might be used in methylene bridge forming reactions. We, of course, tried many reactions between **79C** and **86** under heterogeneous conditions, but to no avail. Concurrent with these failed attempts, we were preparing new glycoluril monomers for expanded **CB[n]** synthesis and therefore **86** was temporarily put on the back burner.[42] Jason and I both had it in the back of our minds that somehow we would make the reaction work. One day in early November 2002, we came up with the idea that CH_3SO_3H might be a good solvent for the cyclization since it is available in anhydrous form and we knew that **86** was soluble in hot sulfuric acid. The heteromeric cyclization reaction between **79C** and **86** worked brilliantly when conducted in CH_3SO_3H at 80 °C for 3 hours, delivering **CB[6]** analog **87** in 78% yield (Scheme 20).[55] Under similar conditions, **82** and **83** were transformed into **CB[6]** analogs **88** and **89** bearing butyl imide and carboxylic acid functional groups on their convex surface. Similarly, a combination of 1_{CO2Et} and **86** delivers **CB[5]** analog **90** albeit in quite modest yield (6%). In contrast, **80CC** and **86** undergo smooth macrocyclization to deliver **CB[7]** analog (±)-**91** in 67% yield.

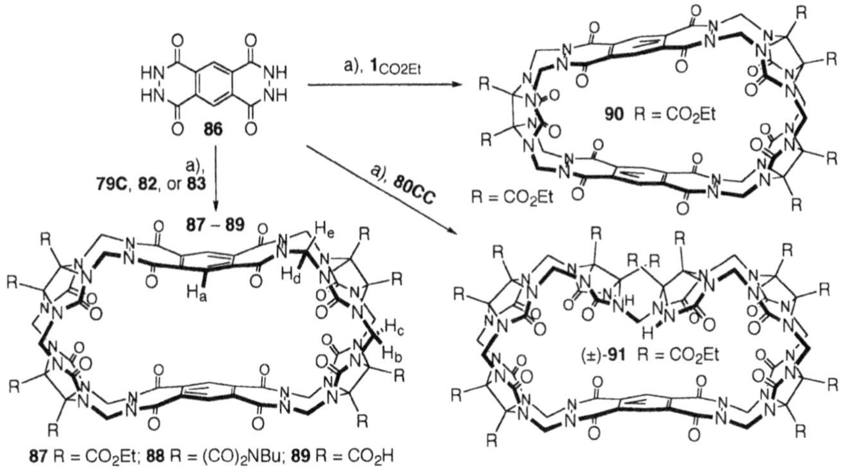

SCHEME 20. Synthesis of **CB**[n] analogs.

Compounds **87 – 91** possess a number of unusual aspects that differentiate them from the known **CB**[n] and suggest they will broaden the range of applications open to the **CB**[n] family. For example, **87 – 91** have remarkable solubility characteristics; **89** is soluble in water whereas **87, 90** and **91** have good solubility in polar organic solvents like $CHCl_3$, CH_3CN, CH_3COCH_3, and CH_3SOCH_3. Second, the cavities of these new **CB**[n] analogs are defined by aromatic rings, enclose significant volumes, and are oblong rather than circular (e.g. **87**: 5.90 × 11.15 × 6.92 Å; **90**: 5.58 Å × 9.75 Å × 6.22 Å; (±)-**91**: 5.71 Å × 11.34 Å × 4.28 Å). The ends of these large cavities potentially define two equivalent binding sites which, just like **C B**[8],[56] may allow for the formation of termolecular complexes. Third, **87 – 91** possess a chromophore within their macrocyclic structure that imparts UV/Vis (λ_{max} = 342 nm) and fluorescence activity (λ_{ex} = 360 nm; λ_{em} = 514nm) to these cycles that may be used as a reporter for recognition properties within their cavities (Figure 4). We expect that the bis(phthalhydrazide) walls will impart electrochemical activity to these macrocycles as well.[57] Fourth, the structure of **CB**[7] analog (±)-**91** is unusual in that it has only a single bis(phthalhydrazide) wall), two free ureidyl NH groups, and one CH_2 group that points directly into the cavity. Compound (±)-**91** is the first chiral member of the **C B**[n] family and is expected to show enantioselective recognition behavior along with communication between the two symmetry equivalent binding sites.

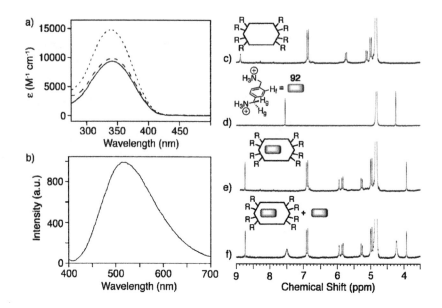

FIGURE 4. a) UV/Vis spectra (CH₃CN) for **87** (- - -), **90** (– – –), and (±)-**91** (——) , b) fluorescence spectrum for **89** (40 μM, 50 mM acetate buffer, pH 4.5), and ¹H NMR spectra for c) **89** (1.2 mM), d) **92** (1.2 mM), e) **89·92** (1.2 mM), and f) **89** (1.2 mM) and **92** (4.8 mM).

We expect that the recognition properties of the new **CB**[n] analogs will retain many features of the **CB**[n] family but will also display novel features as a consequence of their aromatic bis(phthalhydrazide) walls. Indeed, we have preliminary evidence that **89** retains the selectivity of the **CB**[n] family toward ammonium ions due to the high electrostatic potential around its carbonyl groups. In addition, however, **89** is selective for aromatic amines because of its ability to engage in π-π interactions with these guests. Figure 4e shows the NMR spectra recorded for the interaction of **89** with p-xylylenediammonium ion (**92**). Another characteristic of the **CB**[n] family – slow exchange kinetics on the chemical shift timescale – is also retained in **CB**[6] analog **89** (Figure 4f). Currently, we are defining the classes of guests that can be bound within **87 – 91**, investigating the possibility of termolecular complex formation, and studying diastereoselective recognition processes within (±)-**91**.

XII. Conclusions

The remarkable recognition properties of **CB**[6] prompted several groups, including ours, to tackle the synthetic challenges associated with the synthesis of cucurbituril homologs (**CB**[n], n = 5 – 10), cucurbituril derivatives (e.g. Me$_{10}$**CB**[5]), and **CB**[n] analogs. Our approach to this problem relied on the identification of the methylene-bridged glycoluril dimer substructure as the fundamental building block of the **CB**[n] family. Accordingly, we developed efficient homo- and heterodimerization procedures that allow access to methylene-bridged glycoluril dimers bearing substituted o-xylylene walls. These molecular clips possess remarkable recognition properties in water and organic solvents including the ability to distinguish between themselves (enantiomeric self-recognition), their enantiomers (heterochiral recognition), and other molecules present within complex mixtures (self-sorting). Studies of the methylene bridged glycoluril dimer substructure also allowed us to probe the mechanism of **CB**[n] formation to a level not previously possible. These studies: 1) delineated the scope of glycolurils able to undergo **CB**[n] forming reactions, and 2) explained the high yield obtained in **CB**[6] formation, 3) provided evidence for the intermediacy of S-shaped diastereomers during **CB**[n] synthesis and a value of ΔG for the S- to C-shaped equilibrium, and 4) elucidated the mechanism of interconversion of the S- to C-shaped equilibrium. These mechanistic insights have implications for the synthesis of **CB**[n] homologs, derivatives, and analogs. Ultimately, these insights led us to the synthesis of **CB**[5] – **CB**[7] analogs bearing bis(phthalhydrazide) walls. These new **CB**[n] analogs have remarkable features including UV/Vis, fluorescence, and electrochemical activity along with enhanced dimensions. We are currently defining the scope of the recognition properties of these **CB**[n] analogs. Preliminary results suggest that they retain many of the advantageous features of the **CB**[n] family including tight binding, slow kinetics of exchange, and high selectivity. Unlike **CB**[n], these **CB**[n] analogs are soluble in both organic and aqueous solution, which suggests they may expand the range of applications to which the **CB**[n] family may be applied.

Acknowledgments

We are indebted to the contributions of a talented and devoted group of co-workers, named in the references, whose synthetic and mechanistic insights ultimately led to the synthesis of **CB**[n] analogs and congeners. We also thank Dr. Yiu-Fai Lam (NMR) and

Dr. James C. Fettinger (X-ray) for their contributions to this project. We thank the National Institutes of Health (GM61854), the Petroleum Research Fund administered by the American Chemical Society, and the University of Maryland for financial support of the research described in this chapter.

References and Footnotes

1. Behrend, R., Meyer, E., Rusche, F., *Liebigs Ann. Chem.* **1905**, *339*, 1.
2. Freeman, W. A., Mock, W. L., Shih, N. Y., *J. Am. Chem. Soc.* **1981**, *103*, 7367.
3. Kim, J., Jung, I.-S., Kim, S.-Y., Lee, E., Kang, J.-K., Sakamoto, S., Yamaguchi, K., Kim, K., *J. Am. Chem. Soc.* **2000**, *122*, 540.
4. Day, A. I., Arnold, A. P., Blanch, R. J., Snushall, B., *J. Org. Chem.* **2001**, *66*, 8094.
5. Day, A. I., Blanch, R. J., Arnold, A. P., Lorenzo, S., Lewis, G. R., Dance, I., *Angew. Chem., Int. Ed.* **2002**, *41*, 275.
6. Mock, W. L., Shih, N. Y., *J. Org. Chem.* **1983**, *48*, 3618.
7. Mock, W. L., Shih, N. Y., *J. Org. Chem.* **1986**, *51*, 4440.
8. Mock, W. L., Shih, N. Y., *J. Am. Chem. Soc.* **1988**, *110*, 4706.
9. Mock, W. L., Shih, N. Y., *J. Am. Chem. Soc.* **1989**, *111*, 2697.
10. Marquez, C., Hudgins, R. R., Nau, W. M., *J. Am. Chem. Soc.* **2004**, *126*, 5808.
11. Mock, W. L., Irra, T. A., Wepsiec, J. P., Adhya, M., *J. Org. Chem.* **1989**, *54*, 5302.
12. Mock, W. L., Irra, T. A., Wepsiec, J. P., Manimaran, T. L., *J. Org. Chem.* **1983**, *48*, 3619.
13. Mock, W. L., Pierpont, J., *J. Chem. Soc., Chem. Commun.* **1990**, 1509.
14. Buschmann, H. J., Schollmeyer, E., *Textilveredlung* **1994**, *29*, 58.
15. Buschmann, H. J., Schollmeyer, E., *Textilveredlung* **1997**, *32*, 249.
16. Buschmann, H. J., Schollmeyer, E., *Textilveredlung* **1998**, *33*, 44.
17. Jeon, Y.-M., Whang, D., Kim, J., Kim, K., *Chem. Lett.* **1996**, 503.
18. Whang, D., Kim, K., *J. Am. Chem. Soc.* **1997**, *119*, 451.
19. Lee, E., Heo, J., Kim, K., *Angew. Chem., Int. Ed.* **2000**, *39*, 2699.
20. Whang, D., Heo, J., Kim, C.-A., Kim, K., *Chem. Commun.* **1997**, 2361.
21. Lee, J. W., Samal, S., Selvapalam, N., Kim, H.-J., Kim, K., *Acc. Chem. Res.* **2003**, *36*, 621.
22. Mock, W. L. "*Cucurbituril*" in "*Comprehensive Supramolecular Chemistry,*" Ed. Vögtle, F. Pergamon Press: Oxford, 1996, vol. 2, p. 477.
23. Lagona, J., Mukhopadhyay, P., Chakrabarti, S., Isaacs, L., *Angew. Chem., Int. Ed.* **2005**, *44*, in press.
24. Day, A. I., Arnold, A. P., Blanch, R. J., *Molecules* **2003**, *8*, 74.
25. Smeets, J. W. M., Sijbesma, R. P., Niele, F. G. M., Spek, A. L., Smeets, W. J. J., Nolte, R. J. M., *J. Am. Chem. Soc.* **1987**, *109*, 928.
26. Rowan, A. E., Elemans, J. A. A. W., Nolte, R. J. M., *Acc. Chem. Res.* **1999**, *32*, 995.
27. Flinn, A., Hough, G. C., Stoddart, J. F., Williams, D. J., *Angew. Chem., Int. Ed.* **1992**, *31*, 1475.
28. Wu, A., Chakraborty, A., Witt, D., Lagona, J., Damkaci, F., Ofori, M. A., Chiles, J. K., Fettinger, J. C., Isaacs, L., *J. Org. Chem.* **2002**, *67*, 5817.
29. Hoffmann, R., Knoche, W., Fenn, C., Buschmann, H.-J., *J. Chem. Soc., Faraday Trans.* **1994**, *90*, 1507.
30. Ong, W., Gomez-Kaifer, M., Kaifer, A. E., *Org. Lett.* **2002**, *4*, 1791.
31. Sokolov, M. N., Virovets, A. V., Dybtsev, D. N., Gerasko, O. A., Fedin, V. P.,

Hernandez-Molina, R., Clegg, W., Sykes, A. G., *Angew. Chem., Int. Ed.* **2000**, *39*, 1659.

32. Isobe, H., Sato, S., Nakamura, E., *Org. Lett.* **2002**, *4*, 1287.
33. Karcher, S., Kornmuller, A., Jekel, M., *Water Sci. Technol.* **1999**, *40*, 425.
34. Kellersberger, K. A., Anderson, J. D., Ward, S. M., Krakowiak, K. E., Dearden, D. V., *J. Am. Chem. Soc.* **2001**, *123*, 11316.
35. Miyahara, Y., Abe, K., Inazu, T., *Angew. Chem., Int. Ed.* **2002**, *41*, 3020.
36. Sasmal, S., Sinha, M. K., Keinan, E., *Org. Lett.* **2004**, *6*, 1225.
37. Tuncel, D., Steinke, J. H. G., *Macromolecules* **2004**, *37*, 288.
38. Wagner, B. D., Stojanovic, N., Day, A. I., Blanch, R. J., *J. Phys. Chem. B* **2003**, *107*, 10741.
39. Shen, Y., Xue, S., Zhao, Y., Zhu, Q., Tao, Z., *Chin. Science Bull.* **2003**, *48*, 2694.
40. Burnett, C. A., Witt, D., Fettinger, J. C., Isaacs, L., *J. Org. Chem.* **2003**, *68*, 6184.
41. Witt, D., Lagona, J., Damkaci, F., Fettinger, J. C., Isaacs, L., *Org. Lett.* **2000**, *2*, 755.
42. Burnett, C. A., Lagona, J., Wu, A., Shaw, J. A., Coady, D., Fettinger, J. C., Day, A. I., Isaacs, L., *Tetrahedron* **2003**, *59*, 1961.
43. Valdéz, C., Spitz, U. P., Toledo, L. M., Kubik, S. W., Rebek, J. J., *J. Am. Chem. Soc.* **1995**, *117*, 12733.
44. Reek, J. N. H., Kros, A., Nolte, R. J. M., *Chem. Commun.* **1996**, 245.
45. Niele, F. G. M., Nolte, R. J. M., *J. Am. Chem. Soc.* **1988**, *110*, 172.
46. Isaacs, L., Witt, D., *Angew. Chem., Int. Ed.* **2002**, *41*, 1905.
47. Wu, A., Chakraborty, A., Fettinger, J. C., Flowers, R. A., II, Isaacs, L., *Angew. Chem., Int. Ed.* **2002**, *41*, 4028.
48. Reek, J. N. H., Priem, A. H., Engelkamp, H., Rowan, A. E., Elemans, J. A. A. W., Nolte, R. J. M., *J. Am. Chem. Soc.* **1997**, *119*, 9956.
49. Wu, A., Mukhopadhyay, P., Chakraborty, A., Fettinger, J. C., Isaacs, L., *J. Am. Chem. Soc.* **2004**, *126*, 10035.
50. Wu, A., Isaacs, L., *J. Am. Chem. Soc.* **2003**, *125*, 4831.
51. Mukhopadhyay, P., Wu, A., Isaacs, L., *J. Org. Chem.* **2004**, *69*, 6157.
52. Chakraborty, A., Wu, A., Witt, D., Lagona, J., Fettinger, J. C., Isaacs, L., *J. Am. Chem. Soc.* **2002**, *124*, 8297-8306.
53. Day, A. I., Blanch, R. J., Coe, A., Arnold, A. P., *J. Inclusion Phenom. Macrocyclic Chem.* **2002**, *43*, 247-250.
54. Buschmann, H.-J., Fink, H., Schollmeyer, E.; *Ger. Offen.* (Germany): DE 19603377, **1997** [*Chem. Abstr.* **1997**, *127*, 205599].
55. Lagona, J., Fettinger, J. C., Isaacs, L., *Org. Lett.* **2003**, *5*, 3745-3747.
56. Kim, H.-J., Heo, J., Jeon, W. S., Lee, E., Kim, J., Sakamoto, S., Yamaguchi, K., Kim, K., *Angew. Chem., Int. Ed.* **2001**, *40*, 1526-1529.
57. Drew, H. D. K., Pearman, F. H., *J. Chem. Soc.* **1937**, 586-592.

STRATEGIES AND TACTICS IN ORGANIC SYNTHESIS, VOL. 6

Chapter 4

APPLICATION OF SILICON-ASSISTED INTRAMOLECULAR CROSS-COUPLING IN TOTAL SYNTHESIS OF (+)-BRASILENYNE

Scott E. Denmark and Shyh-Ming Yang
Department of Chemistry
University of Illinois, at Urbana-Champaign
Urbana, Illinois 61801

I. Introduction and Background

Red algae and marine organisms that feed on red algae, in particular Laurencia species, have produced various C_{15} non-terpenoid acetogenins containing halogenated medium ring ethers.[1] These metabolites contain different ring sizes such as those found in (+)-laurencin,[2] (+)-prelaureatin,[3] (+)-laurallene,[4] (-)-isolaurallene,[5] and (+)-obtusenyne[6] (Figure 1). Among them, (+)-brasilenyne (**1**), isolated from the digestive gland of a sea hare (*Aplysia brasiliana*) by Fenical *et al.* in 1979,[7] has a novel nine-membered cyclic ether skeleton containing a 1,3–*cis*–*cis* diene unit. The relative and absolute configurations of **1** were established by a single crystal X-ray diffraction analysis.

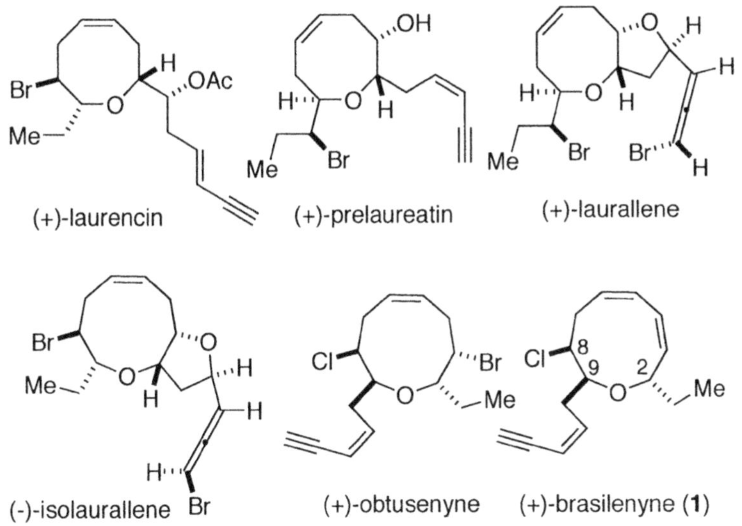

FIGURE 1. Representative C_{15} marine metabolites.

In Nature, many organisms, devoid of physical protection, have evolved the use of toxic and/or noxious organic compounds as defensive agents. For example, sea hares are incapable of evasive maneuvering; yet they lack significant predators. It has been suggested that secondary metabolites, such as 1, are produced and/or concentrated in the digestive gland to act as defensive chemicals. Indeed, the in vivo studies of 1 and (+)-cis-dihydrorhodophytin, a major component from the same natural source, have been shown to be potent antifeedants.[7a]

Investigation of the biosynthesis of cyclic ether metabolites from the Laurencia species reveals that lactoperoxidase (LPO) directly transforms 3Z,6S,7S-laurediol, which occurs in nature as various stereoisomers, into (+)-prelaureatin through a bromonium-ion-induced ether formation.[8] Moreover, (+)-laurallene, (+)-laureatin, and (+)-isolaureatin are assumed to arise from (+)-prelaureatin from a similar biogenetic pathway.[3,9] Therefore, the hypothetical intermediate, Cl-substituted laurediol, has been postulated to produce (+)-obtusenyne by a similar biotransformation and further dehydrobromination to afford (+)-brasilenyne.[7] The speculation is supported by the finding that those compounds have the same configuration at the corresponding C(6) and C(7) centers of Cl-substituted intermediate[10] (Scheme 1).

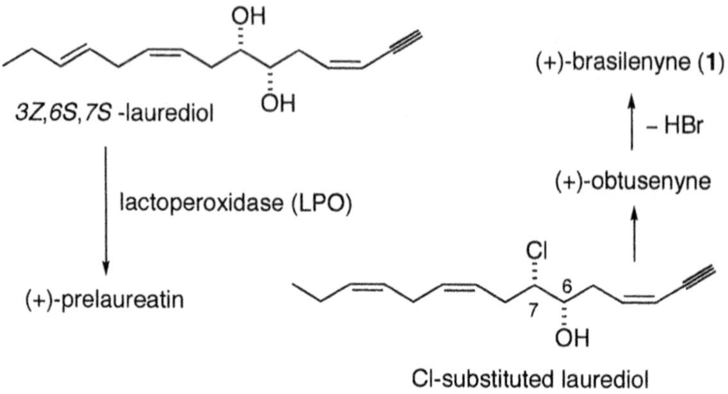

SCHEME 1

The interesting structure of these marine metabolites has stimulated a significant level of effort for construction of oxonin and oxocene ring systems.[11] Recent representative examples include Crimmins' syntheses of (+)-prelaureatin[11b] and (+)-obtusenyne[11e] through an aldol or alkylation/ring-closing metathesis (RCM) sequence, Overman's synthesis of (-)-laurenyne[11j] and (+)-laurencin[11k] by Lewis acid promoted alkene-acetal cyclization, and Holmes' synthesis of (+)-laurencin,[11l] which features a Claisen rearrangement. These methods are particularly well suited for the construction of medium-rings containing a *single* carbon-carbon double bond. The synthetic challenge of the oxonin core of **1**, however, requires the formation of a medium ring bearing a conjugated system. Two recent reports from Negishi *et al*[12] and Isobe *et al*[13] describe viable methods for the synthesis of medium rings that contain a 1,3-diene. These reports feature the cyclic carbopalladation of an allene and the acid-catalyzed cyclization of an acetylene dicobalt complex.

As part of our program on the development of new silicon-based cross-coupling reactions, we have recently demonstrated the synthetic potential of the sequential ring-closing metathesis (RCM)/silicon-assisted intramolecular cross-coupling reaction for constructing medium sized, carbo- and heterocyclic systems bearing a 1,3–*cis*–*cis* diene unit.[14] We soon recognized that (+)-brasilenyne would be an ideal target to illustrate this powerful synthetic method because the coupling process is well suited to generate the oxonin core of **1** with its 1,3-diene. However, this approach introduces several challenges that require additional synthetic manipulations. In contrast to the foregoing methodological studies on simpler systems, the side chain at C(9) and the ethyl group at C(2) presented potential difficulties for the intramolecular coupling process. In addition, the presence of the chlorine-bearing center at C(8) requires the creation of a hydroxyl functional group at C(8) of opposite configuration, which in turn allows the use of a temporary silicon tether for the construction a cyclic alkenylsilyl ether by RCM. The creation of the C(2) stereogenic center is the other critical part of the strategy needed for the synthesis of 1. This problem stimulated the development of a new ring opening reaction of a 1,3-dioxolanone with an acetylenic nucleophile to create the requisite stereogenic center at a propargylic position. In this chapter, we describe in detail our efforts toward the first total synthesis of (+)-brasilenyne by applying these new transformations as key strategic elements.[15]

II. Preliminary Studies

The plan for the synthesis of 1 could be formulated in earnest after the successful methodological development of the tandem RCM/cross-coupling protocol. The intramolecular silicon-assisted cross-coupling reaction provides good generality in the synthesis of medium-sized carbocycles that contain a geometrically defined 1,3-diene (Scheme 2).

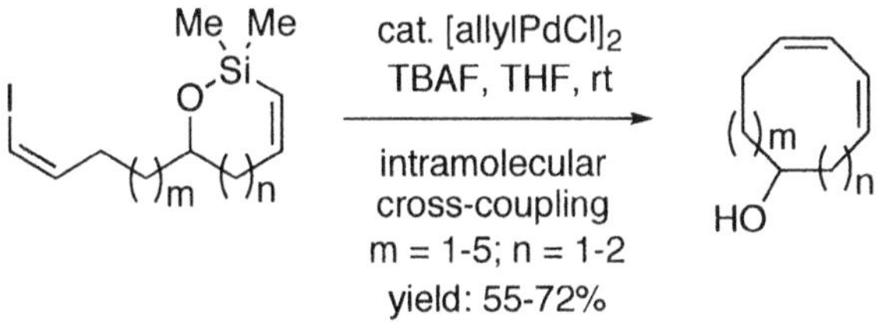

SCHEME 2

To demonstrate the versatility of this process and to provide a relevant model study for the synthesis of 1, we investigated an extension to medium-sized heterocycles. Thus, the diastereomeric silyl ethers 8a-b were selected to test this application by generation of the corresponding 9-membered oxacyclic dienes (Scheme 3). The preparation of 8a-b began with the reduction of pyruvic aldehyde dimethoxy acetal with $NaBH_4$ in MeOH/THF to afford hydroxy acetal 2 (84%). Alkylation of the sodium salt of 2 with propargylic bromide afforded 3 (85%). Conversion of 3 to 5 was achieved by alkyne iodination followed by a cis-reduction of iodoalkyne **4** to iodoalkene **5**. Hydrolysis of the dimethyl acetal was effected by treatment of **5** with p-toluenesulfonic acid in an acetone/H_2O mixture to produce aldehyde **6** in 94% yield. Treatment of aldehyde **6** with allylmagnesium bromide generated hydroxy dienyl iodide **7a-b** in 95% yield as a 56:44 mixture of diastereomers, which were easily separated by silica gel chromatography. Finally, silylation of the alcohols with chlorodimethylvinylsilane furnished **8a-b** in 94% and 95% yield, respectively.

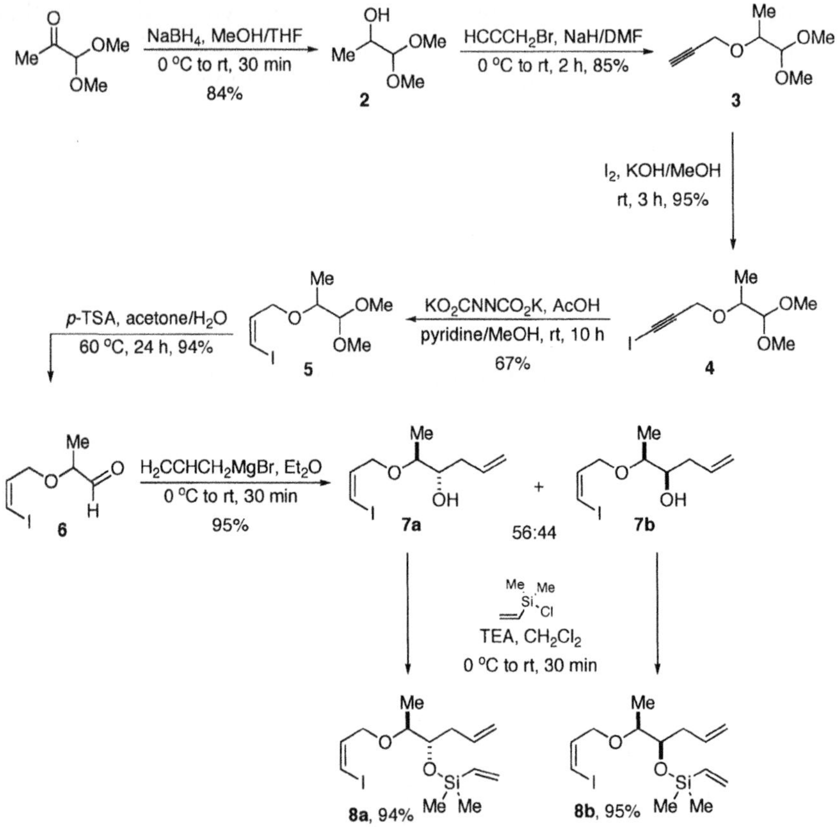

SCHEME 3

With these materials in hand, the ring closing metathesis of **8a** was carried out using 10 mol% of Schrock's molybdenum complex $[(CF_3)_2MeCO]_2Mo(=CHCMe_2Ph)(=NC_6H_3-2,6-i-Pr_2)$ as the catalyst[16] to afford the target siloxane **9a** in 81% yield. Similarly, RCM of **8b** proceeded smoothly to afford cyclic silyl ether **9b** in 80% yield under the same conditions (Scheme 4). Gratifyingly, subsequent exposure of the siloxanes to the optimal conditions established in carbocyclic cases promoted the intramolecular cross coupling effectively.[14] Both diastereomers reacted with equal facility to afford the oxonane dienes **10a** and **10b** (in 72% and 77% yield, respectively) with no difference in rate or efficiency. We felt encouraged by these results since the steric influence of cross-coupling in the synthesis of **1** was one of major issues that concerned us.

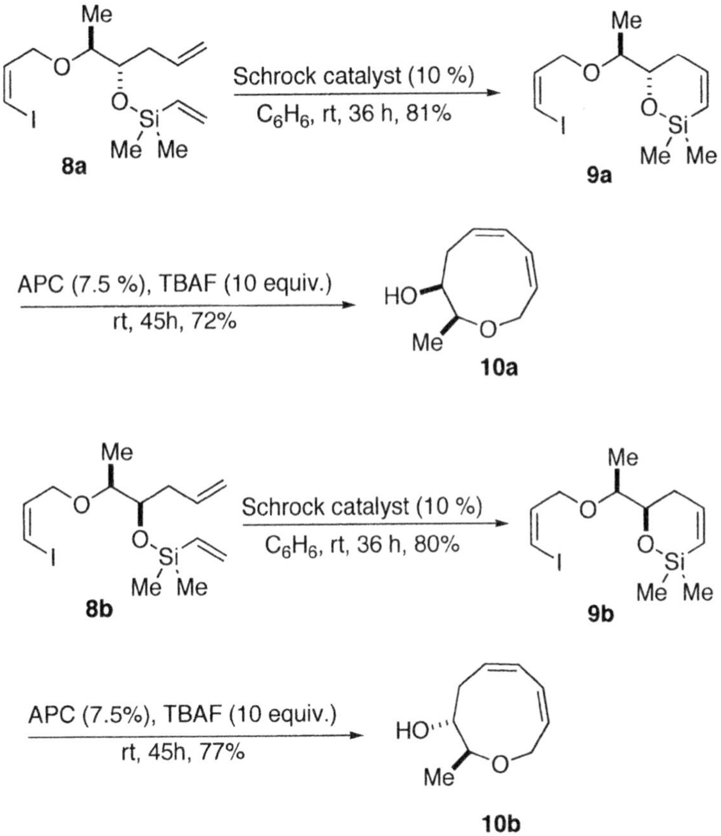

SCHEME 4

III. Synthetic Strategies

The retrosynthetic plan formulated for the synthesis of (+)-brasilenyne is outlined below (Scheme 5). Simplification of the enyne side chain and chloride functionality reduces the challenge to the intermediate **11**, which was projected to arise from palladium-catalyzed, silicon-assisted intramolecular cross-coupling of **12**. By use of the six-membered cyclic siloxane, the hydroxyl group liberated in the cross-coupling is perfectly situated for installation of the chloride at C(8). Cyclic alkenylsilyl ether **12** would arise from diastereoselective allylation of aldehyde **13** and application of ring-closing metathesis (RCM) of the vinyl alkenylsilyl ether derivative. The aldehyde **13**, with a protected primary hydroxy group as well as the geometrically-defined vinyl iodide, could be, without difficulty, elaborated from **14**.

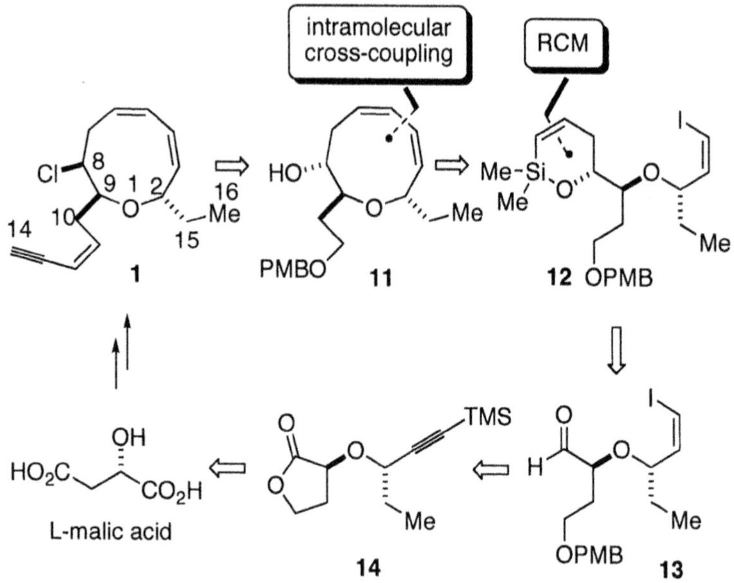

SCHEME 5

The diastereo- and enantioselective synthesis of **14** represented an intriguing synthetic challenge, namely the construction of a doubly branched ether flanked by stereogenic centers (Scheme 6). The straightforward solution to this problem would involve a nucleophilic displacement reaction (path a). Because both enantiomers **15**[17] and **16**[18] are easily available, formation of the doubly branched ether linkage can proceed in either direction. This approach was considered plausible as both hydroxyl groups are activated (flanked by carboxyl or alkynyl groups). However, in recognition of the difficulty of effecting displacements at sterically congested centers, an alternative approach, featuring a diastereoselective ring opening of a 1,3-dioxolanone **17**, was envisioned (path b). This plan called for the Lewis-acid-promoted addition of bis(trimethylsilyl)acetylene to the activated acetal.[19] Thus, the C(2) and C(8) stereocenters were to be installed through a reaction controlled by the stereocenter in the malic acid residue. Either approach is attractive as both **15** and **17** could be easily derived from natural L-(*S*)-malic acid. Therefore, the crucial components in this approach would be the sequential RCM/silicon-assisted intramolecular cross-coupling reaction for construction of the oxonin core structure as well as the diastereoselective installation of a propargylic stereogenic center in key intermediate **14**.

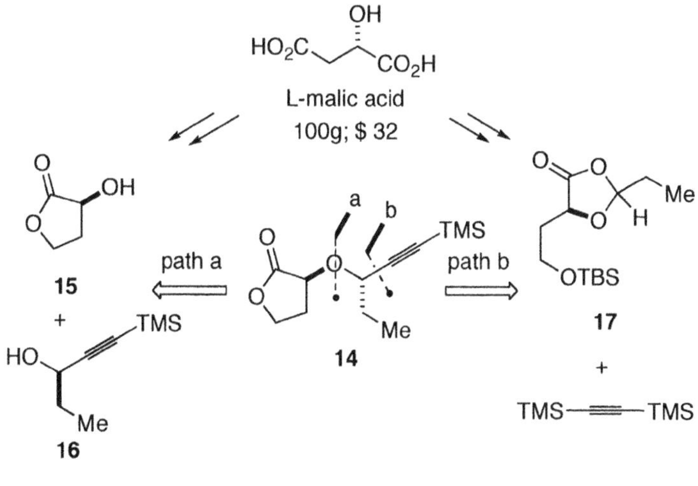

SCHEME 6

IV. Synthesis of (+)-Brasilenyne

A. FIRST APPROACH TO THE DOUBLY BRANCHED ETHER LINKAGE: NUCLEOPHILIC DISPLACEMENT

When this synthesis was initiated in the early months of 2002, we first examined the direct nucleophilic displacement strategy (Scheme 6, path a) to construct the doubly branched ether linkage. To evaluate this possibility, the enantiomerically pure alcohols 15 and 16 were prepared. The (S)-α-hydroxy-γ-butyrolactone 15 is commercially available but can also be easily prepared from L-(S)-malic acid in a three-step sequence in high chemical yield (Scheme 7).[17] Moreover, highly enantiomerically enriched 16 (98.8% ee) was obtained in excellent yield by an efficient transfer hydrogenation reaction of the corresponding ketone 19 in i-PrOH with a catalytic amount of {[(1R,2R)-TsDPEN]Ru(η^6-p-cymene)} as developed by Noyori, et al.[20]

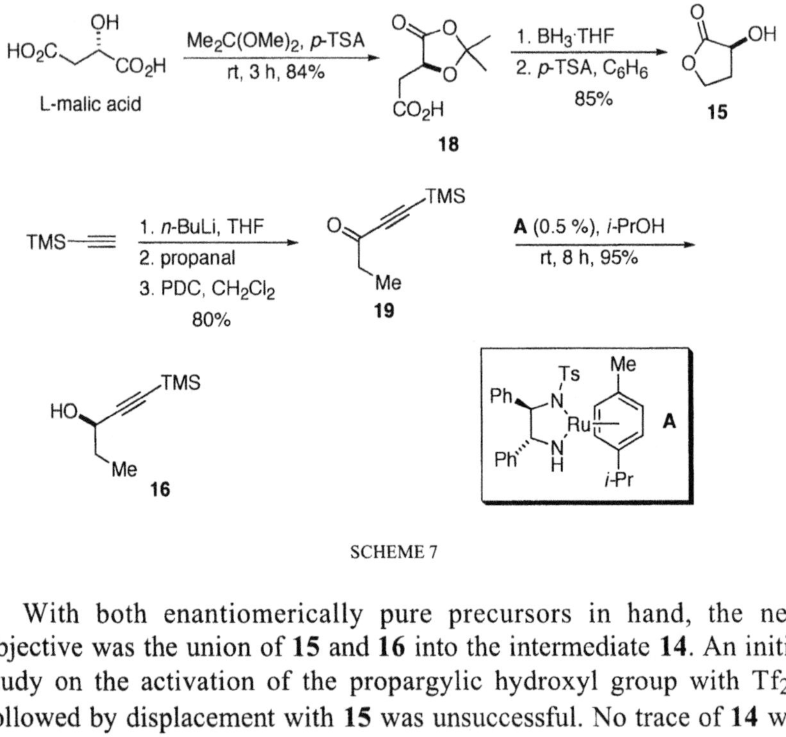

SCHEME 7

With both enantiomerically pure precursors in hand, the next objective was the union of 15 and 16 into the intermediate 14. An initial study on the activation of the propargylic hydroxyl group with Tf$_2$O followed by displacement with 15 was unsuccessful. No trace of 14 was detected and mainly a complex mixture was observed by ^{1}H NMR analysis (Scheme 8). To identify the problem with this transformation, the stable mesylate 20 was prepared by treatment of rac-16 with MsCl and Et$_3$N and was then subjected to ether formation with 15 under various conditions. Unfortunately, no reaction occurred by using (i-Pr)$_2$NEt as the base. Both reactants, 15 and 16, were stable under these and evenharsher conditions (50-60 °C and/or neat). By employing K$_2$CO$_3$ or NaH as the base, silylated compound 21 and desilylated product 22 were observed by

^{1}H NMR analysis.

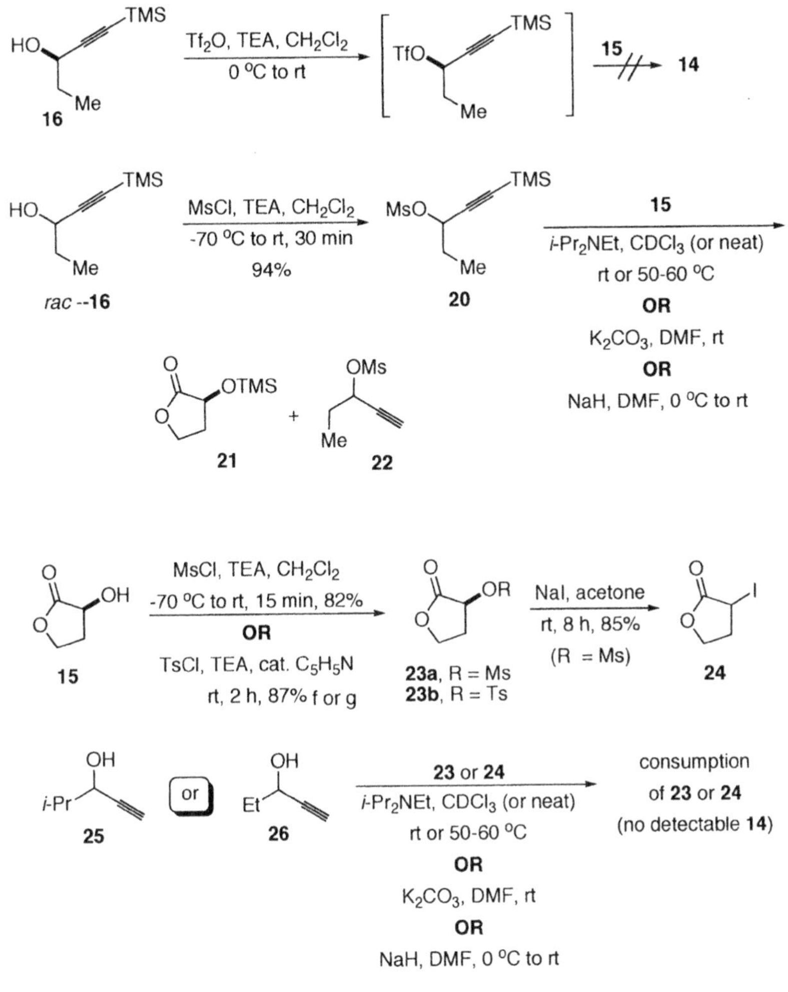

SCHEME 8

We next examined reversing the roles through activation of the hydroxyl group of **15**. Activated substrates **23** and **24** were easily prepared from **15**. Treatment of compound **23** or **24** with propargylic alcohol **25** or **26** using $(i\text{-Pr})_2$NEt led to no reaction. By using K_2CO_3 or NaH as the base, complete consumption of **23** or **24** was observed with no detectable amount of **14**. The formation of doubly branched ether **14** by nucleophilic displacement obviously presented the formidable challenges we anticipated. After a month of unsuccessful attempts, we soon concluded that the problem of elimination with **15** and **16**, the lability of the trimethylsilylacetylenic group, and the imposing steric influence on a direct unification process rendered this strategy non-viable.

B. SECOND APPROACH TO THE DOUBLY BRANCHED ETHER LINKAGE: RING OPENING OF A 1,3-DIOXOLANONE

The failure to create a doubly branched ether by classic substitution reactions necessitated the development of an alternative strategy that could control the introduction of propargylic center from an enantiopure starting material. After considering a number of options, we concluded that the well-known ring opening of chiral acetal templates (promoted by Lewis acids provides) constitutes a potential route to the stereocontrolled creation of chiral *secondary* hydroxyl and/or ethers centers in this molecule (Scheme 9).

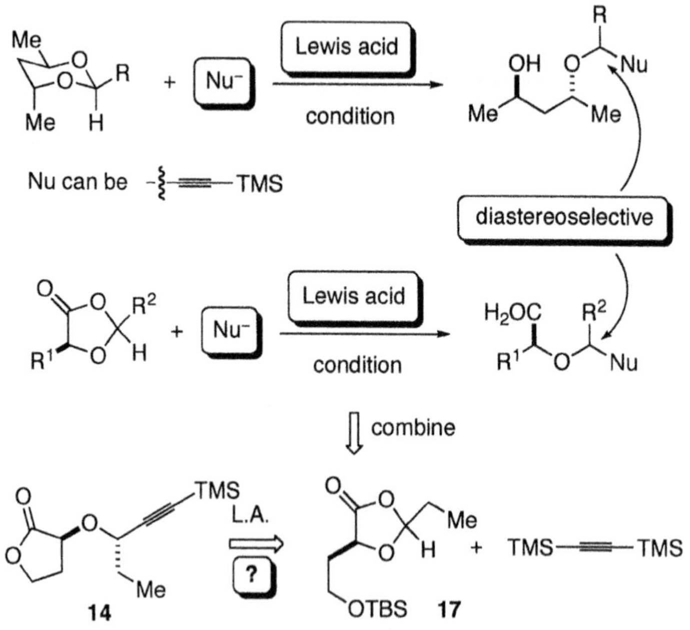

SCHEME 9

Over the past two decades, many nucleophilic organometallic reagents in combination with Lewis acids have been successfully employed in the opening of chiral acetals.[21] The diastereoselectivity of this process strongly depends on the structure of the acetal, the solvent, the Lewis acid, as well as the nucleophilic reagent. Acetylenic organometallic compounds have been employed as nucleophilic reagents to create stereogenic centers at propargylic positions with high diastereoselectivity.[19] Moreover, ring opening of 1,3-dioxolanones with various nucleophilic reagents and Lewis acids have been extensively investigated as well.[22] This method could provide an α-branched ether carboxylate (or ester) in a highly diastereoselective manner. By the combination of these two components, we thought that the diastereoselective ring opening of 1,3-dioxolanones with an acetylenic reagent under activation by a Lewis acid might offer an attractive solution for the preparation of the key intermediate **14** (Scheme 9). However, the ring opening of a 1,3-dioxolanone with acetylenic organometallic reagents was unprecedented. After a moment's reflection, we decided that the 1,3-dioxolanone **17** would be the ideal choice to test this approach because: (1) **17** can be rapidly obtained from L-(S)-malic acid and (2) propanal can act both as the dioxanone protecting group and as a part of the target compound **14** (Scheme 9).

One of the first problems encountered in the implementation of this strategy was stereoselective preparation of the dioxolanone **17**. The condensation of L-(S)-malic acid with propanal in the presence of BF$_3$•Et$_2$O afforded the 1,3-dioxolanone **27** (85%) as an 81:19, *cis/trans* mixture of isomers (Scheme 10).[23] On the basis of literature precedent, high diastereoselectivity was expected in the construction of the 1,3-dioxolanone, which normally favors the thermodynamically more stable *cis* isomer.[23] We noted that the isomers rapidly equilibrated in the presence of acid at room temperature. This ready epimerization suggested that the ring opening process might be stereoconvergent (i.e., independent of the starting material composition). In addition, we surmised that 1,3-dioxolanones would likely react at the S$_N$1 limit of the mechanistic spectrum and therefore not be dependent on the initial diastereomeric mixture.[21h] Thus, we decided to move forward with this *cis/trans* mixture. Selective reduction of the carboxylic acid in **27** using BH$_3$•THF at 0 °C gave alcohol **28** in 82% yield. Subsequently, protection of the primary alcohol with *tert*-butylchlorodimethylsilane and pyridine afforded **17** in 85% yield (ca. 60-65% overall yield in three steps from L-(S)-malic acid). Direct transformation of the *cis/trans* (81:19) mixture of **27** to alcohol **28** and TBS ether **17**, maintained the ratio (*cis/trans* =

83:17). The hydroxy 1,3-dioxolanone intermediate **28** is acid labile and is also unstable toward bases such as Et_3N, $(i\text{-}Pr)_2NEt$, imidazole, and N,N-dimethylaminopyridine, producing (S)-α-hydroxy-γ-butyrolactone **15**. However, by carrying out the reduction of **27** at 0 °C and the protection of **28** using pyridine as the base, we were able to reproducibly perform these transformations on a large scale with comparable yields of isolated products.

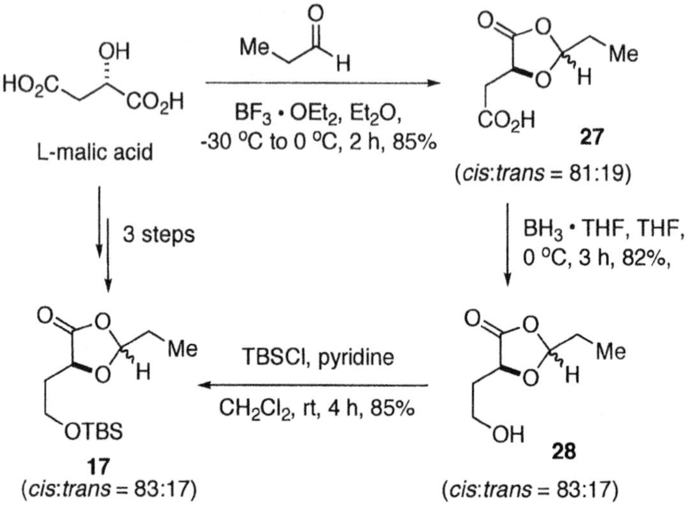

SCHEME 10

With precursor **17** in hand, we studied the installation of the stereogenic center at the propargylic position by a Lewis-acid-mediated ring opening of **17** with bis(trimethylsilyl)acetylene. As mentioned previously, the ring opening of acetal templates with weakly nucleophilic acetylenic reagent, such as bis(trimethylsilyl)acetylene, is expected to occur in the S_N1 region of the mechanistic continuum (i.e. by Lewis acid assisted ring opening prior to nucleophilic attack).[21h] In these cases, the asymmetric induction was rationalized by invoking Cram's rule as applied to the oxocarbenium ion intermediate.[19b] If the mechanism and diastereoselectivity in ring opening of dioxolanone **17** are similar to those operative in the opening of chiral acetals, we anticipated that the desired diastereomer would be obtained. Another beneficial feature of the 1,3-dioxolanone is that the more labile C–OCO bond could facilitate the formation of the reactive oxocarbenium ion intermediate, particularly in

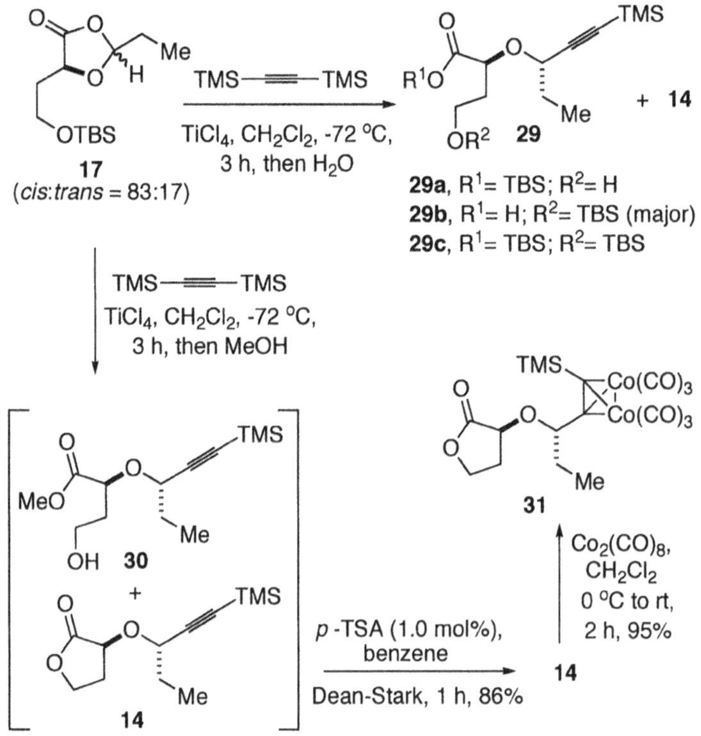

SCHEME 11

resence of stronger Lewis acids such as TiCl$_4$.[22a-b] Therefore, to
ve the rapid equilibration of the isomers and to converge the
ns mixture of dioxolanones **17** into a single diastereomer **14**, TiCl$_4$
mployed as the Lewis acid by adapting a procedure for the ring
ng of acetal templates developed by Johnson *et al.*[19a] However, the
results for the ring opening reaction were disappointing.
licated mixtures were observed in which **29a-c** and **14** were
ned as the products (Scheme 11). Notably, each component was
rised of only a single diastereomer indicating that the ring opening
ss was highly diastereoselective. We also found that the work-up
dure greatly influenced the product distribution. For example, in
to reduce the complexity of the mixture, the silyl group was
ed by quenching the reaction with MeOH. Gratifyingly, the ring
ng of 1,3-dioxolanone **17** proceeded smoothly to afford only two
cts after quenching with MeOH, the desired lactone **14** and the

methyl ester **30**. Furthermore, treatment of the crude reaction mixture with a catalytic amount of *p*-toluenesulfonic acid in refluxing benzene for h gave **14** exclusively in 86% yield as a *single* diastereomer. The *S* configuration of the propargylic position was secured by conversion of **14** to dicobalt complex **31** with $Co_2(CO)_8$ in 95% yield. Slow sublimation of 1 at 65–68 °C (0.1 mmHg) delivered deep red crystals, whose full stereostructure was confirmed by a single crystal X-ray diffraction analysis (Figure 2).[23]

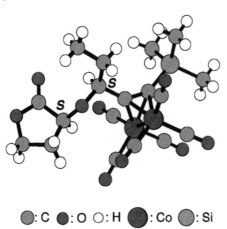

⬤: C ⬤: O ◯: H ⬤: Co ⬤: Si

FIGURE 2. Chem3D image of X-ray crystal structure of cobalt complex **31**.

The production of a single diastereomer **14** from a *cis/trans* mixture of 17 strongly indicated that (1) the mechanism of ring opening of the 1,3-dioxolanone likely proceeded through an oxocarbenium ion intermediate and (2) the diastereoselectivity was effectively controlled by the C(2) stereogenic center of the L-(S)-malic acid residue.[22a-b] A rationale for the convergent diastereoselectivity can be formulated (Figure 3). The equilibration of oxocarbenium ion intermediates **A** and **B** can easily be envisioned in the presence of a strong Lewis acid.[22a-b] To minimize dipole-dipole repulsion, the oxocarbenium ion and the Ti-coordinated carboxylate are arranged in an antiperiplanar conformation. In addition, the ring-opened *E*-oxocarbenium ion **A** is proposed to be thermodynamically favored because of the reduced steric interaction between malic acid residue and the ethyl group. Furthermore, nucleophilic attack from the *Re*-face avoids significant steric repulsion from the bulky R group (CH_2CH_2OTBS), thus affording the desired

though the opening of **17** with bis(trimethylsilylacetylene) was
ssful, it is noteworthy to mention two other approaches that
yed related to ring openings of 1,3-dioxolanones (Scheme 12). On
sis of studies with acetal templates, we expected that an ethyl group
be transferred to the C(2) center with retention of configuration
C(2)-O(3) bond scission upon treatment of 1,3-dioxolanone **32** with
laluminium.[21a-b] In addition, **14** might be constructed from 1,3-
anone **33** by treatment with DIBAL-H. However, these approaches

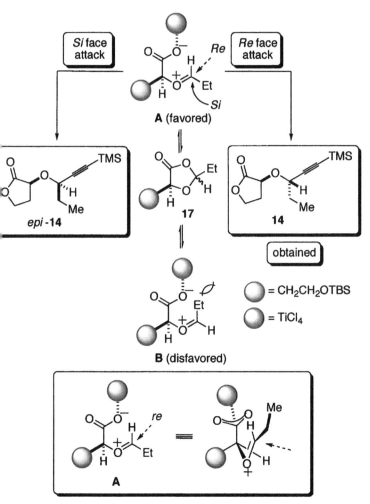

FIGURE 3. Mechanistic and stereochemical rationale for the opening of **17**.

equire the starting 1,3-dioxolanones to be prepared in high
liastereomeric purity. Because the formation of 1,3-dioxolanones bearing
a carboxylic acid residue could not be achieved diastereoselectively, we
elegated these approaches to backup status.

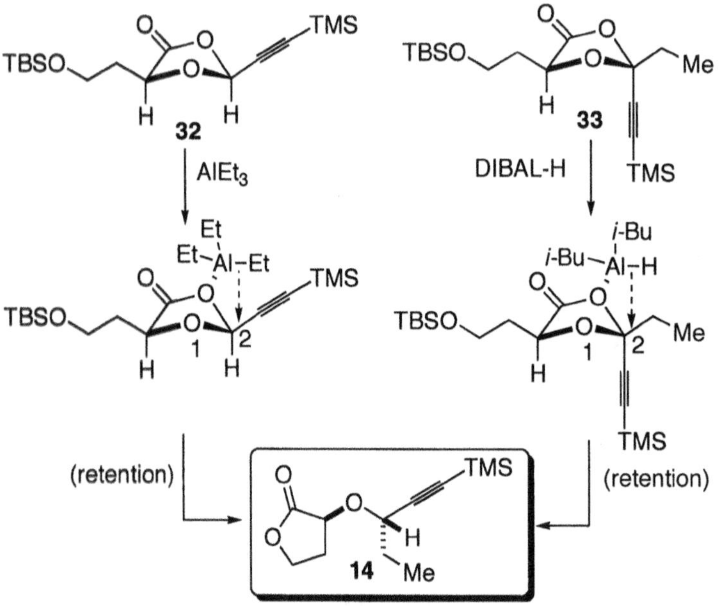

SCHEME 12

C.   INSTALLATION OF THE C(8) STEREOGENIC CENTER: ALLYLBORATION

To construct the oxonin core, a geometrically defined Z-alkenyl
odide was introduced next for the key intramolecular cross-coupling
reaction. Accordingly, conversion of alkynylsilane 14 to alkynyl iodide
34 was efficiently accomplished by treatment of 14 with N-
odosuccinimide and a catalytic amount of silver nitrate in DMF (95%
yield).[24] Furthermore, cis-reduction of 34 with diimide, generated in situ
from potassium azodicarboxylate in AcOH/pyridine, gave the Z-alkenyl
.odide 35 in 80% yield (Scheme 13).[25] The trace amount of over-
reduction product formed could be removed by stirring the crude mixture
n pyridine for 12 h at ambient temperature. Installation of the third
stereogenic center (C(8) in 1) calls for a diastereoselective allylation of an
aldehyde bearing a protected hydroxyl group, such as 37. Thus, reduction

diastereomeric ratio. Unfortunately, all attempts to selectively protect the ring-opened hydroxy aldehyde intermediate with TIPSCl or TBDPSCl were unsuccessful. Only moderate conversion and a small amount of the desired product **37** were obtained after a difficult separation from compound **38**.[26]

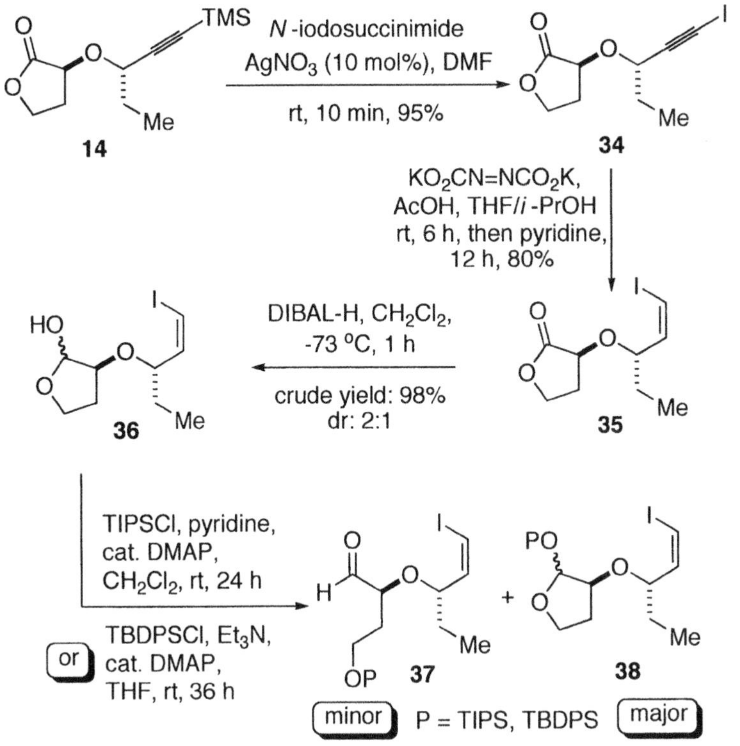

SCHEME 13

Alternatively, we surmised that the lactone **14** could be converted to Weinreb amide **40**, which provided additional options for generation of the homoallylic alcohol **42**. For example, through the influence of an α-alkoxy group, homoallylic alcohol **42** could be obtained by either a chelation-controlled reduction of β,γ-unsaturated ketone **41** or a non-chelation-controlled allylation of **13** (Scheme 14). Thus, elaboration of lactone **35** into amide **40** began by treatment of **35** with N,O-hydroxylamine hydrochloride in the presence of trimethylaluminum to give the amide **39** in 93% yield.[27] Further, protection of the revealed primary hydroxyl group with *p*-methoxybenzyl chloride (PMBCl) under mild conditions (Ag₂O) afforded **40** in 82% yield.[28] Amide **40** could be converted to aldehyde **13** by reduction with DIBAL-H at low temperature (−73 °C, 87% yield)[29] or into β,γ-unsaturated ketone **41** by treatment with allylmagnesium bromide at −40 °C in THF (94% yield).[30]

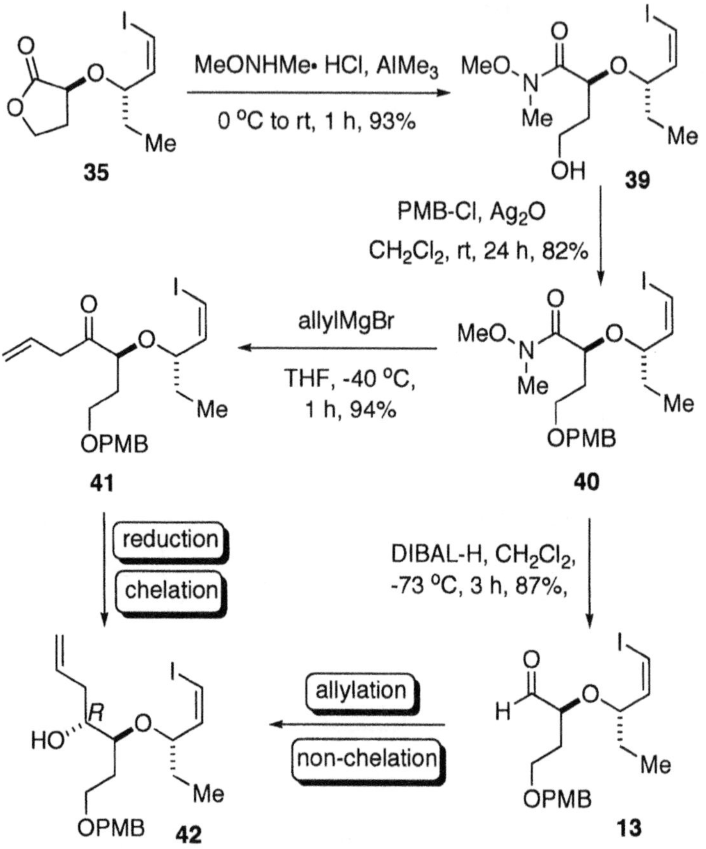

SCHEME 14

With both precursors **13** and **41** in hand, the next critical stage was the introduction of the third stereogenic center in **1**. Initial studies on diastereoselective reduction of the ketone function in **41** into alcohol **42** are compiled in Table 1. The use of DIBAL-H as the reducing agent gave an excellent chemical yield of **42**, but with unsatisfactory diastereoselectivity, slightly favoring the undesired *S*-isomer (Table 1, entry 1). All attempts to improve the selectivity by employing $MgBr_2$ and ZnI_2 were fruitless (entries 2-3).[31] The use of LS-Selectride® led to the preferential formation of the undesired S-isomer (entry 4).

TABLE 1
REDUCTION OF HOMOALLYL KETONE **41**.

entry	conditions	42 yield, % (ratio; $R{:}S)^a$
1	DIBAL–H, THF, –73 °C	93% (43:57)
2	DIBAL–H, $MgBr_2$, CH_2Cl_2, –73 °C	94% (40:60)
3	DIBAL–H, ZnI_2, CH_2Cl_2, –75 °C	92% (48:52)
4	LS-Selectride, THF, –72 °C	– (ca. 15:85)^b

^{a} The numbers correspond to isolated yield and the ratio was determined by ^{1}H NMR integration. ^{b} The product contained small amount of residue of reducing agent.

Next, the influence of the α-alkoxy group on the selective allylation of aldehyde **13** by non-chelation-controlled addition was assayed (Table 2). By using allyltrimethylsilane as the allylating reagent and BF_3•Et_2O as Lewis acid, only a 74:26 ($R{:}S$) ratio of alcohol **42** was obtained in 78% yield (Table 2, entry 1). Increasing the amount of Lewis acid (4.0 equiv) did not have a significant effect on either the yield or the selectivity (entry 2).[32] Allyltributyltin was also employed as the allylation reagent,[33] but poor selectivity was obtained with this reagent as well (entry 3). Direct addition with allylmagnesium bromide in the presence of $ZnCl_2$ gave 91% yield with no stereoselectivity (entry 4).[34]

The inability to accomplish selective, substrate-controlled nucleophilic addition of hydride to ketone **41** or allylating agents to aldehyde **13** suggested the use of reagent-controlled addition with an overwhelming intrinsic bias. The successful generation of **42** was eventually accomplished by employing the chiral allylborane reagent, allylB(l-Ipc)$_2$, developed by Brown et al.[35] Treatment of **13** with the reagent generated in situ from (+)-B-chlorodiisopinocampheylborane [(+)-DIP-Cl] and allylmagnesium bromide at –74 °C afforded **42** in 72% yield and 93:7 diastereoselectivity (entry 5).

TABLE 2
ALLYLATION OF ALDEHYDE **13.**

entry	conditions	**42** yield, % (ratio; $R{:}S$)[a]
1	AllylSiMe$_3$, BF$_3$•Et$_2$O, CH$_2$Cl$_2$, −73 °C	78% (74:26)[b]
2	AllylSiMe$_3$, BF$_3$•Et$_2$O, CH$_2$Cl$_2$, −73 °C	73% (73:27)[c]
3	AllylSnBu$_3$, BF$_3$•Et$_2$O, CH$_2$Cl$_2$, −74 °C	87% (62:38)[d]
4	AllylMgBr, ZnCl$_2$ Et$_2$O, −78 °C	91% (51:49)[e]
5	AllylB(l-Ipc)$_2$, Et$_2$O, −74 °C	72% (93:7)

[a] The numbers correspond to isolated yield and the ratio was determined by ^{1}H NMR integration. [b] AllylSiMe$_3$ (2.0 equiv) and BF$_3$•Et$_2$O (2.0 equiv) were employed. [c] AllylSiMe$_3$ (2.5 equiv) and BF$_3$•Et$_2$O (4.0 equiv) were employed. [d] AllylSnMe$_3$ (2.0 equiv) and BF$_3$•Et$_2$O (2.0 equiv) were employed. [e] AllylMgBr (1.5 equiv) and ZnCl$_2$ (1.75 equiv) were employed.

On the basis of much literature precedent, this reagent has been shown to exhibit a strong intrinsic selectivity bias. Thus, allylation with allylB(l-Ipc)$_2$ derived from (−)-α-pinene, should give the R-configuration at the homoallylic position.[36a] To confirm this stereochemical outcome, the allylboration product **42** was elaborated to lactone **44**, which could be structurally correlated to the known compounds, **45**$_{3,4\text{-}trans}$ and **45**$_{3,4\text{-}cis}$ (Scheme 15).[36] Treatment of **42** with DDQ in CH$_2$Cl$_2$/H$_2$O (19/1) afforded diol **43** in 81% yield. Subsequent oxidation was carried out with a catalytic amount of tetrapropylammonium perruthenate (TPAP) and N-methylmorpholine-N-oxide to furnish lactone **44** in 82% yield.[37] In the ^{1}H NMR spectrum of **44**, the coupling constants of $J_{2,3\text{-}cis}$, $J_{2,3\text{-}trans}$, and $J_{3,4\text{-}trans}$ are 7.0, 3.5, and 2.8 Hz, respectively. These data match well those calculated (7.0, 3.2, and 2.8 Hz) and experimentally determined (6.5, 3.6, 2.8 Hz) for analog **45**$_{3,4\text{-}trans}$. Importantly, strong NOE enhancements between HC(4) and HC(3') were also observed. A 4.7% enhancement of HC(3') was observed when the HC(4) was irradiated, whereas a 7.4% enhancement of HC(4) was observed. These results strongly support the conclusion that HC(4) and the alkenyl iodide residue exist in a *cis* relationship and therefore further confirm the expectation that the R-configuration was obtained in the asymmetric allylboration reaction.

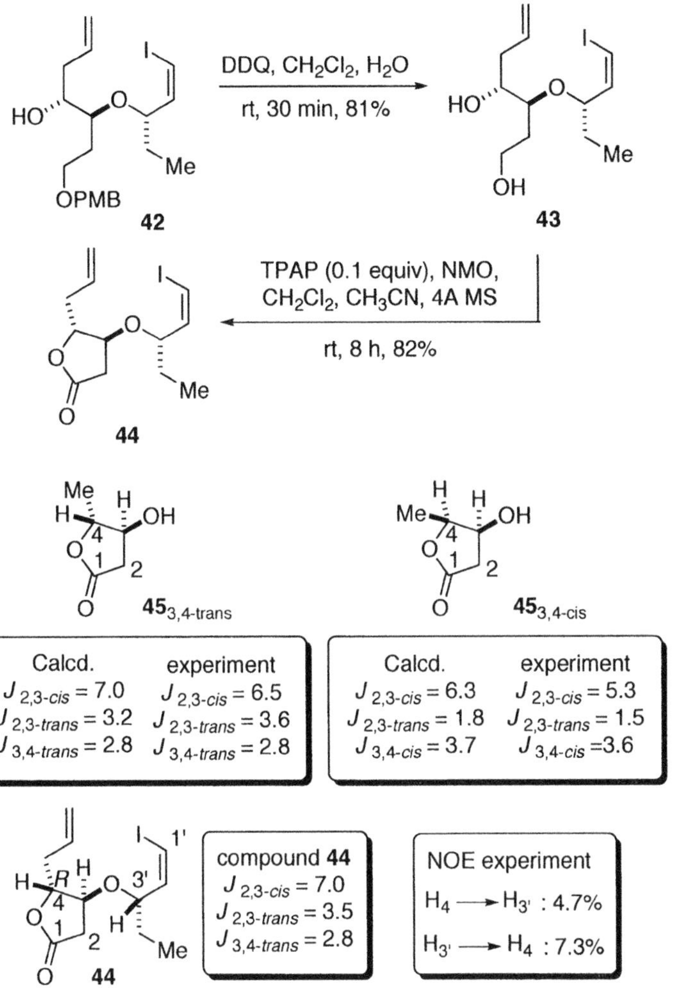

SCHEME 15

Finally, a significant improvement in yield (89%) and selectivity (>97:3) in the allylation could be secured under Mg^{2+}-salt free conditions at -100 °C (Scheme 16).[36b] In the presence of magnesium salts (Table 2, entry 5), the allylboration proceeded less efficiently presumably due to the formation of a stable borate-magnesium complex, which can slow the rate of allylboration. In addition, the level of asymmetric induction was not influenced by the magnesium salts: this effect is attributable solely to the lower reaction temperature (-100 °C vs. -74 °C).[36b]

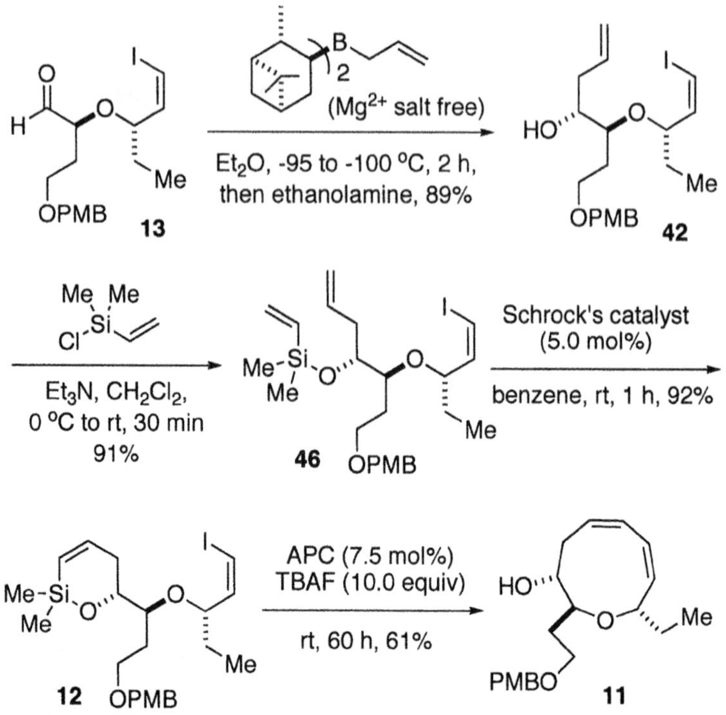

SCHEME 16

D. FORMATION OF THE OXONIN CORE: SEQUENTIAL RING-CLOSING METATHESIS /SILICON-ASSISTED INTRAMOLECULAR CROSS-COUPLING

The preceding synthetic campaign reached its successful conclusion in the installation of stereogenic centers C(2) and C(8) in **1**. We now arrived at a critical junction to construct the oxonin core by the key strategic maneuver, namely the sequential ring-closing metathesis (RCM)/silicon-assisted cross-coupling reaction. This process began by silylation of the secondary alcohol **42** with chlorodimethylvinylsilane to provide vinyl silyl ether **46** in 91% yield (Scheme 16). The silyl vinyl ether was subjected to the action of Schrock's molybdenum complex $[(CF_3)_2MeCO]_2Mo(=CHCMe_2Ph)(=NC_6H_3-2,6-i-Pr_2)$ to effect the RCM reaction.[16] By using a 5.0 mol% catalyst loading in benzene at room temperature, the ring-closure went to completion efficiently within 1 h to afford **12** in excellent yield (92%). With precursor **12** in hand, the crucial intramolecular cross-coupling reaction was carried out under optimized conditions established previously; 7.5 mol% of [allylPdCl]₂ (APC) as the catalyst and 10 equiv of 1.0 M TBAF solution as the activator using a syringe-pump addition of **12**.[14] The intramolecular cross-coupling process proceeded smoothly to afford the corresponding nine-membered ether **11** in 61% yield from a highly linear structure with little constraint. Surprisingly, no significant inhibition by either the bulky side chain or the ethyl group was observed.

E.   SIDE CHAIN ELABORATION AND COMPLETION OF THE SYNTHESIS

With all the stereocenters installed and the oxonin core constructed, we were confident that the synthesis of **1** was nearly in hand. To complete this enterprise, elaboration of the enyne side chain as well as introduction of the chloride functionality was required (Scheme 17). To avoid the potential elimination of hydrogen chloride, the installation of the halide was chosen to be the final step. Transformation of the hydroxyethyl group into the enyne side chain began by protection of the C(8) hydroxyl group with TBSOTf using pyridine and a catalytic amount of DMAP to afford **47** in 88% yield. Further, deprotection of the PMB group with DDQ in CH₂Cl₂/H₂O (19/1)[38] gave **48** efficiently in 84% yield. Oxidation of **48** with Dess-Martin periodinane[39] afforded aldehyde **49** in 83% yield (61% overall yield from coupling product **11**).

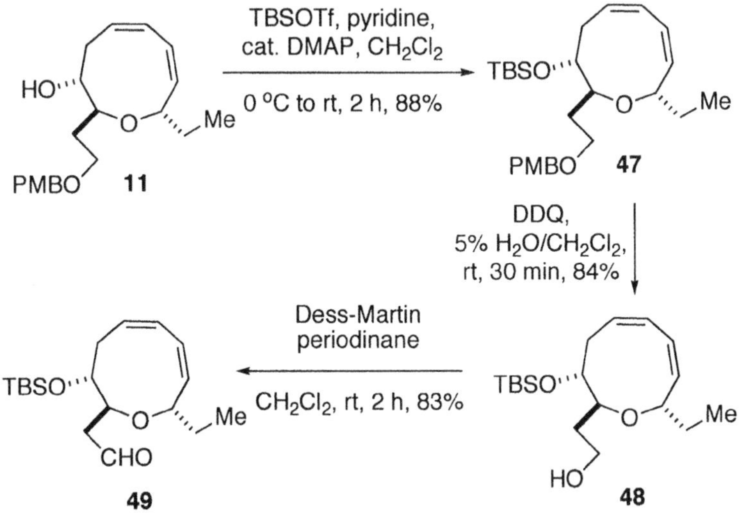

SCHEME 17

Three approaches were considered to generate geometrically defined Z-enyne side chain as described in Scheme 18. First, the desired Z-enyne could be prepared in one step using a Peterson-type olefination (center line). Second, a two-step sequence, which features the Stork-Wittig reaction to generate Z-alkenyl iodide followed by Sonogashira coupling, has been employed to install the Z-enyne functionality in several marine natural product syntheses (top line).[11b,c] Finally, a three-step sequence, introduced by Murai in their synthesis of obtusenyne was evaluated as well (bottom line).[11d]

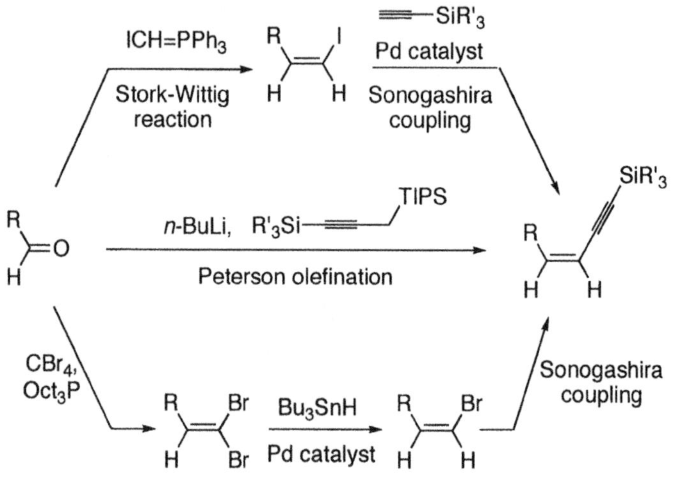

SCHEME 18

For simplicity we settled on the Peterson-type olefination to introduce the final carbon-carbon bonds. To obtain high geometric selectivity, the bulky 1,3-bis(triisopropylsilyl)propyne recommended by Corey et al. was employed[40]. Thus, treatment of **49** with lithiated 1,3-bis(triisopropyl)propyne at low temperature followed by slow warming to room temperature afforded the enyne product **50** in 83% yield with ca. 6:1 Z/E geometric selectivity (Scheme 19). Attempts to improve the geometric selectivity by employing $MgBr_2$ were unsuccessful.[41] Elimination of the β-alkoxy group to form an α,β-unsaturated aldehyde was observed by ^1H NMR analysis [diagnostic signals: 9.55 (d, $J = 8.0$, 1H, CHO); 6.78 (dd, $J = 15.5, 4.5$, 1 H, H_β); 6.27 (dd, $J = 16.0, 8.0$, 1 H, H_α)]. The coordination of β-alkoxy group with Mg^{2+} apparently facilitates the elimination process.

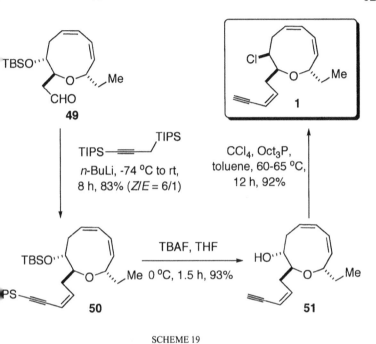

SCHEME 19

ιe geometrically homogeneous enyne (Z)-**50** can be easily isolated
ιca gel chromatography. Next, removal of the TBS and TIPS groups
ι 1.0 M solution of TBAF in THF afforded hydroxy enyne **51** in
yield. Finally, introduction of the (8S)-chloride by inversion of
ιydroxy group proceeded in 92% yield using $CCl_4/(n\text{-}Oct)_3P$ in
ιe as described by Murai et al.[11m,42] thus completing the first total
esis of (+)-brasilenyne **1** (230 mg). The synthetic sample was
cal in all respects [mp 37–38 °C (lit. 36–37 °C), ^{1}H NMR, ^{13}C
, IR, and $[\alpha]_D^{24}$ +228.0 (c = 1.08, CHCl₃); lit. +216 (c = 0.017,
₃)] to that reported for natural (+)-brasilenyne.[7a]

onformational Analysis

ι part of the synthetic studies towards (+)-brasilenyne, we carried
ιtailed conformational analyses of several of the 9-membered ring
 that were synthesized along the way. In view of the limited number
ιdies that address the conformational preferences for highly
ιtuted medium ring ethers, we wish to describe some interesting
ιations as an epilog to this chapter.

SCOTT E. DENMARK AND SHYH-MING YANG

We anticipated that the substitution pattern would have significant influence on the conformation of the ring in solution. Therefore, both molecular mechanics calculations (MM2) and NOE studies were performed. The two lowest energy conformations calculated for **10a** (Scheme 4) are shown in Figure 4. The most striking feature is the orthogonal disposition of the two double bonds. Conformation 1 possesses an oxonin core structure that has the ether oxygen up, the C(5)-C(6) double bond up, and the C(7)-C(8) double bond down. Conformation 2 possesses a different orientation with the ether oxygen up, the C(5)-C(6) double bond down, and the C(7)-C(8) double bond up, respectively. These conformations are nearly equal in energy differing by only 0.06 kcal/mol. Interestingly, viewing the modified Newman projection along the C(9)-O-C(2) vector reveals a staggered structure in Conformation 1 whereas an eclipsed structure is found in Conformation 2. The results from MM2 calculations of the conformations of **10b** are similar to those found for **10a** (Figure 5).

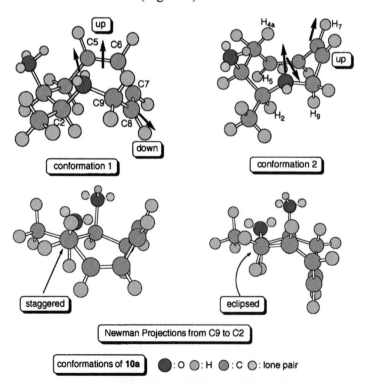

FIGURE 4. Chem3D representations of molecular mechanics (MM2) calculations on **10a**.

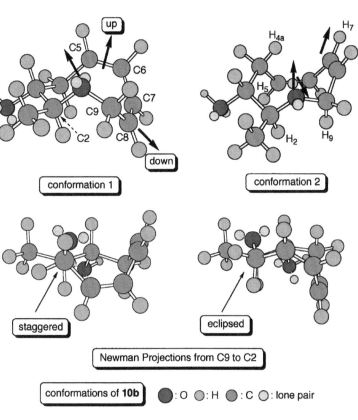

FIGURE 5. Chem3D image of molecular mechanics (MM2) calculations on **10b**.

tensive NOE experiments from **10a** display a strong enhancement
) between HC(2) and HC(9) which would be consistent with either
rmation, but more so for conformation 2 (Figure 6).   More
tantly, a 1.9% enhancement of HC(7) by irradiation of $H_aC(4)$, a
enhancement of HC(5) by irradiation of HC(2) as well as a 2.1%
cement of HC(5) by irradiation of HC(3) were also observed for
ound **10a**. Despite the small magnitude of these enhancements, the
s suggest that the conformation of **10a** in solution more closely
es the lower energy calculated Conformation 2 (Figure 4).
nfortunately, the results from NOE experiments with **10b** were not
sive as the chemical shifts of HC(2) and HC(3) of **10b** are nearly
ident. However, the strong enhancement (13.4%) of HC(2) by
ation of HC(9) implies that the conformation of **10b** is very likely
r to that of **10a**.

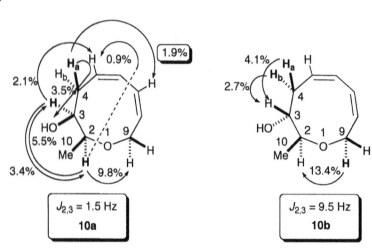

FIGURE 6. NOE experiments on **10a** and **10b**.

NOE experiments were also performed with compounds **11, 48, 49, 1** and **1** and the results are compiled in Figure 7. Interestingly, the strong enhancement between HC(2) and H$_b$C(4), HC(2) and HC(8), HC(3) and HC(9) as well as H$_b$C(4) and HC(7) were observed in those compounds. The results strongly support the conclusion that the oxonin core structure of those compounds in solution are represented by Conformation 1 of **10b** (Figure 5), which has the ether oxygen up, the C(5)-C(6) double bond up, and the C(7)-C(8) double bond down.

To clarify the significance of these NOE observations, modified conformations of **11, 48, 49**, and **51** derived from Conformation 1 of **10b** are presented in Figure 8. Apparently, the preferred conformation has changed from Conformation 2 (eclipsed) to Conformation 1 (staggered) upon replacement of H$_b$C(9) by an ethyl group. This likely arises from the increased steric interference. In addition, the ring oxygen becomes slightly twisted away from the oxonin core when the C(3) hydroxyl group is protected with a bulky TBS group (see Newman projection and top view). This twist causes HC(3) and HC(9) to point into the oxonin core while HC(2) and HC(8) point slightly away from the oxonin core (bottom of Figure 8). This conclusion was further supported by observation of increased NOE enhancements (from ca. 5% to 12%) between HC(3) and HC(9) and the decreased NOE enhancements (from ca. 5% to 2%) between HC(2) and HC(8), in **48** and **49**, respectively. The inversion of C(3) in the change from a hydroxyl group to a chloride maintained the conformation of oxonin core. These NOE studies show

e size of substituents have a significant effect on the conformation
nin ring.

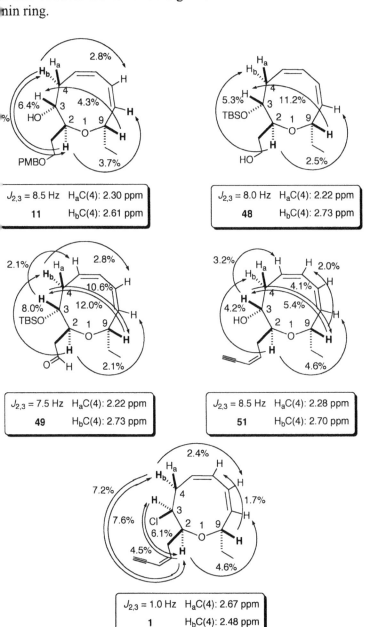

$J_{2,3}$ = 8.5 Hz $H_aC(4)$: 2.30 ppm
11 $H_bC(4)$: 2.61 ppm

$J_{2,3}$ = 8.0 Hz $H_aC(4)$: 2.22 ppm
48 $H_bC(4)$: 2.73 ppm

$J_{2,3}$ = 7.5 Hz $H_aC(4)$: 2.22 ppm
49 $H_bC(4)$: 2.73 ppm

$J_{2,3}$ = 8.5 Hz $H_aC(4)$: 2.28 ppm
51 $H_bC(4)$: 2.70 ppm

$J_{2,3}$ = 1.0 Hz $H_aC(4)$: 2.67 ppm
1 $H_bC(4)$: 2.48 ppm

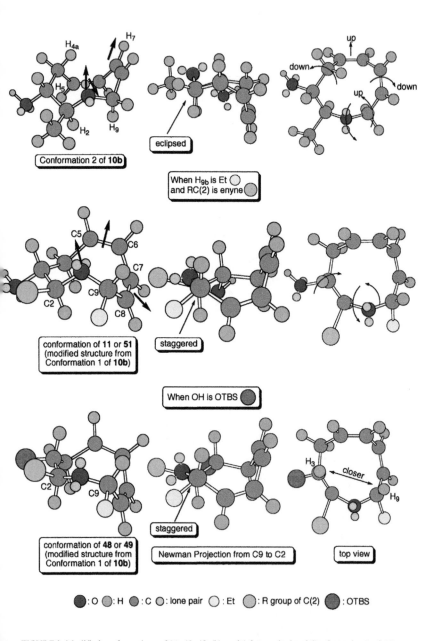

FIGURE 8. Modified conformations of **11**, **48**, **49**, **51**, and **1** from calculated Conformation 1 of **10b**.

VI.   Summary

The overall synthetic scheme for the total synthesis of (+)-brasilenyne is summarized in Scheme 20. Despite the difficulty of introducing the doubly branched ether linkage by nucleophilic displacement from **15** and **16**, several transformations benefited from this sterically congested skeleton. For instance, the branched ether at propargylic position of **34** (C(2) of **1**) suppressed the occurrence of over-reduction in a *cis*-reduction of acetylene group with diimide (Scheme 13).[26] The desired product **35** was obtained in good yield (80%) with only trace amount of over-reduction product. Furthermore, the branched ether flanking the vinyldimethylsilyl ether unit of **46** (C(9) of **1**) provides an important conformational constraint, which facilitates the RCM process. The cyclic alkenylsilyl ether **3** was obtained in excellent yield (92%) within 1 h using only 5 mol% of catalyst (Scheme 16). In a comparison to previous investigations, substrates with a methylene group adjacent to vinyldimethylsilyl ether, the RCM required a higher catalyst loading (8.0 mol%) and a longer reaction time (9-24 h) to reach completion. Fortunately, the intramolecular cross-coupling reaction to form the oxonin core was minimally influenced by the steric hindrance of from the branched ether in contrast to previous model studies (Scheme 4).[14]

Additionally, in the elaboration of lactone **35** into key intermediate **11**, mild conditions were employed to preserve the sensitive Z alkenyl iodide unit. The completion of synthesis from coupling product **11** was focused on the elaboration of the enyne side chain as well as the introduction of the chloride functionality. These transformations proceeded extremely well with an exception that the Peterson-type olefination reaction provided only modest geometric selectivity (Z/E = 6:1).

In conclusion, an efficient synthesis of the marine natural product (+)-brasilenyne has been accomplished in a linear approach (19 steps, 5.1%) from commercially available L-(S)-malic acid. The significant features of the strategy include: (1) a highly diastereoselective ring-opening of 1,3-dioxolanone promoted by $TiCl_4$ for a novel creation of stereogenic center at a propargylic position, and (2) the successful application of the sequential RCM/silicon-assisted intramolecular cross-coupling reaction for construction of medium-sized ether ring with 1,3-*cis*,*cis*-diene unit. The reagent controlled asymmetric allylboration was employed to generate a homoallylic alcohol with excellent diastereoselectivity, which in turn perfectly set the introduction of the (8S)-chloride function.

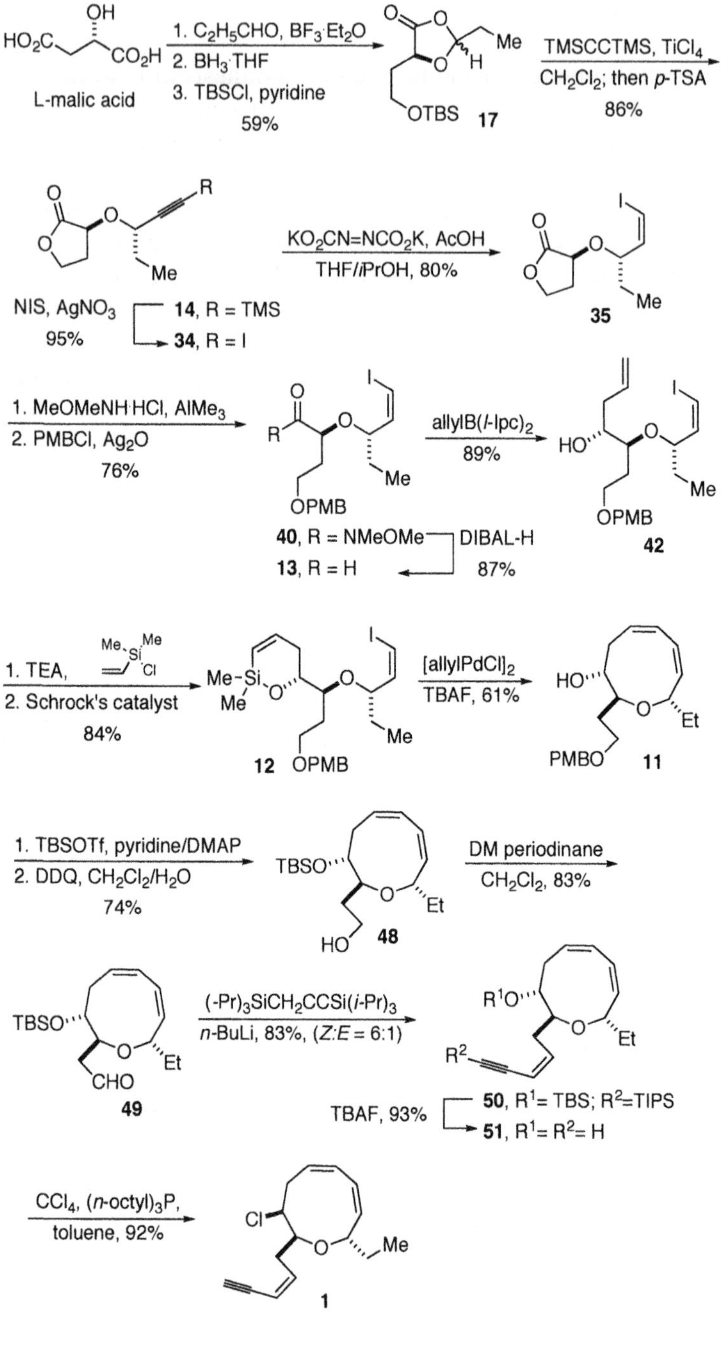

HO2C
OH
CO2H
L-malic acid
1. C2H5CHO, BF3·Et2O
2. BH3·THF
3. TBSCl, pyridine
59%
OTBS
17
Me
H
TMSCCTMS, TiCl4
CH2Cl2; then p-TSA
86%
R
Me
NIS, AgNO3
95%
14, R = TMS
34, R = I
KO2CN=NCO2K, AcOH
THF/iPrOH, 80%
35
1. MeOMeNH·HCl, AlMe3
2. PMBCl, Ag2O
76%
OPMB
40, R = NMeOMe
13, R = H
DIBAL-H
87%
allylB(l-Ipc)2
89%
HO
42
1. TEA,
2. Schrock's catalyst
84%
Me-Si
12
[allylPdCl]2
TBAF, 61%
Et
PMBO
11
1. TBSOTf, pyridine/DMAP
2. DDQ, CH2Cl2/H2O
74%
TBSO
48
DM periodinane
CH2Cl2, 83%
CHO
49
(-Pr)3SiCH2CCSi(i-Pr)3
n-BuLi, 83%, (Z:E = 6:1)
R1O
R2
TBAF, 93%
50, R1= TBS; R2=TIPS
51, R1= R2= H
CCl4, (n-octyl)3P,
toluene, 92%
Cl
1

SCHEME 20

Acknowledgements

We are grateful to the National Institutes of Health (GM63167-01A1) for generous financial support. We are grateful to Professor William Fenical (UCSD) for providing IR, ^{1}H NMR, and ^{13}C NMR spectra of natural (+)-brasilenyne.

References and Footnotes

1. (a) Erickson, K. L. In *Marine Natural Products*; Scheuer, P. J., Ed.; Academic Press: New York, 1983; vol. 5, p. 131-257. (b) Faulkner, D. J. *Nat. Prod. Rep.* **2001**, *18*, 1.

2. Irie, T., Suzuki, M., Masamune, T., *Tetrahedron Lett.* **1965**, *6*, 1091.

3. Fukuzawa, A., Takasugi, Y., Murai, A., *Tetrahedron Lett.* **1991**, *32*, 5597.

4. Fukuzawa, A., Kurosawa, E. *Tetrahedron Lett.* **1979**, *20*, 2797.

5. Kurata, K., Furusaki, A., Suehiro, K., Katayama, C., Suzuki, T., *Chem. Lett.* **1982**, 1031.

6. (a) King, T. J., Imre, S., Oztunc, A., Thomson, R. H., *Tetrahedron Lett.* **1979**, *20*, 1453. (b) Howard, B. M., Schulte, G. R., Fenical, W., *Tetrahedron* **1980**, *36*, 1747.

7. (a) Kinnel, R.B., Dieter, R. K., Meinwald, J., Engen, D. V., Clardy, J., Eisner, T., Stallard, M. O., Fenical, W., *Proc. Natl. Acad. Sci. USA* **1979**, *76*, 3576. (b) Fenical, W., Sleeper, H. L., Paul, V. J., Stallard, M. O., Sun, H. H., *Pure and Appl. Chem.* **1979**, *51*, 1865.

8. (a) Ishihara, J., Kanoh, N., Murai, A., *Tetrahedron Lett.* **1995**, *36*, 737. (b) Fukuzawa, A., Aye, M., Nakamura, M, Tamura, M., Murai, A., *Chem. Lett.* **1990**, 1287. (c) Kurosawa, E., Fukuzawa, A., Irie, T., *Tetrahedron Lett.* **1972**, *13*, 2121. For transformation of 3*E*,6*R*,7*R*-laurediol to deacetyl-laurencin by LPO, see: (d) Fukuzawa, A., Aye, M., Murai, A., *Chem. Lett.* **1990**, 1579. (e) Irie, T., Suzuki, M., Masamune, T., *Tetrahedron* **1968**, *24*, 4193.

9. In vitro studies have shown that laureatin and isolaureatin are derived from *Z*-prelaureatin by LPO and that laurallene is derived from *E*-prelaureatin by bromoperoxidase (BPO), see: Ishihara, J., Shimada, Y., Kanoh, N., Takasugi, Y., Fukuzawa, A., Murai, A., *Tetrahedron* **1997**, *53*, 8371.

10. Norte, M, Gonzalez, A. G., Cataldo, F., Rodriguez, M. L., Brito, I., *Tetrahedron* **1991**, *47*, 9411.

11. For a review of the construction of medium-ring ethers, see: (a) Elliott, M. C., *Contemp. Org. Synth.* **1994**, *1*, 457. For total syntheses of (+)-prelaureatin, see: (b) Crimmins, M. T., Tabet, E. A., *J. Am. Chem. Soc.* **2000**, *122*, 5473. (c) Fujiwara, K., Souma, S.-i., Mishima, H., Murai, A., *Synlett* **2002**, 1493. For total syntheses of (+)-obtusenyne, see: (d) Fujiwara, K., Awakura, D., Tsunashima, M., Nakamura, A., Honma, T., Murai, A,. *J. Org. Chem.* **1999**, *64*, 2616. (e) Crimmins, M. T., Powell, M. T., *J. Am. Chem. Soc.* **2003**, *125*, 7592. For total syntheses of (-)-isolaurallene, see: (f) Crimmins, M. T., Emmitte, K. A., Choy, A. L., *Tetrahedron* **2002**, *58*, 1817. (g) Crimmins, M. T., Emmitte, K. A., *J. Am. Chem. Soc.* **2001**, *123*, 1533. For total syntheses of (+)-laurallene, see: (h) Saitoh, T., Suzuki, T., Sugimoto, M., Hagiwara, H., Hoshi, T., *Tetrahedron Lett.* **2003**, *44*, 3175. Also see

reference 10b. For a total synthesis of natural (+)-laurenyne, see: (i) Boeckman, Jr., R. K., Zhang, J., Reeder, M. R., *Org. Lett.* **2002**, *4*, 3891. For a total synthesis of (-)-laurenyne, see: (j) Overman, L. E., Thompson, A. S., *J. Am. Chem. Soc.* **1988**, *110*, 2248. For representative examples of total syntheses of (+)-laurencin, see: (k) Bratz, M., Bullock, W. H., Overman, L. E., Takemoto, T., *J. Am. Chem. Soc.* **1995**, *117*, 5958. (l) Burton, J. W., Clark, J. S., Derrer, S., Stork, T. C., Bendall, J. G., Holmes, A. B., *J. Am. Chem. Soc.* **1997**, *119*, 7483. (m) Tsushima, K., Murai, A. *Tetrahedron Lett.* **1992**, *33*, 4345. (n) Crimmins, M. T., Emmitte, K. A., *Org. Lett.* **1999**, *1*, 2029. For a total synthesis of (+)-isolaurepinnacin, see: (o) Berger, D., Overman, L. E., Renhowe, P. A., *J. Am. Chem. Soc.* **1997**, *119*, 2446.

12. Ma, S., Negishi, E.-i., *J. Am. Chem. Soc.* **1995**, *117*, 6345.

13. (a) Yenjai, C., Isobe, M., *Tetrahedron* **1998**, *54*, 2509. (b) Hosokawa, S., Isobe, M., *Tetrahedron Lett.* **1998**, *39*, 2609. (c) Kira, K., Isobe, M., *Tetrahedron Lett.* **2000**, *41*, 5951.

14. (a) Denmark, S. E., Yang, S.-M., *J. Am. Chem. Soc.* **2002**, *124*, 2102. (b) Denmark, S. E., Yang, S.-M., *Tetrahedron* **2004**, *60*, 9395.

15. Previous reports: (a) Denmark, S. E., Yang, S.-M., *J. Am. Chem. Soc.* **2002**, *124*, 15196. (b) Denmark, S. E., Yang, S.-M., *J. Am. Chem. Soc.* **2004**, *126*, 12432.

16. The Schrock molybdenum complex is commercially available (Strem) and can be prepared according to the reported procedure with consistent purity and reactivity, see: (a) Fox, H. H., Yap, K. B., Robbins, J., Cai, S., Schrock, R. R., *Inorg. Chem.* **1992**, *31*, 2287. (b) Schrock, R. R., Murdzek, J. S., Bazan, G. C., Robbins, J., DiMare, M., O'Regan, M., *J. Am. Chem. Soc.* **1990**, *112*, 3875. (c) Oskam, J. H., Fox, H. H., Yap, K. B., McConville, D. H., O'Dell, R., Lichtenstein, B. J., Schrock, R. R., *J. Organomet. Chem.* **1993**, *459*, 185. (d) Fox, H. H., Lee, J.-K., Park, L. Y., Schrock, R. R., *Organometallics* **1993**, *12*, 759.

17. For the preparation of **15**, see: (a) Green, D. L. C., Kiddle, J. J., Thompson, C. M., *Tetrahedron* **1995**, *51*, 2865. (b) White, J. D., Hrnciar, P. *J. Org. Chem.* **2000**, *65*, 9129. (c) Mulzer, J., Mantoulidis, A., Ohler, E., *J. Org. Chem.* **2000**, *65*, 7456.

18. For the preparation of **16**, see: (a) Allevi, P., Ciuffreda, P., Anastasia, M., *Tetrahedron: Asymmetry* **1997**, *8*, 93. (b) Niwa, S., Soai, K., *J. Chem. Soc., Perkin Trans. 1* **1990**, 937. (c) Mukaiyama, T., Suzuki, K., *Chem. Lett.* **1980**, 255. (d) Mori, M., Nakai, T., *Tetrahedron Lett.* **1997**, *38*, 6233.

19. To the best of our knowledge, the ring opening of a 1,3-dioxolanone with bis(trimethylsilyl)acetylene is unprecedented. Ring opening of acetal templates with silylacetylenic compounds promoted by Lewis acid have been reported, see: (a) Johnson, W. S., Elliott, R., Elliott, J. D., *J. Am. Chem. Soc.* **1983**, *105*, 2904. (b) Yamamoto, Y., Nishii, S., Yamada, J.-i., *J. Am. Chem. Soc.* **1986**, *108*, 7116. (c) Rychnovsky, S. D., Dahanukar, V. H., *J. Org. Chem.* **1996**, *61*, 7648.

20. (a) Haack, K.-J., Hashiguchi, S., Fujii, A., Ikariya, T., Noyori, R., *Angew. Chem. Int. Ed.* **1997**, *36*, 285. (b) Matsumura, K., Hashiguchi, S., Ikariya, T., Noyori, R., *J. Am. Chem. Soc.* **1997**, *119*, 8738. The air stable, purple Ru-complex was easily prepared from $(1R,2R)$-1,2-diphenyl-N-(4-toluenesulfonyl)-ethylenediamine with $[RuCl_2(\eta^6\text{-}p\text{-cymene})]_2$ in 87% yield.

21. For some representative examples of ring opening of acetal templates, see: (a) Mori, A., Ishihara, K., Arai, I., Yamamoto, H., *Tetrahedron* **1987**, *43*, 755. (b) Ishihara, K., Hanaki, N., Yamamoto, H., *J. Am. Chem. Soc.* **1993**, *115*, 10695. (c) Lindell, S. D., Elliott, J. D., Johnson, W. S., *Tetrahedron Lett.* **1984**, *25*, 3947. (d) Harada, T, Nakamura, T., Kinugasa, M., Oku, A., *J. Org. Chem.* **1999**, *64*, 7594. (e) Bartlett, P. A., Johnson, W. S., Elliott, J. D., *J. Am. Chem. Soc.* **1983**, *105*, 2088. (f) Elliott, J. D., Choi, V. M. F., Johnson, W. S., *J. Org. Chem.* **1983**, *48*, 2294. (g) For examples using acetylenic organometallic reagents, see reference 18. For mechanism studies, see: (h) Denmark, S. E., Almstead, N. G., *J. Am. Chem. Soc.* **1991**, *113*, 8089. (i) Denmark, S. E., Almstead, N. G., *J. Org. Chem.* **1991**, *56*, 6485. (j) Denmark, S. E., Willson, T. M., *J. Am. Chem. Soc.* **1989**, *111*, 3475. (k) Mori, I., Ishihara, K., Flippin, L. A., Nozaki, K., Yamamoto, H., Bartlett, P. A., Heathcock, C. H., *J. Org. Chem.* **1990**, *55*, 6107. (l) Sammakia, T., Smith, R. S., *J. Org. Chem.* **1992**, *57*, 2997.

22. For some representative examples of ring opening of 1,3-dioxolanones, see: (a) Mashraqui, S. H., Kellogg, R. M., *J. Org. Chem.* **1984**, *49*, 2513. (b) Heckmann, B., Mioskowski, C., Yu, J., Flack, J. R., *Tetrahedron Lett.* **1992**, *33*, 5201. (c) Heckmann, B., Mioskowski, C., Lumin, S., Flack, J. R., Wei, S., Capdevila, J. H., *Tetrahedron Lett.* **1996**, *37*, 1425. (d) Heckmann, B., Mioskowski, C., Bhatt, R. K., Flack, J. R., *Tetrahedron Lett.* **1996**, *37*, 1421.

23. The crystallographic coordinates of the dicobalt complex have been deposited with the Cambridge Crystallographic Data Centre, deposition no. CCDC-195245.

24. Nishikawa, T., Shibuya, S., Hosokawa, S., Isobe, M., *Synlett* **1994**, 485.

25. For modified procedures, see: (a) Nicolaou, K. C., Marron, B. E., Veale, C. A., Webber, S. E., Serhan, C. N., *J. Org. Chem.* **1989**, *54*, 5527. (b) Chavez, D. E., Jacobsen, E. N., *Angew. Chem. Int. Ed.* **2001**, *40*, 3667.

26. (a) Kazuta, Y., Matsuda, A., Shuto, S., *J. Org. Chem.* **2002**, *67*, 1669. (b) Yau, E. K., Coward, J. K., *J. Org. Chem.* **1990**, *55*, 3147.

27. For a modified procedure, see: Nemoto, H., Nagamochi, M., Ishibashi, H., Fukumoto, K., *J. Org. Chem.* **1994**, *59*, 74.

28. Bouzide, A., Sauve, G., *Tetrahedron Lett.* **1997**, *38*, 5945.

29. For modified procedures, see: (a) Wernic, D., DiMaio, J., Adams, J., *J. Org. Chem.* **1989**, *54*, 4224. (b) Evans, D. A., Ratz, A. M., Huff, B. E., Sheppard, G. S., *J. Am. Chem. Soc.* **1995**, *117*, 3448.

30. For modified procedures, see: (a) Guanti, G., Banfi, L., Riva, R., *Tetrahedron* **1995**, *51*, 10343. (b) Berts, W. B., Luthman, K., *Tetrahedron* **1999**, *55*, 13819.

31. (a) For the use of MgBr$_2$, see reference 31a. (b) For using ZnI$_2$, see: Solladie, G., Hanquet, G., Rolland, C. ,*Tetrahedron Lett.* **1999**, *40*, 177.

32. (a) Danishefsky, S. J., Deninno, M. P., Phillips, G. B., Zelle, R. E., *Tetrahedron* **1986**, *42*, 2809. (b) Veloo, R. A., Wanner, M. J., Koomen, G.-J., *Tetrahedron* **1992**, *48*, 5301.

33. Heck, G. E., Boden, E. P., *Tetrahedron Lett.* **1984**, *25*, 265.

34. Scarlato, G. R., DeMattei, J. A., Chong, L. S., Ogawa, A. K., Lin, M. R., Armstrong, R. W., *J. Org. Chem.* **1996**, *61*, 6139.

35. (a) Brown, H. C., Bhat, K. S., Randad, R. S., *J. Org. Chem.* **1989**, *54*, 1570. (b) Racherla, U. S., Brown, H. C., *J. Org. Chem.* **1991**, *56*, 401.
36. Jaime, C., Segura, C., Dinares, I., Font, J., *J. Org. Chem.* **1993**, *58*, 154.
37. (a) Ley, S. V., Norman, J., Griffith, W. P., Marsden, S. P., *Synthesis* **1994**, 639. (b) Burke, S. D., Austad, B. C., Hart, A. C., *J. Org. Chem.* **1998**, *63*, 6770.
38. Horita, K., Yoshioka, T., Tanaka, T., Oikawa, Y., Yonemitsu, O., *Tetrahedron* **1986**, *42*, 3021.
39. Dess, D. B., Martin, J. C., *J. Org. Chem.* **1983**, *48*, 4155.
40. Corey, E. J., Rucker, C., *Tetrahedron Lett.* **1982**, *23*, 719.
41. Yamakado, Y., Ishiguro, M., Ikeda, N., Yamamoto, H., *J. Am. Chem. Soc.* **1981**, *103*, 5568.
42. The high yield secured here was particularly satisfying as slightly lower yields (ca. 60-75%) have been obtained from 8- and 9-membered ring ethers containing a single carbon-carbon double bond. Competitive elimination of hydrogen chloride has been observed in these cases. See references 11e and 11i.

STRATEGIES AND TACTICS IN ORGANIC SYNTHESIS, VOL. 6

Chapter 5

SAMARIUM(II) PROMOTED KETYL OLEFIN CYCLIZATIONS APPLIED TO THE TOTAL SYNTHESES OF (–)-STEGANONE AND (+)-ISOSCHIZANDRIN

Gary A. Molander and Kelly M. George
Department of Chemistry
University of Pennsylvania
Philadelphia, Pennsylvania 19104

I. Introduction

In 1980, Professor Henri Kagan of the Université de Paris-Sud (Orsay) published his landmark paper regarding the use of samarium(II) iodide (SmI_2) in organic synthesis.[1] In that contribution, Kagan and coworkers had outlined a wealth of simple transformations involving the reduction and reductive coupling reactions of a variety of organic functional groups. At the time, the senior author (GM) was a postdoctoral research associate in the laboratories of Professor Barry M. Trost, then at the University of Wisconsin, Madison. Artie Romero, a graduate student and laboratory partner in Trost's group at the time, had read Kagan's article and questioned whether this exotic reagent possessed any real advantages over the myriad of extant reductants already at the synthetic chemist's disposal.

Kagan's paper was read and dutifully noted. Over the next few months, ideas began to ferment about the potential of this unusual reductant. After gaining a position at the University of Colorado, Boulder, a research program had to be initiated. However, not knowing whether the SmI_2 chemistry would be the focus of intense study in the

Kagan group and worried that other, more established, groups could quickly take advantage of this initial report, we hesitated to commit resources to investigating its potential. Nothing would be worse for an assistant professor than starting a program and being scooped at every step. At any rate, during the first year at Colorado there were no graduate students in the group to undertake this research, and so attention was initially focused on getting other, perhaps less risky, projects initiated.

During the ensuing year, things changed. Kagan's initial report had apparently not incited a rush of new research and two new graduate students (Jeff Etter and Paul Zinke) had joined the group. Consequently, we decided that the time was right to begin our studies in lanthanide chemistry. Although nearly two years had passed before we decided to explore this area, we were lucky enough to be able to make some contributions to this field.

We first explored the intramolecular Barbier reaction of haloalkyl-substituted ketones. Although this transformation had first been described in 1902[2] and numerous reductants had been utilized (e.g., magnesium, lithium, sodium, organolithiums, organocuprates, low-valent nickel species and chromium(II) salts, to name a few[2-18]) no general solution to the problem had been found. The synthesis of six-membered rings by this process was particularly problematic.[16] We postulated that a homogeneous reagent with the proper reduction potential was needed to solve this problem, and Sm(II) reagents appeared to fit the bill. Indeed, SmI_2 worked well (Eq. 1). Consequently, our first contribution on this topic and in Sm(II) chemistry appeared in 1984.[19]

$$\text{(1)}$$

Our initial efforts in Barbier cyclization reactions[20] were rapidly followed by investigations on the reduction of vinyl oxiranes[21] and the reduction of α-heterosubstituted carbonyls.[22,23] Armed with our experiences in this exciting new area of research, in 1986 we began a lengthy odyssey to examine the use of SmI_2 in the construction of carbocycles and heterocycles utilizing ketyl-olefin cyclizations.

II. Samarium(II) Promoted Ketyl Olefin Cyclizations

One of the limitations of the intramolecular Barbier reaction of ω-haloalkyl β-keto ester substrates, as developed in our laboratory, was the inability to utilize secondary alkyl halides. Under the aprotic conditions utilized, a retroaldol reaction transpired that led to undesired transformations of the intermediate generated (Eq. 2).

$$\text{(2)}$$

We hoped to circumvent this shortcoming by developing an intramolecular ketyl-olefin cyclization that would transpire under protic conditions. Our early work on the reductive coupling of ω-haloalkyl β-keto esters[24] and indeed Kagan's initial mechanistic work,[25] had suggested that ketyl radical anions were intimately involved in the Barbier process. Although this postulate subsequently came under intense scrutiny,[26] for these particular substrates this paradigm is still undoubtedly correct.[27] Believing this to be the case, we embarked on a study of the use of SmI$_2$ in mediating intramolecular ketyl-olefin cyclizations.[28-30]

Studies of similar reactions under a variety of conditions had been previously reported. Thus, electroreductive, photoreductive, as well as metal-induced ketyl-olefin cyclizations had all been explored prior to our investigations.[31-37] Many of these cyclization reactions using simple unsaturated ketones took place with reasonably high diastereoselectivity at two stereocenters (Eq. 3). This feature of the transformation was ascribed to favorable secondary orbital interactions between the developing methylene radical center and the alkyl group of the ketyl,[33,35,38] and/or to electrostatic interactions in the transition state leading to product.[31,34,38]

$$\text{(3)}$$

the information gained at the end of the process was often negligible. Recognizing this, and becoming increasingly frustrated, Lauren did what any good graduate student should do – she took the matter into her own hands and essentially designed her own thesis.

Rather than study the cyclization of abstract substrates with little or no application to real-world problems, Lauren decided to showcase the method by synthesizing several interesting natural products. She had independently developed approaches to three molecules, and laid these out for approval. All three (spartadienedione, (-)-steganone and (+)-isoschizandrin) were exceptionally well conceived, and she embarked on their construction. Detailed below are the successful approaches to the latter two that eventually derived from her designs.

III. Synthesis of (-)-Steganone

(-)-Steganone (**1**, Figure 1) was first isolated in 1973,[57] and by 1976 two total syntheses were reported.[58,59] Over the next two decades, the natural product remained a popular target not only because of its unique structure, which became a platform for asymmetric biaryl synthesis, but also because of the discovery of some congeners' inhibition of tubulin polymerization in both *in vivo* and *in vitro* studies.[28,60-64] Before embarking upon a synthesis of (-)-steganone, two key structural challenges must be addressed: (1) the biaryl system possessing the correct axial chirality and, (2) the fused 8-5 ring system containing the requisite stereochemistry.

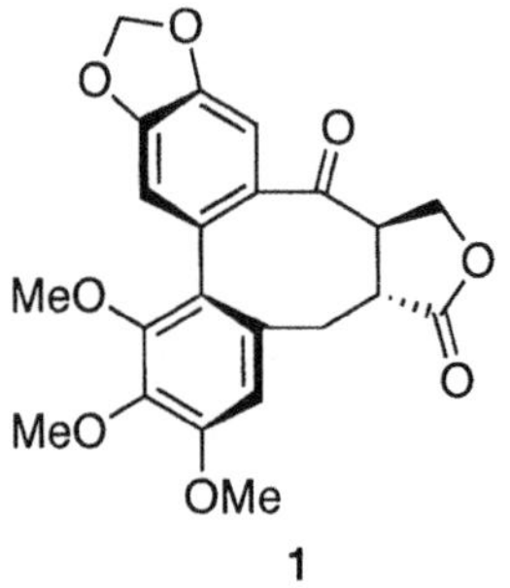

FIGURE 1. (-)-Steganone.

Our plan was to utilize the elegant strategy of Uemura and coworkers to establish the chirality of the biaryl, followed by a simultaneous construction of both the eight- and five-membered rings of steganone utilizing our SmI_2-promoted ketyl-olefin radical cyclization method (Eq. 14). In doing so, increased efficiency would be incorporated into the construction of this interesting target.

Although at the time we began there had been several reported total syntheses of (-)-steganone, none featured a radical ring closure to form the eight-membered ring, and in all cases the formation of the five- and eight- membered rings were separated by several transformations.[65-67] We saw the 8-5 ring system as a unique opportunity to explore our SmI_2-promoted-8-endo ketyl-olefin cyclization (Eq. 14). Earlier studies in our research group had shown that substrates possessing olefins that were substituted with electron-withdrawing groups often provide excellent yields in the SmI_2-promoted 8-endo ketyl-olefin cyclization. [68,69] Specifically, the butenolide moiety, resulting from a standard retrosynthetic disconnection, would be an excellent precursor poised to undergo a reductive coupling with a samarium ketyl radical anion generated from the aryl carbaldehyde unit.

In terms of electronics, aldehyde ketyls (particularly those derived from aromatic aldehydes) usually do not react as well as ketones in ketyl couplings because of their lower-lying SOMO.[70] Electron-donating groups, such as those found in our envisioned precursor, would facilitate coupling over simple reduction by raising the SOMO of the aldehyde.[71] Additionally, we felt that the selectivity[71-74] of SmI_2 would allow exclusive reduction of the aromatic aldehyde to the ketyl radical in the presence of the butenolide and the product lactone. Finally, the conformational bias imposed by the biaryl unit would aid in the cyclization because of the prearrangement of four of the eight carbons present in the 8-membered ring.

With a plan in mind, Lauren wisely chose to work through a racemic version of the synthesis. It was important to understand the reaction details further before undertaking an enantioselective synthesis, particularly because the stereochemical outcome of the 8-endo cyclization product was difficult to predict. Lauren envisioned biaryl (±)-3 as an early intermediate, a structure that resembles the enantiopure biaryls (4,5) synthesized by Uemura[66] and Meyers[65], respectively (Figure 2). Initial investigations of an Ullmann coupling between 6-bromopiperonal and 2-iodo-3,4,5-trimethoxybenzyl alcohol protected as a TBDMS ether proved to be problematic as it provided only a 28% yield of the cross-coupled product using 2 equivalents of the aryl iodide. After several unsuccessful attempts to optimize the Ullmann reaction, Lauren finally opted instead to explore a Suzuki coupling between an arylboronic acid or ester and a suitable aryl iodide, a reaction that is well documented[75] and most successful with either Pd or Ni catalysts in sterically hindered biaryl couplings.[76,77]

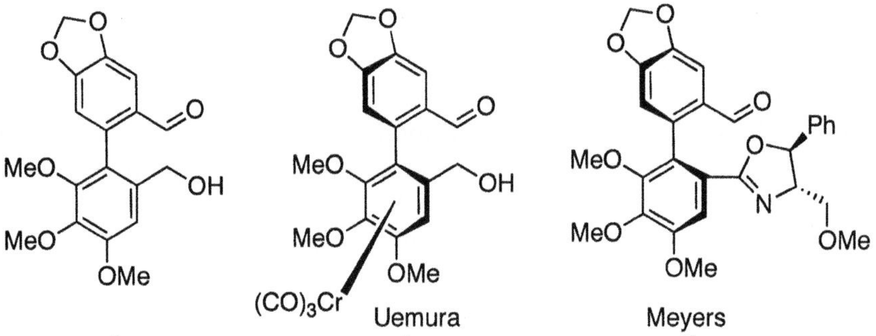

FIGURE 2

Upon investigating the literature, Lauren noted that arylboronic acids substituted with electron-withdrawing groups often experience extensive protodeboronation and poor yields under the standard Suzuki conditions that utilize aqueous Na_2CO_3 as a base.[78,79] A solution to this problem lies in either the use of a milder base[30,80] or nonaqueous conditions.[81] Therefore, Lauren chose to employ $NaHCO_3$ as a base. The arylboronic acid derivative of 6-bromopiperonal was coupled with the suitably elaborated aryl iodide to provide the desired biaryl (±)-3 in moderate yields (Eq. 15).

The resulting benzyl alcohol (±)-**3** was then converted to the corresponding benzyl bromide, (±)-**8**, after reacting with MsCl and triethylamine in CH_2Cl_2 and subsequent displacement of the mesylate with LiBr in acetone (Scheme 2). The benzyl bromide was then poised to

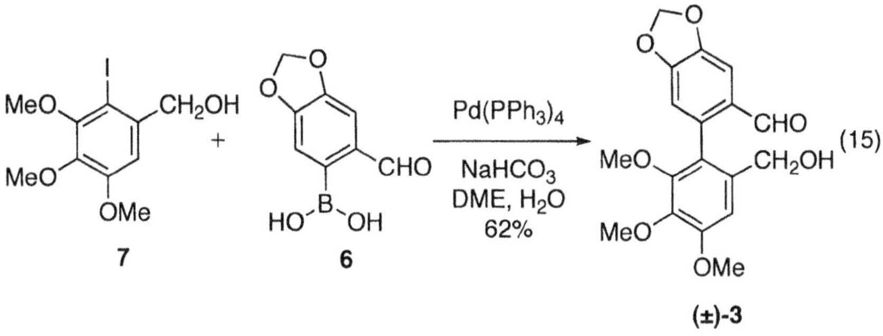

undergo a Pd-catalyzed Stille coupling with 3-(tributylstannyl)-(5H)-furan-2-one to install the requisite butenolide moiety.[82] The desired transformation occurred between the stannylated butenolide and the benzyl bromide at 80 °C with $Pd_2(dba)_3$ and trifurylphosphine (Pd:L 1:2)

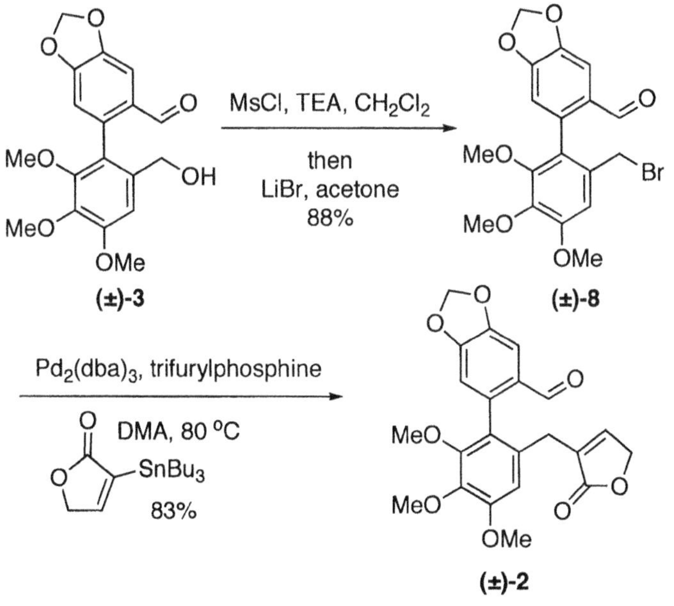

SCHEME 2

as the catalyst (Scheme 2).[83] A minor amount of the fully conjugated product was obtained as a result of either thermal- or palladium-catalyzed isomerization.

With the coupling precursor in hand, (±)-2 was treated with 2.2 equivalents of SmI_2 in THF/HMPA at 0 °C using t-BuOH as a proton source. Two diastereomeric products, each of the picro-series (8-5 cis ring system), were formed in a 3:1 ratio (Scheme 3). As Scheme 3 shows, the oxygen of the reacting ketyl assumes the pseudoequatorial position and the resulting stereochemistry of the adjacent stereocenter can be rationalized by attack on one of the diastereotopic faces of the butenolide (one example is depicted). Protonation of the samarium(III) enolate from the convex face provides the two diastereomers. Epipicrosteganol, **B**, was the minor diastereomer formed and its structure was confirmed by comparison with literature spectral data.[84] The major diastereomer was previously unknown and required an X-ray crystal structure to support the

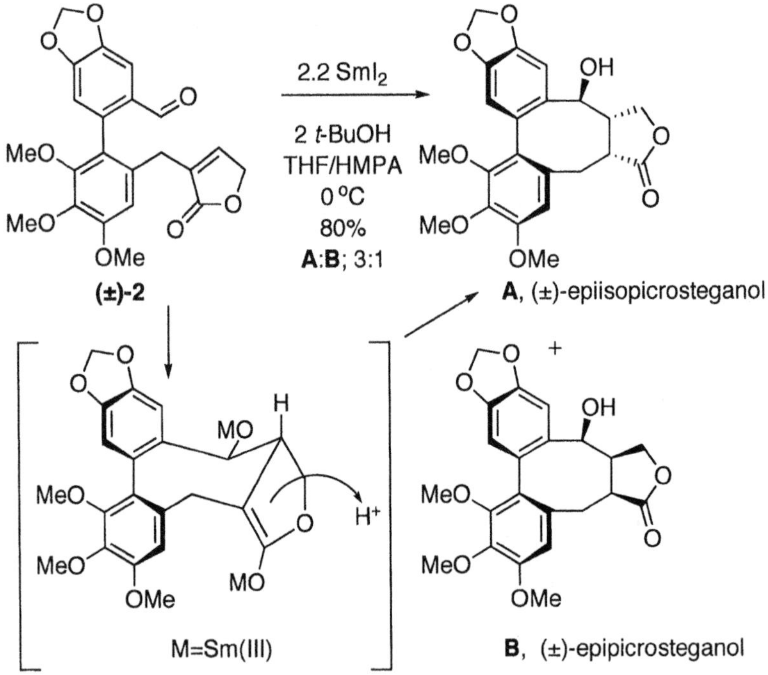

SCHEME 3

structural assignment. Upon confirming that the structure was indeed correct, **A** was named epiisopicrosteganol in keeping with the nomenclature established for these systems.

Unfortunately, oxidation of **A** *or* **B** yielded an equilibrating pair of atropisomeric ketones, (±)-picrosteganone and (±)-isopicrosteganone (Eq. 16).

A or **B** ———————→   (16)
Dess-Martin periodinane

(±)-isopicrosteganone   (±)-picrosteganone

We were confident that the final transformation of the picrosteganones to steganone would proceed smoothly due to the precedented thermodynamic driving force inherent within the system.[85] Notably, steganone is the only diastereomer possible that allows planar alignment of the ketone carbonyl with the neighboring arene ring system. This key structural arrangement is supported by the chemical shift of the ortho aromatic proton ($\delta 7.53$ ppm for steganone vs. δ 6.71 ppm for isosteganone) and by the carbonyl stretching frequency (1667 cm^{-1} vs 1707 cm^{-1} for isosteganone).[85] Exploitation of this thermodynamic driving force has been frequently used in total syntheses of steganone. For example, isosteganone was known to be converted to steganone by either of two different conditions: (1) heating to reflux a solution of isosteganone in xylenes, or (2) heating to reflux a solution of isosteganone in MeOH or EtOH in the presence of NaOAc (Eq. 17). The

xylenes, Δ
or
NaOAc, EtOH, Δ   (17)

isosteganone   steganone

former set of conditions was the only known method to convert nonracemic isosteganone to enantioenriched steganone with retention of the absolute stereochemistry α to the lactone carbonyl.[86]

Unfortunately, when picrosteganone was heated to reflux in xylenes, there was no evidence of product and only starting material was recovered (Eq. 18).

$$(\pm)\text{-picrosteganone} \xrightarrow{\text{xylenes, } \Delta} \hspace{-1.2em}/\hspace{-0.6em}/ \quad (18)$$

(±)-steganone

Luckily, reaction of (±)-picrosteganone under base-promoted conditions (NaOAc in EtOH) provided (±)-steganone in a 90% yield (Eq. 19). Overall, the racemic synthesis of steganone was achieved quite efficiently in 6 steps from known compounds.

$$(\pm)\text{-picrosteganone} \xrightarrow[90\%]{\substack{\text{3 NaOAc} \\ \text{EtOH, } \Delta}} \quad (19)$$

(±)-steganone

The difference between the behavior of picrosteganone and isosteganone under the suggested equilibration conditions may be explained by the need for picrosteganone to undergo enolate formation. The isomerization of isosteganone to steganone under both sets of conditions was postulated to involve a retro-Michael reaction (generating an exo-methylene ketone) followed by a Michael reaction of the resultant carboxylic acid on the enone. However, it is also possible that the isomerization of isosteganone to steganone in hot xylenes occurs by a simple thermal-promoted rotation about the biaryl axis, which is known to occur in related systems and is a process that is facilitated by the presence of sp^2-hybridized carbons at the ortho positions of the biaryl system. In the case of picrosteganone, this rotation about the biaryl axis yields only starting material: the two atropisomers of the picro series. Therefore, base-promoted conditions are required, and we believe that the retro-Michael/Michael mechanism is operating. Formation of the enolate, followed by β-elimination to give the carboxylate, affords a common intermediate enone from picrosteganone or isosteganone.

With racemic steganone in hand, Lauren then turned her attention toward obtaining a single atropisomer in the enantioselective synthesis of (-)-steganone. At the time, there were two suitable options found in the literature for generating an enantioselective biaryl intermediate.[65, 66, 87] Despite its sensitivity to oxidative conditions, Uemura's Cr^o arene complex seemed especially appealing because of its success in being used alongside the SmI_2 reducing agent.[88,89] Although there was some precedent for attack of the chromium-complexed arene by radicals, there were also examples in which the arene remained untouched and the chromium complex only served as a stereocontrol element.[65,66,87] It was our goal to exploit the latter paradigm.

The main difficulties associated with constructing the biaryl moiety as a single atropisomer early in the synthesis lies within the arene substitution pattern. The thermal isomerization barrier for a steganone precursor that carries only three ortho substituents is extremely sensitive to the exact identity of those substituents. Specifically, when one of the three groups (e.g., a formyl moiety) adds only a small amount to the inversion barrier and a second substituent (e.g., a methoxy group) is also relatively small, the barrier to inversion renders the biaryl stereochemically labile even at 0 °C.[65,66,87] For example, at least one of three intermediates in the Meyers' synthesis of steganone resulted in some loss of biaryl stereochemistry. In that case, even with careful handling (e.g., low temperature column chromatography), the ee of the product did not reflect the initial success of the biaryl formation. We believed that the chromium-arene complex would serve as a tool for both forming and protecting the desired biaryl atropisomer and thus we set out to explore this route.

Using Uemura's method, the enantiomerically pure chromium-complexed biaryl **4** was generated from the enantiomerically pure Cr-carbonyl complex **9** in five steps from 3,4,5-trimethoxybenzaldehyde.[66] Subsequent $NaBH_4$-mediated reduction of the benzaldehyde proceeded smoothly to provide the corresponding alcohol. This chromium-complexed alcohol underwent a Suzuki-Miyaura coupling with 2-formyl-4,5-methylenedioxyboronic acid in the presence of Pd(PPh$_3$)$_4$ and aqueous Na$_2$CO$_3$ to provide the desired biaryl in good crude yields. As shown in Scheme 4, the desired diastereomer derives from an initially formed conformational isomer. One can rationalize that rotation about the biaryl axis under the reaction conditions would afford the desired diastereomer, **4**. Because some decomplexed product was obtained under the reaction conditions and upon purification, the resulting alcohol was converted directly to the corresponding benzyl halide (Cl or Br), which was purified and stored without decomplexation or decomposition. Lauren attempted to couple the benzyl chloride under a variety of metal-catalyzed Stille coupling conditions without much success. The bromide provided a reasonable alternative.

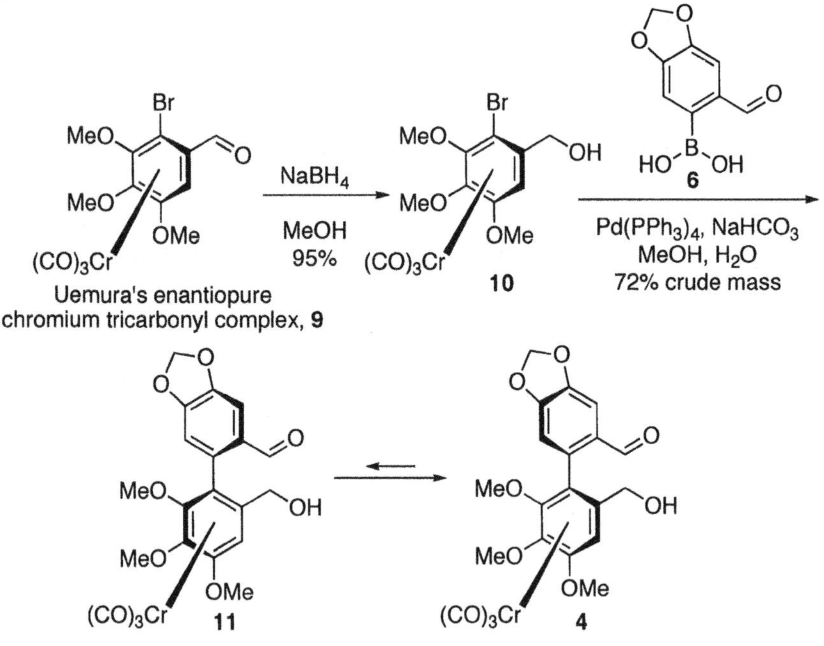

SCHEME 4

It was at this point that Magnus Rönn and Yvan LeHuérou assumed their roles as postdoctoral associates in our laboratories and worked to complete the enantioselective total synthesis of (-)-steganone. To obtain the more reactive benzyl bromide coupling partner in the enantioselective synthesis, methanesulfonic anhydride was required to avoid formation of an intermediate chloride mixture, which could neither be separated from, nor converted to, the bromide with excess LiBr. Utilizing optimized reaction conditions, both aryl halides were transformed to the desired chromium-complexed SmI$_2$ precursor **13**, although the benzyl bromide indeed provided much higher yields when compared to the chloride (Scheme 5). Specifically, AsPh$_3$ was found to be the optimal ligand of choice because of its weak coordination to the metal.

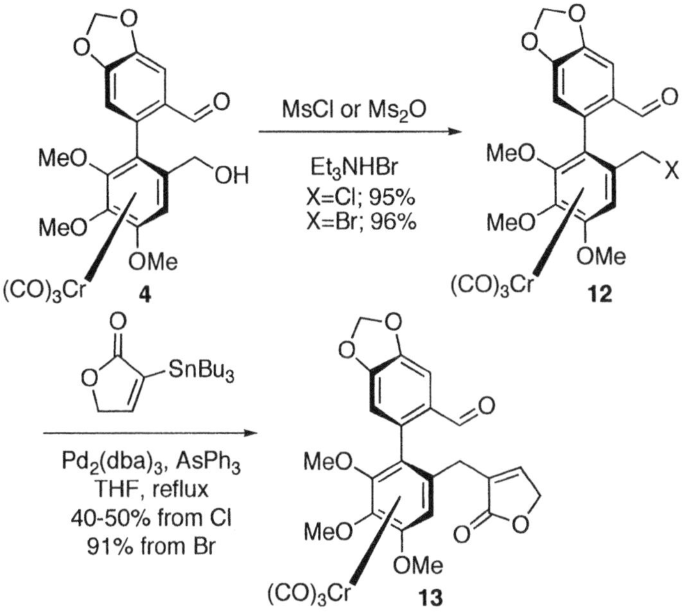

SCHEME 5

To our surprise and delight, the SmI$_2$-promoted ketyl-olefin cyclization of the chromium-complexed biaryl intermediate **13** afforded a single diastereomer **14** in 65–75% yield, in contrast to the uncomplexed, racemic substrate which provided two diastereomeric products (Scheme 6). Overall, the relative stereochemistry of the biaryl and ring juncture stereocenters in the Cr-complexed product corresponded to the major diastereomer observed in the racemic synthesis.

Subsequent oxidation of the chromium-complexed isopicrosteganol **14** with PCC/NaOAc in CH$_2$Cl$_2$ simultaneously deprotected/decomplexed the chromium species and oxidized the carbinol to the corresponding ketone (Scheme 7). Although Dess-Martin periodinane was utilized in

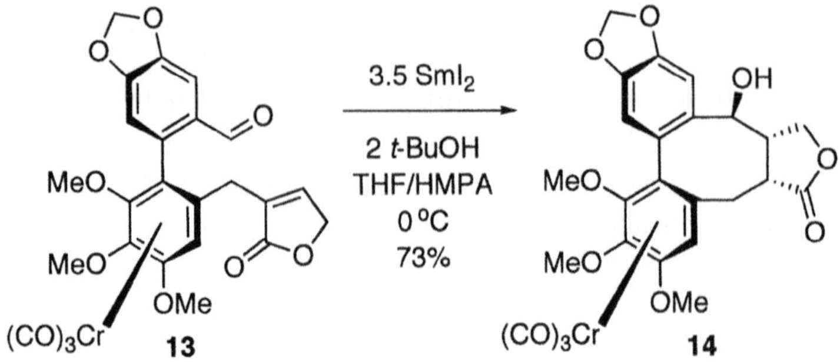

SCHEME 6

the racemic synthesis, an inseparable impurity was formed when these conditions were applied in the enantioselective approach, and thus attempts to utilize this reagent were ultimately abandoned. Again, oxidation to the ketone provided a mixture of atropisomers in a 2:1 ratio.

Yvan now faced the challenge of equilibrating the stereocenter α to the ketone to provide the trans stereochemistry at the ring juncture and the requisite driving force to obtain the correct atropisomer. As alluded to previously, in one of the earlier syntheses of steganone, isosteganone was converted to the desired natural product simply by heating neat or in xylenes; however, this transformation only required rotation about the biaryl bond.[86] Additionally, the use of NaOAc in boiling EtOH had been employed only in the racemic series, and thus the effect of its application on the various enantiomerically enriched isomers of steganone was unknown. Utilizing the refluxing NaOAc/EtOH conditions employed in our nonracemic synthesis led to an 87/13 mixture of (-)-steganone to (+)-steganone by chiral HPLC, demonstrating a loss of ee by equilibration of the stereocenter α to the lactone. The absolute stereochemistry at this center is set relative to the biaryl during the SmI_2-promoted cyclization and is ultimately necessary to relay all of the stereochemical information back to the remaining stereocenters in the final equilibration step.

After investigating the literature, Yvan found a paper from the Still[90] laboratories describing a similar equilibration. With only a few milligrams of material left and the situation thus quite tense, Yvan proceeded very carefully to heat the mixture of atropisomers with DBU in THF at reflux. Fortunately, these conditions worked beautifully, leading to the successful transformation of picrosteganone to (-)-steganone in

82% yield with an optical purity >99%.

The enantioselective synthesis of (-)-steganone was thus completed within 11 steps from commercially available material. A highlight was the construction of the medium-sized ring via the Sm(II)-promoted ketyl-olefin coupling reaction, which, before the advent of the ring-closing metathesis reaction, provided one of the best methods available for the formation of eight-membered rings.[91-95]

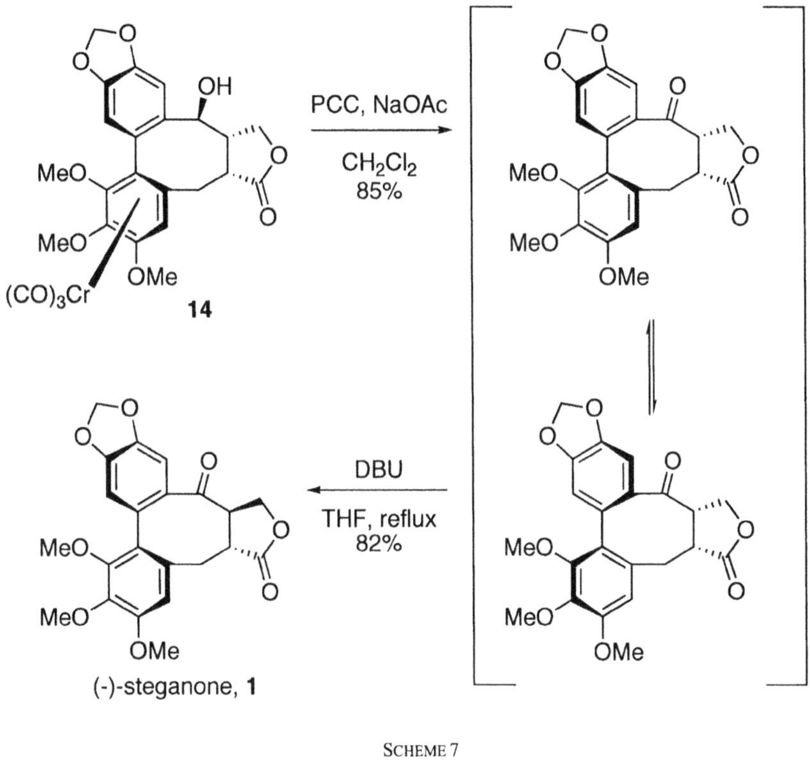

SCHEME 7

IV. Synthesis of (+)-Isoschizandrin

(+)-Isoschizandrin **15** represents one of nearly forty dibenzocyclooctadiene lignans isolated from the fruit of *Schizandra chinensis*, a creeping vine native to northern China.[29,64,96,97] The extracts from this lignan-rich plant have been used in both Chinese and Japanese traditional medicine as an antitussive, and several lignans isolated from these extracts are thought to exhibit antirheumatic and antihepatotoxic activity. (+)-Isoschizandrin, a minor component of the extract, displays anti-ulcer activity in rats.[98,99] Although these lignans exhibit notable biological activity, our principal synthetic interest in this family of natural products was their unique dibenzocyclooctadiene structure and the potential to demonstrate our samarium(II) iodide promoted 8-endo ketyl olefin radical cyclization. This key method was envisioned to provide the eight-membered ring, the required functionality, and the correct stereochemistry present in (+)-isoschizandrin in a single transformation (Scheme 8).[68] As already addressed, a general approach to dibenzocyclooctadienes was first demonstrated in our synthesis of (-)-steganone.[100]

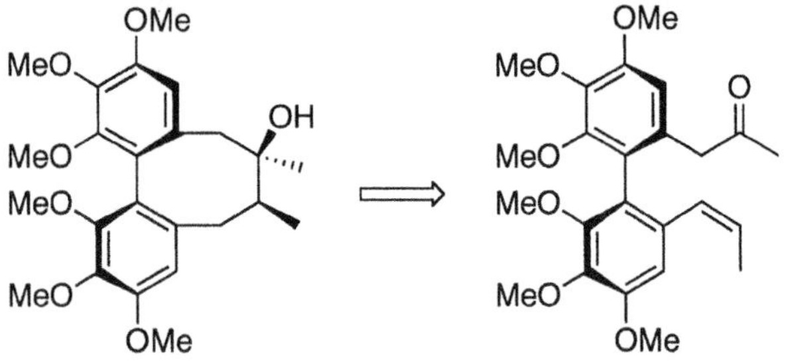

(+)-isoschizandrin, **15**   **(+)-16**

SCHEME 8

To date, only three research groups have communicated syntheses of isoschizandrin. The Meyers group[101] first reported the total synthesis of (–)-isoschizandrin, thereby establishing the absolute stereochemistry of the natural product. This was soon followed by a racemic synthesis of (±)-isoschizandrin from the laboratories of Tobinaga.[102] More recently, Tanaka and coworkers reported the first asymmetric total synthesis of (+)-isoschizandrin.[97,103]

Retrosynthetic analysis to incorporate a samarium(II) iodide-promoted 8-endo ketyl-olefin coupling suggests structure (+)-16 as a direct precursor to isoschizandrin 15 (Scheme 9). Previous studies in our group demonstrated that samarium(II)-promoted 8-endo ketyl-olefin cyclizations were greatly enhanced by placing an electron-withdrawing or π-conjugating substituent directly on the olefin. Increasing the radical-SOMO/alkene-LUMO interactions in the transition state favors the cyclization product formation.[68,100] In the present synthesis, the aryl group thus provided a useful structural element in facilitating this type of cyclization.

To obtain the penultimate keto-olefin **(+)-16**, we initially envisioned a Pd-mediated coupling reaction between a Z-propenyl organometallic and keto-triflate **(+)-17** to install the necessary carbon functionality in the ortho position. Construction of the tetraorthosubstituted biaryl moiety must proceed with special regard to its atropisomerism, because the 8-endo cyclization sets the remaining stereocenters relative to the biaryl geometry. Generally, three or four ortho substituents are required to make asymmetric biaryl synthesis a viable consideration. Few biaryl couplings succeed when three or four ortho substituents are present,[76] and fewer still control the resultant biaryl geometry.[87,104,105] Meyers' pioneering method, which utilizes a chiral oxazoline auxiliary to facilitate nucleophilic aromatic substitution of a methoxy group, provides yields in the 60-90% range with good to excellent levels of diastereoselection (up to 96% de) and has been used successfully in several total syntheses.[106,107] This protocol has the advantage of forming the biaryl and establishing the stereochemistry in a single step.

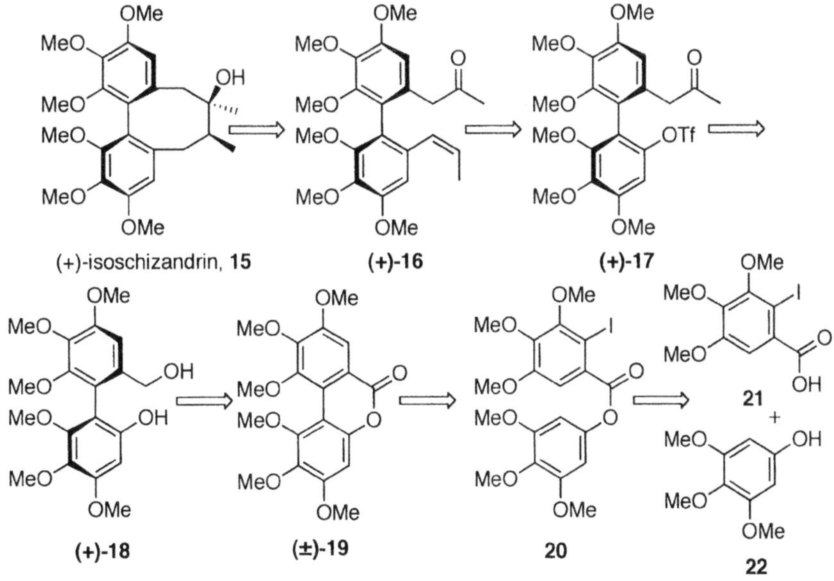

(+)-isoschizandrin, **15**   **(+)-16**   **(+)-17**

(+)-18   **(±)-19**   **20**

SCHEME 9

A novel method developed by Bringmann and coworkers, by contrast, controls the formation of the biaryl juncture and atropisomerism separately. We chose to apply this latter strategy to our system to provide diol **(+)-18** as a single atropisomer from benzocoumarin lactone **(±)-19** via a dynamic kinetic resolution. Lactone (±)-19 could be derived from halobenzoate ester **20** via a palladium-mediated coupling reaction. Further disconnection of the iodo-ester precursor reveals DCC coupling partners 3,4,5-trimethoxyphenol **22** and 2-iodo-3,4,5-trimethoxybenzoic acid **21**.

Our first synthetic approach was performed racemically by reducing a biaryl lactone intermediate to a diol, with an eye toward applying Bringmann's method to achieve an asymmetric synthesis (Scheme 10). First, iodination of 3,4,5-trimethoxybenzoic acid followed by DCC coupling with 3,4,5-trimethoxyphenol **22** provided aryl halobenzoate ester

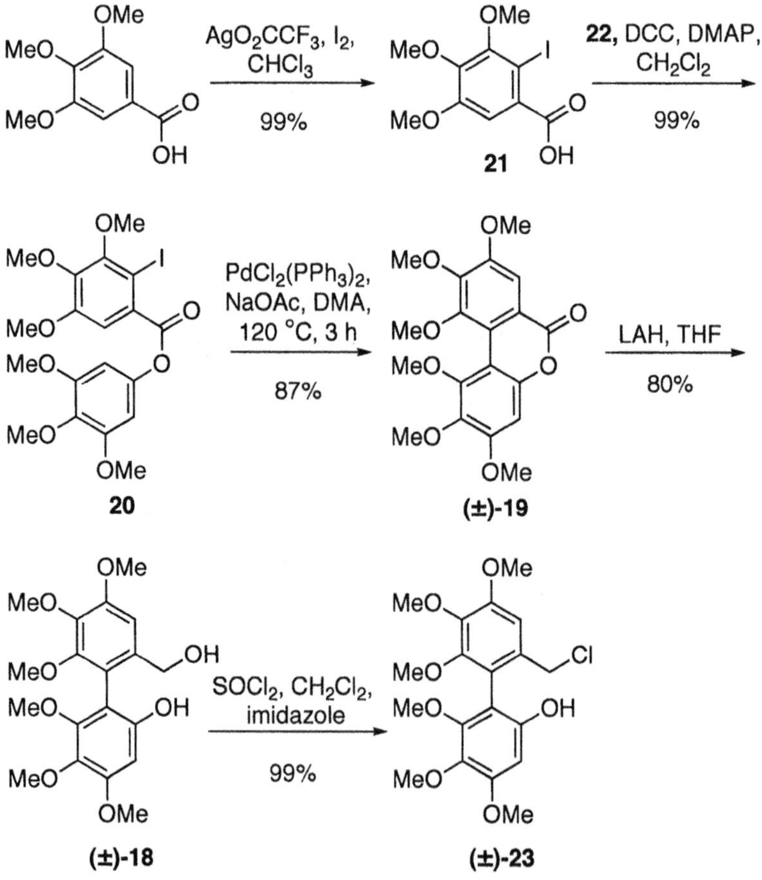

SCHEME 10

20. Intramolecular coupling of **20** afforded benzocoumarin lactone **19** that was reduced with LAH to form the racemic diol **18**.

Lauren had demonstrated previously in our laboratories that the atropoenantioselective reduction of benzocoumarin lactone (±)-**19** with 4 equivalents of BH$_3$·THF in the presence of excess (R)-2-methyl-CBS-oxazaborolidine (3 equiv) provided the nonracemic diol (+)-**18** in 95% yield and 95.6% ee by chiral HPLC.[51] Additionally, the racemate crystallized readily. Filtration of the crystallized racemate followed by concentration of the mother liquor afforded >90% yield of the enantiomerically pure material. This provided entrée to an asymmetric synthesis of the target via a dynamic kinetic resolution. Notably, the scale of this reaction was ten-fold higher than that reported by Bringmann, demonstrating its viability in total synthesis. Nevertheless, for initial investigations LAH reduction was employed to provide diol **18**, which was then utilized to explore subsequent transformations. Benzyl chloride (±)-**23** was then prepared by reacting the racemic diol with SOCl$_2$ (Scheme 10).

Acylation of the corresponding phenol **23** (Scheme 11) provided a substrate capable of undergoing a samarium(II) iodide promoted acyl

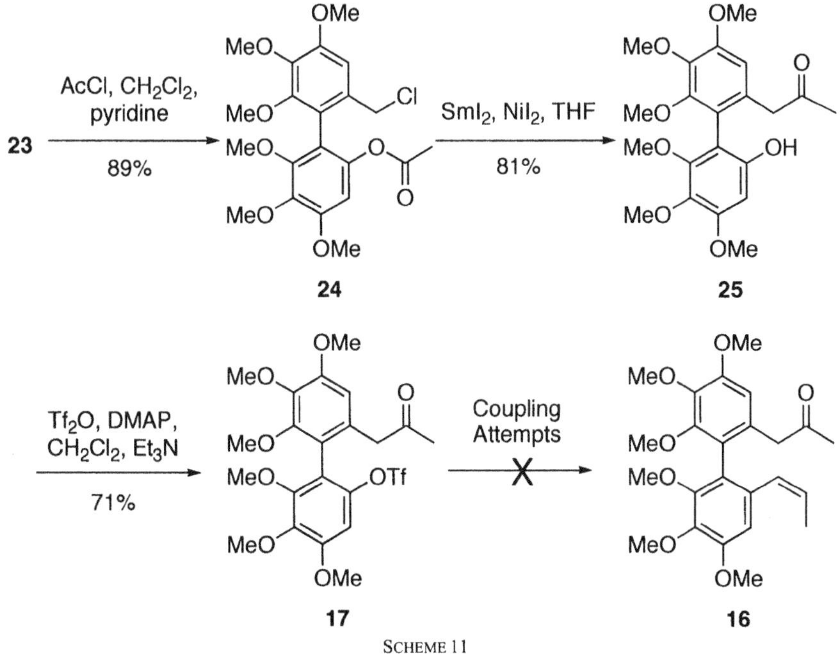

SCHEME 11

transfer reaction.[108] The use of NiI_2[109] as an additive in this reaction provided the desired phenolic ketone, **25**, but required high dilution (0.007 M). Finally, conversion of keto-phenol **25** with Tf_2O afforded the desired coupling precursor, **17**.[28,62,66]

It is well known that transition metal-catalyzed carbon-carbon bond formations are extraordinarily difficult with sterically hindered, electron-rich aryl triflates.[110,111] However, just such a transformation was required to install the Z-propenyl fragment necessary for the samarium-mediated cyclization. Lauren's initial attempts to achieve this goal afforded predominantly detriflated material and capricious reactions leading to loss of biaryl enantiopurity.

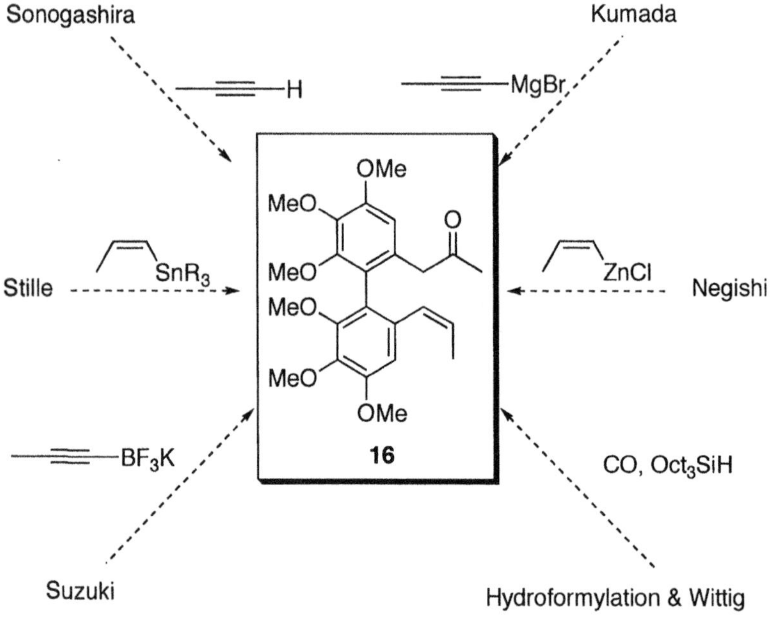

SCHEME 12

When Kelly George first joined the laboratory in the summer of 2001, she was presented with the challenge to find a way to improve this step in order to obtain the penultimate intermediate **16** necessary for the key Sm(II) iodide cyclization. Consequently, numerous cross-coupling strategies were tested (utilizing both *cis*-propenyl and 1-propynyl organometallics) to convert **17** to **16** or a suitable precursor (Scheme 12). Unfortunately, after nine months of investigating an exhaustive number of

ols, the desired product was never obtained in a reproducible
n. Although the coupling strategies worked well on a simple model
1, overall these conditions led predominantly to detriflated or
posed material on the real system.

potential solution was to utilize a nonafluorobutanesulfonate
late), which is known to be less prone to oxygen–sulfur cleavage, a
deleterious side reaction in the original coupling.[112,113] With this
n mind, keto-nonaflate **26** was also explored in a variety of
ium-mediated reactions. Again in this case, the desired product
not be accessed. The X-ray crystal structure of keto-nonaflate **26**
led evidence of the steric hindrance in this system and this,
ned with the electron-rich nature of the electrophile, explained the
f successful coupling (Figure 3). Finally, Kelly determined that a
ent approach was needed to arrive at the desired samarium
rsor, **16**, and designed a new route that addressed the key
chemical and functional group considerations.

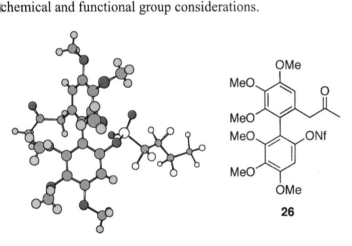

FIGURE 3

owing that the atropoenantioselective reduction of benzocoumarin
e (±)-**19** provided an excellent pathway to a single atropisomer, we
t to apply a similar strategy to a seven-membered lactone. We
ed during the painstaking coupling attempts that it was
rdinarily difficult to install a C-C bond in one of the ortho positions
biaryl system when the other three were already substituted. Kelly
ed that the two carbon ortho substituents needed to be in place as
as possible with the opportunity to homologate each functional

group independently. By altering our approach, both carbon ortho substituents would now be poised for further chemistry without the need to couple a carbon unit into a sterically hindered, electron rich position. Upon investigating the literature, Kelly was excited to learn that Bringmann and coworkers had just developed such chemistry to obtain enantiomerically enriched biaryls via the kinetic resolution[114-117] of conformationally stable seven-membered biaryl lactones using a chiral oxazaborolidine.[118] We sought to apply this chemistry to our synthesis to obtain a single atropisomer and a novel and more versatile biaryl framework.

With this new approach in mind, the revised retrosynthetic analysis would utilize two different Wittig reactions to homologate the carbon side chains, providing ketoolefin (+)-16 (Scheme 13). Further disconnection revealed hydroxy aldehyde (+)-27 which could arise from a selective DIBALH reaction of lactone (+)-28. Applying Bringmann's method, lactone (+)-28 would be enantiomerically enriched via a kinetic resolution of a racemic lactone mixture. Racemic lactone (±)-28 could arise from a

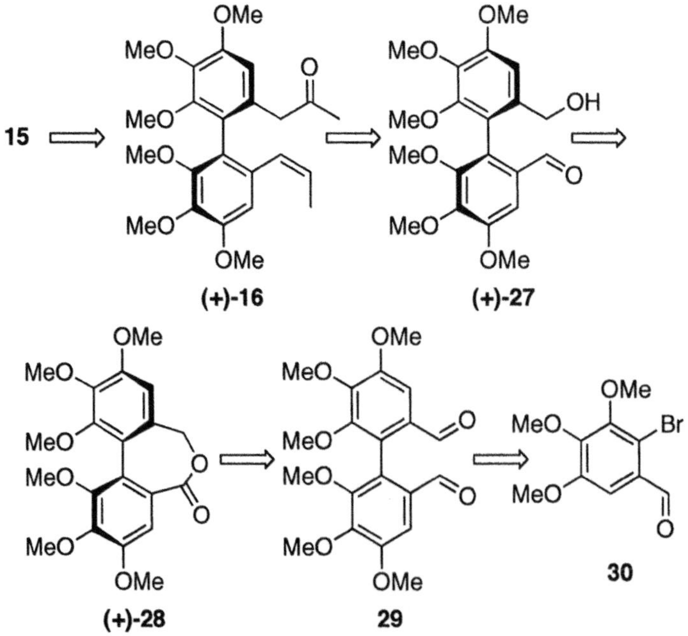

SCHEME 13

Cannizzaro reaction of **29** followed by DCC coupling. Finally, dialdehyde **29** could be accessed via a standard Ullmann coupling of the corresponding haloaldehyde **30**.

Our revised approach began with the bromination of 3,4,5-trimethoxybenzaldehyde to provide the corresponding bromoaldehyde **30**[119] (Scheme 14). Using a standard Ullmann coupling protocol, bromoaldehyde **30** was heated with activated Cu in DMF to provide the desired biaryl dialdehyde in good yields.[104,119-121] Treatment of **29** with KOH in boiling ethanol[122-125] afforded the hydroxy acid, which was directly lactonized with DCC to produce the desired racemic seven-membered lactone (±)-**28**. This lactone was poised to undergo a kinetic

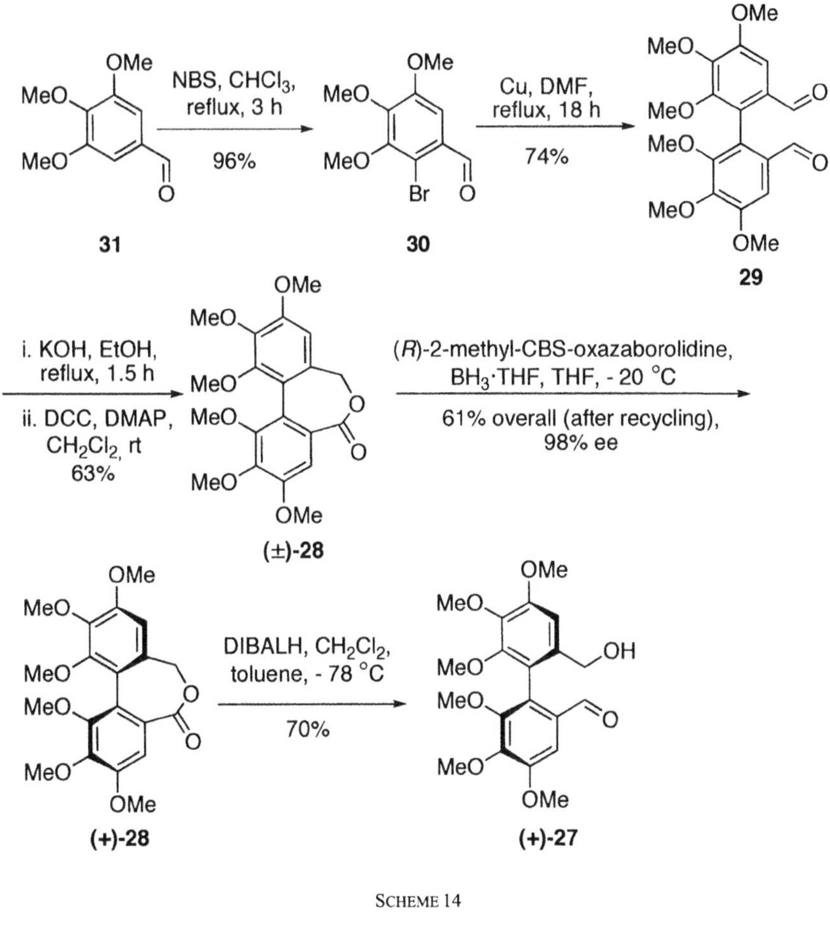

SCHEME 14

resolution using Bringmann's method, and in the event the desired lactone (+)-28 was obtained in a 45% yield (out of a possible 50%) with 98% ee as determined by chiral HPLC.

The undesired diol (–)-32 was recycled by first oxidizing to the dialdehyde with Dess-Martin's reagent[126] (Scheme 15). After heating the dialdehyde in acid to racemize the material, (±)-29 was subjected to the Cannizzaro/DCC conditions providing racemic lactone (±)-28 in a 60% yield. Performing this kinetic resolution/recycling process provided (+)-28 in an overall yield of 61% (98% ee).

DIBALH reduction of the enantiomerically enriched lactone at –78 °C afforded aldehyde (+)-27 (Scheme 14). Using ethyltriphenylphosphonium bromide, aldehyde (+)-27 was then converted to the desired Z-propenyl fragment via a Wittig reaction in a 2:1 (Z:E) ratio (Scheme 16). After testing several protecting group strategies, the TIPS ether improved the

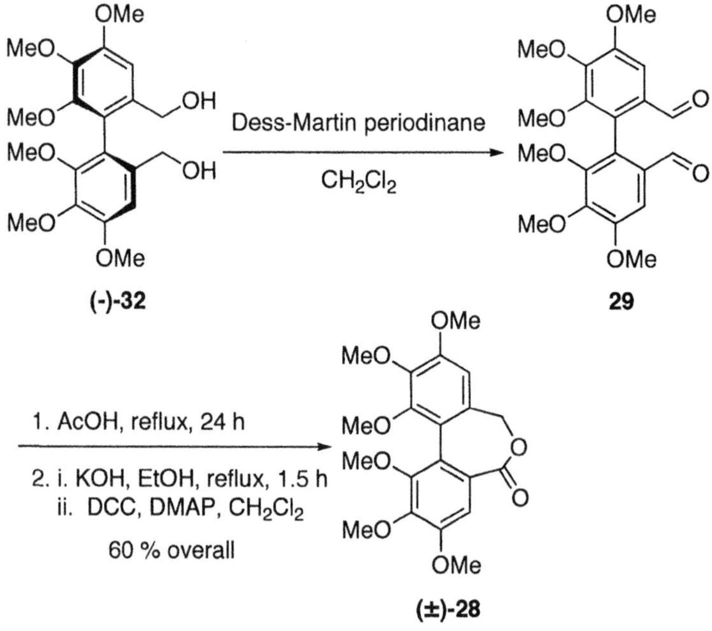

SCHEME 15

Z:E selectivity to 4:1 (85% yield) and also enabled the geometric isomer ratio to be more easily enriched to >10:1 (35% yield) via MPLC. The enriched material was then carried through to the end of the synthesis.

synthesis. After removing the TIPS group using TBAF, benzyl alcohol
(+)-34 was then oxidized using Dess-Martin periodinane to provide
unsaturated aldehyde (+)-35. Treatment of aldehyde (+)-35 with the ylide
derived from (α-methoxyethyl)-triphenylphosphonium chloride[127]
afforded an enol ether that was carefully hydrolyzed to alkenyl ketone
(+)-16 without olefin isomerization (Z-E).

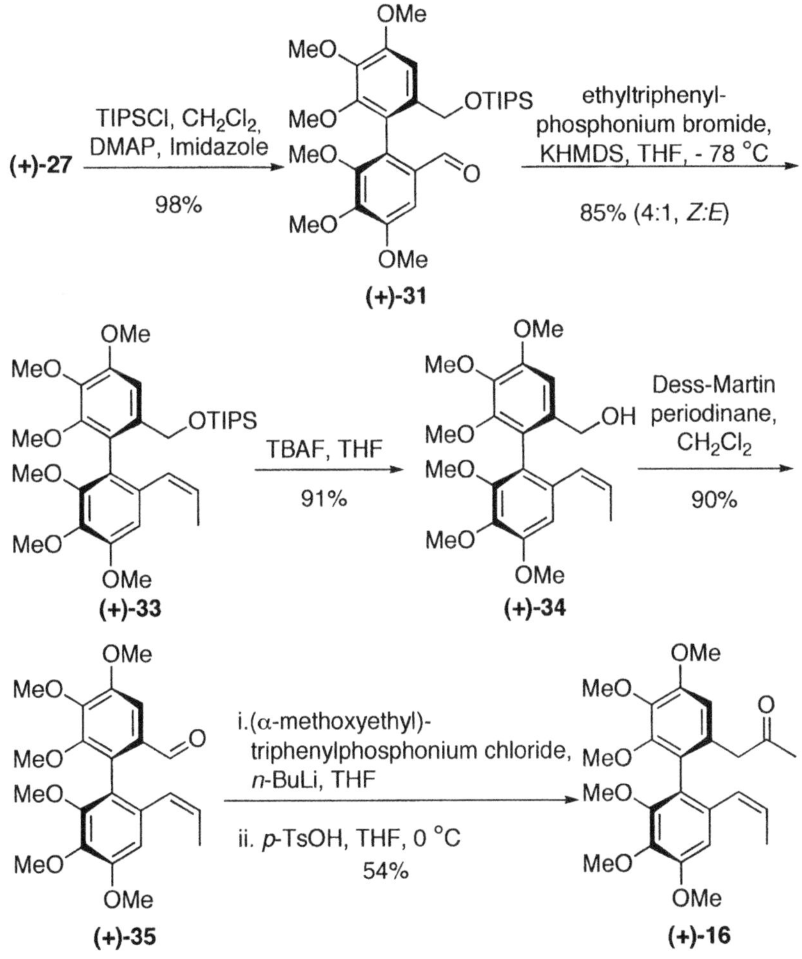

SCHEME 16

With the desired substrate in hand, keto-olefin 16 was treated with 2.2
equivalents of samarium(II) iodide and HMPA[128,129] in THF along with 2

equivalents of the proton source, *t*-BuOH (Eq. 20). We were delighted to find that the natural product was obtained with a dr >18:1 and with no loss of ee (98%).[130] The identity of the target compound was confirmed by comparison with reported analytical data.[97,103]

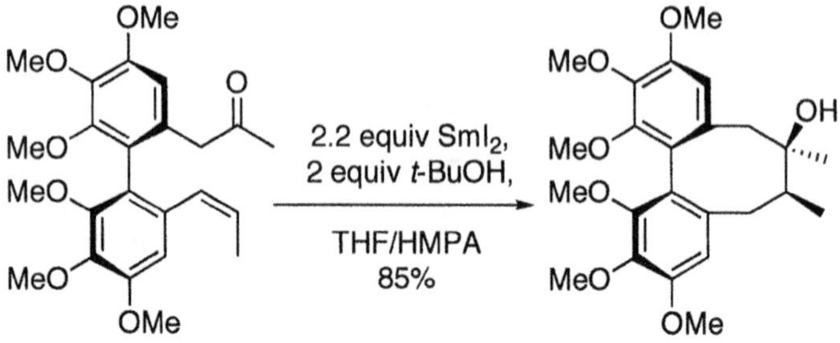

(+)-16   (+)- isoschizandrin, **15**

It should be noted that there are several important factors contributing to the high yield and selectivity of the samarium(II) iodide promoted

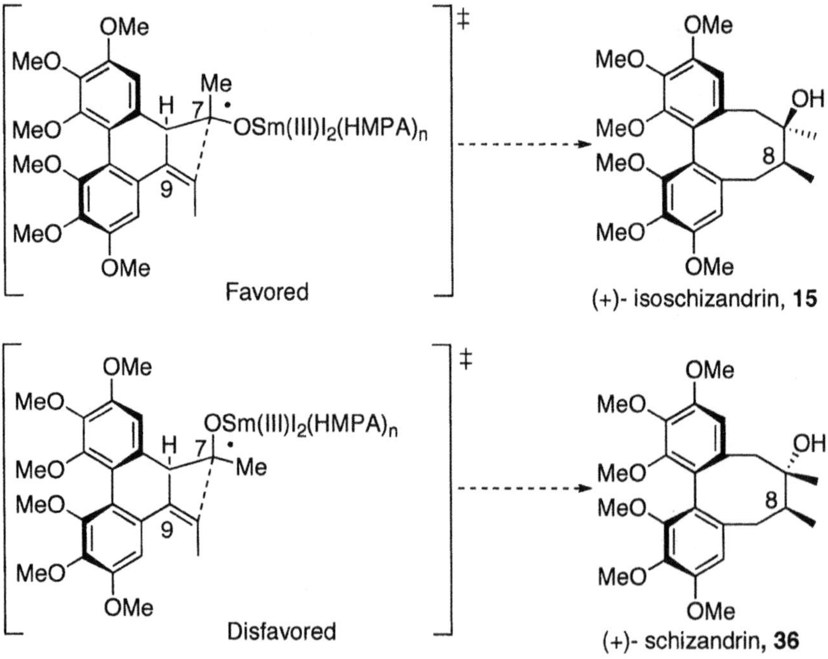

SCHEME 17

coupling (Scheme 17). In terms of the yield, we have alluded previously to the favorable SOMO/LUMO interactions engendered by an aryl unit on the olefin. Additionally, the biaryl motif likely enhances the rate of the 8-endo cyclization by partially ordering at least four of the ring's eight carbons, thereby reducing the entropic demand on the reaction and further enhancing ring closure.

Access to the correct relative stereochemistry of isoschizandrin is realized through three separate stereochemical control elements (Scheme 17). First, the stereochemistry at C-8 is set by the Z olefin geometry, which was obtained via a standard Wittig reaction. Conformational restrictions placed on dibenzocyclooctadiene ring systems are generally limited to the twist-boat-chair (TBC) and the twist-boat (TB) limiting conformations.[131] The stereodefined Z-alkene excludes the TB structure. We therefore expected carbons 6, 7, 8, and 9 to assume the chair-like transition state geometry of the TBC conformation.

The second source of stereochemical control is a result of the alkoxy samarium moiety assuming a pseudoequatorial orientation. Scheme 17 depicts how the alkoxy samarium(III) substituent of the slightly pyramidalized[132-136] ketyl radical anion at C-7 may take on either a pseudoaxial or pseudoequatorial orientation. Tight coordination of the ketyl oxygen to the Sm(III) species, adorned with up to four HMPA ligands in its coordination sphere,[54,137] is more imposing than the methyl group to the pseudoaxial hydrogen at C-9 and to the pendant aromatic ring.

Third, the diastereotopic facial selectivity between the olefin and the ketyl is determined by the biaryl stereochemistry. This stereocontrol element, set earlier via a kinetic resolution of racemic lactone (±)-**28**, provided an enantiomerically enriched atropisomer (98% ee) that was carried throughout the synthesis without loss of stereochemical integrity.

V. Conclusion

Samarium diiodide is often featured as a mild, chemoselective reagent for simple functional group transformations employing complex substrates.[47, 48, 138] Therefore, the flexibility introduced in the 8-endo cyclization by varying the olefin substitution can be applied to the construction of many eight-membered ring-containing natural products, creating the rings and installing the attendant functionality stereoselectively in a single synthetic operation.

Overall, the syntheses of (-)-steganone and (+)-isoschizandrin illustrate the potential for the samarium(II) iodide-promoted reductive coupling to achieve rapid and efficient access to the dibenzocyclooctadiene lignans. Specifically, a stereocontrolled 8-endo ketyl-olefin coupling produced (+)-isoschizandrin in good yield with excellent stereo- and regioselectivity. Additionally, this method allowed the direct construction of the 8-5 fused ring system of (-)-steganone without the need for multiple functional group manipulations.

Acknowledgments

We thank our colleagues listed in the references below who have contributed so much to the success of this program. In particular, we acknowledge the dedicated efforts of Drs. Lauren Gilchrist-Monovich, Magnus Rönn, and Yvan LeHuérou, without whose effort the projects described herein would not have been possible. Last, but certainly not least, we thank the National Institutes of Health for their steadfast support throughout the years.

References and Footnotes

1. Girard, P., Namy, J. L., Kagan, H., *J. Am. Chem. Soc.* **1980**, 2693.
2. Zelinsky, N.; Moser, A., *Chem. Ber.* **1902**, *35*, 2684.
3. Zelinsky, N. D., Elagina, N. V. *Dokl. Akad. Nauk USSR* **1952**, *86*, 1117.
4. Barton, D. H. R., Robinson, C. H., *Proc. Chem. Soc., London* **1961**, 207.
5. House, H. O., Riehl, J. J., Pitt, C. G., *J. Org. Chem.* **1965**, *30*, 650.
6. Leroux, Y., *Bull. Soc. Chim. Fr.* **1968**, 359.
7. Hamon, D. P. G., Sinclair, R. W., *J. Chem. Soc., Chem. Commun.* **1968**, 890.
8. Danishefsky, S., Dumas, D., *J. Chem. Soc., Chem. Commun.* **1968**, 1287.
9. Felkin, H., Gault, Y., Rossi, G., *Tetrahedron* **1970**, *26*, 3761.
10. Mirrington, R. M., Schmalzl, K. J., *J. Org. Chem.* **1972**, *37*, 69.
11. Teisseire, P., Pesnelle, P., Corbies, B., Plattier, M., Manpetit, P., *Recherches* **1974**, *19*.
12. Blomberg, C., Hartog, F. A., *Synthesis* **1977**, 18-30.
13. Stojanac, N., Stojanac, Z., White, P. S., Valenta, Z., *Can. J. Chem.* **1979**, *57*.
14. Dadson, W. M., Money, T., *Can. J. Chem.* **1980**, *58*, 2524.
15. Trost, B. M., Coppola, B. P., *J. Am. Chem. Soc.* **1983**, *104*, 6879.
16. Crandall, J. K., Magaha, H. S., *J. Org. Chem.* **1982**, *47*, 5368.
17. Corey, E. J., Kuwajima, I., *J. Am. Chem. Soc.* **1970**, *92*, 395.
18. Corey, E. J., Narasida, M., Hiraoka, T., Ellison, R. A., *J. Am. Chem. Soc.* **1970**, *92*, 396.
19. Molander, G. A., Etter, J. B., *Tetrahedron Lett.* **1984**, *25*, 3281.
20. Molander, G. A., Etter, J. B., *J. Org. Chem.* **1986**, *51*, 1778.
21. Molander, G. A., La Belle, B. E., Hahn, G., *J. Org. Chem.* **1986**, *51*, 5259.
22. Molander, G. A., Hahn, G., *J. Org. Chem.* **1986**, *51*, 2596.
23. Molander, G. A., Hahn, G., *J. Org. Chem.* **1986**, *51*, 1135.
24. Molander, G. A., Etter, J. B., Zinke, P. W., *J. Am. Chem. Soc.* **1987**, *109*, 453.
25. Kagan, H., Namy, J. L., Girard, P., *Tetrahedron* **1981**, *37*, 175.

26. Curran, D. P., Fevig, T. L., Jasperse, C. P., Totleben, M. J., *Synlett* **1992**, 943.
27. Molander, G. A., McKie, J. A., *J. Org. Chem.* **1991**, *56*, 4112.
28. Wickramaratne, D. B. M., Pengsuparp, T., Mar, W., Chai, H. B., Chagwedera, T. E., Beecher, C. W. W., Farnsworth, N. R., Kinghorn, A. D., Pezzuto, J. M., Cordell, G. A., *J. Nat. Prod.* **1993**, *56*, 2083.
29. Whiting, D. A., *Nat. Prod. Rep.* **1987**, *4*, 499.
30. Wright, S. W., Hageman, D. L., McClure, L. D., *J. Org. Chem.* **1994**, *59*, 6095.
31. Shono, T., Nishiguchi, I., Ohmizu, H., Mitani, M., *J. Am. Chem. Soc.* **1978**, *100*, 545.
32. Fox, D. P., Little, R. D., Baizer, M. M., *J. Org. Chem.* **1985**, *50*, 2202.
33. Karviv-Miller, E., Mahachi, T. J., *Tetrahedron Lett.* **1985**, *26*, 4591.
34. Belotti, D., Cossy, J., Pete, J. P., Portella, C., *Tetrahedron Lett.* **1985**, *26*, 4591.
35. Pradham, S. K., Kadam, S. R., Kolhe, J. N., Radhakrishnan, T. V., Sohani, S. V., Thaker, V. B., *J. Org. Chem.* **1981**, *46*, 2622.
36. Corey, E. J., Pyne, S. G., *Tetrahedron* **1983**, *24*, 2821.
37. Ikeda, T., Yue, S., Hutchinson, C. R., *J. Org. Chem.* **1985**, *50*, 5193.
38. Beckwith, A. L. J., *Tetrahedron* **1981**, *37*, 3073.
39. Molander, G. A., Kenny, C., *Tetrahedron Lett.* **1987**, *28*, 4367.
40. Molander, G. A., Kenny, C., *J. Am. Chem. Soc.* **1989**, *111*, 8236.
    Molander, G. A., Etter, J. B., *J. Am. Chem. Soc.* **1987**, *109*, 6556.
42. Molander, G. A., Etter, J. B., Harring, L. S., Thorel, P. J., *J. Am. Chem. Soc.* **1991**, *113*, 8036.
43. Molander, G. A., Etter, J. B., *J. Org. Chem.* **1987**, *52*, 3942.
44. Molander, G. A., Harring, L. S., *J. Org. Chem.* **1989**, *54*, 3525.
45. Molander, G. A., Kenny, C., *J. Org. Chem.* **1988**, *53*, 2132.
46. Molander, G. A., Harring, L. S., *J. Org. Chem.* **1990**, *55*, 6171.
47. Molander, G. A., Harris, C. R., *Chem. Rev.* **1996**, *96*, 307.
48. Molander, G. A., Harris, C. R., *Tetrahedron* **1998**, *54*, 3321.
49. Molander, G. A., McKie, J. A., *J. Org. Chem.* **1995**, *60*, 872.
50. Molander, G. A., McKie, J. A., *J. Org. Chem.* **1992**, *57*, 3132.
51. Monovich, L. G., Samarium(II) Iodide Promoted 8-Endo Ketyl-Olefin Couplings: New Methods in Total Synthesis. Ph.D. Thesis, University of Colorado at Boulder, Boulder, 1998.
52. Beckwith, A. L. J., Schiesser, C. H., *Tetrahedron* **1985**, *41*, 3925.
53. Sowinski, A. F., Whitesides, G. M., *J. Org. Chem.* **1979**, *44*, 2369.
54. Shabangi, M., Robert A. Flowers, I., *Tetrahedron Lett.* **1997**, *38*, 1137.
55. Prasad, E., Knettle, B. W., Flowers, R. A., II, *J. Am. Chem. Soc.* **2004**, *126*, 6891.
56. Still, W. C., *Tetrahedron* **1981**, *37*, 3981.
57. Kupchan, S. M., Britton, R. W., Ziegler, M. F., Gilmore, C. J., Restivo, R. J., Bryan, R. F., *J. Am. Chem. Soc.* **1973**, *95*, 1335.
58. Kende, A. S., Liebeskind, L. S., *J. Am. Chem. Soc.* **1976**, *98*, 267.
59. Hughes, L. R., Raphael, R. A., *Tetrahedron Lett.* **1976**, 1543.
60. Tomioka, K., Ishiguro, T., Mizuguchi, H., Komeshima, N., Koga, K., Tsukagoshi, S., Tsuruo, T., Tashiro, T., Tanida, S., Kishi, T., *J. Med. Chem.* **1991**, *34*, 54.
61. Wang, R. W.-J., Rebhun, L. I., Kupchan, S. M., *Cancer Res.* **1977**, *37*, 3071.
62. Zavala, F., Guenard, D., Robin, J.-P., Brown, E., *J. Med. Chem.* **1980**, *23*, 546.
63. Dhal, R., Brown, E., Robin, J.-P., *Tetrahedron* **1983**, *39*, 2787.

64. Ayres, D. C., Loike, J. D., *Chemistry & Pharmacology of Natural Products: Lignans, Chemical, Biological, and Chemical Properties*. Cambridge University Press: Cambridge, 1990.

65. Meyers, A. I., Flisak, J. R., Aitken, R. A., *J. Am. Chem. Soc.* **1987**, *109*, 5446.

66. Uemura, M., Daimon, A., Hayashi, Y., *J. Chem. Soc., Chem. Commun.* **1995**, *19*, 1943.

67. Ziegler, F. E., Fowler, K. W., Sinha, N. D., *Tetrahedron Lett.* **1978**, 2767.

68. Molander, G. A., McKie, J. A., *J. Org. Chem.* **1994**, *59*, 3186.

69. Dinesh, C. U., Reissig, H.-U., *Angew. Chem., Int. Ed. Engl.* **1999**, *38*, 789.

70. Kan, T., Nara, S., Ito, S., Matsuda, F., Shirahama, H., *J. Org. Chem.* **1994**, *59*, 5111.

71. Molander, G. A., McWilliams, J. C., Noll, B. C., *J. Am. Chem. Soc.* **1997**, *119*, 1265.

72. Kamochi, Y., Kudo, T., *Tetrahedron Lett.* **1991**, *32*, 3511.

73. Kamochi, Y., Kudo, T., *Tetrahedron* **1992**, *48*, 4301.

74. Inanaga, J., Sakai, S., Handa, Y., Yamaguchi, M., Yokoyama, Y., *Chem. Lett.* **1991**, 2117.

75. Benbow, J. W., Martinez, B. L., *Tetrahedron Lett.* **1996**, *37*, 8829.

76. Stanforth, S. P., *Tetrahedron* **1998**, *54*, 263.

77. Johnson, M. G., Foglesong, R. J., *Tetrahedron Lett.* **1997**, *38*, 7001.

78. Oh-e, T., Miyaura, N., Suzuki, A., *Synlett* **1990**, 221.

79. Miyaura, N., Suzuki, A., *Synth. Commun.* **1981**, *11*, 513.

80. Gronowitz, S., Hornfeldt, A.-B., Yang, Y.-H., *Chem. Scr.* **1986**, *26*, 383.

81. Watanabe, H., Miyaura, N., Suzuki, A., *Synlett* **1992**, 207.

82. Hollingworth, G. J., Sweeney, J. B., *Tetrahedron Lett.* **1992**, *33*, 7049.

83. Farina, V., Krishnan, B., *J. Am. Chem. Soc.* **1991**, *113*, 9585.

84. Tomioka, K., Ishiguro, T., Iitaka, Y., Kogo, K., *Chem. Pharm. Bull.* **1986**, *34*, 1501.

85. Kende, A. S., Liebeskind, L. S., Kubiak, C., Eisenberg, R., *J. Am. Chem. Soc.* **1976**, *98*, 6389.

86. Robin, J.-P., Gringore, O., Brown, E., *Tetrahedron Lett.* **1980**, *21*, 2709.

87. Kamikawa, T., Watanabe, T., Uemura, M., *J. Org. Chem.* **1996**, *61*, 1375.

88. Schmaltz, H.-G., Siegel, S., Schwarz, A., *Tetrahedron Lett.* **1996**, *37*, 2947.

89. Taniguchi, N., Hata, T., Uemura, M., *Angew. Chem., Int. Ed. Engl.* **1999**, *38*, 1232.

90. Still, W. C., Murata, S., Revial, G., Yoshihara, K., *J. Am. Chem. Soc.* **1983**, *105*, 625.

91. Fukuzawa, S.-I., Iida, M., Nakanishi, A., Fujinami, T., Sakai, S., *J. Chem. Soc., Chem. Commun.* **1987**, 920.

92. Fukuzawa, S.-I., Nakanishi, A., Fujinami, T., Sakai, S., *J. Chem. Soc., Perkin Trans. 1* **1988**, 1669.

93. Enholm, E. J., Satici, H., Trivellas, A., *J. Org. Chem.* **1989**, *54*, 5841.

94. Enholm, E. J., Trivellas, A., *J. Am. Chem. Soc.* **1989**, *111*, 6463.

95. Enholm, E. J., Trivellas, A., *Tetrahedron Lett.* **1989**, *30*, 1063.

96. Whiting, D. A., *Nat. Prod. Rep.* **1985**, *2*, 191.

97. Tanaka, M., Mukaiyama, C., Mitsuhashi, H., Maruno, M., Wakamatsu, T., *J. Org. Chem.* **1995**, *60*, 4339.

98.   Ikeya, Y., Taguchi, H., Mitsuhashi, H., Takeda, S., Kase, Y., Aburada, M., *Phytochemistry* **1988**, *27*, 569.
99.   Ikeya, Y., Sugama, K., Okada, M., Mitsuhashi, H., *Phytochemistry* **1991**, *30*, 975.
100.   Monovich, L. G., Le Huerou, Y., Roenn, M., Molander, G. A., *J. Am. Chem. Soc.* **2000**, *122*, 52.
101.   Warshawsky, A. M., Meyers, A. I., *J. Am. Chem. Soc.* **1990**, *112*, 8090.
102.   Takeya, T., Ohguchi, A., Tobinaga, S., *Chem. Pharm. Bull.* **1994**, *42*, 438.
103.   Tanaka, M., Itoh, H., Mitsuhashi, H., Maruno, M., Wakamatsu, T., *Tetrahedron: Asymmetry* **1993**, *4*, 605.
104.   Bringmann, G., Walter, R., Weirich, R., *Angew. Chem., Int. Ed. Engl.* **1990**, *29*, 977.
105.   Hayashi, T., Niizuma, S., Kamikawa, T., Suzuki, N., Uozumi, Y., *J. Am. Chem. Soc.* **1995**, *117*, 9101.
106.   Meyers, A. I., Himmelsbach, R. J., *J. Am. Chem. Soc.* **1985**, *107*, 682.
107.   Moorlag, H., Meyers, A. I., *Tetrahedron Lett.* **1993**, *34*, 6989.
108.   Molander, G. A., McKie, J. A., *J. Org. Chem.* **1993**, *58*, 7216-27.
109.   Machrouhi, F., Hamann, B., Namy, J. L., Kagan, H., *Synlett* **1996**, *7*, 633.
110.   Ritter, K., *Synthesis* **1993**, 735.
111.   Stang, P. J., Hanack, M., Subramanian, L. R., *Synthesis* **1982**, 85.
112.   Rottlander, M., Knochel, P., *J. Org. Chem.* **1998**, *63*, 203.
113.   Han, X., Stoltz, B. M., Corey, E. J., *J. Am. Chem. Soc.* **1999**, *121*, 7600.
114.   Bringmann, G., Hinrichs, J., *Tetrahedron: Asymmetry* **1997**, *8*, 4121.
115.   Bringmann, G., Hinrichs, J., Pabst, T., Henschel, P., Peters, K., Peters, E.-M., *Synthesis* **2001**, 155.
116.   Bringmann, G., Hartung, T., *Angew. Chem., Int. Ed. Engl.* **1992**, *31*, 761.
117.   Bringmann, G., Hartung, T., *Tetrahedron* **1993**, *49*, 7891.
118.   Corey, E. J., Bakshi, R. K., Shibata, S., *J. Am. Chem. Soc.* **1987**, *109*, 7925.
119.   Ziegler, F. E., Chliwner, I., Fowler, K., Kanfer, S. J., Kuo, S. J., Sinha, N. D., *J. Am. Chem. Soc.* **1980**, 790.
120.   Fanta, P. E., *Synthesis* **1974**, 9.
121.   Degnan, A. P., Meyers, A. I., *J. Am. Chem. Soc.* **1999**, *121*, 2762.
122.   Abbaszadeh, M. R., Bowden, K., *J. Chem. Soc., Perkin Trans. 2* **1990**, 2081.
123.   Kobayashi, S., Senoo, F., Kihara, M., Sakata, K., Miura, A., *Chem. Pharm. Bull.* **1971**, *19*, 1262.
124.   Kobayashi, S., Mineo, S., Kihara, M., Tagashira, S., *Chem. Pharm. Bull.* **1976**, *24*, 2191.
125.   Kobayashi, S., Kihara, M., Hashimoto, T., Shingu, T., *Chem. Pharm. Bull.* **1976**, *24*, 716.
126.   Dess, D. B., Martin, J. C., *J. Am. Chem. Soc.* **1991**, *113*, 7277.
127.   Coulson, D. R., *Tetrahedron Lett.* **1964**, *45*, 3323.
128.   Inanaga, J., Ishikawa, M., Yamaguchi, M., *Chem. Lett.* **1987**, 1485.
129.   Otsubo, K., Kawamura, K., Inanaga, J., Yamaguchi, M., *Chem. Lett.* **1987**, 1487.
130.   Molander, G. A., George, K., Monovich, L., *J. Org. Chem.* **2003**, *68*, 9533-9540.
131.   Eliel, E. L., Wilen, S. H., *Stereochemistry of Organic Compounds*. Wiley-Interscience: New York, 1994.
132.   Davies, A. G., Neville, A. G., *J. Chem. Soc., Perkin Trans. 1* **1992**, 163.
133.   Russell, G. A., Stevenson, G. R., *J. Am. Chem. Soc.* **1971**, *93*, 2432.
134.   Bennett, J. E., Mile, B., *J. Chem. Soc. A* **1968**, 298.

instances. Thus, while the intellectual challenge of complex synthesis has been largely met, the logistical issues associated with multistep synthesis remain a substantial barrier to the acquisition of adequate quantities of materials for evaluation of their properties.

There are at least two strategies for dealing with this problem. The first is tactical and requires refinement of each and every step of a multi-step synthesis to a much higher level than is typically the case. In this way, even a thirty-step synthesis could deliver a useful quantity of a complex natural substance of interest, but the investment of time and resources in such a labor-intensive enterprise would be beyond what most individuals (and granting agencies) are willing to commit. A second approach is to reduce the number of steps in a synthesis, or at least the number of steps where material-depleting operations such as chromatographic separations, protection-deprotection, etc., are likely to occur. This option is attractive, not only because it is more likely to bring the synthesis to a successful conclusion within a reasonable time frame, but also because it encourages new strategic thinking about synthesis. Multicomponent reactions and multi-step sequences that can be contrived in the absence of protecting groups are illustrations of this "reductionist" approach to synthesis, but there are few examples in the literature of natural products synthesis where these strategic elements play the major role.

Our forays into macrolide synthesis that began with methynolide and continue today have succumbed on many occasions to the logistical difficulties associated with excessive length and/or low-yielding reactions. In some cases, the synthesis was saved by one or more strategic revisions to the plan, occasionally inspired by an adroit observation at the bench but quite often by a desperate "try-it-and-see" approach. Our synthesis of polycavernoside A described in the pages that follow illustrates both of these paradigms. In presenting the synthesis as it actually unfolded, we hope the reader will agree with us that not all ideas that don't work are bad ones. The important lesson to be learned from a successful synthesis is not that a particular sequence of steps can be made to conclude at a designated end point, but that many seemingly good ideas tested against a resistant adversary will be found wanting. Multistep synthesis is the ultimate sifting ground for sorting ideas; some will be discarded, perhaps because they are poorly conceived for the context in which they were intended, but others will prevail and, in the best case scenario, will become permanently woven into the broad tapestry of the craft.

II. Polycavernoside A: A Lethal Metabolite from *Polycavernosa tsudai*

Edible seaweeds provide an important dietary component for the peoples of the Western Pacific islands and these nutritious algae rarely give cause for concern. However, in early 1991, thirteen people were unexpectedly poisoned after consuming small quantities of a normally innocuous seaweed purchased from their local food market near Tanguisson Beach, Guam.[1] The type of seaweed eaten by those afflicted, the red alga *Polycavernosa tsudai* (formerly *Gracilaria edulis*), is widely consumed in the area and had not previously been associated with reports of toxicity. The symptoms of the seaweed poisoning were of a severe and neurotoxic nature, and three of the affected individuals developed cardiac arrhythmia and later died. Preliminary toxicological studies on samples of the ingested alga identified a substance that shared some of the characteristics of the virulent marine toxin palytoxin. An acetone extract of 2.6 kg of *P. tsudai* collected soon after the outbreak had occurred was successfully fractionated with multiple chromatographies guided by mouse bioassay.[2] Two substances, polycavernosides A and B, were isolated from the algal extract that elicited very similar symptoms in the rodent model to those manifested by the human victims of the Tanguisson Beach intoxication.

Polycavernoside A (1)

Skillful application of a barrage of analytical techniques by Yasumoto and co-workers at Tohoku University enabled elucidation of the planar structure of polycavernoside A (1) from the minute quantity (0.4 mg) of purified isolate.[2] The lethal metabolite was revealed to be a glycosidic macrolide possessing an unusual tricyclic aglycon, formally derived from a partially dehydrated and unsaturated 3,5,7,13,15-pentahydroxy-9,10-dioxotricosanoic acid. The aglycon was linked at C5 to a disaccharide moiety composed of heavily methylated xylose and fucose sugar units. The identity of polycavernoside B, while clearly related to polycavernoside A, could not be established at that time due to insufficient material. The following year, Yasumoto and his team made one further collection of *P. tsudai* from Tanguisson Beach and discovered three more polycavernoside congeners. NMR spectral analysis indicated that these new compounds, together with polycavernoside B, shared the same aglycon as polycavernoside A, but differed slightly in the carbohydrate portion.[3]

The production of polycavernosides in *P. tsudai* proved to be a transient phenomenon and the natural supply of these metabolites vanished before chemical and biological studies could be concluded. With access to further samples of the polycavernosides denied, the precise cause of the fatal human poisoning in Guam remains shrouded in mystery. Fortunately, Yasumoto had garnered enough structural detail of polycavernoside A to make its recreation by purely synthetic means a definite possibility. In this context, it was recognized that total synthesis would not only help to map out the relative and absolute stereochemical configuration for the natural product, but might also serve to re-establish a supply of the toxin, vital for a better understanding of its threat to human health.

A desire to take on this challenge prompted us, in 1994, to embark on what turned out to be a lengthy synthetic chemistry program directed at tackling the complex molecular architecture presented by polycavernoside A. Completion of the total synthesis ultimately required over a decade and, as described below, involved many diversions and cul-de-sacs along the way.[4] Our own efforts towards polycavernoside A were aided by seminal contributions made concurrently in the same area by the groups of Professor Akio Murai[5] at Hokkaido University in Japan and Professor Leo Paquette[6,7] at Ohio State University.

III. First Generation Strategy

At the outset of our studies, reliable information concerning the likely stereostructure of polycavernoside A was sketchy at best. After defining the gross topology of the toxin, and the all *trans* nature of the conjugated triene, Yasumoto and co-workers went on to assign the relative configurations for four essentially separate stereochemical domains within the glycoside. It was ascertained that the terminal sugar moiety was of a fucopyranose type and that the inner carbohydrate residue was a xylopyranose derivative, but the diastereochemical relationship between these two sugars, and their absolute configurations were unknown. The relative stereochemistry of the tetrahydropyran containing part of the aglycon (C1-C9) was established, as was the relationship between

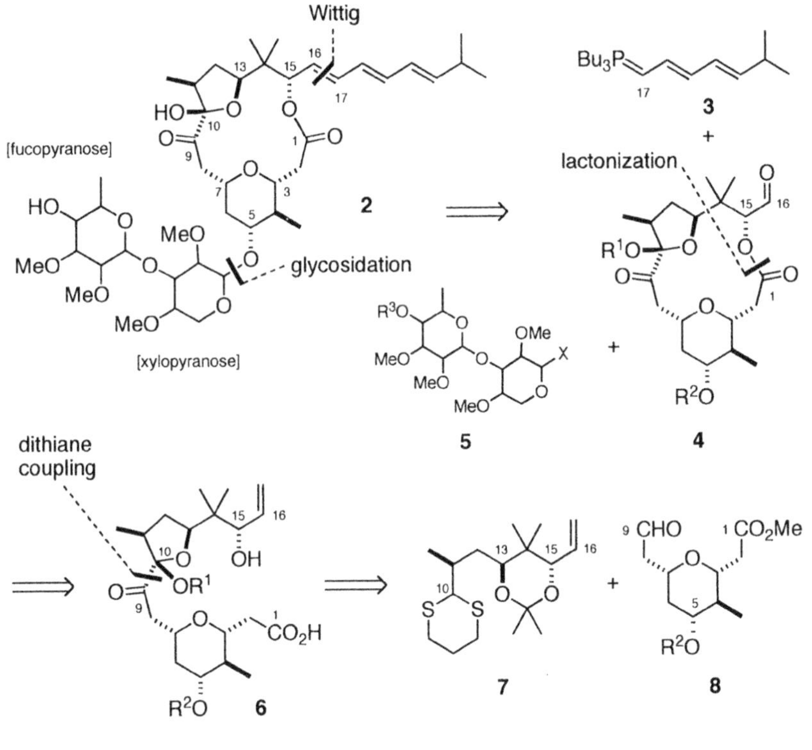

SCHEME 1. First generation strategy for the synthesis of a likely polycavernoside A stereoisomer (2).

stereogenic centers of the upper tetrahydrofuran derived segment (C10-C16). However, it was not clear how these two domains related to each other, nor how aglycon stereochemistry was related to the disaccharide moiety. The early work of Yasumoto had thus narrowed the stereostructure of polycavernoside A down to one of sixteen possibilities. Any synthetic plan directed at such an incompletely defined target must offer enough flexibility to enable the production of any one of the possible isomers. Our first generation strategy for the synthesis of polycavernoside A was designed to be sufficiently malleable to cope with this demand (Scheme 1).

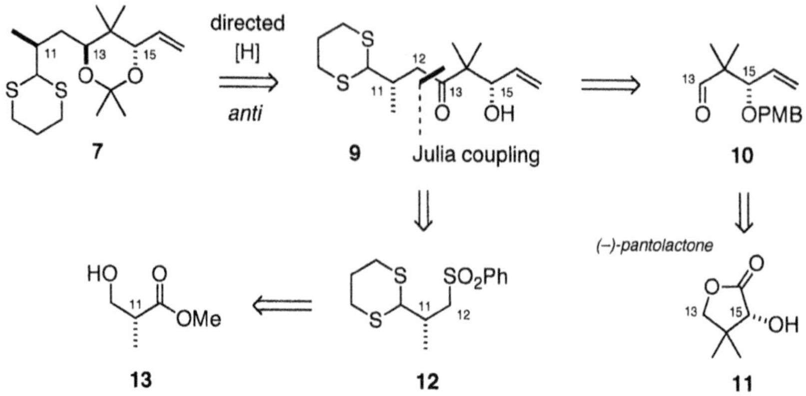

SCHEME 2. Retrosynthetic analysis of C10-C16 dithiane fragment.

Late stage attachment of the trienyl side-chain at C16 was envisioned *via* a Wittig reaction (or a similar construct) between a C16 aldehyde aglycon core fragment **4** and a suitable phosphorane **3**. Glycosidation onto this platform, which might conceivably precede or follow triene installation, would require an appropriately dressed glycosyl donor **5**, the exact nature of which was not clear at the outset. Yasumoto had suggested a tentative stereochemical assignment for the aglycon portion of polycavernoside A and it was this configuration we initially targeted in **4**.[8] The aglycon core **4** would be forged from two halves, dithiane **7**, representing C10-C16, and tetrahydropyranyl aldehyde **8**, which was to provide atoms C1-C9. Assembly of these two fragments of roughly equal complexity was envisioned by addition of the dithiane anion of **7** to aldehyde **8** to form the C9-C10 bond. Adjustment of the oxidation level and dithiane removal from the adduct would, according to our plan, lead to an appropriate precursor **6** for lactonization to the aglycon core **4**. A terminal olefin at C16, carried through the synthesis from its inception, would serve as the latent form of an aldehyde to be used for subsequent fabrication of the triene.

It was imperative that our syntheses of fragments **7** and **8** would allow for their stereoselective generation in either enantiomeric series, and since our plan for the synthesis of the C10-C16 dithiane fragment **7** relied on utilization of readily available substances from the chiral pool (Scheme

2), these materials would need to be easily accessible in both antipodal forms. The so-called Roche esters, 3-hydroxy-2-methylpropionates, provide extremely versatile building blocks for polyketide synthesis and both enantiomorphs of these bacterial metabolites are commercially available. We elected to employ one of these esters, methyl (R)-$(-)$-3-hydroxy-2-methylpropionate (13), as the source of carbon atoms C10-C12 of 7 and the methyl bearing stereogenic center at C11. The hydroxyl-bearing stereogenic center at C15 and the geminal dimethyl group at C14 were to be provided by $(-)$-pantolactone (11). Again, our choice of this starting material was predicated not only on its close structural resemblance to the desired sub-target, but also on its commercial availability in both dextro- and levorotatory forms. The union of these two carefully selected starting materials, 13 and 11, was envisioned *via* Julia coupling between their respective derivatives, sulfone 12 and aldehyde 10. Following the fusion of 12 and 10, a ketone would be introduced at C13 prior to reductive removal of the phenyl sulfone. The issue of stereoselectivity in the sulfone anion addition reaction was therefore of no consequence. To complete the skipped stereotriad embedded within 7, a free hydroxyl group at C15 would be exploited to effect *anti* selective reduction of the C13 ketone in 9.

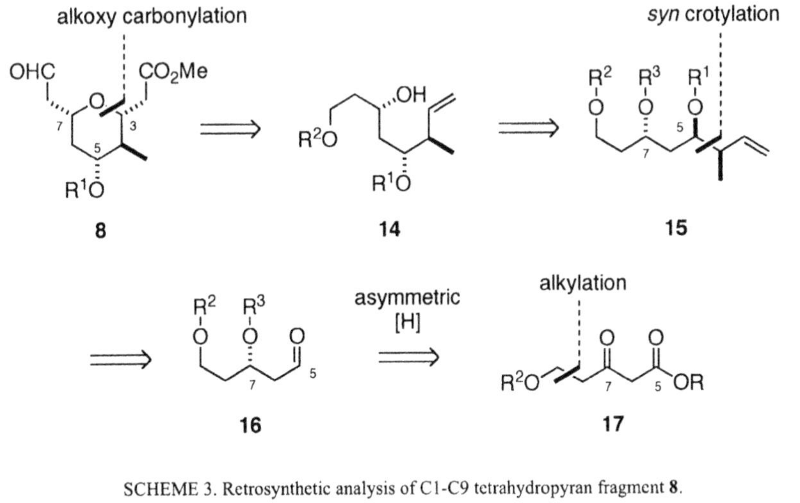

SCHEME 3. Retrosynthetic analysis of C1-C9 tetrahydropyran fragment 8.

The route we initially envisioned for the C1-C9 tetrahydropyran fragment 8 was conceptually very different from our plan for dithiane 7 (Scheme 3). Rather than employ starting materials from the chiral pool, the various stereochemical elements of **8** would be created using a mix of substrate control and asymmetric synthetic methods. If the need arose, access to the opposite enantiomorph of **8** would not present a problem with such an approach. Diastereoselective closure of the oxane ring was envisioned *via* Pd(II) mediated alkoxy carbonylation of an acyclic precursor, hydroxy alkene **14**. This rather unusual method for constructing tetrahydropyrans had only limited precedent at the time and was an exciting prospect in the context of this synthetic venture. Reorientation of **14** clearly revealed a *syn* crotylation retron, and application of the appropriate antithetic operation led to aldehyde **16**. The single stereogenic center present in this material, which would eventually reside at C7, was in principle easy to install enantioselectively by asymmetric hydrogenation of β-ketoester **17**.

With the strategy for our assault on polycavernoside A now fully formed, our focus shifted to its execution in the laboratory.

IV. Synthesis of a C10-C16 Dithiane Fragment

Initial progress was swift and the two subunits required for the C10-C16 dithiane fragment, sulfone **12** and aldehyde **10**, were easily prepared. Ester **13** was first converted to the mono-protected diol **18** by silylation followed by ester reduction (Scheme 4). The phenyl sulfone auxiliary was next installed in two steps by a Mitsunobu-like thioether formation with diphenyl disulfide and tributylphosphine[9] followed by oxidation with Oxone®. The resulting sulfone **19** was desilylated and the liberated hydroxyl group converted to an aldehyde with Swern's procedure.[10] Subunit **12** was completed by formation of the dithiane from aldehyde **20** under standard conditions.

The aldehyde subunit **10** was prepared in straightforward fashion from (–)-pantolactone (**11**) (Scheme 5). Exhaustive reduction of this lactone with lithium aluminum hydride gave triol **21** in 71% yield. This could be easily isolated providing that an aqueous work-up was avoided. The 1,2,4-triol moiety of **21** now presented an interesting problem in hydroxyl group differentiation that is more commonly encountered in carbohydrate chemistry. It is well known that ketones will preferably acetalize with 1,2,4-triols to form 1,2-dioxolanes, whereas aldehydes typically give 1,3-dioxanes under the same conditions. As expected, and with precedent provided by the work of Lavallée,[11] transacetalization of the dimethyl acetal of *p*-anisaldehyde with triol **21** gave the 1,3-dioxanyl *p*-methoxybenzylidene acetal **22** rather than an alternative 1,2-dioxolane

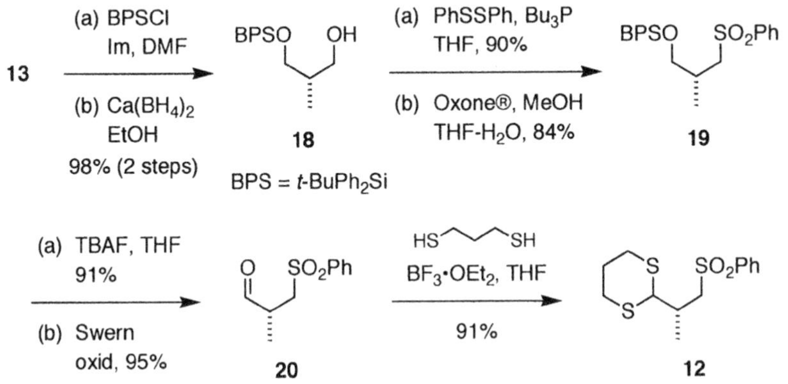

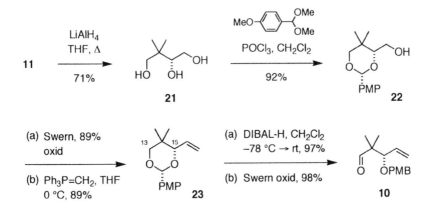

SCHEME 4. Synthesis of sulfone subunit **12**.

product. The remaining hydroxyl substituent of **22** was converted to a
vinyl group by Wittig methylenation after Swern oxidation in 79% yield.
We next employed a well-known tactic in protecting group chemistry to
advance **23** toward allylic ether **10**. Treatment of six-membered cyclic
benzylidene acetals with diisobutylaluminum hydride (DIBAL-H)
typically results in selective cleavage of one of the two carbon oxygen
bonds, the site of hydride attack being determined by the oxygen atom
that is more available for complexation with the Lewis acidic reductant.
In the case of **23**, the accessible oxygen atom resides at C13, and DIBAL-
H reduction afforded a primary alcohol in which the secondary hydroxyl

SCHEME 5. Synthesis of aldehyde subunit **10**.

group at C15 was now protected as its *p*-methoxybenzyl (PMB) ether. A second Swern oxidation gave the completed subunit **10**.

It was hoped that dithianyl sulfone **12** would function as a unique double-ended nucleophilic synthon, selective manipulation being made possible in principle by the greater acidifying effect of the sulfonyl moiety. Lithiation of **12** with *n*-butyllithium did indeed occur adjacent to the sulfone group, and addition of aldehyde **10** to the metallate gave the desired hydroxy sulfone **24** as an inconsequential mixture of four diastereoisomers. The next part of our plan called for conversion of **24** to the corresponding ketone; however, we were unable to find a method to effect this normally trivial transformation. Attempted oxidation of **24** with a variety of reagents either returned unreacted starting material or resulted in decomposition. The recalcitrance of this oxidation was ascribed to the hindered environment surrounding C13, which bears neopentyl substitution to one side and is flanked by three contiguous methine carbons on the other.

This obstacle was easily surmounted by making recourse to the sterically less encumbered sulfone **19** (Scheme 6). Julia coupling, as before with aldehyde **10**, produced hydroxy sulfone **25** in near quantitative yield as a mixture of four diastereoisomers. As we had hoped, alcohol **25** proved more amenable to oxidation, and its conversion to a ketone was accomplished in excellent yield with the Dess-Martin periodinane.[12] Reduction of the resultant α-ketosulfone with samarium(II) iodide[13] gave 93% of an isomerically homogenous ketone from which the PMB ether was removed by treatment with DDQ in wet dichloromethane. Substrate directed reduction of β-hydroxy ketone **26** with tetramethylammonium triacetoxyborohydride, a reagent originally developed by Evans and Carreira,[14] gave a diol (**27**) containing all three stereogenic centers required for the targeted polycavernoside A fragment **7**. Acetonide formation from *anti*-1,3-diol **27** was followed by desilylation and Swern oxidation to give aldehyde **28**, a compound that would later take on greater significance in this enterprise. Finally,

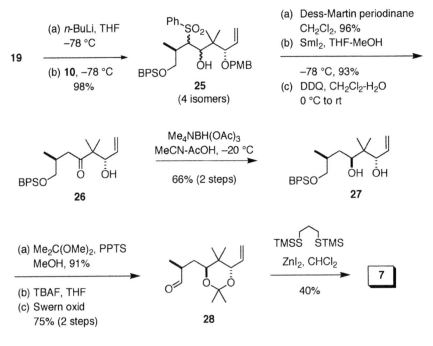

SCHEME 6. Synthesis of the C10-C16 dithiane fragment 7.

treatment of **28** with the bistrimethylsilyl thioether derivative of propane-1,3-thiol in the presence of zinc iodide, gave dithiane **7**.

With the latent nucleophilic C10-C16 aglycon fragment **7** now in hand, we directed our attention to preparing its electrophilic coupling partner, originally envisioned as the C1-C9 tetrahydropyran aldehyde **8**.

V. Intramolecular Alkoxy Carbonylation of 6-Hydroxy-1-octenes

The tetrahydopyran nucleus is found in many classes of secondary metabolites, and a variety of methods are now available for the synthesis of this saturated heterocycle.[15] As one might expect, cyclization strategies to oxanes are manifold. However, relatively few of these approaches are able to generate heavily substituted tetrahydropyrans with complete stereocontrol.

One particular method for tetrahydropyran synthesis attracted our attention as a possible means to prepare a C1-C9 fragment of polycavernoside A. In 1990, Semmelhack had reported that Pd(II) mediated intramolecular alkoxy carbonylation of 6-hepten-2-ol (29) gave cis-2,6-disubstituted tetrahydropyran 30 as a single diastereoisomer.[16] Semmelhack's precedent held great promise for our projected synthesis of a fragment 8, but the effect of substituents in the interconnecting chain between hydroxyl and alkenyl moieties might have on cyclization were essentially unknown. Analogous reactions leading to tetrahydrofurans had been independently studied by Semmelhack[17] and Liotta,[18] and in these cases the effects of chain substitution were generally favorable. The prospects for a successful route to 8 thus seemed excellent, and, eager to apply this construct within the context of the total synthesis of polycavernoside A, we set about preparing an appropriate cyclization precursor, which we formulated as octenol 39 (Scheme 7).[19]

The synthesis of 39 commenced with isobutyl acetoacetate (31), which was doubly deprotonated by addition of sodium hydride followed by n-butyllithium in THF. Alkylation of the resulting diendiolate with benzyl chloromethyl ether occurred at the more reactive peripheral site and gave β-keto ester 32 in moderate yield. Hydrogenation of 32 with Noyori's versatile Ru(II)-(S)-BINAP system[20] yielded hydroxy ketone 33 with excellent enantioselectivity as predicted from a close literature precedent.[21] Following silylation of the hydroxyl group in 33, controlled reduction of the resulting ester gave aldehyde 34 in 92% yield. Homoallylic alcohol 38 was next prepared from this carbonyl compound using Brown's reagent-controlled crotylboration method.[22] Following described procedures, cis-2-butene (35) was metallated with the Schlosser base, and the resulting (Z)-crotyl metal species was treated with the levorotatory antipode of B-methoxydiisopinocampheylborane (36), itself derived from (+)-pinene. The ensuing K → B transmetallation process afforded cis crotylborane 37, which effected syn crotylation of aldehyde 34 to afford 38 with moderate diastereoselectively. Silylation of the new secondary hydroxyl group in 38 was followed by removal of the more labile TBS ether in acidic ethanol and gave octenol **39** in seven steps from **31**.

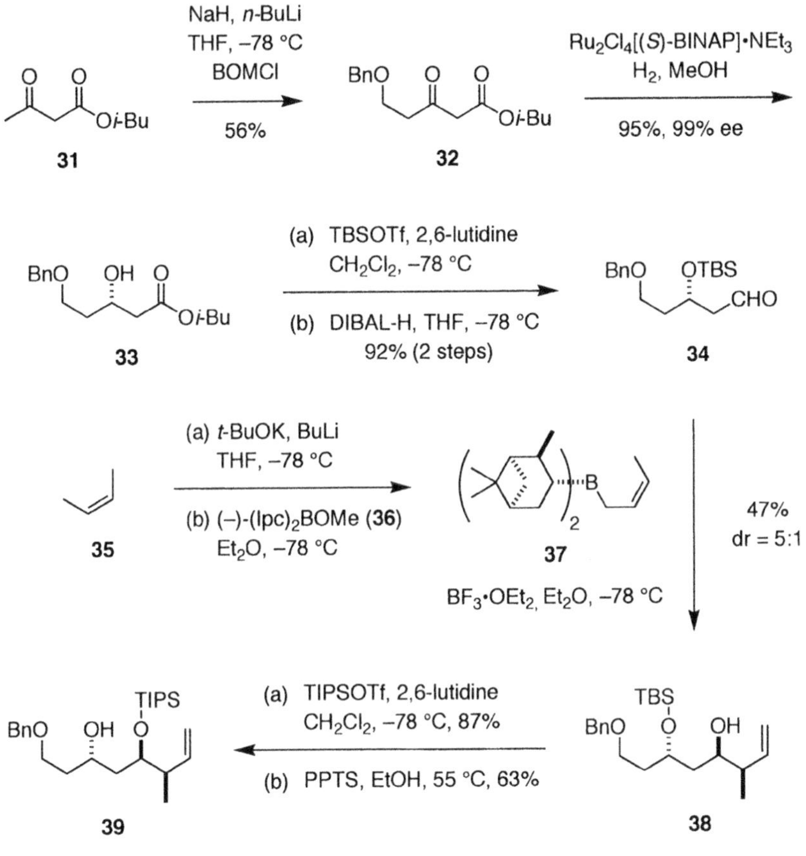

SCHEME 7. Synthesis of alkoxycarbonylation precursor **39**.

Alkoxycarbonylative cyclization of **39** was attempted in methanol under an atmosphere of CO in the presence of a catalytic quantity of $PdCl_2$ with $CuCl_2$ as stoichiometric oxidant (Table 1). To our surprise, not even a trace of the desired tetrahydropyran **40** was formed. Three additional 6-hydroxy-1-octenes, **41**, **43**, and **45**, were prepared from **34** by using the appropriate isomer of crotylborane **37** in each case.[22] Under identical reaction conditions to those previously tested with **39**, alkoxycarbonylation of these isomers gave varying amounts of the expected 2,6-*cis*-tetrahydropyran products (Table 1). Evidently, cyclization efficiency was highly dependent on the configuration of the hydroxyoctene substrate. Thus, in the two extremes, the process worked well for **45**, giving 61% of tetrahydropyran **46**, but failed completely for **39**, the substrate with the appropriate stereochemistry for elaboration to polycavernoside A!

TABLE 1
Aldoxycarbonylative cyclization of 6-hydroxy-1-octenes[a]

substrate	product	yield
39	**40**	0%
41	**42**	20%
43	**44**	40%
45	**46**	61%

[a]Reaction conditions: $PdCl_2$ (25 mol%), $CuCl_2$, CO (1 atm), MeOH, rt.

At first glance, the results in Table 1 seem counterintuitive. The substrate that gives the highest yield, **45**, leads to the tetrahydropyran that should be least stable (2 axial substituents), whereas **39** would have produced a tetrahydropyran **40** with all substituents equatorial. A likely explanation for the observed outcome can be found by considering the Pd complexed alkenes **47** and **48** that precede cyclization in these two examples. With the large *exo* Pd substituent positioned pseudoequatorially in a bisected conformation of the alkene, an eclipsing interaction with the methyl group is present in **47** that is absent in **48**. This line of reasoning suggested that relocating the methyl group to a site more remote from the double bond would diminish the steric encumbrance that prevented intramolecular alkoxycarbonylation of **39** and that therefore should lead to successful cyclization. To test this hypothesis, an alternative hydroxy octene **49** was prepared along similar lines to **39**. If cyclization from **49** were to occur, a tetrahydropyran of identical configuration to **40** would be produced, but with side-chain functionality at C2 and C6 reversed. When pivalate **49** was exposed to $PdCl_2/CuCl_2$ in the presence of CO and MeOH, cyclization was indeed successful but gave the expected oxane **50** in only 45% yield.

Unfortunately, the way forward from **50** was judged to be too circuitous to make this a practical route to a major fragment of polycavernoside A, and we therefore abandoned it in favor of a different concept. However, we had demonstrated that alkoxycarbonylation can be a viable method for the preparation of certain highly substituted tetrahydropyrans, and we later exploited this methodology more successfully in a different context.[23]

VI. Synthesis of a C1–C9 Aldehyde Fragment and Failure of Dithiane Coupling

Our exploration of the intramolecular palladium(II)-catalyzed alkoxy carbonylation reaction had yielded useful insights into the scope of this cyclization process, which turned out to be highly dependent on subtle conformational effects. Ironically, the only stereochemical configuration of an appropriately substituted oxane that we could not directly access using this methodology was that required for a synthesis of

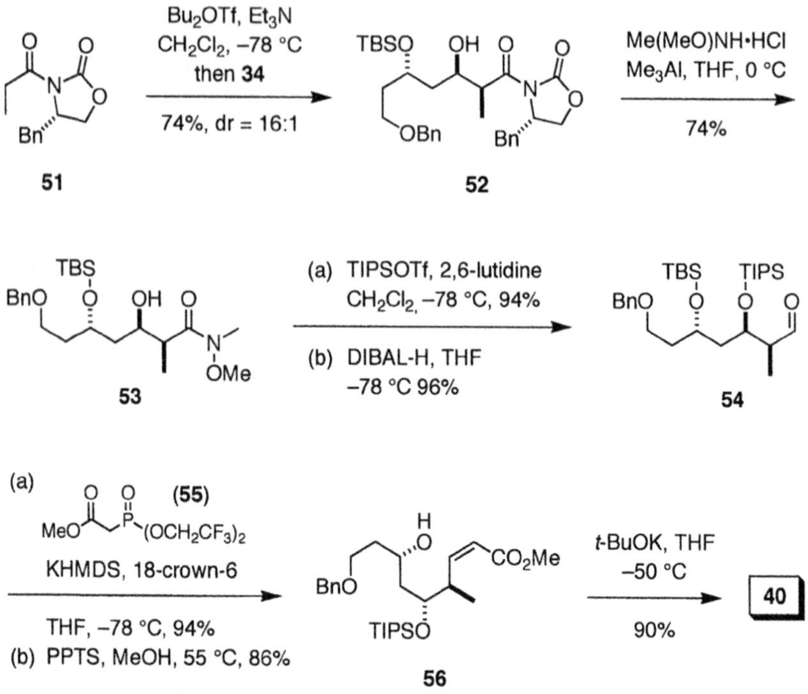

SCHEME 8. Synthesis of tetrahydropyran **40** by a conjugate addition reaction.

polycavernoside A! Fortunately, an alternative route to this compound presented itself that now merited serious consideration. Oxan-2-ylacetates, such as **40**, have often been prepared by intramolecular alkoxide conjugate addition of ω-hydroxy-α,β-unsaturated esters, and this well-precedented approach was adopted as our contingency plan (Scheme 8).

A return was made to aldehyde **34**, and its reaction with the (Z)-boron enolate of the (S)-phenylalanine derived oxazolidinone **51** gave the expected *syn* aldol adduct **52** in 74% yield and with high diastereoselectivity.[24] Conversion of imide **52** to Weinreb amide **53** was effected by the aluminum amide reagent derived from trimethylaluminum and N-methoxymethylamine hydrochloride.[25] It is worth noting that this useful transamidation reaction is only applicable to Evans aldol adducts which are still in possession of a free β-hydroxyl group. As a consequence, protection of alcohol **53** was deferred until after its genesis from **52**. Following silylation of **53**, reduction of the resulting hydroxamate with DIBAL-H gave a differentially protected trihydroxy aldehyde **54** in 96% yield. Aldehyde **54** was homologated to a (*Z*)-enoate with the Still-Gennari phosphonoacetate **55**,[26] before selective removal of the TBS ether in mildly acidic methanol afforded our cyclization precursor, ω-hydroxy α,β-unsaturated ester **56**. Treatment of **56** with potassium *tert*-butoxide in THF gave the desired tetrahydropyran **40** as the only detectable isomer in 90% yield. The stereochemical outcome of this cyclization was anticipated from a closely related precedent provided by Murai[27] and is that predicted by a chair-like transition state **57** with all substituents equatorially disposed (Figure 1). The (*Z*)-configuration of **56** ensures that severe 1,3-allylic strain would develop in an alternative transition state **58** leading to the unwanted 2,6-*trans* tetrahydropyran. Banwell has examined similar intramolecular conjugate addition reactions and has defined the role that alkene geometry plays in determining product oxane stereochemistry.[28]

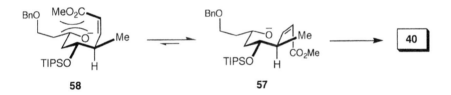

FIGURE 1. Stereochemical model for cyclization of **56**.[27,28]

A suitable C1-C9 fragment **60** required for our projected coupling with dithiane **7** was obtained in two steps from **40** (Scheme 9). Debenzylation of **40** by hydrogenation over Pearlman's catalyst, followed by Dess-Martin oxidation of the resulting primary carbinol, gave the crystalline aldehyde **60** in 73% overall yield. An X-ray crystallographic analysis of **60** furnished unambiguous proof of configuration for the four stereogenic centers installed in this tetrahydropyran. With both key fragments, **7** and **60**, now in hand, we began the task of joining them together through construction of a C9-C10 bond.

1,3-Dithiane metallate alkylation was introduced by Corey and Seebach in 1965 and has served well as a method for forging carbon-carbon bonds at a masked carbonyl group by nucleophilic displacement.[29] The method, an example of an umpolung process, has found extensive use for advanced fragment assembly in target-directed synthesis. Contributions from the Smith research group at the University of Pennsylvania, in particular, have served to demonstrate the power of this

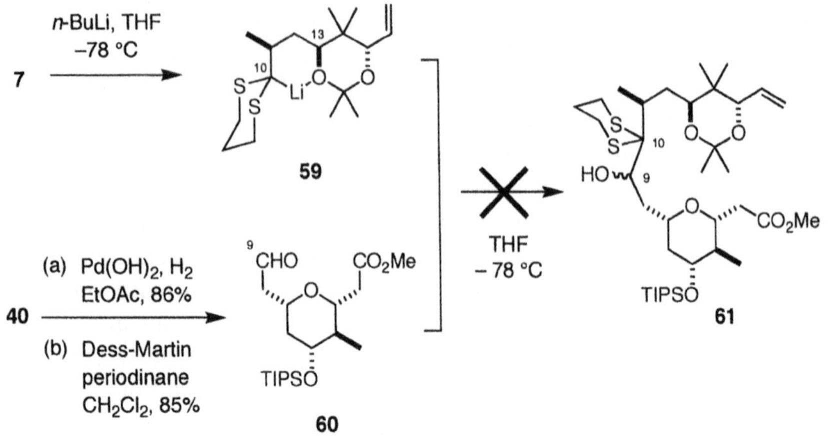

SCHEME 9. Failure of attempted coupling between **7** and **60**.

versatile technique for the fabrication of very complex natural products.[30] Thus, the pedigree of our proposed method for fragment linkage was undeniable. In the event we were to discover that the synthesis of **61** *via* addition of lithiated dithiane **59** to aldehyde **60** was not going to be viable for reasons associated with **59**'s refusal to perform as a reactive nucleophile (Scheme 9). We attribute the lack of reactivity of **59** to the formation of a stable six-membered lithium chelate between C10-Li and C13-OR. Within such a metal complex, the carbanionic site of **59** would not only experience steric encumbrance from its local environment, but would also suffer depressed nucleophilicity. Whether or not this rationalization was actually correct, the fact remained that the C9-C10 bond of polycavernoside A could not be formed as originally intended. Not for the first or last time, a revised synthetic plan was called for.

VII. A Revised Plan: Fragment Role Reversal and an Antipodean Target

In 1995, the mist surrounding the absolute stereostructure for polycavernoside A began to clear. From commercially available samples of D-fucose, L-fucose, and D-xylose, Murai and co-workers prepared both possible diastereoisomers of methyl 2,3-di-*O*-methyl-α-fucopyanosyl-(1→3)-2,4-di-*O*-methyl-β-xylopyranose glycoside. Of the two glycosides, only the ^{1}H NMR spectrum for the disaccharide derived from L-fucose and D-xylose compared well with appropriate sub-regions of a spectrum obtained from natural polycavernoside A.[31] With the question of carbohydrate relative stereochemistry probably settled, both enantiomers of a truncated C1-C9 polycavernoside A aglycon model compound were separately glycosidated with the L-fucosyl-D-xylosyl disaccharide.[32] Spectral comparison of the two resulting diastereomeric partial structures with the natural product itself revealed only one good match, establishing the stereochemical connection between disaccharide and aglycon tetrahydropyranyl stereochemical domains. The final piece of the puzzle was put in place when an NOE between the methine proton at C11 of polycavernoside A and one of the diastereotopic protons at C8 was detected. Observation of this key NOE enabled spatial and configurational connectivity between tetrahydropyran and tetrahydrofuran moieties within the aglycon to be identified. The absolute configuration for polycavernoside A could also be tentatively proposed at this time since xylose sugars found in marine biota are almost always of the D-form. The complete stereostructure for polycavernoside A as originally suggested by Murai and co-workers on the basis of these arguments,[31,32] and illustrated above as 1, was later proved to be correct by total synthesis.

The astute reader will notice that up to this point we had been pursuing the incorrect aglycon enantiomer for polycavernoside A. With our dithiane coupling plan now in tatters, we decided to change course and head for the antipode of our original aglycon fragment 4. We also elected to reverse the reactivity roles for C1-C9 and C10-C16 fragments in search of a better construct with which to form the pivotal C9-C10 bond. The addition of a vinyl chromium reagent to an aldehyde,[33] known as the Nozaki-Hiyama-Kishi (NHK) reaction, presented itself as an excellent vehicle for our purpose (Scheme 10). The requisite C10-C19 aldehyde fragment ent-28 necessary to bring this new strategy to fruition had been synthesized before, albeit in the antipodal form, en route to dithiane 7. An appropriate precursor of the C1-C9 organochromium reagent, vinyl bromide 62, was also available from the enantiomer of a previously synthesized material, aldehyde 60.

Union of ent-28 and 62 by bond formation between C9 and C10 was anticipated by the combined action of chromous chloride and nickel(II) chloride, the usual prescription for a successful NHK reaction. Advancement of NHK coupling product 63, representing atoms C1-C16 of polycavernoside A, to a suitable lactonization precursor 65 was projected to be a relatively straightforward operation. As it turned out, this aspect of the new plan would not be without difficulties (vide infra). The terminal alkenes present at both C9 and C16 in this new series of compounds would serve as latent carbonyl equivalents, their exposure hopefully taking place at a later stage by oxidative scission.

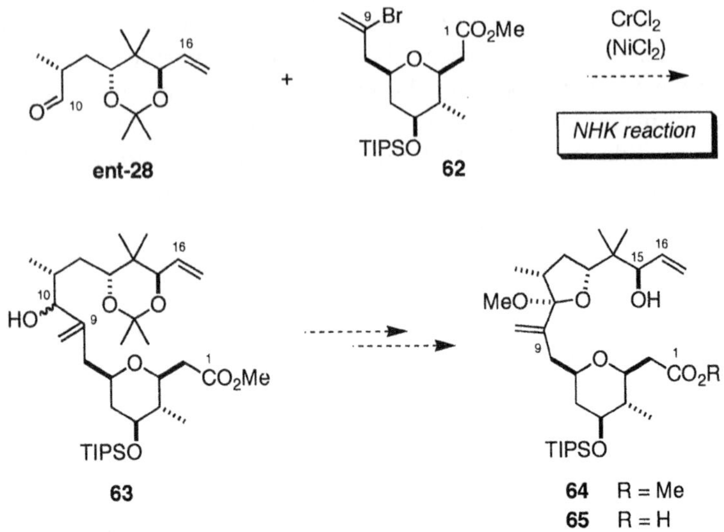

SCHEME 10. A revised coupling strategy.

VIII. Fragment Union by a Nozaki-Hiyama-Kishi (NHK) Reaction

As a first step towards our revised fragment assembly plan, we prepared aldehyde **ent-28** from the ester **ent-13** and (+)-pantolactone (**ent-11**) exactly as we had previously prepared its enantiomer **28**. However, we felt that our existing route to aldehyde **60**, the enantiomer of which would be required to prepare **62**, could be substantially improved in terms of its efficiency. Accordingly, a shorter and more practical synthesis of this substance was developed that incorporated a less cumbersome protecting group strategy (Scheme 11).

Asymmetric allylation of aldehyde **66**, available in two steps from 3-hydroxypropionitrile,[34] with allyl borane **67** gave the expected homoallylic alcohol **68** in good yield and with acceptable enantioselectivity. This material was converted uneventfully to a doubly protected dihydroxy aldehyde **69** and thence to **70** along lines similar to our synthesis of **56** from **34** (see Scheme 8). The aldol reaction employed within this five step sequence provided the means to remove vestiges of the minor enantiomer of **69** introduced from the allylation of **66**. In an

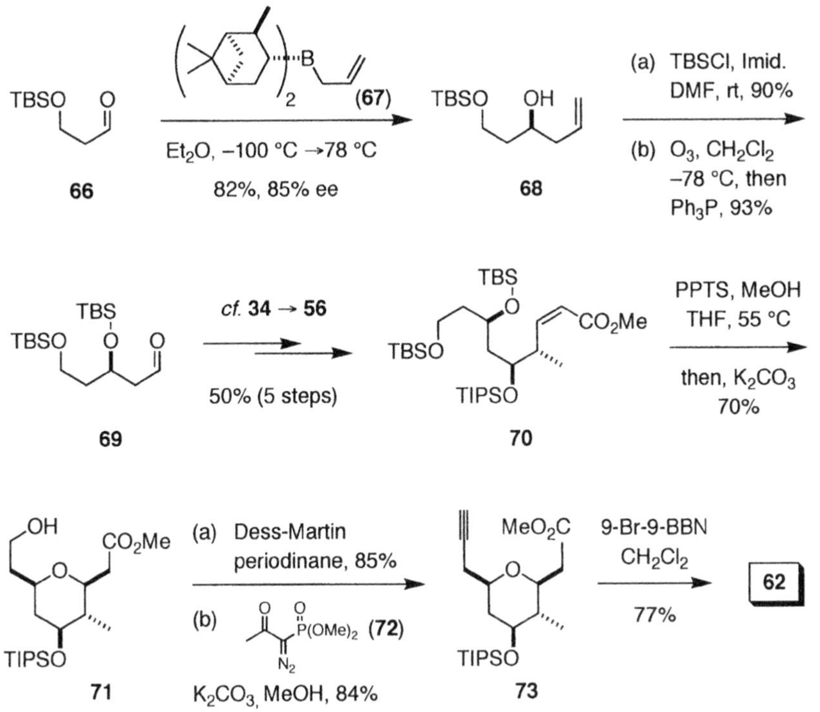

SCHEME 11. Synthesis of vinyl bromide 62.

efficient one-pot process, both TBS groups were removed simultaneously from 70 in acidic methanol, basification of the reaction medium with potassium carbonate inducing immediate cyclization of the intermediate dihydroxy enoate. In this manner, oxane 71 was obtained directly from 70 in 70% yield. By contrast, the use of potassium *tert*-butoxide in THF to effect cyclization of isolated dihydroxy enoate, akin to the method used earlier in the preparation of 40 from 56, led to oligomerization of the product 71 *via* intermolecular esterification. By performing the cyclization in methanol, any accumulation of oligomers of 71 is prevented since these adducts would be broken down and the monomer reconstituted by basic methanolysis. Final oxidation of 71 with Dess-Martin periodinane gave **ent-60** in 85% yield, identical in all respects to previously synthesized enantiomeric material except for the sign of its optical rotation.

Preparation of the new C1-C9 fragment, vinyl bromide 62, was accomplished in just two operations from aldehyde **ent-60**. Treatment of

ent-60 with Ohira's reagent (**72**),[35] followed by bromoboration of the resulting alkyne **73** with 9-bromo-9-borabicyclo[3.3.1]nonane[36] gave **62** in 65% overall yield. Both fragments required for the projected NHK reaction, **ent-28** and **62**, were now in hand and it was not without apprehension that we approached the critical process of their union. After all, the coupling of **59** and **60** had seemed a perfectly reasonable proposition before experiment proved otherwise. Would fragment linkage by the NHK reaction fare any better?

Nozaki and Hiyama were among the first investigators to demonstrate the facile oxidative addition of Cr(II) into carbon-halogen bonds in aprotic dipolar solvents. They further showed that the resulting organochromium(III) reagents are competent nucleophiles capable of highly chemoselective addition to aldehydes.[37,38] However, it was noticed in the early developmental stage of this chemistry that the addition of alkenyl halides to aldehydes was difficult to reproduce. Yields depended critically on the source of chromous chloride used in the reaction. Takai[39] and Kishi[40] independently discovered that samples of $CrCl_2$ that gave best results in this reaction were those of supposedly inferior quality; in fact, rigorously purified $CrCl_2$ was quite ineffective! Subsequent analysis of batches of $CrCl_2$ that promoted successful NHK reactions revealed traces of contamination by nickel salts.[39] It was established that Ni(II) catalyzes the formation of the carbon-Cr(III) bond, and the addition of alkenyl halides to aldehydes is now routinely conducted with $CrCl_2$ deliberately doped with a small quantity of $NiCl_2$. The nickel-catalyzed variant of this organochromium chemistry, referred to today as the Nozaki-Hiyama-Kishi (NHK) reaction, is highly reproducible and has been applied successfully in some truly challenging scenarios.[33] The metal insertion step is compatible with a wide variety of functional groups which makes the NHK reaction particularly well suited to the linkage of very complex molecular fragments. A lasting testament to the power of the NHK reaction remains Kishi's landmark synthesis of palytoxin,[41] an accomplishment that provided much of the impetus for development of the method in the first place.[39]

The NHK transform proved equal to the task we demanded of it. When a mixture of **ent-28** and **62** was reacted with a large excess of $CrCl_2$ and a sub-stoichiometric quantity $NiCl_2$ in DMF, the desired C1-C16 polycavernoside A fragment **63** was produced as an equal mixture of two epimers in 79% yield (Scheme 12). We were delighted by this result and not at all disappointed, or surprised, by the lack of diastereoselectivity we encountered in this pivotal step of the synthesis. Despite its many favorable attributes, the NHK reaction is notoriously non-stereoselective. The formation of two epimers of alcohol **63** is inconsequential here since they converged on a single enone **74** after a high yielding oxidation with Dess-Martin periodinane.[12]

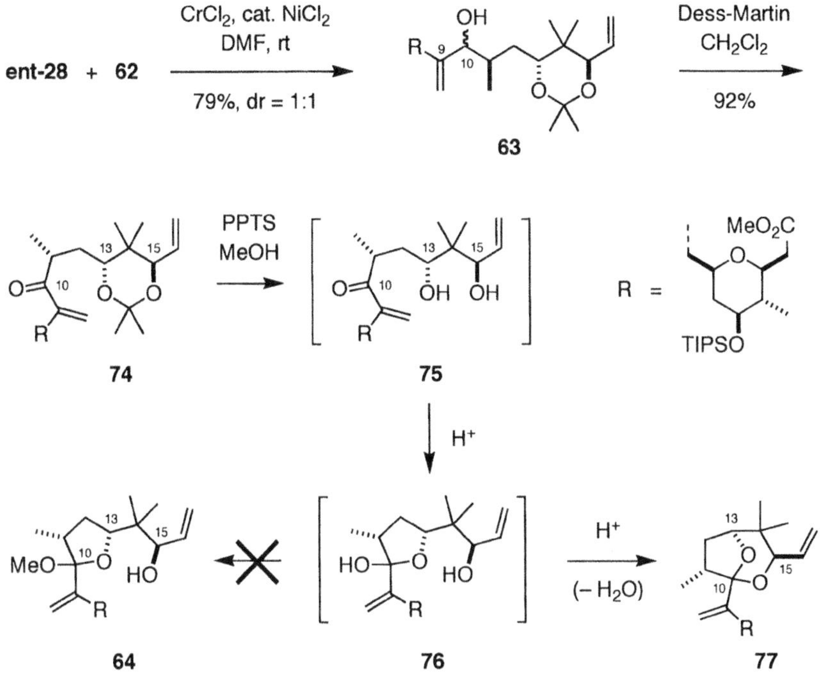

SCHEME 12. NHK coupling of **ent-28** and **62** and attempted advancement of **63**.

At this point, a part of our plan we had taken for granted failed utterly. The latent hydroxyl groups at C13 and C15 of **74** had been mutually protected by an acetonide; thus, one carbinol could not be revealed without exposing the other. Cognizant of this fact, we had counted on the two hydroxyl groups naturally differentiating themselves when liberated in the presence of a C10 ketone. The C13-OH would form a five-membered hemiketal with the proximal carbonyl group (and thence methyl ketal **64**) it was supposed, leaving C15-OH free for subsequent lactonization. To our surprise, *in situ* generation of **75** by treatment of a methanolic solution of **74** with PPTS was followed by rapid *double* cyclization to bicyclic ketal **77**. None of the desired methyl ketal **64** was found! Any chance of reopening the bicyclic ketal of **77** by acidic hydrolysis seemed remote since it would require harsh conditions unlikely to be compatible with other functionality contained within this relatively sensitive structure. The facile cyclodehydration of **76** to form **77**, which was presumably promoted by a Thorpe-Ingold effect induced by the geminal dimethyl substituents at C14, had thwarted yet another assault on polycavernoside A, and it appeared we had entered a cul-de-sac from which there was no obvious exit strategy.

IX. The Quest for a Differentially Protected C10-C16 Aldehyde Fragment

It was clear upon reflection that any keto diol related to **75** was likely to experience spontaneous and irreversible dehydration to form a stable bicyclic ketal. A logical solution to this problem was to employ a C10-C16 aldehyde fragment in the NHK reaction in which the hydroxyl groups at C13 and C15 were orthogonally protected. If the coalescence of such an aldehyde with **62** and subsequent introduction of a ketone at C10 could be carried out, the hydroxyl group at C13 could be selectively liberated and a methoxy tetrahydrofuran installed by acidic methanolysis, as in **78** to **79** (Scheme 13). With a stable form of the hemiketal in place, the hydroxyl group at C15 could perhaps be exposed, avoiding acidic conditions in the process, to give **64**. But how could this be accomplished without lengthening the synthesis?

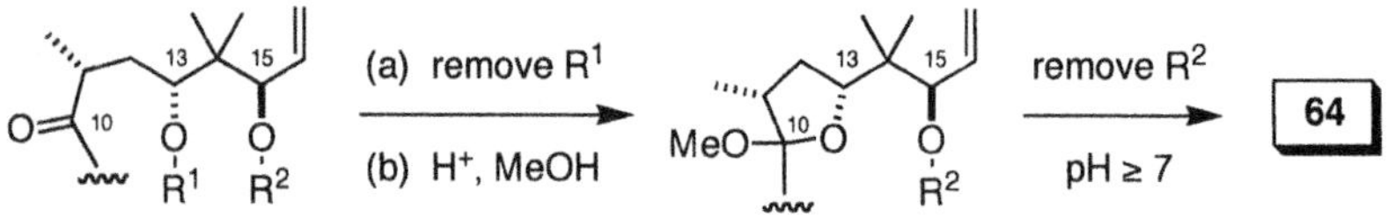

SCHEME 13. Strategic use of orthogonal protecting groups at C13-OH and C15-OH.

Our existing route to the C10-C16 aldehyde fragment was clearly not appropriate for this new plan. Reduction of β–hydroxyketone **ent-26** with triacetoxyborohydride proffered a 1,3-diol intermediate (**ent-27**) with no obvious means available for distinguishing the two secondary carbinol moieties. On the other hand, the Evans-Hoveyda variant of the classical Tishchenko reduction would provide a method to effect diastereoselective reduction of **ent-26** while at the same time allowing differentiation of the C13 and C15 hydroxyl groups.[42] According to the Evans-Tishchenko reduction protocol, a β-hydroxyketone **80** is treated with an aldehyde and a catalytic quantity of SmI$_2$ (Scheme 14). Transfer of hydride from the resulting Sm(III) bound hemiacetal to the proximal chelated ketone in **81** would yield an *anti*-1,3-diol derivative **83** in which the stereodirecting hydroxyl group finds itself converted to an ester.

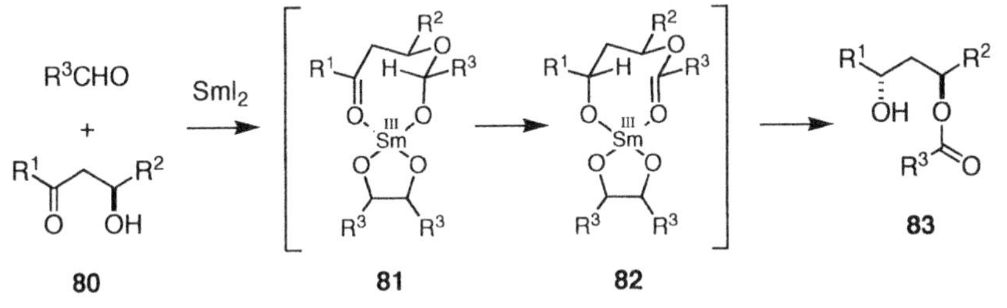

SCHEME 14. The Evans-Hoveyda variant of the Tishchenko reduction.

Evans-Tishchenko reduction of **ent-26** with either *p*-nitrobenzaldehyde or benzaldehyde itself afforded benzoates **84** and **85**, respectively, in good yield and with complete diastereoselectivity in each case. Interestingly, the same reaction, when attempted with *p*-anisaldehyde, gave none of the expected *p*-methoxybenzoate **86**. Failure of this reaction may reflect the lower propensity of *p*-anisaldehyde to form hemiacetals owing to its poor electrophilicity. Alternatively, the *in situ* generation of a Sm(III) "pinacolate" catalyst from SmI$_2$ and *p*-anisaldehyde may be hampered by the relatively high reduction potential of this electron rich aldehyde.

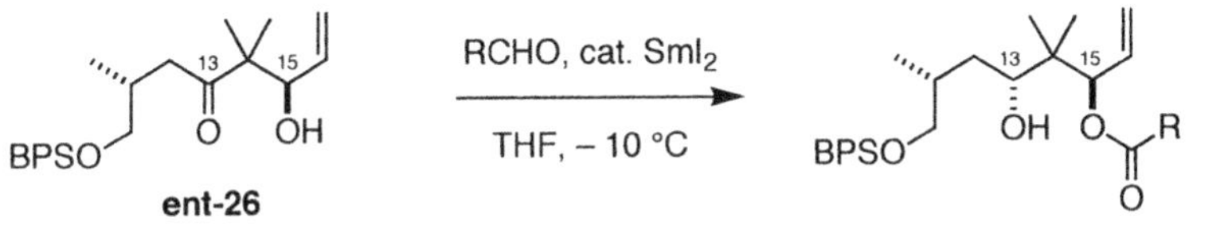

84, R = 4-NO$_2$C$_6$H$_4$ 80%
85, R = C$_6$H$_5$ 60%
86, R = 4-MeOC$_6$H$_4$ 0%

In any event, the available benzoates, **84** and **85**, were further converted to orthogonally protected dihydroxyaldehydes **87**, **88** and **89** by a simple three step reaction sequence involving etherification at C13-OH, desilylation at C10 and oxidation of the free hydroxyl group. Our efforts to prepare these substrates were in vain, however, since none of the three aldehydes gave acceptable yields of a C1-C16 allylic alcohol in their

NHK coupling reactions with vinyl bromide **62**. Indeed, treatment of nitrobenzoate **87** with **62** and CrCl$_2$/NiCl$_2$ resulted in complete annihilation of the starting materials. In this case, it is almost certain that the nitroarene in **87** was reduced by the low valent transition metal reagents.

| aldehyde | 87 | 88 | 89 |
| yield in NHK reaction with **62** | 0% | 29% | 28% |

It was hoped that interference from redox behavior in the pivotal NHK coupling steps could be suppressed if easily reducible ester protecting groups were absent from the aldehyde fragment. To this end, acetate **90** was prepared from **ent-26** by Evans-Tishchenko reaction with acetaldehyde, albeit in a disappointingly low yield (45%). The seemingly straightforward task of transforming **90** into an aldehyde **94** suitable for subsequent coupling with **62** serves to illustrate the vexing surprises that can attend "simple" protecting group manipulations. Desilylation of **90** with TBAF was followed by exhaustive silylation with TBSOTf and then mono-desilylation with hydrogen fluoride pyridine complex. These three operations were expected to have taken the course: **90** → **91** → **92** → **93** (Scheme 15). Indeed, a cursory examination of the spectral data obtained for the intermediate compounds seemed to support this sequence of events. However, Dess-Martin oxidation of the final carbinol product from the above set of reactions clearly gave an enone, later determined to be **98**, and not the expected aldehyde **94**. Evidently, the basic fluoride anion of TBAF had initiated acyl migration in which the acetyl moiety of **90** was transferred from its original site to the newly exposed primary hydroxyl group at C10. If this is an intramolecular transfer, as we believe, it must proceed through a rather unusual nine-membered transition state. Thus, the true product of TBAF-mediated desilylation of **90**, acetate **95**, had actually been advanced to **97** by way of **96** and thence to enone **98**.

From the foregoing results, it can be concluded that our plan to employ a differentially protected C10-C16 dihydroxy aldehyde in the synthesis of

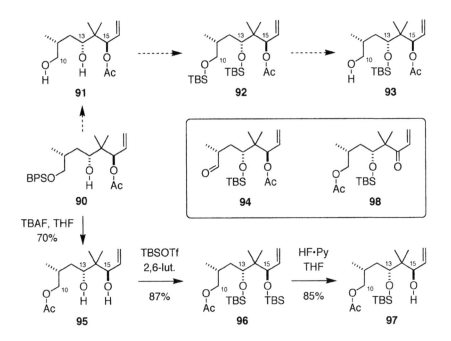

SCHEME 15. The unexpected course of a protecting group manipulation sequence.

the aglycon was only partially successful. The aldehydes that were most easily prepared, benzoates **87** to **89**, turned out to be largely incompatible with the previously validated NHK assembly protocol. Conversely, the attempted preparation of an aldehyde that might offer compatibility with this pivotal transformation was thwarted by an unanticipated rearrangement. Although these problems were not insurmountable, it was felt that any plausible solution would necessitate a circuitous sequence of protection and deprotection steps. The prospect of lengthening our route to polycavernoside A, particularly with such a tedious series of operations, was not an attractive one so, in a moment of desperation, we decided to throw caution to the winds and abandon the selective hydroxyl protection idea altogether. In this way, we moved forward to consider a much bolder strategy, hoping that we could count on the axiom "fortune favors the brave."

X. Regioselective Macrolactonization of a Trihydroxycarboxylic Acid

In retrospect, our design for a seco acid precursor to the aglycon core of polycavernoside A in which the C13 hydroxyl group was protected while that at C15 was free to participate in lactonization, was unnecessarily complicated. The new strategy was simple, perhaps naïve, but we would forgo the use of protecting groups for hydroxyl functions at C13 and C15 and wager on a seco acid precursor undergoing regioselective macrolactonization at the C15-OH. In fact, we took this idea one stage further and prepared a trihydroxy acid, (10*S*)-99, from one of the two easily separable epimers of NHK adduct 63. With three unprotected hydroxyl groups, (10*S*)-99 could conceivably lactonize to give any combination of 9-, 12-, or 14-membered lactones, but cognizant of the fact that a preponderance of the natural macrolides are 14- and 16-membered lactones,[43] we gambled on there being a propensity for 99 to close to the largest of the three possible cycles. Of course, nature fabricates these structures under enzymic control; nevertheless, there is evidence to suggest their seco acid precursors may possess an intrinsic bias for forming these particular ring sizes.

The Yamaguchi-Yonemitsu protocol (Scheme 16) is arguably one of the most versatile methods for macrolactonization and this was selected for the cyclodehydration of (10*S*)-99.[44] Following the prescribed recipe of Yamaguchi, trihydroxy acid (10*S*)-99 was first treated with 2,4,6-trichlorobenzoyl chloride (100) in the presence of triethylamine and the resulting mixed anhydride was transferred *via* a syringe pump into a refluxing toluene solution of the catalyst DMAP (102). This second stage of the process ensures that a competent acyl donor, an acyl pyridinium cation (103), is generated at high dilution, thus favoring intramolecular macrocyclization over competing oligomerization pathways. With considerable relief and a little self-congratulation, we found that the lactonization of (10*S*)-99 was completely selective for the 14-membered lactone 104, which was produced in a very acceptable 75% yield. No

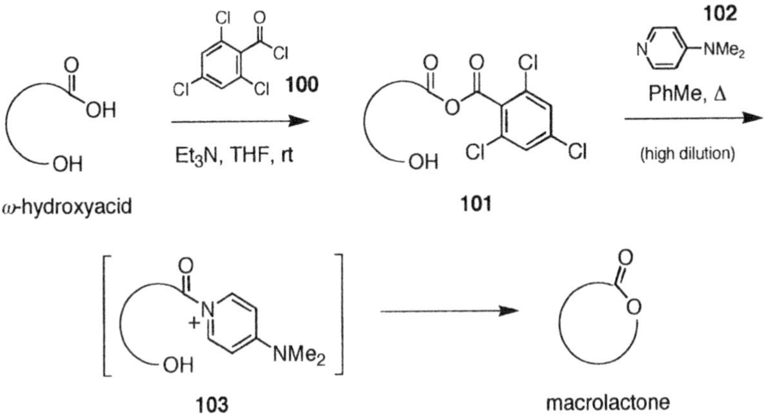

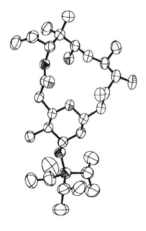

SCHEME 16. The Yamaguchi-Yonemitsu macrolactonization.

trace of 9- and 12-membered lactones resulting from alternative cyclization pathways at the C10 and C13 hydroxyl groups appeared in the course of the reaction. Lactone **104** crystallized as good quality needles, and X-ray crystallographic analysis confirmed its structure in all stereochemical detail (Figure 2).

To gain some insight into the remarkable regioselectivity exhibited by

FIGURE 2. X-ray crystal structure of **104**.

the lactonization of (10*S*)-**99**, a conformational study of the likely pre-lactonization complexes was conducted at the PM3 level of theory (Table

2). Attack on the putative acyl DMAP adduct of (10S)-**99** along the Burgi-Dunitz approach trajectory was considered for each of the hydroxyl groups at C10, C13 and C15. The reactive conformers considered, **107**, **108**, and **109**, represent crude transition state mimics for lactonization and their relative energies should give an indication of $\Delta\Delta G^{\ddagger}$ between the three reaction pathways. The conformer **107** leading to the observed 14-membered lactone **104** was found to be more than 8 kcal mol^{-1} lower in energy than alternative assemblies leading to either 9- or 12-membered lactones (both *re* and *si* face attack on the C1 carboxyl group was considered, but only the lower energy conformers are illustrated).

TABLE 2
Lactonization of (10S)-**99** as Predicted by Molecular Modeling

Product (yield)	**104** (75%)	**105** (0%)	**106** (0%)
Conformation of lactonization precursor	**107**	**108**	**109**
ΔE (kcal mol^{-1})	0.0	9.1	8.8

We were pleased by the good agreement between our experimental observations and our rationale from modeling the lactonization of (10S)-**99**, an outcome that instilled some faith in our theoretical treatment of the problem. However, it remained to be seen whether regioselective lactonization of the epimeric trihydroxy acid (10R)-**99** could also be

TABLE 3
Yamaguchi Lactonization of (10R)-99 and Results of Molecular Modeling

110	111	112
ΔE(kcal mol^{-1}) 0.0	12.1	8.6

achieved. This material was prepared in 87% yield from (10R)-**63**, the remnant epimer from the NHK coupling between **ent-28** and **62**, by an identical route to that used for the synthesis of (10S)-**99** from (10S)-**63**. Modeling studies were carried out as before on the pre-lactonization complex from (10R)-**99**, and the results again indicated that a selective cyclization in favor of the 14-membered lactone should be expected (Table 3). However, lactonization of (10R)-**99** gave only 46% of the 14-membered lactone, **epi-104**, together with 16% of the 12-membered lactone, **epi-105**. No 9-membered lactone was isolated. Thus, although our modeling exercise was clearly not infallible, it had predicted in the case of (10S)-**99** an outcome that was in broad agreement with experiment. Our gamble on regioselective lactonization had paid off and both **104** and **epi-104** could now be regarded promising candidates for further elaboration towards polycavernoside A.

XI. Completion of the Aglycon Core: A Formal Synthesis of Polycavernoside A

We now had what seemed to be an excellent opportunity to move forward with the products of macrolactonization, **104** and **epi-104**, towards an aglycon fragment of polycavernoside A suitable for attaching the disaccharide at C5-OH and installing the conjugated triene at C16. This task necessitated a series of functional group adjustments: ketone

formation was required at C10 and the alkenes serving as latent carbonyl moieties at C9 and C16 needed to be cleaved by oxidative scission. The most direct route to achieving the desired overall transformation met with an immediate snag. Selective oxidation of the allylic alcohol in **104** with barium manganate gave enone **113** in good yield, but this keto alcohol refused to be coaxed into forming a hemiketal. To make matters worse, double ozonolysis of **113** gave a transient tricarbonyl compound in which the hydroxyl group at C13 preferably engaged in cyclization with the aldehyde generated at C16 rather than with either of the ketone carbonyls at C9 and C10. Once more we seemed to have entered a cul-de-sac since there would be obvious difficulties in differentiating between the masked aldehyde at C16 and the highly reactive 1,2-diketone moiety in diketo lactol **114**.

Fortunately, a brief detour from **104** enabled us to circumvent this problem, which was remedied by delaying installation of the C10 ketone in favor of first revealing the C9 and C16 carbonyl groups. To this end, diol **104** was exhaustively silylated and the resulting bis-silyl ether was ozonolyzed to give keto aldehyde **115** in 77% overall yield (Scheme 17). Differentiation of the ester, ketone, and aldehyde carbonyl groups contained within **115** does not present selectivity issues and this intermediate was regarded as an ideal platform from which to launch completion of our synthesis of the aglycon.

At the time we reached this advanced stage of our work, the research groups of Murai and Paquette had already completed their own syntheses of polycavernoside A.[5,6] Both groups had independently chosen the same aglycon core en route to polycavernoside A and each had successfully demonstrated the conversion of their compound, vinyl iodide **117**, to the

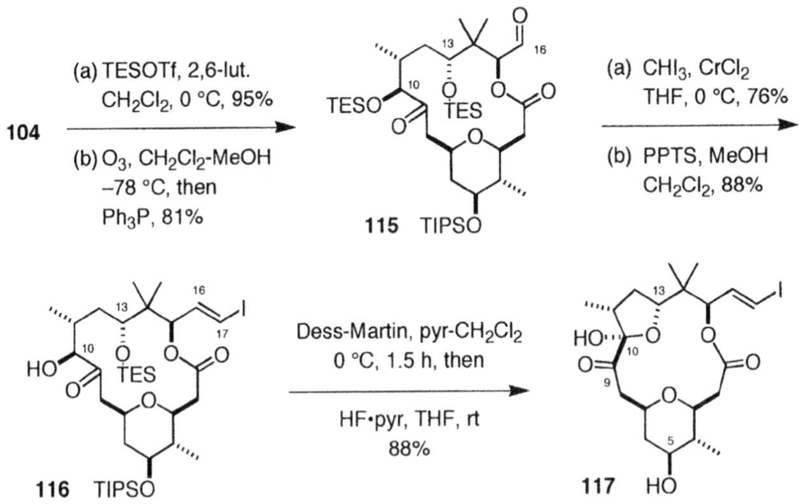

104

(a) TESOTf, 2,6-lut.
CH$_2$Cl$_2$, 0 °C, 95%

(b) O$_3$, CH$_2$Cl$_2$-MeOH
−78 °C, then
Ph$_3$P, 81%

115   TIPSO

(a) CHI$_3$, CrCl$_2$
THF, 0 °C, 76%

(b) PPTS, MeOH
CH$_2$Cl$_2$, 88%

116   TIPSO

Dess-Martin, pyr-CH$_2$Cl$_2$
0 °C, 1.5 h, then

HF·pyr, THF, rt
88%

117   HO

SCHEME 17. Synthesis of the Murai-Paquette iodide **117**.

marine toxin. Our competitors' end-game strategy was undeniably an attractive one (*vide infra*) and we elected to first complete a formal synthesis of polycavernoside A by merging our aldehyde **115** with the Murai-Paquette iodide **117**. The desired convergence was achieved in just three steps beginning with a Takai reaction between **115** and the geminal dichromium reagent generated *in situ* from iodoform and chromous chloride.[45] The C10 hydroxyl group of the resulting *trans* vinyl iodide was unmasked selectively with acidic methanol to afford hydroxy ketone **116** in 67% overall yield from **115**. Finally, alcohol **116** was converted transiently to an unstable 1,2-diketone with Dess-Martin periodinane before global desilylation was effected by the direct addition of hydrogen fluoride-pyridine complex to the reaction. This profitable one-pot process delivered the target iodide **117** in 88% yield from **116**. The same sequence of five transformations, when applied to **epi-104**, also gave **117** as expected. Spectral data (IR, ^{1}H NMR, and ^{13}C NMR) for **117** were in complete agreement with those previously reported,[5,6] and our optical rotation measurement, $[\alpha]_D^{22}$ −36 (c 0.09, CHCl$_3$), matched closely with that obtained by Murai, $[\alpha]_D^{22}$ −38 (c 0.12, CHCl$_3$).[5]

Our independent synthesis of iodide **117** provided additional support to Murai and Yasumoto's assignment of structure for the aglycon of polycavernoside A, which was based largely upon NOE data.[5] The perfect correspondence between material prepared by our route and that previously obtained by Paquette and Murai was significant since the stereochemical attribution made to our lactone **104** had been firmly secured by an X-ray crystallographic analysis. The assignment of absolute configuration to the stereogenic centers at C3, C5, C7, C11, C13, and C15 within the aglycon of polycavernoside A was now irrefutable. While it is true we were somewhat late arriving at **117**, we derived some satisfaction from the fact that our synthesis of this compound, which can also be used to make bioactive analogs of the polycavernosides,[46] was both significantly shorter and higher yielding overall than the previously reported preparations (Table 4). The superior efficiency of our route can be attributed in large part to our dispensing with protecting group chemistry associated with macrolactonization.

TABLE 4
Summary of Syntheses of Polycavernoside A of Aglycon Fragment 117

	Murai	Paquette	White
No. of steps to a C1-C9 fragment	21	16	14
No. of steps to a C10-C16 fragment	21	18	14
Longest linear sequence to 117	32	29	22
Overall yield of 117	2.6%	1.5%	4.7%

XII. Coup de Grace: The Realization of Polycavernoside A

With the aglycon subunit **117** in hand, our synthesis of polycavernoside A was for all practical purposes complete. Nevertheless, for aesthetic reasons as well as reassurance that no hidden obstacles lurked in the attachment of the glycoside unit or in elaboration of the triene side chain, we decided to advance **117** to the natural product itself. We carried out this conversion along very similar lines to those used by both Murai and Paquette in their syntheses of polycavernoside A.[5,6,46] Thus, activation of glycosyl donor **118** with *N*-bromosuccinimide in the presence of aglycon acceptor **117** gave the desired β-glycoside as a single anomer, albeit in low yield. Subsequent removal of the benzyl ether protecting group from the disaccharide moiety was a prelude to the final step, Stille coupling of dienylstannane **120**[6] with vinyl iodide **119**. The latter pair of transformations proceeded without difficulty, and we had in our possession at last a small quantity of the elusive toxin that had resulted in such dire consequences ten years earlier in Guam. The polycavernoside A we prepared in our laboratory was authenticated by comparison of its ^{1}H NMR spectrum with that from natural material.

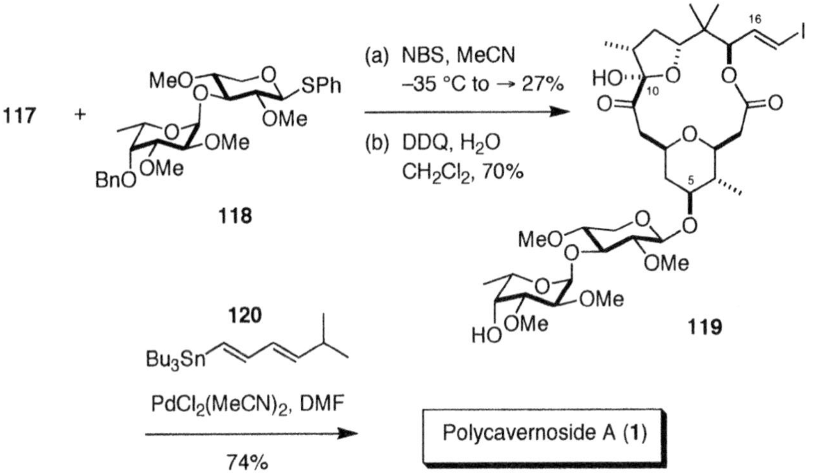

SCHEME 18. Conclusion of the polycavernoside A synthesis by glycosidation and Stille coupling.

XIII. Conclusion

Despite a 30-year history in our laboratory of tangling with the problems and perplexities of macrolide synthesis, it is clear from the foregoing account that we have not yet learned everything there is to know about the field. The polycavernoside synthesis not only produced obstacles we had not anticipated, but it also presented issues that we had not previously confronted in any of our macrolide work. One of those issues, the challenge of orchestrating macrolactonization of a seco carboxylic acid so that it could lead only to a 14-membered ring and no other came close to defeating us. In the end, a "let-the-chips-fall-where-they-may" attitude paid dividends beyond our best expectations; the lesson learned was that an overindulgence in protecting groups can work to the detriment of a sound plan. We readily confess we had not predicted the high level of selectivity seen in this key step, nor were we confident that some other nefarious process would not intervene in the putative lactonization of our trihydroxy acid. An after-the-fact conformational study of the seco acid precursor to lactonization using a molecular mechanics computation provides a plausible rationale for the outcome, but we hasten to add that we have been deceived in other circumstances by computations that had led us to believe we could predict a reaction pathway when in fact we were well wide of the mark.

Perhaps the most enduring lesson to be learned from our polycavernoside effort is that a key ingredient for success in complex synthesis is a flexible mindset. Of course, there must be collateral flexibility within the logistical framework of the synthesis so that it can accommodate the inevitable revisions that become necessary as the synthesis progresses. This flexibility is evident in our assembly of the two major subunits of the polycavernoside aglycon. With the failure of our first plan, there was the choice of trying to force a stubborn reaction that was intrinsically unwilling to yield the product we wanted, or of reversing the roles of the two subunits by interchanging electrophilic and nucleophilic partners. We chose the latter option and were rewarded.

The final word on polycavernoside A has probably not been written, either from its perspective as a viciously toxic but now extinct natural product or from the angle of total synthesis. The molecule is one that engenders substantial interest from both sides, and it is virtually certain that a future synthetic chemist will rise to the challenge of finding a different, perhaps more efficient route to the compound than ours.

Acknowledgments

We are indebted to the several co-workers who participated in our polycavernoside A synthesis. Cindy Browder, Jian Hong, Pavel Nagornyy, Lonnie Robarge and Duncan Wardrop made critical experimental contributions, Alex Yokochi determined the X-ray crystal structure of **104**, and Christopher Lincoln carried out the molecular modeling studies. Financial support was provided by the National Institute of General Medical Sciences through grant GM-50574.

References and Footnotes

1. Haddock, R. L., Cruz, O. L. T., *The Lancet* **1991**, 338, 195.
2. Yotsu-Yamashita, M., Haddock, R. L., Yasumoto, T., *J. Am. Chem. Soc.* **1993**, 115, 1147.
3. Yotsu-Yamashita, M., Seki, T., Paul, V. J., Naoki, H., Yasumoto, T., *Tetrahedron Lett.* **1995**, 36, 5563.
4. White, J. D., Blakemore, P. R., Browder, C. C., Hong, J., Lincoln, C. M., Nagornyy, P. A., Robarge, L. A., Wardrop, D. J., *J. Am. Chem. Soc.* **2001**, 123, 8593.
5. Fujiwara, K., Murai, A., Yotsu-Yamashita, M., Yasumoto, T., *J. Am. Chem. Soc.* **1998**, 120, 10770.

6. Paquette, L. A., Barriault, L., Pissarnitski, D., Johnston, J. N., *J. Am. Chem. Soc.* **2000**, 122, 619.
7. Paquette, L. A., *Chemtracts* **2002**, 15, 345.
8. Yotsu-Yamashita, M., Yasumoto, T., Haddock, R. L. In *Conference Proceedings of the 34th Symposium on the Chemistry of Natural Products*, Tokyo, 1992.
9. Walker, K. A. M., *Tetrahedron Lett.* **1977**, 4475.
10. Mancuso, A. J., Huang, S.-L., Swern, D., *J. Org. Chem.* **1978**, 43, 2480.
11. Lavallée, P., Ruel, R., Grenier, L., Bissonnette, M., *Tetrahedron Lett.* **1986**, 27, 679.
12. Dess, D. B., Martin, J. C., *J. Am. Chem. Soc.* **1991**, 113, 7277.
13. Molander, G. A., Hahn, G., *J. Org. Chem.* **1986**, 51, 1135.
14. Evans, D. A., Chapman, K. T., Carreira, E. M., *J. Am. Chem. Soc.* **1988**, 110, 3560.
15. Boivin, T. L. B., *Tetrahedron* **1987**, 43, 3309.
16. Semmelhack, M. F., Kim, C., Zhang, N., Bodurow, C., Sanner, M., Dobler, W., Meier, M., *Pure & Appl. Chem.* **1990**, 62, 2035.
17. Semmelhack, M. F., Zhang, N., *J. Org. Chem.* **1989**, 54, 4483.
18. McCormick, M., Monahan, R., Soria, J., Goldsmith, D., Liotta, D., *J. Org. Chem.* **1989**, 54, 4485.
19. White, J. D., Hong, J., Robarge, L. A., *Tetrahedron Lett.* **1999**, 40, 1463.
20. Noyori, R., Ohkuma, T., Kitamura, M., Takaya, H., Sayo, N., Kumobayashi, H., Akutagawa, S., *J. Am. Chem. Soc.* **1987**, 109, 5856.
21. Loubinoux, B., Sinnes, J.-L., O'Sullivan, A. C., Winkler, T., *Tetrahedron* **1995**, 51, 3549.
22. Brown, H. C., Bhat, K. S., *J. Am. Chem. Soc.* **1986**, 108, 5919.
23. White, J. D., Kranemann, C. L., Kuntiyong, P., *Org. Lett.* **2001**, 3, 4003.
24. Gage, J. R., Evans, D. A., In *Organic Syntheses*, John Wiley & Sons: New York 1993; Collect. Vol. 8, p 339.
25. Levin, J. I., Turos, E., Weinreb, S. M., *Synth. Commun.* **1982**, 12, 989.
26. Still, W. C., Gennari, C., *Tetrahedron Lett.* **1983**, 24, 4405.
27. Fujiwara, K., Amano, S., Oka, T., Murai, A., *Chem. Lett.* **1994**, 2147.
28. Banwell, M. G., Bui, C. T., Pham, H. T. T., Simpson, G. W., *J. Chem. Soc., Perkin Trans. 1* **1996**, 967.
29. Corey, E. J., Seebach, D., *Angew. Chem. Int. Ed. Engl.* **1965**, 4, 1075.
30. A. B. Smith, I.; Condon, S. M.; McCauley, J. A.; J. L. Leazer, J.; Leahy, J. W.; R. E. Maleczka, J., *J. Am. Chem. Soc.* **1997**, 119, 947.
31. Fujiwara, K., Amano, S., Murai, A., *Chem. Lett.* **1995**, 191.
32. Fujiwara, K., Amano, S., Murai, A., *Chem. Lett.* **1995**, 855.
33. Fürstner, A., *Chem. Rev.* **1999**, 99, 991.
34. Kozikowski, A. P., Stein, P. D., *J. Org. Chem.* **1984**, 49, 2301.
35. Ohira, S., *Synth. Commun.* **1989**, 19, 561.
36. Hara, S., Dojo, H., Takinami, S., Suzuki, A., *Tetrahedron Lett.* **1983**, 24, 731.
37. Okude, Y., Hirano, S., Hiyama, T., Nozaki, H., *J. Am. Chem. Soc.* **1977**, 99, 3179.
38. Takai, K., Kimura, K., Kuroda, T., Hiyama, T., Nozaki, H., *Tetrahedron Lett.* **1983**, 24, 5281.
39. Takai, K., Tagashira, M., Kuroda, T., Oshima, K., Utimoto, K., Nozaki, H., *J. Am. Chem. Soc.* **1986**, 108, 6048.
40. Jin, H., Uenishi, J.-I., Christ, W. J., Kishi, Y., *J. Am. Chem. Soc.* **1986**, 108, 5644.

41. Armstrong, R. W., Beau, J.-M., Cheon, S. H., Christ, W. J., Fujioka, H., Ham, W.-H., Hawkins, L. D., Jin, H., Kang, S. H., Kishi, Y., Martinelli, M. J., McWhorter, W. W., Mizuno, M., Nakata, M., Stutz, A. E., Talamas, F. X., Taniguchi, M., Tino, J. A., Ueda, K., Uenishi, J.-I., White, J. B., Yonaga, M., *J. Am. Chem. Soc.* **1989**, 111, 7525.
42. Evans, D. A., Hoveyda, A. H., *J. Am. Chem. Soc.* **1990**, 112, 6447.
43. Norcross, R. D., Paterson, I., *Chem. Rev.* **1995**, 95, 2041.
44. Inanaga, J., Hirata, K., Saeki, H., Katsuki, T., Yamaguchi, M., *Bull. Chem. Soc. Jpn.* **1979**, 52, 1989.
45. Takai, K., Nitta, K., Utimoto, K., *J. Am. Chem. Soc.* **1986**, 108, 7408.
46 Barriault, L., Boulet, S. L., Fujiwara, K., Murai, A., Paquette, L. A., Yotsu-Yamashita, M., *Bioorg. Med. Chem. Lett.* **1999**, 9, 2069.

STRATEGIES AND TACTICS IN ORGANIC SYNTHESIS, VOL. 6

Chapter 7

FIRST TOTAL SYNTHESIS OF SEVERAL NATURAL PRODUCTS BASED ON ALKYNE-Co₂(CO)₆ COMPLEXES

Chisato Mukai
Division of Pharmaceutical Sciences, Graduate School of Natural Science and Technology, Kanazawa University
Kakuma-machi, Kanazawa 920-1192, Japan

I. Introduction

The alkyne-Co₂(CO)₆ complexes **1** are the binuclear cluster complexes of the acetylenic derivatives with the hexacarbonyldicobalt moiety.[1] These complexes can be readily prepared by treatment of alkynes with commercially available octacarbonyldicobalt [Co₂(CO)₈] and can regenerate the parent triple bond functionality under some mild oxidation conditions. Two synthetically very useful reactions have so far been developed by taking advantage of the characteristic properties of the alkyne-Co₂(CO)₆ complexes **1**; one is so-called Nicholas reaction[2] and the other is so-called Pauson-Khand reaction (Scheme 1).[3] The alkyne-Co₂(CO)₆ complexes **1** possessing a hydroxyl group or its equivalent at carbon β- to alkyne moiety (propargyl alcohol derivatives) could easily provide the corresponding stable carbocationic charge at propargyl position upon treatment with acid. This cationic species **2** can be captured by various nucleophiles resulting in, after decomplexation, a novel propargylation without the formation of allenic by-products (Nicholas reaction). On the other hand, the Pauson-Khand reaction is regarded as a formal [2+2+1] cyclization of alkyne-$Co_2(CO)_6$ complexes **1**, alkene, and carbon monoxide, leading to cyclopentenone frameworks **3** in one operation. In particular, the intramolecular version of this intriguing reaction is now well recognized as one of the most reliable methods for construction of bicyclo[3.3.0]octenone as well as bicyclo[4.3.0]nonenone derivatives.

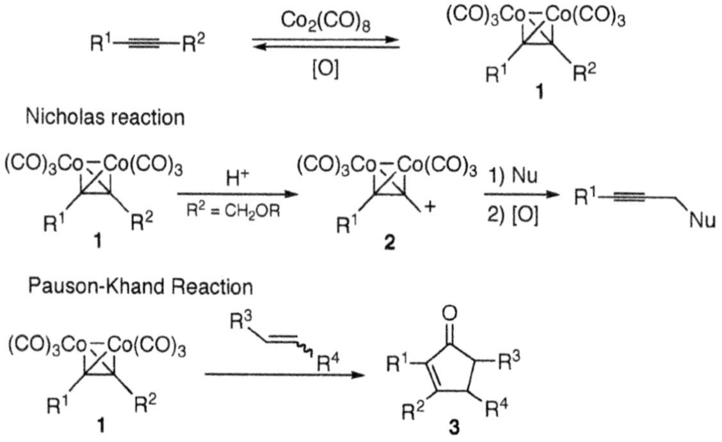

SCHEME 1

In this account, we describe three first total synthesis of natural products, (+)-secosyrins 1 and 2,[4] (-)-ichthyothereol,[5] and (±)-8β-hydroxystreptazolone[6] by making the most of the significant features of the alkyne-$Co_2(CO)_6$ complexes **1** [propargyl cation stabilizing ability (Nicholas reaction), formal [2+2+1] cycloaddition (Pauson-Khand reaction), and their inherent steric bulkiness).

II. First Total Synthesis of (+)-Secosyrins 1 and 2

In 1993, syringolides 1 and 2 (**4** and **5**), novel nonproteinaceous low molecular weight metabolites, were isolated from *Pseudomonas syringae* pv. Tomato (Figure 1).[7] These oxygen-rich tricyclic compounds **4** and **5** have been found to be produced by bacteria expressing avirulence gene D and to elicit a hypersensitive reaction in soybean plants carrying the resistance gene Rpg4. Because of biological interest as well as the intriguing oxygen-rich tricyclic framework of these elicitors, several reports[8] on the total synthesis of these oxygen-rich tricyclic natural products 4 and 5 have been recorded. Furthermore, two years after the first isolation of 4 and 5, Sims and co-workers[9] reported the isolation and structure elucidation of the additional four related natural products having the simpler oxacycles, two dioxaspiro derivatives, secosyrins 1 and 2 (6 and 7), and two butenolide derivatives, syributins 1 and 2 (8 and 9). Although the newly isolated four natural products 6-9 are not active elicitors, in sharp contrast to structurally more complex syringolides 4 and 5, isolation of the former is particularly of interest since they would provide some information and suggestions for understanding the biosynthetic pathway of the entire family of compounds.

The paper[9] from Sims' laboratories in 1995 prompted us to investigate a total synthesis of secosyrins 1 and 2 (6 and 7),[4] because we envisaged that secosyrins 1 and 2 (6 and 7) could be important synthetic intermediates, which might be transformed into more complex syringolides 1 and 2 (4 and 5) through intramolecular aldol-type condensation, according to the retrobiosynthetic process proposed by Sims.[9] In addition, it might be anticipated that intramolecular Michael-type reaction of the deacylated congener of syributins 1 and 2 (8 and 9) affords secosyrins 1 and 2 (6 and 7). Thus, by taking synthesis of all types of β-lactone natural products 4-9 into account according to the above-mentioned scenario, the simplest syributins 1 and 2 (8 and 9) were chosen as the first our target molecules.

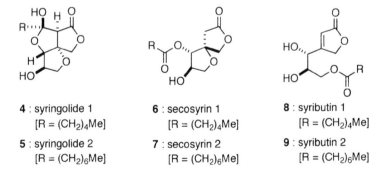

4 : syringolide 1
[R = (CH₂)₄Me]

5 : syringolide 2
[R = (CH₂)₆Me]

6 : secosyrin 1
[R = (CH₂)₄Me]

7 : secosyrin 2
[R = (CH₂)₆Me]

8 : syributin 1
[R = (CH₂)₄Me]

9 : syributin 2
[R = (CH₂)₆Me]

FIGURE 1. Oxygen-rich natural products isolated from *P. syringae* pv. tomato.

A. TOTAL SYNTHESIS OF (+)-SYRIBUTINS 1 AND 2

The simple retrosynthetic analysis of **8** and **9** indicated that two chiral centers of D-tartrate must be incorporated into the target natural products. Thus, diisopropyl D-tartrate became the common starting material for this investigation.

The acetonide **10**,[10] derived from diisopropyl D-tartrate, was oxidized under Swern conditions to give the aldehyde, which was subsequently exposed to a Horner-Emmons reaction with ethyl (diethylphosphono)-acetate producing (*E*)-ethyl ester derivative **11** in 85% yield (Scheme 2). Michael reaction of **11** with nitromethane as the C_1-unit in the presence of diazabicyclo[5.4.0]undec-7-ene (DBU)[11] furnished **12** in 80% yield. Compound **12** could be obtained as a single stereoisomer and be free from other diastereoisomer. However, the newly generated stereogenic center at the β position of **12** was not determined, since it would be disappear in a later step. Transformation of nitromethane group of **12** to an aldehyde functionality was realized under Nef conditions (KOH, $KMnO_4$)[12] and subsequent treatment of the resulting aldehyde derivative **13** with sodium borohydride ($NaBH_4$) effected reduction and spontaneous lactone formation leading to the β-lactone derivative **14** in 72% yield from **12** as a single stereoisomer.

Introduction of double bond at the β,β positions of **14** was the next subject. We first attempted conventional selenoxide chemistry[13] for conversion of **14** into **15**. Thus, treatment of **14** with lithium bis(trimethylsilyl)amide (LHMDS) and phenylselenyl chloride gave the corresponding β-phenylselenylated derivative, which was then oxidized with aqueous hydrogen peroxide to afford the desired **15** in rather lower yield (40%). Alternatively and more efficiently, **15** was obtained in 74% yield when **14** was first converted into the corresponding trimethylsilyl (TMS) enol ether derivative (LHMDS/TMSCl), which was subsequently exposed to palladium diacetate [$Pd(OAc)_2$] and benzoquinone.[14] The final step required for the synthesis of the target natural products **8** and **9** was deprotection and acylation. Thus, desilylation of **15** with tetra-*n*-butylammonium fluoride (TBAF) and hydrofluoric acid gave the primary alcohol **16**, which was acylated with hexanoyl chloride and octanoyl chloride to provide **17** and **18** in 81% and 85% yield, respectively. Finally (+)-syributin 1 (**8**)[4b] was obtained in 72% yield from **17** under acidic conditions. Similar treatment of **18** gave (+)-syributin 2 (**9**)[4b] in 69% yield. During our investigation on total synthesis of these compounds, Honda[8f] completed the first total synthesis of (+)-syributin 1 (**8**) based on

Sharpless asymmetric dihydroxylation of an alkenyl butenolide derivative.

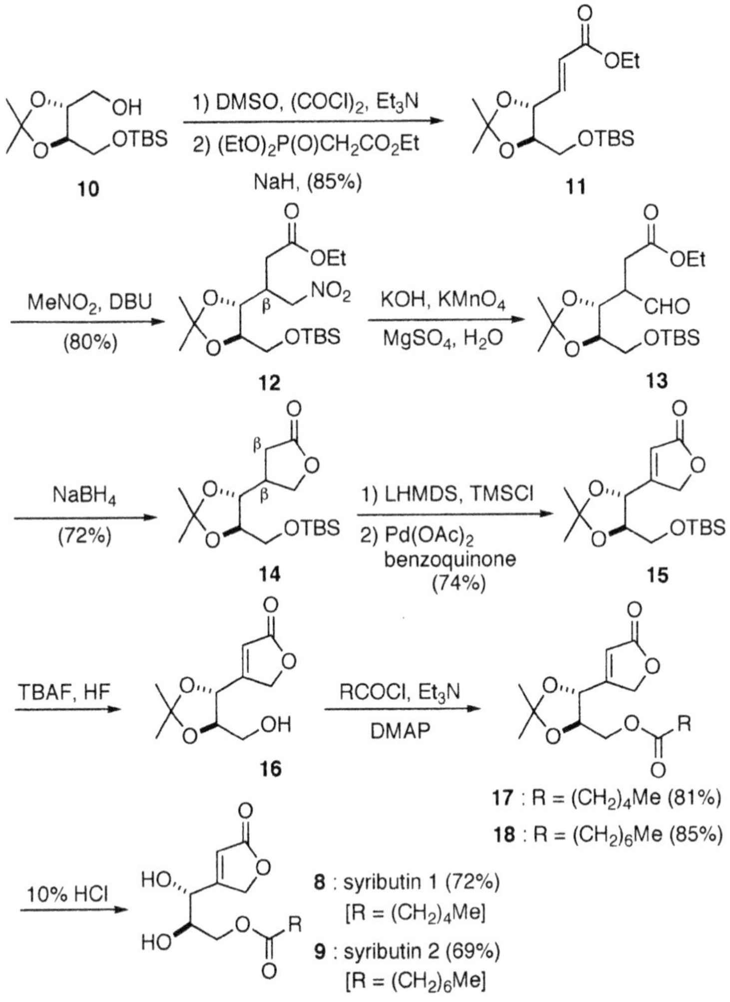

SCHEME 2

Since the simplest target molecules, syributins 1 and 2 (**8** and **9**), could be easily prepared, we next investigated transformation of the simplest monocyclic butenolide derivatives into the more complex dioxaspiro skeletons according to the original scenario. Acetonide and silyl protecting groups on the three hydroxyl groups of the butenolide derivative **15** were removed by exposure to 10% hydrochloric acid in methanol to give **19** in 90% yield. Several basic conditions (NaH/THF, LHMDS/THF, 'BuOK/'BuOH, 'Pr₂NEt/CH₂Cl₂) as well as acidic conditions (*p*-toluenesulfonic acid/THF, trifluoroacetic acid/THF, zinc iodide/THF) were employed for Michael-type ring-closing reaction of **19** leading to the dioxaspiro framework **20**, a basic skeleton of secosyrins 1 and 2 (**6** and **7**) (Scheme 3). However, the desired product **20** could never be obtained from the reaction mixture. We only observed an intractable reaction mixture or no reaction. We thus discovered that the intramolecular Michael-type reaction of the butenolide **19** to the dioxaspiro compound **20** seemed to be much harder than imagined.

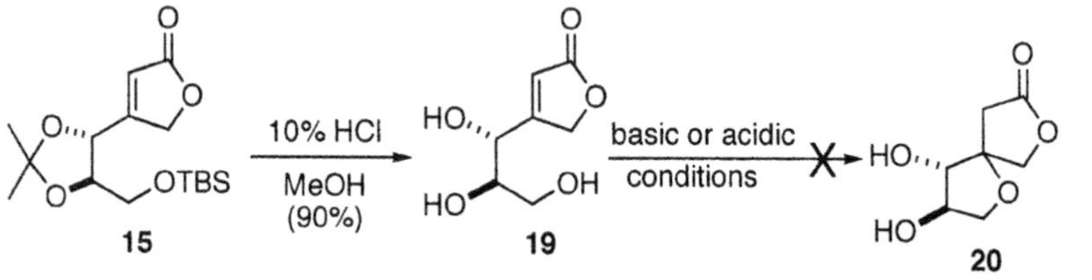

SCHEME 3

The triol derivative **19** was found to be an unsuitable substrate for Michael-type ring-closing reaction. Therefore, we next investigated the direct preparation of secosyrin 1 (**6**) from the hexanoyloxy derivative **21**. Compound **21** was synthesized in 63% yield from **19** by successive acetonide formation[15] and acylation. Exposure of **21** to deacetonization conditions (20% hydrochloric acid/THF), however, was shown to give syributin 1 (**8**) in 99% yield instead of the desired secosyrin 1 (**6**), due to acyl migration (Scheme 4). Deacetonized derivative **22** with hexanoyl moiety on the secondary hydroxyl group could be detected on TLC, but it was never isolated in more than trace quantities. These observations indicated that acyl migration from a secondary hydroxyl group to a primary one must be much faster than the Michael-type ring-closing reaction, resulting in the exclusive formation of **8**. Similar unsuccessful results on conversion of monocyclic butenolides to the corresponding dioxaspiro ones under several conditions were reported by Honda.[8f] Because our first scenario involving transformation of simpler butenolide natural products into the more complex dioxaspiro compounds was shown to be difficult, we tried to devise an alternative pathway for synthesis of the dioxaspiro natural products, secosyrins 1 and 2 (**6** and **7**).

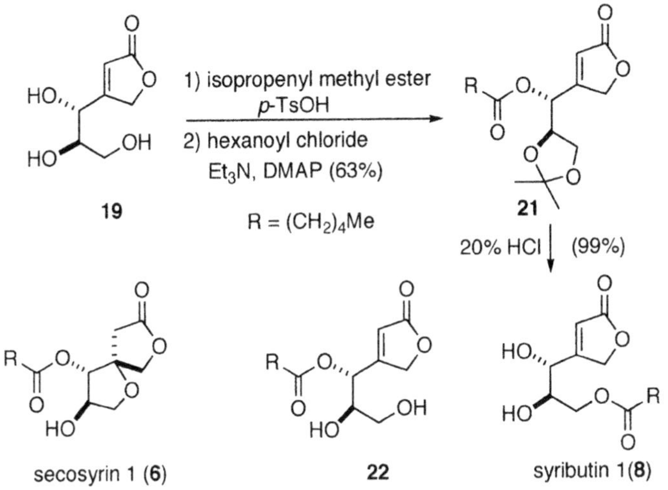

SCHEME 4

B. TOTAL SYNTHESIS OF (+)-SECOSYRINS 1 AND 2

The most significant point for synthesis of secosyrins 1 and 2 (**6** and **7**) must be the stereoselective construction of the quaternary center. Our first attempt at construction of the quaternary center of **6** and **7** by the intramolecular Michael-type reaction of butenolides was fruitless. We anticipated that a compound having an oxygen functionality at the β position such as **23** would react with the nucleophile (C$_2$-unit) via the chelation model transition state **A** (five-membered transient ring) leading to the desired adduct **24**, which has the same stereochemistry as that of the basic skeleton of **6** and **7** (e.g. compound **25**, Scheme 5). However, the possibility of production of the epimer of **24** arising from the transition state **B** (six-membered transient ring) due to chelation with the β oxygen functionality and/or a nonchelating transition state could not be ruled out at this stage.

With the above consideration in mind, we first prepared the ketone derivative **28** with an acetylenic moiety as a C$_1$-unit. The known hydroxyl compound **26**,[16] derived from diisopropyl D-tartrate, was consecutively oxidized under Swern conditions and exposed to lithium phenylacetylide to afford an 80% yield of **27**, which was again oxidized under Swern conditions to provide **28** in 90% yield (Scheme 6). Stereoselective

introduction of the C_2-unit was achieved when **28** was treated with the
lithium enolate derived from *tert*-butyl thioacetate in THF at −78 °C, to
furnish **29** in 94% yield as a sole product.

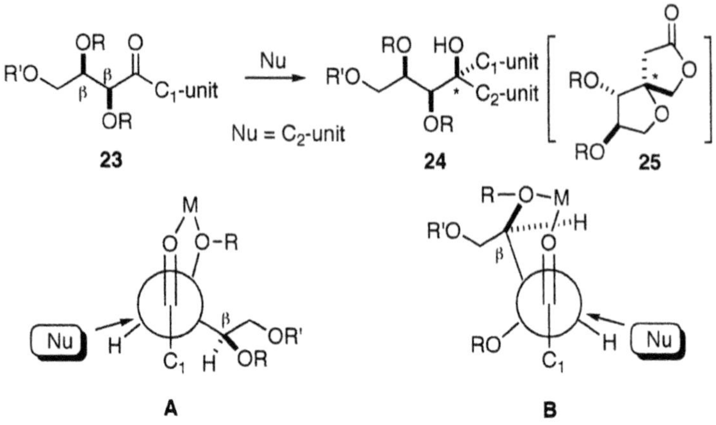

SCHEME 5

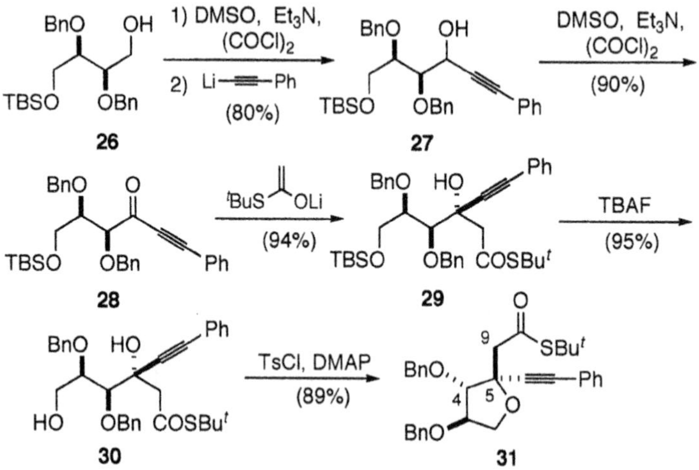

SCHEME 6

We expected that addition of the enolate would have proceeded via
transition state **A** (Scheme 5) and resulted in the formation of our desired
product, although configuration of the newly generated quaternary carbon
center of **29** was uncertain at this stage.

The stereochemistry of **29** was determined by chemical transformation to the tetrahydrofuran derivative **31** by conventional means. Desilylation of **29** with TBAF gave the primary hydroxyl derivative **30** in 95% yield. This was subsequently treated with p-toluenesulfonyl chloride (TsCl), Et₃N, and (dimethylamino)pyridine (DMAP) to afford **31** in 89% yield. An NOE experiment with **31** revealed 6.5% enhancement between C₄-H and C₉-H.[17] This observation strongly indicated that the configuration of the newly constructed quaternary carbon center of **29** was not as same as the target molecules **6** and **7**.

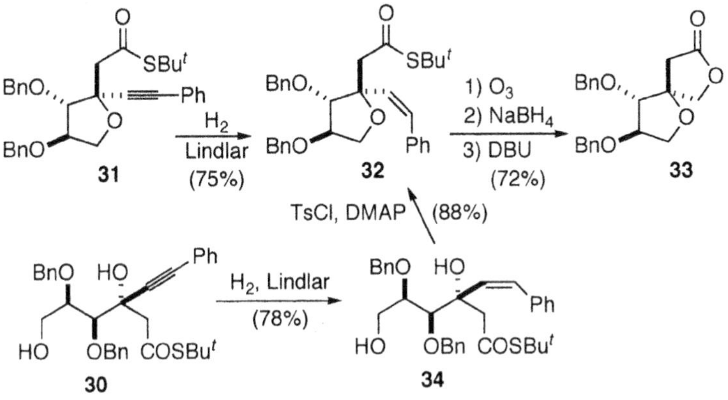

SCHEME 7

In addition, the newly generated quaternary carbon center of **29** was unambiguously established as follows. Hydrogenation of **31** in the presence of Lindlar catalyst in ethyl acetate gave the *cis*-olefin **32** in 75% yield (Scheme 7). Upon successive treatment with ozone, NaBH₄, and DBU, **32** underwent oxidative cleavage of its olefinic portion, reduction, and a ring-closing reaction to afford spiro compound **33** in 72% yield. Compound **32** was also prepared from **30** by half-reduction (**34**, 78%) and a ring-closing reaction (88%). X-ray crystallographic analysis of **33** unambiguously demonstrated its relative and absolute stereochemistry as depicted in Scheme 7, thereby showing that **29** had the undesired stereochemistry at the quaternary carbon.

Contrary to our expectation, the reaction of the ketone derivative **28** with the lithium enolate of *t*-butyl thioacetate exclusively produced the adduct **29**, which possessed the wrong stereochemistry for our purposes (Scheme 6). Thus, we had to develop an efficient way to invert the tertiary hydroxyl group at the propargyl center of **29** or **30**. The Mitsunobu reaction[18] is well known as one of the most reliable methods for inversion of configuration of hydroxyl group. However, in the case of **29**, it seemed to be much difficult to invert the tertiary hydroxyl group at the activated position (propargyl position) in a satisfactory yield. Alternatively, we envisaged that the propargyl cation species,[2] derived from the cobalt-complexed **30**, would be intramolecularly captured by a primary hydroxyl group through transition state **C** rather than the transition state **D**, the former leading to a compound with the desired stereochemistry (Figure 2). Transition state **D** would suffer from serious nonbonding interactions between the fairly bulky alkyne-$Co_2(CO)_6$ portion and benzyloxy functionality. This is not the case for the transition state **C**, where unfavorable but rather weak nonbonding interaction of the benzyloxy group with the thioacetate residue might be the only nonbonding interaction in existence. Formation of the transition state **C**, therefore, would be expected to be preferred over the transition state **D**.

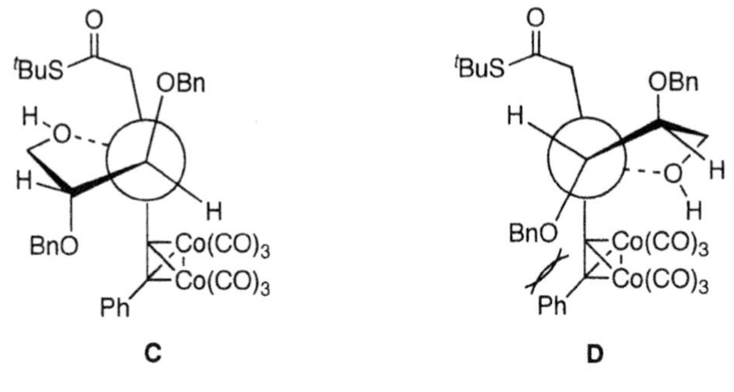

FIGURE 2. Capture of a propargylic cation species by a terminal hydroxyl group.

Treatment of **30** with $Co_2(CO)_8$ in diethyl ether at room temperature gave the corresponding alkyne-$Co_2(CO)_6$ complex **35** in 98% yield (Scheme 8). Exposure of **35** to $BF_3 \cdot OEt_2$ in methylene chloride at room temperature effected ring-closing reaction with inversion of the configuration at the propargyl center to provide tetrahydrofuran derivative **36** and a small amount of its epimer, which were subsequently demetalated by treatment with cerium(IV) ammonium nitrate (CAN) to afford **37** in 74% yield along with its epimer **31** in 15% yield. It should be mentioned that an NOE experiment with **37** showed no enhancement between C_4-H and C_9-H,[17] in sharp contrast to the case of **31** (6.5% enhancement). Thus, efficient inversion of the propargyl carbon center

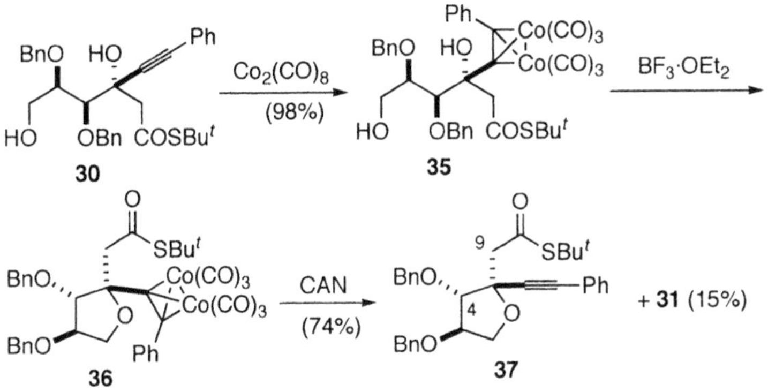

SCHEME 8

with concomitant tetrahydrofuran ring formation was realized.

With the proper stereochemistry set in **37** possessing the requisite quaternary carbon center in hand, the stage was set for completing a first total synthesis of secosyrins 1 and 2 (**6** and **7**). According to the procedure developed for transformation of **31** into **33**, tetrahydrofuran derivative **37** was converted to **39** (75%) through *cis*-olefin derivative **38** (78%) (Scheme 9). Debenzylation of **39** under hydrogenolysis conditions in the presence of 10% Pd-C produced diol **40** in 90% yield.

Selective monoacylation proceeded when **40** was exposed to hexanoic anhydride in THF in the presence of Et$_3$N and DMAP to provide (+)-secosyrin 1 (**6**)[4] in 70% yield along with diacylated compound **41** in 25% yield. By changing the acylating reagent from hexanoic anhydride to octanoic anhydride, the diol **40** produced (+)-secosyrin 2 (**7**)[4] and **42** in 78% and 15% yield, respectively. Similarly, hydrogenolysis of **33** provided a 94% yield of **43**, which was subsequently converted to (+)-5-*epi*-secosyrin 1 (**44**) in 70% yield, together with the corresponding diacylated compound (20%).

In conjunction with our first plan, which involved transformation of the simpler dioxaspiro framework of secosyrins into more complex tricyclic skeleton of syringolides, several basic conditions were screened. Unfortunately, base treatment of secosyrin 1 (**6**) caused easy and quantitative conversion into the simpler butenolide derivative, syributin 1 (**8**), through a retro-Michael-type reaction (via compound **45**) accompanied by migration of the hexanoyl group from the secondary to the primary hydroxyl group (Scheme 10). This result is in good agreement with the biogenetic route proposed by Sims.[7] The desired aldol-type condensation between C_9-position and hexanoyl group was never observed.

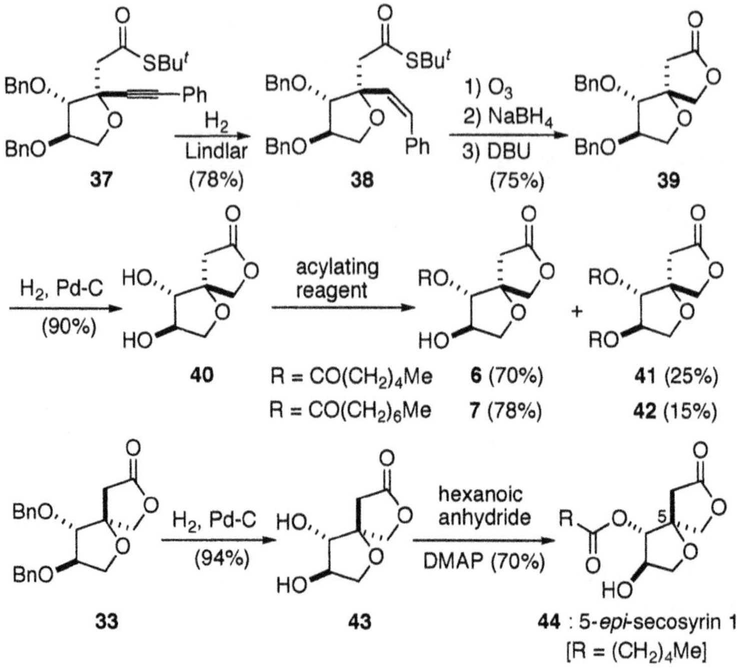

SCHEME 9

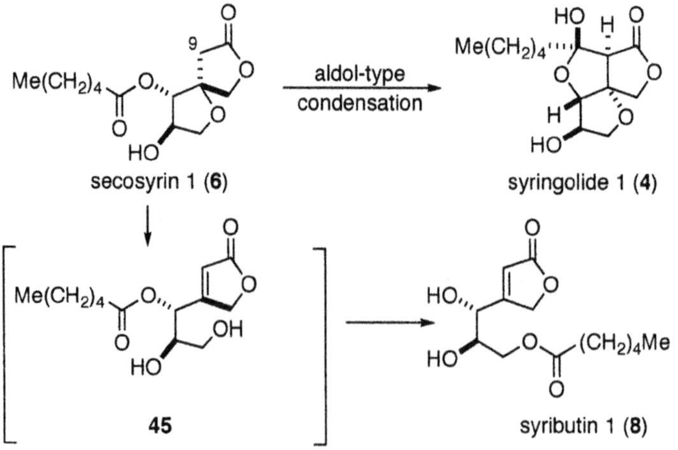

SCHEME 10

Thus, we have succeeded in not only the first total synthesis of (+)-secosyrins 1 and 2[19] but also the total synthesis of (+)-syributins 1 and 2 from a common starting material, diisopropyl D-tartrate. The crucial step for the stereocontrolled total synthesis of (+)-secosyrins 1 and 2 is the stereoselective tetrahydrofuran ring formation by taking advantage of inherent properties of alkyne-$Co_2(CO)_6$ complex. Although direct transformation of simpler syributins into more complex syringolides via secosyrins by a Michael-type ring-closing reaction and aldol-type condensation could not be realized (yet!), the first total syntheses of secosyrins 1 and 2 have unambiguously established their structures.

III. First Total Synthesis of (-)-Ichthyothereol

In 1965, (-)-ichthyothereol (46) and its acetate 47[20,21] (Figure 3) were isolated from the leaves and flowers of *Dahlia coccinea* as well as from the leaves of *Ichthyother terminals*, the latter of which have long been known to be used as a fish poison by the natives of the Lower Amazon Basin.[20,22] The crude extracts of the leaves of *Ichthyother terminals* had been found to be extremely poisonous not only to fish, but also to mammals.[22] The effects in dogs were typically convulsant, similar to those of picrotoxin, indicating bulbar (i.e., affecting the medulla oblongata) action. Minute quantities of either ichthyothereol (46) or its acetate 47 were extremely toxic to the fish *Lebistes reticulatus*, confirming that these triyne derivatives were likely at least in part responsible for the toxicity of the leaves of *Ichthyother terminals*.[20,22] These two compounds were also shown to kill mice[20] when injected

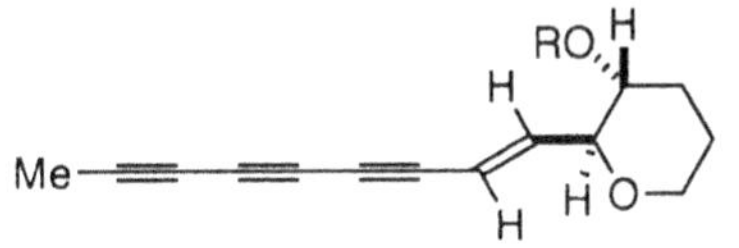

(-)-ichthyothereol (46) : R = H
(+)-ichthyothereol acetate (47): R = Ac

FIGURE 3. Tetrahydropyran natural products from *Ichthyother terminals*.

intraperitoneally in doses of 1 mg in olive oil. The gross structure including the relative stereochemistry of compounds 46 and 47 was determined by ¹H NMR analysis. The absolute configuration of two chiral centers of 46 and 47 was first tentatively deduced on the basis of the optical rotatory dispersion curve of the substituted tetrahydropyran-3-one derivatives prepared by oxidation of perhydroichthyothereol. Finally, the absolute configuration of **46** and **47** was unambiguously established by chemical transformation of **46** into the (-)-bis(2,4-dinitrobenzoate) of *trans*-3-hydroxy-2-hydroxymethyltetahydropyran.[23]

Surprisingly, no publications dealing with the total synthesis of toxic ichthyothereol (**46**) and/or its acetate **47** have yet appeared, despite the fact that more than thirty years have now elapsed since their first isolation in 1965. We describe here the first total synthesis of (-)-ichthyothereol (**46**) and (+)-ichthyothereol acetate (**47**) on the basis of the endo, ring-closing reaction of an alkyne-$Co_2(CO)_6$ complex.

A. RETROSYNTHESIS OF (-)-ICHTHYOTHEREOL

Recent efforts from our laboratories[24] led to a highly stereoselective, as well as stereospecific, novel endo, ring-closing reaction of alkyne-$Co_2(CO)_6$ complexes, leading to the formation of oxacycles. For instance, treatment of alkyne derivatives possessing a *trans*-epoxy group at the β position (*trans*-**48**) with $Co_2(CO)_8$ gave the corresponding alkyne-$Co_2(CO)_6$ derivatives, *trans*-**49**. These were then exposed to $BF_3 \cdot OEt_2$ at $-78°C$ to give the *cis*-tetrahydropyran derivatives, *cis*-**50**, in a highly

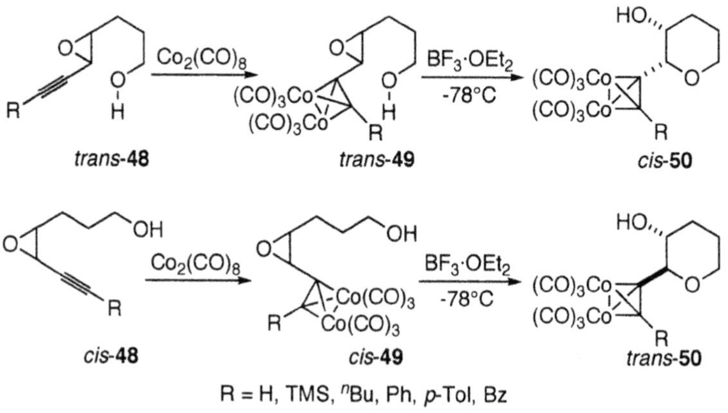

R = H, TMS, ^{n}Bu, Ph, *p*-Tol, Bz

SCHEME 11

stereoselective manner (*cis* : *trans* = 99~91 : 1~9) in high yields (Scheme 11).[24a,b,d] Similar high stereoselectivity was observed when the starting *cis* epoxides (*cis*-**48**) were exposed to the standard ring-closing conditions to provide the corresponding *trans*-tetrahydropyran derivatives, *trans*-**50**, (*cis* : *trans* = 0~3 : 100~97) in high yields via *cis*-**49**. No tetrahydrofuran derivatives (exo products) could be detected in the reaction mixture.[24a,b,d] Thus, this intriguing ring-closing reaction was shown to proceed in an endo fashion with *retention* of the configuration at the propargyl position. It should be mentioned that this method could be applied to the stereoselective preparation of the corresponding tetrahydrofuran congeners.[24c]

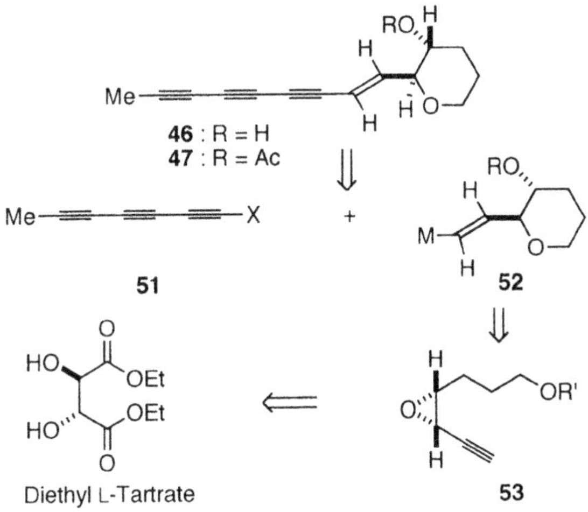

46 : R = H
47 : R = Ac

51

Diethyl L-Tartrate

52

53

SCHEME 12

On the basis of the newly developed method for the stereoselective and stereospecific construction of the 2-ethynyl-3-hydroxytetrahydro-pyran framework, our retrosynthetic analysis of (-)-ichthyothereol (**46**) and its acetate **47** is outlined in Scheme 12. The first carbon-carbon bond disconnection of **46** would be made between the triyne moiety and the (*E*)-olefin part leading to the triyne species **51** and the *trans*-3-hydroxy-2-vinyltetrahydropyran derivative **52**. The latter would be prepared from the optically active *cis*-epoxide **53** through an endo ring-closing reaction of the Co₂(CO)₆-complexed *cis*-epoxy-alkyne compound **53**, which would be obtained by taking advantage of the two contiguous chiral centers of diethyl L-tartrate. Thus, diethyl L-tartrate became the starting material for this program. The triyne counterpart **51** (X=SnBu₃) for the coupling reaction would be obtained from hex-1,4-diyn-3-one by combination of Tykwinski's triyne synthesis[25] and transformation of the silyl group to a stannyl one by Buchwald's procedure.[26]

B. CONSTRUCTION OF OPTICALLY ACTIVE *CIS*-EPOXIDE

The optically active *cis*-epoxide **60**, the key starting material for construction of the carbon framework of **46** and **47**, was prepared as depicted in Scheme 13. The diol **54** was easily obtained from diethyl L-tartrate according to Saito's procedure.[27] Activation of the primary hydroxyl group of **54** by tosylation was followed by base treatment, affording the epoxide **55** in 63% yield. Transformation of **55** into the vinyl derivative **56** was realized as follows. Addition of the acetylide, prepared from trimethylsilylacetylene and butyllithium (^{n}BuLi), to **55** in

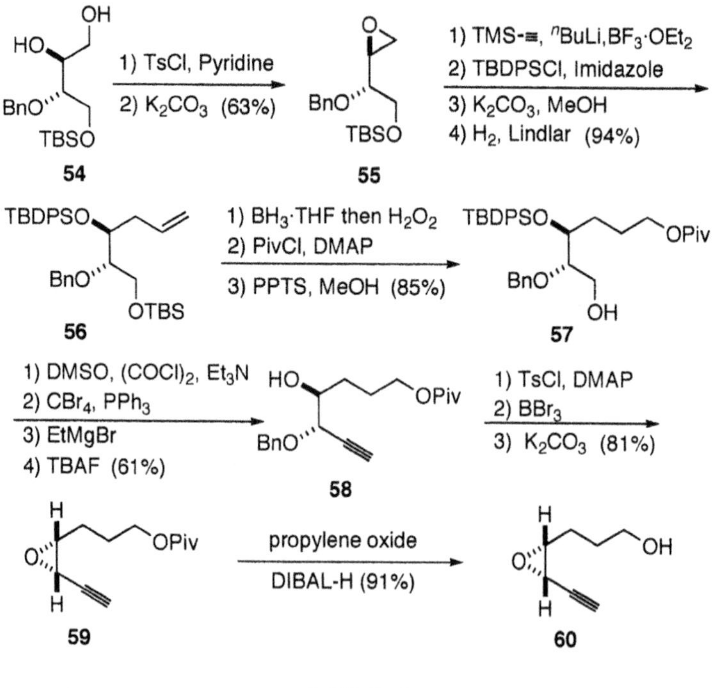

SCHEME 13

the presence of $BF_3 \cdot OEt_2$[28] furnished the homopropargyl alcohol. This was protected with a *tert*-butyldiphenylsilyl (TBDPS) group, and then the terminal TMS group was removed. The resulting alkyne derivative was semi-hydrogenated over a Lindlar catalyst to give **56** in 94% overall yield. Upon successive exposure to $BH_3 \cdot THF$ and hydrogen peroxide, compound **56** underwent hydroboration-oxidation to produce the primary alcohol. Protection of the primary hydroxyl moiety with a pivaloyl group and desilylation under acidic conditions provided **57** in 85% yield. Compound **57** was oxidized under Swern conditions to give the labile aldehyde, which was subsequently exposed to Corey's dibromoolefination conditions[29] and ethylmagnesium bromide[30] to produce, after TBAF treatment, the alkyne derivative **58** in 61% overall yield. Treatment of **58** with TsCl at room temperature and debenzylation with BBr$_3$ at −78°C afforded the hydroxyl compound, epoxidation of which was then accomplished by exposure to K$_2$CO$_3$ in MeOH yielding the epoxy derivative **59** in 81% overall yield.

Conversion of **59** into **60** was somewhat troublesome. Treatment of **59** with ethylmagnesium bromide, methylmagnesium bromide or methyllithium in several kinds of solvents under standard conditions gave an intractable mixture that included the desired **60**. When two equivalents of diisobutylaluminum hydride (DIBAL-H) in THF was employed, **59** gave a rather clean mixture mainly consisting of **60**, the epoxy ring-opened products of **60**, and the starting **59**. We envisaged that DIBAL-H might have reacted with the pivaloyl group faster than with the epoxy moiety of **59** leading to the production of **60**. However, **60** seems susceptible to further reduction by DIBAL-H. Thus, the addition of another epoxy compound in the reaction mixture would avoid the over-reaction of **60**. The epoxy derivative that would be used for the above purpose must be one with less steric hindrance compared to the epoxy moiety of **59** and **60**. In other words, epoxy derivatives more reactive than **59** and **60** toward DIBAL-H would be favorable for this purpose. Therefore, we selected propylene oxide as a scavenger of excess of DIBAL-H. As a result, reaction of **59** with DIBAL-H (5.0 equiv.) in the presence of excess propylene oxide (20 equiv.) [31] at −78°C proceeded very cleanly to provide **60** as the sole isolable product in 91% yield. The substrate for the endo cyclization was thus obtained.

C. TOTAL SYNTHESIS OF (-)-ICHTHYOTHEREOL

According to the procedure[24a-d] for the endo cyclization of the racemic epoxyalkyne derivatives, optically active *cis*-epoxide **60** was converted to the corresponding dicobalthexacarbonyl species, which was then treated with a catalytic amount of BF$_3$·OEt$_2$ in CH$_2$Cl$_2$ at −78°C to afford the *trans*-tetrahydropyran derivative **61** in 87% yield (Scheme 14). Demetalation of **61** with CAN was followed by protection of the secondary hydroxyl functionality with a silyl group to afford **62**[32] in 87% yield. Upon consecutive treatment with tri-*n*-butyltin hydride in the

presence of a palladium catalyst at –78°C and iodine at room temperature,
62 was consecutively hydrostannylated,[33] and iodinated via a stannyl-
iodide exchange reaction,[34] resulting in the formation of the (E)-iodovinyl
derivative **63** in 83% yield.

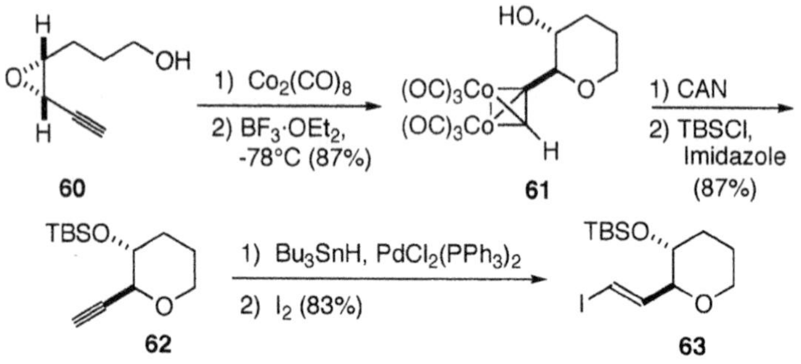

SCHEME 14

Our next endeavors were then directed toward the preparation of
counterpart **66** for the coupling reaction. Trimethylsilylpropynal (**64**) was
reacted with propargylmagnesium bromide to give the alcohol, which was
subsequently oxidized with manganese dioxide. The resulting diynone
derivative was then converted into the dibromoolefin **65** in 45% overall
yield under the standard conditions (Scheme 15).[29] By taking advantage
of the method developed by Tykwinski,[25] **65** was exposed to ^{n}BuLi in
hexane at –78 °C, furnishing 1-trimethylsilyl-1,3,5-heptatriyne (**66**) in
76% yield.

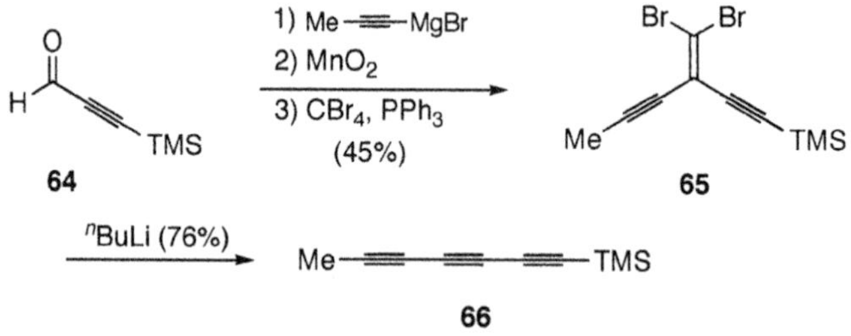

SCHEME 15

With ready access to **66**, we were ready to consider the completion of
the construction of the carbon skeleton of **46** and **47** by the palladium-

catalyzed coupling reaction under Stille reaction conditions. Transmetalation of the silyl group of **66** to the corresponding stannyl one was realized by Buchwald's procedure.[26] Thus, treatment of **66** with bis(tributyltin)oxide in THF in the presence of a catalytic amount of TBAF provided the crude stannyl derivative **67** (Scheme 16). The Stille coupling of this compound with **63** under the standard conditions (5 mol% of PdCl$_2$(PPh$_3$)$_2$ in THF at room temperature)[35] proceeded without difficulty to produce the coupling product **68** in 95% yield. The final phase of this route involved simple chemical modifications of **68**. Desilylation of **68** with TBAF in THF at room temperature afforded (-)-ichthyothereol (**46**)[5] in a quantitative yield. Acetylation of (-)-**46** by the conventional means provided (+)-**47** in 88% yield.

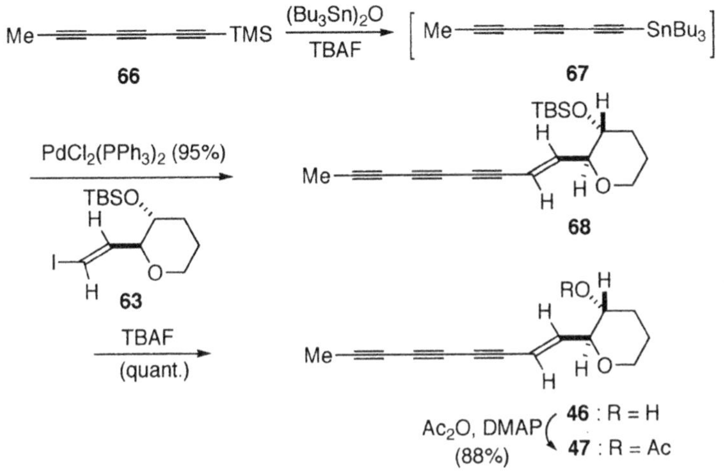

SCHEME 16

We have completed the first total synthesis of (-)-ichthyothereol (**46**) and its acetate **47** by the palladium-catalyzed coupling reaction between the (*E*)-iodoolefin **63** and the triyne derivative **66**. The iodoolefin **63** was prepared from commercially available diethyl L-tartrate through the Co$_2$(CO)$_8$-mediated endo cyclization of the optically active *cis*-epoxy-alkyne derivative **60** in a highly stereoselective manner.

IV. FIRST TOTAL SYNTHESIS OF (±)-8β-HYDROXYSTREPTAZOLONE

Streptazolin (**69**) was first isolated from a culture of *Streptomyces* *viridochromogenes* by Drautz *et al.* in 1981.[36] This lipophilic neutral tricyclic compound has been shown to possess antibiotic and antifungal activities.[37] Tang *et al.*[38] recently reported the isolation of some of the novel and more oxidized streptazolin-related natural products, 8β-hydroxystreptazolone (**70**),[39] 4,12-epoxystreptazolin (**71**),[39] and 9β-hydroxystreptazolin (**72**).[39] In addition to these, the streptazolin-dimer (**73**)[39] together with streptazolin (**69**) and related known compounds[40] as secondary metabolites from *Streptomyces spec.* and *viridochromogenes* were isolated, all via chemical screening (Figure 4).

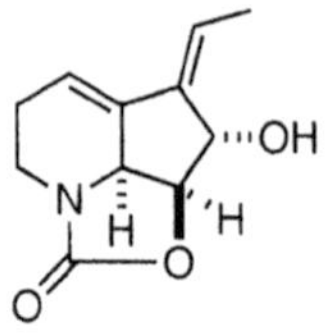

streptazolin (**69**)

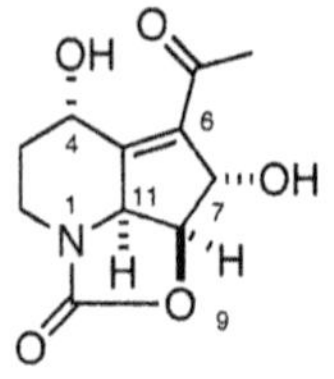

8β-hydroxy-streptazolone (**70**)

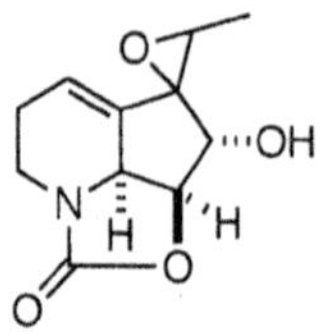

4,12-epoxy-streptazolin (**71**)

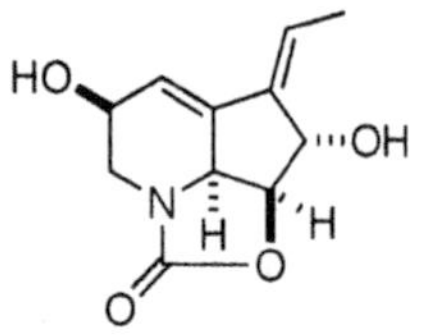

9β-hydroxystreptazolin (**72**)

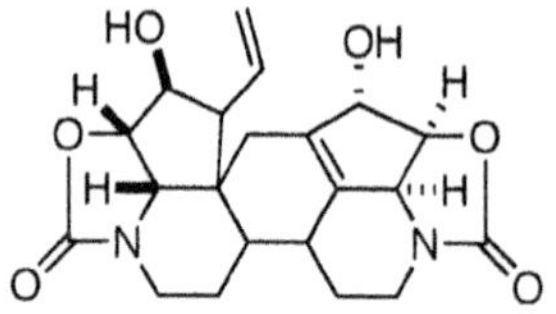

streptazolin-dimer (**73**)

FIGURE 4. Streptazolin and related natural products.

Because of its unique structural features as well as its promising biological activity profile, streptazolin (**69**), having the common basic skeleton of the streptazolin analogues, has been synthesized by four groups.[41-44] The total synthesis of **69** in racemic form was first reported by Kozikowski and Park[41] through the intramolecular nitrile oxide [2+3]-cycloaddition for construction of the azabicyclo[4.3.0] framework. Overman and Flann[42] completed the first enantioselective synthesis of **69** starting from L-tartrate via a ring-closing reaction between an N-acyliminium cation and a vinylsilane species for construction of the azabicyclo[4.3.0] framework. Kibayashi and co-workers[43] also reported an enantioselective synthesis of **69** starting from L-tartrate by taking advantage of a palladium-mediated ring-closing reaction. Comins and Huang[44] recently revealed an asymmetric synthesis of **69** based on a chiral auxiliary-mediated asymmetric preparation of a dihydropyridone derivative, followed by palladium-mediated construction of the tricyclic framework. All these syntheses of streptazolin (**69**) used a stepwise procedure for the construction of the 9-oxa-1-azatricyclo[6.2.1.0^{5,11}]undecan-10-one skeleton.

A. RETROSYNTHESIS OF TRICYCLIC BASIC SKELETON

We have devoted considerable attention to the total synthesis of streptazolin and related natural products. A general retrosynthetic analysis for streptazolin and its analogues is outlined in Scheme 17. A common structural feature of these natural products[36-38,40-45] is the 7-hydroxy-6-substituted-9-oxa-1-azatricyclo[6.2.1.0^{5,11}]undecan-10-one framework **74**.[39] Therefore, we envisaged the tricyclic core framework, namely, 4-hydroxy-6-(C$_2$ unit)-9-oxa-1-azatricyclo[6.2.1.0^{5,11}]undec-5-ene-7,10-dione **75**,[39] being the key intermediate for further elaboration leading to the synthesis of various streptazolin-related natural products. The tricyclic skeleton **75** might be directly constructed by the intramolecular Pauson-Khand reaction of the 2-oxazolone derivative **76**, which has a suitable heptynyl moiety on the nitrogen atom. To the best of our knowledge, no

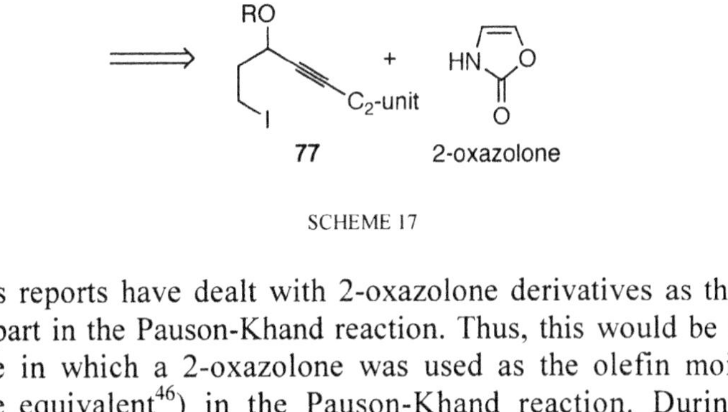

SCHEME 17

previous reports have dealt with 2-oxazolone derivatives as the olefin counterpart in the Pauson-Khand reaction. Thus, this would be the first example in which a 2-oxazolone was used as the olefin moiety (an enamine equivalent[46]) in the Pauson-Khand reaction. During these ongoing studies, Magnus[46a] and Pérez-Castells[46b] independently reported examples of the Pauson-Khand reaction of an enamine equivalent. The 2-

oxazolone-alkyne derivative **76**, a substrate for the Pauson-Khand reaction, can be prepared from the coupling reaction between the heptynyl iodo derivative **77** and 2-oxazolone.

On the basis of the above simple retrosynthetic analysis, (±)-8β-hydroxystreptazolone (**70**)[38,39] was chosen as our first target natural product in this investigation. We describe in detail the results relating (i) the development of the stereoselective and direct construction of the 9-oxa-1-azatricyclo[6.2.1.0^{5,11}]undec-5-ene-7,10-dione skeleton (e.g. **75**) via the intramolecular Pauson-Khand reaction of 2-oxazolone derivatives and (ii) its successful application to the first total synthesis of (±)-8β-hydroxystreptazolone (**70**).

B. PAUSON-KHAND REACTION OF 2-OXAZOLONE DERIVATIVES

Since the targeted streptazolin-related natural products have a C_2-unit at the C_6-position,[39] we first prepared the 2-oxazolone-alkyne derivative **82** with the simplest C_2-unit, an ethyl group, at the triple bond terminus to not only identify suitable ring-closing conditions, but also to determine the level of stereoselectivity that could be expected in the intramolecular Pauson-Khand reaction. Thus, the 2-oxazolone-alkyne derivative **82**, required for the intramolecular Pauson-Khand reaction in our retrosynthesis, was easily prepared from the known alcohol **78** by conventional means, as shown in Scheme 18. Oxidation of **78** was

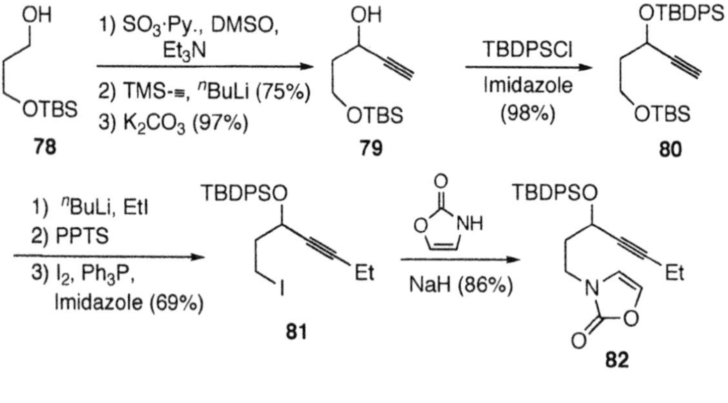

SCHEME 18

followed by addition of the acetylide derived from trimethylsilylacetylene to afford the adduct in 75% yield. The terminal silyl group of the adduct

was removed by base treatment to afford **79** (97%),[47] the secondary
hydroxyl group of which was then protected with a TBDPS group to give
80 in 98% yield. Introduction of an ethyl group at the triple bond terminus
of **80** was followed by desilylation and iodination to furnish the iodo
derivative **81** in 69% overall yield. The coupling reaction between **81** and
2-oxazolone proceeded, upon treatment with NaH in DMF, to produce **82**
in 86% yield.

We began the intramolecular Pauson-Khand reaction of the 2-
oxazolone derivative by treating **82** with Co$_2$(CO)$_8$ in diethyl ether at
room temperature to give the corresponding cobalt-complexed **82**, which
was heated in acetonitrile[48] without a promoter to give only a trace
amount of **83**. When the cobalt-complexed **82** was exposed to
trimethylamine *N*-oxide dihydrate (TMANO·2H$_2$O, 3.5 equiv)[49] in THF at
room temperature for 3h, the desired tricyclic compound **83** was obtained
in 37% yield as the sole isolable product (Scheme 19). The relative
stereochemistry of **83** was assigned on the basis of ^{1}H NMR spectral data.
Compound **83** was converted into the acetate derivative **84** in 79% yield
by treatment with TBAF and then acetic anhydride. The NOE study of **84**
showed no enhancement between C$_4$-H and C$_{11}$-H, whereas the methylene
protons of the C$_6$-ethyl group were enhanced by 4.2% upon irradiation at
C$_4$-H. This observation was in good accord with the prediction based on
analysis of the molecular model.

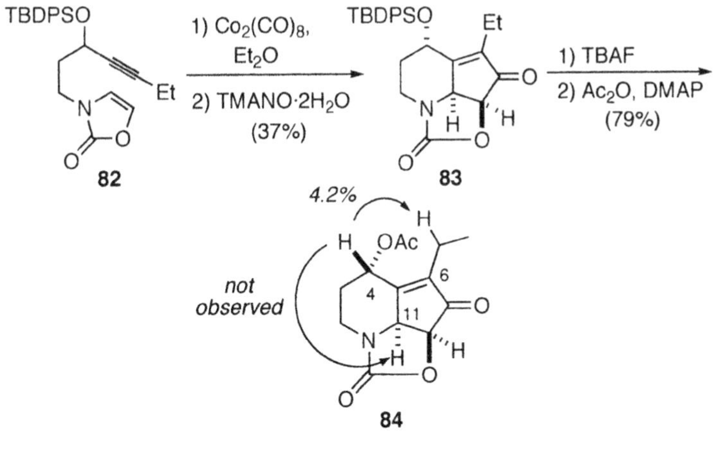

SCHEME 19

Although the chemical yield (37%) of **83** was not satisfactory, we could synthesize stereoselectively the desired tricyclic product **83**, having the common core framework of the target natural products, from the 2-oxazolone-alkyne derivative **82**, in a straightforward manner. The mechanism for the stereoselective formation of **83** is uncertain, but it might be tentatively rationalized on the basis of the mechanistic hypothesis proposed by Magnus (Figure 5).[50] Two plausible cobalt-metallocyclic intermediates **85** and **86**, derived from the cobalt-complexed **82**, would collapse to the desired **83** and its C_4-epimer **87**. In the cobalt-metallocycle **86** leading to **87**, the bulky siloxy group at the C_4-position would have a nonbonding interaction with the C_2-carbon appendage (ethyl group) at the triple bond terminus due to a kind of 1,3-pseudodiaxial relationship on the sterically congested concave face of the transient cobaltabicyclic structure; therefore, a serious unfavorable interaction should occur. This would not be the case in the intermediate **85** where the siloxy group and the C_2-substituent (ethyl group) have a *trans* alignment. As a result, the cyclization pathway via the intermediate **85** would be preferred over that via **86**.

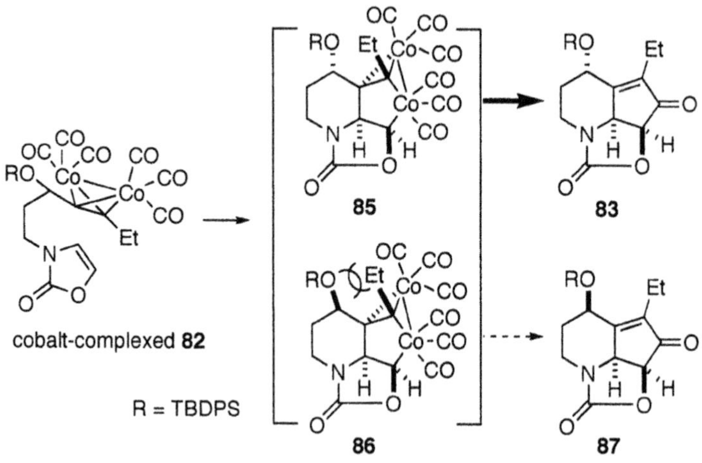

FIGURE 5. Proposed mechanism for the Pauson-Khand reaction.

We next sought to transform the carbonyl functionality of the tricyclic compound **83** into a hydroxyl group with the same relative stereochemistry as that of the target natural products (Scheme 20). Thus, reduction of **83** with $NaBH_4$ in the presence of $CeCl_3$ gave the alcohol **88** in a quantitative yield, which was then acetylated under conventional conditions to provide **89** in 99% yield. The NOE study of **89** (11.4%

enhancement between C_7-H and C_8-H as well as 12.6% enhancement between C_8-H and C_{11}-H) clearly showed that the newly generated stereogenic center (C_7-position) was not the same as that of streptazolin (**69**) and its related natural products. Inversion of the configuration at the C_7-position was realized by exposure of **88** to Mitsunobu conditions[18] (*p*-nitrobenzoic acid, diethyl azodicarboxylate, and triphenylphosphine), which led in 94% yield to the exclusive formation of **90** with the inverted stereochemistry at the C_7-position.

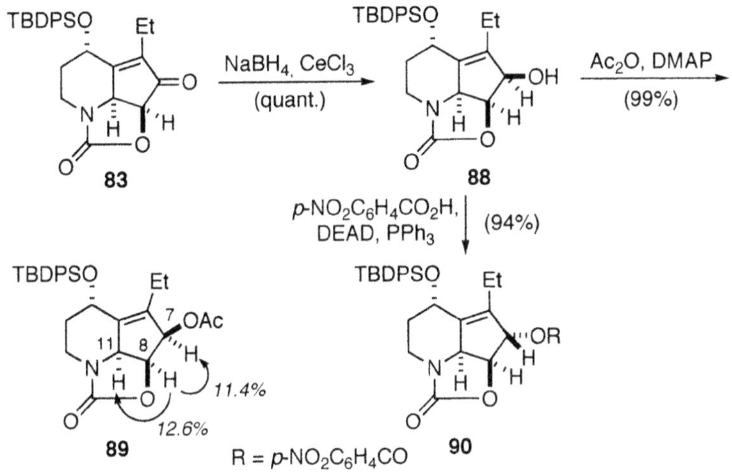

SCHEME 20

We could now develop a procedure for constructing the tricyclic framework of streptazolin (**69**) and its related natural products, in which all of the stereogenic centers of 8β-hydroxystreptazolone (**70**) are constructed in a stereocontrolled fashion. Prior to turning our efforts to the total synthesis of the target natural product, the most significant issue remaining to be solved at this point was to optimize the ring-closing conditions for the conversion of the 2-oxazolone derivative **82** into the tricyclic compound **83**. The first effort was directed toward determining the influence of the protecting group on the secondary propargylic hydroxyl group in the Pauson-Khand reaction. Compound **82** was converted into four related 2-oxazolone derivatives **91a-d** by conventional means. With four newly synthesized precursors for the Pauson-Khand reaction in hand, we next investigated the ring-closing reaction of these compounds under the conditions described for the preparation of **83**. As can be seen in Scheme 21, the desired tricyclic skeletons **84**, and **92a,c,d** were stereoselectively formed, but no improvement of chemical yield was achieved by changing the protecting group on the hydroxyl functionality. On the basis of the observation in Scheme 21 and the result of the reaction of **82**, it might be concluded that a bulky protection group would be favorable in this transformation. Interestingly, it took somewhat longer (20 h) to consume the starting material when the acetyl congener **91b** was submitted to the standard conditions, presumably due to the electron-withdrawing nature of the acetoxy group at the propargylic position.

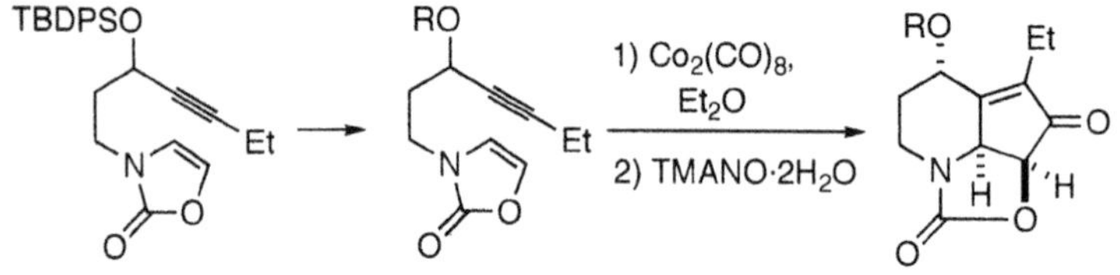

82

91a : R = H
91b : R = Ac
91c : R = TBS
91d : R = TIPS

92a : R = H (20%)
84 : R = Ac (14%)
92c : R = TBS (14%)
92d : R = TIPS (29%)

SCHEME 21

We next screened various Pauson-Khand conditions by using compound **82**. Table 1 summarizes several typical results. The Pauson-Khand reaction of **82** with TMANO·$2H_2O$ (3.5 equiv)[49] in refluxing CH_2Cl_2 instead of THF at room temperature gave a lower yield (23%, entry 1). Increasing of the amount of trimethylamine N-oxide dihydrate (TMANO·$2H_2O$) to 4.5 equiv in refluxing CH_2Cl_2 afforded **83** in an improved yield of 55% (entry 2). Another *N*-oxide promoter, *N*-methylmorpholine *N*-oxide (NMO)[51] in CH_2Cl_2 at room temperature produced **83** in 38% yield (entry 3), similar to that obtained with TMANO·$2H_2O$ in THF. Sugihara's method[52] of using amines or sulfides as the promoter did not improve the yield of **83** (entries 4 to 6). However, the procedure developed by Pérez-Castells[53] was found to be effective for our purposes. Thus the cobalt-complexed 2-oxazolone derivative **82** was exposed to anhydrous TMANO (4.5 equiv) and 4Å molecular sieves in toluene at −10°C for 12 h to furnish **83** in 60% yield (entry 7). This transformation was reproducible, and **83** was consistently obtained in the designated yield. We had developed an efficient method for the construction of 4-hydroxy-6-(C_2-unit)-9-oxa-1-azatricyclo[6.2.1.$0^{5,11}$]

undec-5-ene-7,10-dione (e.g., **75** in Scheme 17), a possible common synthetic intermediate for streptazolin (**69**) and its related natural products, in a highly stereoselective manner based on the intramolecular Pauson-Khand reaction of the 2-oxazolone derivative in acceptable yields.

TABLE 1.

Pauson-Khand reaction of 2-oxazolone derivative **82**

Entry	Promoter	Equiv.	Solvent	Temp.	Time	Yield (%)
1	TMANO·2H₂O	3.5	CH₂Cl₂	reflux	3.5 h	23
2	TMANO·2H₂O	4.5	CH₂Cl₂	reflux	5.5 h	55
3	NMO	4.5	CH₂Cl₂	rt	20 h	38
4	ʲPrSMe	3.5	ClCH₂CH₂Cl	reflux	45 min	16
5	ⁿBuSMe	3.5	ClCH₂CH₂Cl	reflux	1 h	14
6	Cyclohexylamine	3.5	ClCH₂CH₂Cl	reflux	30 min	11
7	TMANO/4Å MS	4.5	toluene	-10 °C	12 h	60

C. TOTAL SYNTHESIS OF 8β-HYDROXYSTREPTAZOLONE

As shown in Figure 4, the target natural product, 8β-hydroxystreptazolone (**70**), has an acetyl group at the C_6-position.[39] Therefore, the 2-oxazolone derivative **95** possessing a vinyl group at the triple bond terminus was chosen as the starting material,[54] because a vinyl group is well known to be easily converted to an acetyl group under the Wacker oxidation conditions (Scheme 22).[55] The 2-oxazolone-enyne derivative **95** was prepared from **80** via **93** and **94** by conventional means. Exposure of **95** to ring-closing conditions, optimized in Table 1 (entry 7),[53] produced exclusively the desired tricyclic compound **96** in 50% yield. According to the procedure described for the transformation of **83** to **90**, compound **98** was obtained in high yield from the keto derivative

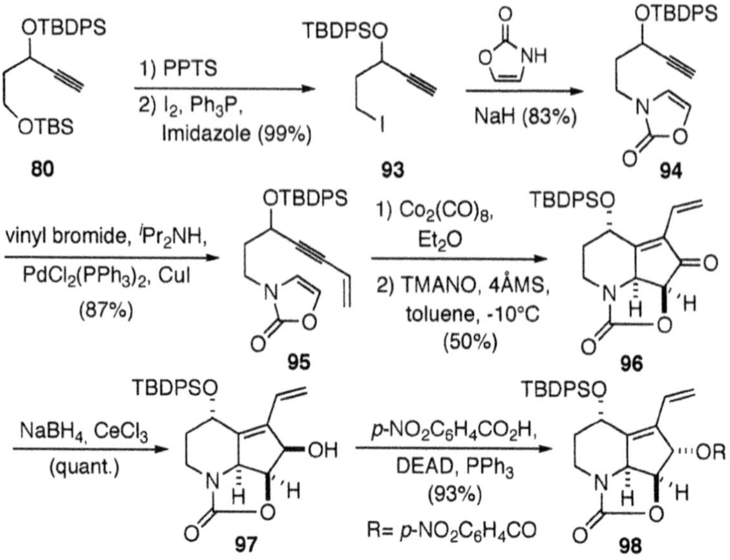

SCHEME 22

96 by stereoselective reduction (compound **97**) followed by the Mitsunobu reaction. However, it soon became apparent that the vinyl moiety of **98** could not be converted into the acetyl group under the Wacker conditions.[55] A careful inspection of the ^{1}H NMR spectrum of the crude mixture from the Wacker reaction disclosed the following observations; (i) disappearance of vinyl protons, (ii) no peaks due to the acetyl group, and (iii) a lower field-shifted proton presumably due to an aldehyde functionality. Judging from this observation, we tentatively concluded that oxidative carbon-carbon bond cleavage of the vinyl group must have occurred, although no pure compounds could be isolated from the reaction mixture.

We next addressed the introduction of the 1-hydroxyethyl group as an acetyl precursor at the triple bond terminus of the 2-oxazolone derivative.[54] Treatment of **94** with acetaldehyde in the presence of NaHMDS afforded **99** in 83% yield as a mixture of two diastereoisomers, which were exposed to $Co_2(CO)_8$ under the Pauson-Khand conditions to produce the corresponding tricyclic compound **100** in 51% yield (Scheme 23).

Protection of the secondary hydroxyl moiety of **100** with the TBS group was followed by stereoselective reduction of the carbonyl

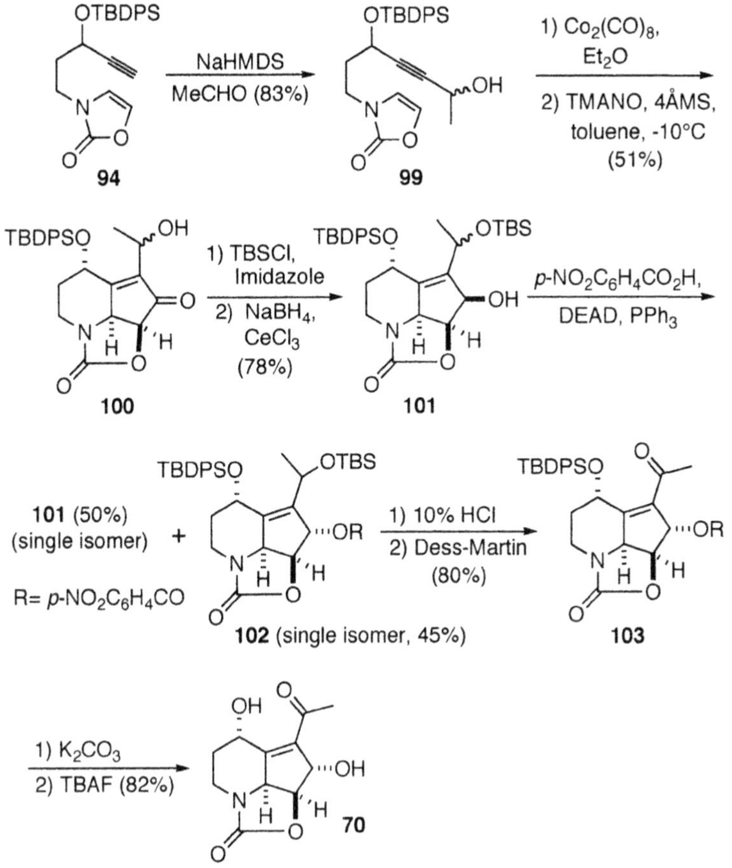

SCHEME 23

functionality to furnish **101** in 78% yield. According to our model studies, inversion of the C$_7$-hydroxyl group of **101** (a mixture of two diastereoisomers) was carried out under Mitsunobu conditions at room temperature to give the inverted product **102** in 45% yield along with 50% yield of the recovered starting material **101**. Complete consumption of the starting material **101** could not be achieved even at higher reaction temperature. Although substrate **101** for the inversion reaction was used as a mixture of two diastereoisomers, the recovered **101** and the inverted product **102** were both apparently a single stereoisomer, the stereochemistries of which were undetermined. This result obviously indicated that one of two diastereoisomers of **101** reacted readily with the

Mitsunobu reagents to provide the desired **102**, while the other isomer was inactive toward those reagents. The chemical yield of **102** (45%) was not good enough, nevertheless we took this compound forward to the final stage of the synthesis of **70**. Removal of the TBS group of **102** under acidic conditions gave the corresponding hydroxyl derivative, which was then oxidized with Dess-Martin periodinane to afford **103** in 80% yield. Finally, successive treatment of **103** with potassium carbonate and TBAF produced (±)-8β-hydroxystreptazolone (**70**)[6] in 82% yield.

The first total synthesis of (±)-**70** from 2-oxazolone was thus accomplished. One half of the synthetic intermediate **101**, however, could not be converted into **70**. This step made this total synthesis less efficient. Therefore, an alternative procedure was necessary to transform **101** into **102**. One of two diastereoisomers of **101** did not react with the Mitsunobu reagent, which might indicate steric congestion by the bulky TBDMS group, although this was unclear.

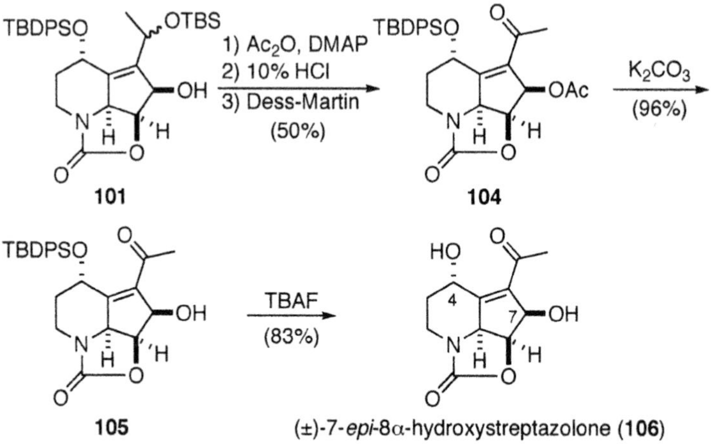

101 **104** **105**

(±)-7-*epi*-8α-hydroxystreptazolone (**106**)

SCHEME 24

We envisaged constructing an essential acetyl functionality at the C_6-position prior to inverting the stereochemistry at the C_7-position. Thus, compound **101** was converted into the corresponding C_6-acetyl derivative **105** via **104** as shown in Scheme 24. Treatment of **105** with the Mitsunobu reagents under various conditions, however, led to the recovery of the starting material **105**.[56] The intramolecular hydrogen bonding between the C_6-acetyl carbonyl oxygen and the C_7-hydroxyl group presumably caused fairly low reactivity toward the Mitsunobu reagents. It should be mentioned here that desilylation of **105** with TBAF afforded (±)-7-*epi*-8β-hydroxystreptazolone (**106**)[6b,39] in 83% yield, although its isolation has not been reported.

On the basis of these results, we decided to change the protecting group on the secondary hydroxyl group of the tricyclic compound from the bulky TBS group to the less sterically hindered methoxymethyl (MOM) group (Scheme 25). Protection of the hydroxyl group of the tricyclic compound **100** with the MOM group was followed by stereoselective reduction to give **107** in 90% yield. The Mitsunobu reaction[18] of **107**, which required 60°C for completion of the reaction,[57] afforded the inverted products, which were subsequently exposed to concentrated HCl and Dess-Martin periodinane to produce **103** in 65% overall yield.[6b] Thus, the transformation of **100** into **103** could be improved by changing the protecting group on the secondary hydroxyl moiety.

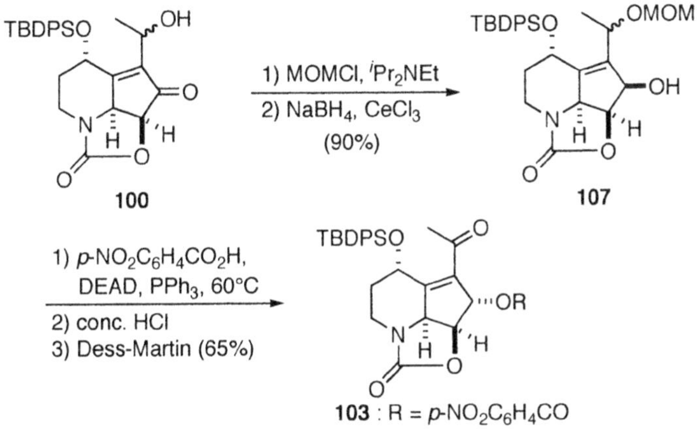

SCHEME 25

We have developed a novel and efficient procedure for constructing 7-hydroxy-6-substituted-9-oxa-1-azatricyclo[6.2.1.0^{5,11}]undec-5-ene-7,10-diones by the intramolecular Pauson-Khand reaction of the 2-oxazolone species with the required alkyne appendages. In addition, by taking advantage of this newly developed method, we have achieved the first total synthesis of (±)-8β-hydroxystreptazolone (**70**)[6] in a highly stereoselective manner. Synthesis of (±)-7-*epi*-8β-hydroxystreptazolone (**106**)[6b] was also achieved. Since the tricyclic compound **100** has the entire carbon framework and suitable functionalities for further elaborations, it should be a versatile intermediate for the synthesis of streptazolin (69) and its related natural products.

V. SUMMARY

We have completed the first total syntheses of three types of natural products: secosyrins, ichthyothereol, and 8β-hydroxystreptazolone. Application of the unique properties of alkyne-$Co_2(CO)_6$ complexes as a crucial step made these total syntheses possible. The facile preparation of alkyne-$Co_2(CO)_6$ complexes as well as their intriguing properties will make them useful for other total synthesis of other natural products.

Acknowledgements

First of all, I want to thank emeritus professor Miyoji Hanaoka for our long-term collaboration. I also wish to thank my highly dedicated group of co-workers, Dr. Izumi Nomura, Dr. Sameh M. Moharram, and Naoki Miyakoshi for their intellectual, creative, and experimental contributions to the overall success of the work described herein.

References and Footnotes

1. For examples, see: (a) Collman, J. P., Hegedus, L. S., Norton, J. R., Finke, R. G. *"Principles and Applications of Organotransition Metal Chemistry,"* University Science Book: Mill Valley, CA, 1987. (b) Went, M., *Adv. Organomet. Chem.* **1997**, *41*, 69.

2. For, leading reviews, see: (a) Nicholas, K. M., *Acc. Chem. Res.* **1987**, *20*, 207. (b) Caffyn, A. J. M., Nicholas, K. M. " *Comprehensive Organic Chemistry II*," Eds. Abel, E. W., Stone, F. G. A., Wilkinson, G. Pergamon, Oxford, 1995, vol. 12, p. 685. (c) Fletcher, A. J., Christie, S. D. R., *J. Chem. Soc., Perkin Trans. 1* **2000**, 1657. (d) Green, J. R., *Curr. Org. Chem.* **2001**, *5*, 809. (e) Teobald, B. J., *Tetrahedron* **2002**, *58*, 4133.

3. For leading reviews, see: (a) Pauson, P. L. *"Organometallics in Organic Synthesis. Aspects of a Modern Interdisciplinary Field"* Eds. de Meijere, A., tom Dieck, H., Springer: Berlin, 1988; pp 233. (b) Schore, N. E., *Chem. Rev.* **1988**, *88*, 1081. (c) Schore, N. E. *Org. React.* **1991**, *40*, 1. (d) Schore, N. E. *"Comprehensive Organic Synthesis,"* Ed. Trost, B. M. Pergamon: Oxford, 1991; Vol. 5, pp 1037. (e) Schore, N. E. *"Comprehensive Organometallic Chemistry II,"* Eds. Abel, E. W., Stone, F. G. A., Wilkinson. G. Elsevier: New York, 1995; Vol. 12, p 703. (f) Frühauf, H.-W., *Chem. Rev.* **1997**, *97*, 523. (g) Jeong, N. *"Transition Metals in Organic Synthesis,"* Eds. Beller, H., Bolm, C. Wi ley-VCH: Weinheim, 1998, Vol. 1, pp 560. (h) Geis, O., Schmalz, H.-G., *Angew. Chem. Int. Ed. Engl.* **1998**, *37*, 911. (i) Chung, Y. K., *Coord. Chem. Rev.* **1999**, *188*, 297. (j) Brummond, K. M., Kent, J. L., *Tetrahedron* **2000**, *56*, 3263.

4. (a) Mukai, C., Moharram, S. M., ,Hanaoka, M. *Tetrahedron Lett.* **1997**, *38*, 2511.

(b) Mukai, C., Moharram, S. M., Azukizawa, S., Hanaoka, M., *J. Org. Chem.* **1997**, *62*, 8095.

5. Mukai, C., Miyakoshi, N., Hanaoka, M., *J. Org. Chem.* **2001**, *66*, 5875.

6. (a) Nomura, I., Mukai, C. ,*Org. Lett.* **2002**, *4*, 4301. (b) Nomura, I. Mukai, C., *J. Org. Chem.* **2004**, *69*, 1803.

7. (a) Smith, M. J., Mazzola, E. P., Sims, J. J., Midland, S. L., Keen, N. T., Burton, V., Stayton, M. M., *Tetrahedron Lett.* **1993**, *34*, 223. (b) Midland, S. L., Keen, N. T., Sims, J. J., Midland, M. M., Stayton, M. M. Burton, V., Smith, M. J., Mazzola, E. P., Graham, K. J., Clardy, J., *J. Org. Chem.* **1993**, *58*, 2940.

8. (a) Wood, J. L., Jeong, S., Salcedo, A,., Jenkins, J., *J. Org. Chem.* **1995**, *60*, 286. (b) Kuwahara, S., Moriguchi, M., Miyagawa, K., Konno, M., Kodama, O., *Tetrahedron Lett.* **1995**, *36*, 3201. (c) Kuwahara, S., Moriguchi, M., Miyagawa, K., Konno, M., Kodama, O., *Tetrahedron* **1995**, *32*, 8809. (d) Henschke, J. P., Rickards, R. W., *Tetrahedron Lett.* **1996**, *37*, 3557. (e) Ishihara, J., Sugimoto, T., Murai, A., *Synlett* **1996**, 335. (f) Honda, T., Mizutani, H., Kanai, K., *J. Org. Chem.* **1996**, *61*, 9374.

9. Midland, S. L., Keen, N. T., Sims, J. J., *J. Org. Chem.* **1995**, *60*, 1118.

10. Savage, I., Thomas, E. J., *J. Chem. Soc., Chem. Commun.* **1989**, 717.

11. Ono, N., Kamimura, A., Kaji, A., *Synthesis* **1984**, 226.

12. Steliou, K., Poupart, M.-A., *J. Org. Chem.* **1985**, *50*, 4971.

13. Reich, H. J., Renga, J. M., Reich I. L., *J. Am. Chem. Soc.* **1975**, *97*, 5434.

14. Ito, Y., Hirao, T., Saegusa, T., *J. Org. Chem.* **1978**, *43*, 1011.

15. Fanton, E., Gelas, J., Horton, D., *J. Chem. Soc., Chem. Commun.* **1980**, 21.

16. Mukai, C., Moharram, S. M., Kataoka, O., Hanaoka, M., *J. Chem. Soc., Perkin Trans. 1* **1995**, 2849.

17. Numbering for a 1,7-dioxaspiro[4.4]nonan-8-one skeleton was used for convenience.

18. Mitsunobu, O., *Synthesis* **1981**, 1.

19. After our first total synthesis of secosyrins 1 and 2, three groups have reported their syntheses. (a) Yu, P., Yang, Y., Zhang, Z. Y., Mak, T. C. W., Wong, H. N. C.. *J. Org. Chem.* **1997**, *62*, 6359. (b) Carda, M., Castillo, E., Rodríguez, S., Falomir, E., Marco, J. A., *Tetrahedron Lett.* **1998**, *39*, 8895. (c) Donohoe, T. J., Fisher, J. W., Edwards, P. J., *Org. Lett.* **2004**, *6*, 465.

20. Chin, C., Jones, E. R. H., Thaller, V., Aplin, R. T., Durham, L. J., Cascon, S. C., Mors, W. B., Tursch, B. M., *Chem. Commun.* **1965**, 152.

21. Several reports on isolation of ichthyothereol and its acetate were also recorded: (a) Gorinsky, C., Templeton, W., Zaidi, S. A. H., *Lloydia* **1973**, *36*, 352. (b) Czerson, H., Bohlmann, F., Stuessy, T. F., Fischer, N. H., *Phytochemistry* **1979**, *18*, 257. (c) Bohlmann, F., Borthakur, N., King, R. M., Robinson, H., *Phytochemistry* **1982**, *21*, 1793. (d) Lam, J., Christensen, L. P., Thomasen, T., *Phytochemistry* **1991**, *30*, 515.

22. Cascon, S. C., Mors, W. B., Tursch, B. M., Aplin, R. T., Durham, L. J., *J. Am. Chem. Soc.* **1965**, *87*, 5237.

23. Chin, C., Cutler, M. C., Jones, E. R. H., Lee, J., Safe, S., Thaller, V., *J. Chem. Soc. (C)* **1970**, 314.

24. (a) Mukai, C., Ikeda, Y., Sugimoto, Y., Hanaoka, M. *Tetrahedron Lett.* **1994**, *35*, 2179. (b) Mukai, C., Sugimoto, Y., Ikeda, Y., Hanaoka, M., *Tetrahedron Lett.* **1994**, *35*, 2183. (c) Mukai, C., Sugimoto, Y., Ikeda, Y., Hanaoka, M., *J. Chem. Soc., Chem. Commun.* **1994**, 1161. (d) Mukai, C., Sugimoto, Y., Ikeda, Y.,

Hanaoka, M.. *Tetrahedron* **1998**, *54*, 823. (e) Mukai, C., Sugimoto, Y., Miyazawa, K., Yamaguchi, S., Hanaoka, M., *J. Org. Chem.* **1998**, *63*, 6281. (f) Mukai, C., Yamaguchi, S., Sugimoto, Y., Miyakoshi, N., Kasamatsu, E., Hanaoka, M., *J. Org. Chem.* **2000**, *65*, 6761. (g) Mukai, C., Yamaguchi, S., Kim, I. J., Hanaoka, M., *Chem. Pharm. Bull.* **2001**, *49*, 613.

25. Eisler, S., Tykwinski, R. R., *J. Am. Chem. Soc.* **2000**, *122*, 10736.
26. Warner, B. P., Buchwald, S. L., *J. Org. Chem.* **1994**, *59*, 5822.
27. Saito, S., Kuroda, A., Tanaka, K., Kimura, R., *Synlett* **1996**, 231.
28. Yamaguchi, M., Hirao, I., *Tetrahedron Lett.* **1983**, *24*, 391.
29. Corey, E. J., Fuchs, P. L., *Tetrahedron Lett.* **1972**, 3769.
30. *"*BuLi was found to be ineffective in this case.
31. An excess of propylene oxide (bp 34°) can be easily removed by evaporation.
32. Martín *et al* reported an alternative method for the preparation of 2-ethynyl-3-hydroxytetrahydropyrane derivatives: Alvarez, E., Pérez, R., Rico, M., Rodríguez, R. M., Suárez, M. C., Martín, J. D., *Synlett* **1996**, 1082.
33. Zhang, H. X., Guibé, F., Balavoine, G., *J. Org. Chem.* **1990**, *55*, 1857.
34. Barth, W., Paquette, L. A., *J. Org. Chem.* **1985**, *50*, 2438, and references cited therein.
35. Stille, J. K., *Angew. Chem. Int. Ed. Engl.* **1986**, *25*, 508, and references cited therein.
36. Drautz, H., Zähner, H., Kupfer, E., Keller-Schierlein, W., *Helv. Chim. Acta* **1981**, *64*, 1752.
37. (a) Grabley, S., Kluge, H., Hoppe, H.-U., *Angew. Chem. Int. Ed. Engl.* **1987**, *26*, 664. (b) Grabley, S., Hammann, P., Kluge, H., Wink, J., Kricke, P., Zeeck, A., *J. Antibiot.* **1991**, *44*, 797. (c) Grabley, S., Hammann, P., Thiericke, R., Wink, J., Philipps, S., Zeeck, A., *J. Antibiot.* **1993**, *46*, 343.
38. Tang, Y.-Q.; Wunderlich, D.; Sattler, I.; Grabley, S.; Feng, X.-Z.; Thiericke, R. Abstract of Division of Organic Chemistry. *223rd ACS National Meeting*, Apr 7-12, Orlando; American Chemical Society: Washington, DC, 2002, p 341.
39. Tang et al called compound **70** 8β-hydroxystreptazolone, compound **71** 4,12-epoxystreptazolin, compound **72** 9β-hydroxystreptazolin, and compound **73** streptazolin-dimer.[38] According to the IUPAC nomenclature system, (±)-**70** should be described as (*4R**,*7R**,*8R**,*11R**)-6-acetyl-4,7-dihydroxy-9-oxa-1-azatricyclo-[6.2.1.0^{5,11}]undec-5-en-10-one. This numbering system is used for the tricyclic compounds.
40. Puder, C., Loya, S., Hizi, A., Zeeck, A., *J. Nat. Prod.* **2001**, *64*, 42.
41. (a) Kozikowski, A. P., Park, P., *J. Am. Chem. Soc.* **1985**, *107*, 1763. (b) Kozikowski, A. P., Park, P., *J. Org. Chem.* **1990**, *55*, 4668.
42. Flann, C. J., Overman, L. E., *J. Am. Chem. Soc.* **1987**, *109*, 6115.
43. Yamada, H., Aoyagi, S., Kibayashi, C., *J. Am. Chem. Soc.* **1996**, *118*, 1054.
44. Huang, S., Comins, D. L., *Chem. Commun.* **2000**, 569.
45. (a) Cossy, J., Pévet, I., Meyer, C., *Synlett* **2000**, 122. (b) Cossy, J., Pévet, I., Meyer, C., *Eur. J. Org. Chem.* **2001**, 2841.
46. Recently, the carbamate and amide functionalities were used as an enamine equivalent in the Pauson-Khand reaction. (a) Magnus, P., Fielding, M. R., Wells, C., Lynch, V., *Tetrahedron Lett.* **2002**, *43*, 947. (b) Domínguez, G., Casarrubios, L., Rodríguez-Noriega, J., Pérez-Castells, J., *Helv. Chim. Acta* **2002**, *85*, 2856.
47. (a) Takano, S., Sugihara, T., Ogasawara, K., *Tetrahedron Lett.* **1991**, *32*, 2797. (b) Pearson, W. H., Postich, M. J., *J. Org. Chem.* **1994**, *59*, 5662.

48. (a) Hoye, T. R., Suriano, J. A., *J. Org. Chem.* **1993**, *58*, 1659. (b) Chung, Y. K., Lee, B. Y., Jeong, N., Hudecek, M., Pauson, P. L., *Organometallics* **1993**, *12*, 220.
49. Jeong, N., Chung, Y. K., Lee, B. Y., Lee, S. H., Yoo, S.-E., *Synlett* **1991**, 204.
50. (a) Magnus, P., Principe, L. M., *Tetrahedron Lett.* **1985**, *26*, 4851. (b) Magnus, P., Exon, C., Albaugh-Robertson, P., *Tetrahedron* **1985**, *41*, 5861.
51. Shambayati, S., Crowe, W. E., Schreiber, S. L., *Tetrahedron Lett.* **1990**, *31*, 5289.
52. (a) Sugihara, T., Yamada, M., Ban, H., Yamaguchi, M., Kaneko, C., *Angew. Chem. Int. Ed. Engl.* **1997**, *36*, 2801. (b) Sugihara, T., Yamada, M., Yamaguchi, M., Nishizawa, M., *Synlett* **1999**, 771.
53. Pérez-Serrano, L., Casarrubios, L., Domínguez, G., Pérez-Castells, J., *Org. Lett.* **1999**, *1*, 1187.
54. The Pauson-Khand reaction of the 2-oxazolone-alkyne derivative, possessing an acetyl group at the triple bond terminus was not investigated, because it appeared difficult to differentiate the resulting two carbonyl functionalities (the C$_6$-acetyl group and the C$_7$-carbonyl moiety) of the tricyclic compound.
55. For example, see, (a) Smidt, J., Hafner, W., Jira, R., Sieber, R., Sedlmeier, J., Sabel, A., *Angew. Chem.* **1962**, *74*, 93. (b) Tsuji, J., *Synthesis*, **1984**, 369. (c) Molander, G. A., Cameron, K. O., *J. Am. Chem. Soc.* **1993**, *115*, 830, and references cited therein.
56. In some cases, a trace amount of the inverted compound could be detected.
57. The reaction did not occur at room temperature.

STRATEGIES AND TACTICS IN ORGANIC SYNTHESIS, VOL. 6

Chapter 8

TOTAL SYNTHESIS OF MYRIAPORONES 1, 3, AND 4

Richard E. Taylor, Kristen N. Fleming, and Brian R. Hearn
Department of Chemistry and Biochemistry and
The Walther Cancer Research Center
251 Nieuwland Science Hall
University of Notre Dame
Notre Dame, IN 46556

I. The Foundation: Tedanolide and the Myriaporones

Despite not being obvious from the title of this chapter, our work in this area was initiated by interest in a complex marine polyketide isolated over twenty years ago. Tedanolide is a potent cytotoxic macrolide isolated from the Caribbean sponge *Tedania ignis*.[1] Tedanolide demonstrated potent cytotoxicity against cancer cell lines, but detailed biological studies were hampered by material supply issues. Despite significant synthetic interest,[2] only 13-desoxytedanolide[3] has been prepared to date and it is not clear if these remarkable success stories will lead to an increased understanding of its interesting biological activity.[4]

In 1995, Rinehart reported the isolation of the myriaporones which, in addition to their structural similarity to the C10-C23 region of tedanolide, have an IC_{50} = 100 ng/mL in L-1210 cells.[5] Unfortunately, the small amount of these materials isolated from the natural source was insufficient for the unambiguous assignment of the C5 and C6 stereogenic centers as well as more comprehensive biological studies. In 1996, a patent from the same group disclosed the structures of myriaporones 1 and 2.[6] Unfortunately, the C5 stereochemistry was not assigned for either compound. The structure of Myriaporone 2, with its "fragmented" epoxide, foreshadowed the acid sensitivity of synthetic intermediates and difficulties experienced by many of the synthetic groups targeting tedanolide including the challenges uncovered in our own group.

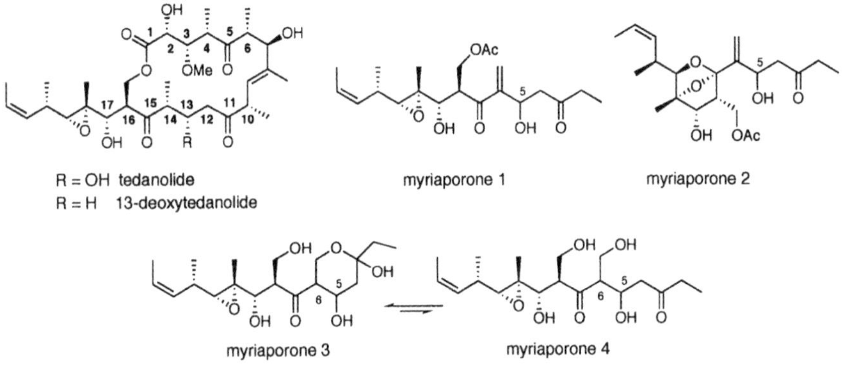

R = OH tedanolide
R = H   13-deoxytedanolide

myriaporone 1

myriaporone 2

myriaporone 3

myriaporone 4

SCHEME 1

The remarkable structural similarities between the myriaporones and tedanolide suggest an obvious evolutionary relationship between the organisms (PKS genes) that produce such elegant chemical entities. Moreover, we wondered if their biological activity could be related as well and whether synthetic work towards the simpler of the two, the myriaporones, might lead to a greater understanding of both and an opportunity for new leads in cancer chemotherapy. Despite these interesting themes, only limited synthetic studies had been carried out on the myriaporones.[7]

We began the myriaporone project in 1996 with a primary goal of designing and executing efficient and elegant total syntheses of these natural products. To this end, we identified obvious synthetic challenges and proposed solutions that we felt offered high probabilities for success. Where possible, we aimed to incorporate strategies applicable to the overall goals of our program and tactics that would highlight recent advances in chemical synthesis.

II. An Issue of Timing: A Strategy for the Synthesis of the Myriaporones

Upon visual inspection of the myriaporone scaffold (Scheme 1), several synthetic challenges present themselves. Each myriaporone contains multiple β-hydroxy ketone subunits that may be susceptible to retro-aldol fragmentation or β-elimination. Hydroxypropionate structural units, found at C8-C9 of myriaporone 1 and C6-C7 and C8-C9 of myriaporones 3 and 4, had received little attention from the synthetic community at that time and would likely prove difficult to construct. Finally, the presence of the C10-C11 trisubstituted epoxide presented the most significant challenge as early incorporation would render subsequent intermediates prone to acid-catalyzed degradation. Alternatively, a late-stage epoxidation might fail to produce the desired diastereomer. In spite of these challenges, we fully expected two enthusiastic first year graduate students, Jeff Ciavarri and Brian Hearn, to rapidly complete a myriaporone total synthesis as the first of many achievements to be highlighted in their theses. Blinded by an excusable bit of new investigator overzealousness, we were obviously a bit naïve.

Based on earlier work by Sharpless, our preferred epoxidation strategy involving a late-stage oxidation seemed unlikely to result in the desired *syn*-C9-C10 epoxy alcohol.[8] In 1979, the Sharpless group proposed a model to rationalize the diastereoselectivity of epoxidations of secondary allylic alcohols containing vinyl methyl substituents as in **1** (Scheme 2). Vanadium-catalyzed, alcohol-directed epoxidations of substrates such as **1** generated *anti*-epoxy alcohols **3** in 98:2 selectivity (R = *n*-propyl). This observation was rationalized by comparing the two possible rotamers, **A** and **B**, which maintain the optimal 50° dihedral angle between the olefin and the alcohol. In rotamer **A**, the steric interaction of the vinyl methyl and *n*-propyl R substituent limits the formation of the *syn*-diastereomer **2** required for the synthesis of the myriaporones. Such steric repulsions are alleviated in rotamer **B**; however, this particular pathway generates the *anti*-diastereomer **3**. Sharpless also evaluated *m*-CPBA epoxidations of substrate **1**, and the reaction was found to be virtually unselective as ~1:1 mixture of **2** and **3** was obtained. Based on these results, we decided to initially pursue a synthetic route that incorporated the C10-C11 epoxide prior to the formation of the C9 stereocenter.

Although we were less than thrilled by the notion of maintaining a potentially reactive epoxide throughout our entire synthesis, we viewed this challenge as an opportunity to investigate new reactions. Relatively little was then known about the diastereoselectivity of nucleophilic additions to α-methyl epoxyaldehydes, and we thought this area would provide significant opportunities for scientific contributions. Before we could even begin to investigate stereochemical issues, however, we needed to identify an appropriate reaction for the formation of the C8-C9 hydroxypropionate.

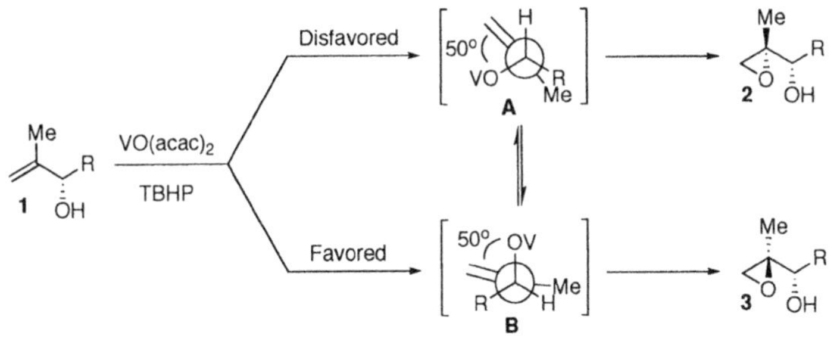

SCHEME 2

As shown in Scheme 3, classic approaches to β-hydroxy ketones are often ineffective for the formation of hydroxypropionates. Standard aldol reactions are complicated by the presence of the β-alkoxide of **4**. Enolization of **4** typically results in facile β-elimination, thus preventing aldol addition and formation of hydroxypropionate **7**. This approach was later successfully applied to the myriaporone problem by Loh through the use of boron enolates.[2g] Likewise, relatively few allylation reactions are appropriate for hydroxypropionate synthesis as the required organometallic species **8** would also be prone to elimination.

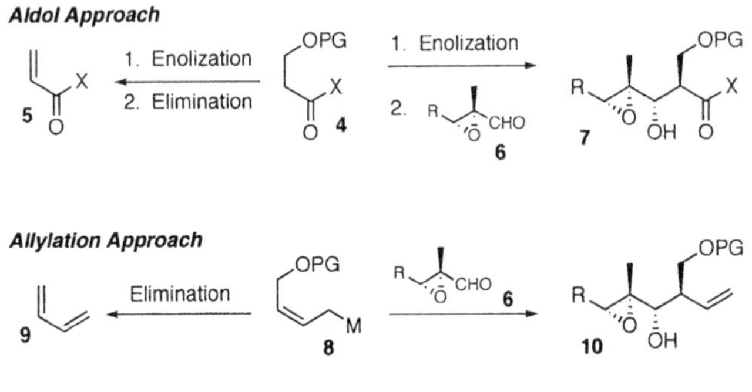

SCHEME 3

III. C9-C13: The Most Elegant Reaction is One That Works Well.

Our synthesis of the C9-C13 aldehyde of the myriaporones is detailed in Scheme 4. This route is characterized by a series of standard reactions that produces the desired intermediates with good selectivity and in high yields. Initially, we investigated alternative reaction sequences that result from less obvious disconnections; however, each of these approaches was limited by low yield or poor scalability. Since the initial objective was to prepare reasonable quantities of the myriaporones (exemplified eloquently by a group motto, "Just make it"), we decided that the most elegant route involves reactions that are selective, high yielding and scalable, not simply less obvious.

Beginning with commercially available methyl-(S)-2-hydroxymethyl propionate 11, aldehyde 12 was synthesized via a three-step protection, reduction and oxidation sequence (Scheme 4). Although a number of methods for the installation of the C10-C11 olefin were evaluated,[7c] the most reliable route utilized a Wittig olefination for the elaboration of aldehyde 12 to unsaturated ester 13. After reduction of the methyl ester, a Sharpless asymmetric epoxidation[8] provided epoxy alcohol 14, which was then oxidized with catalytic TEMPO to provide the aldehyde 15. This route was capable of producing the desired intermediate 15 as a single diastereomer on multigram scale.

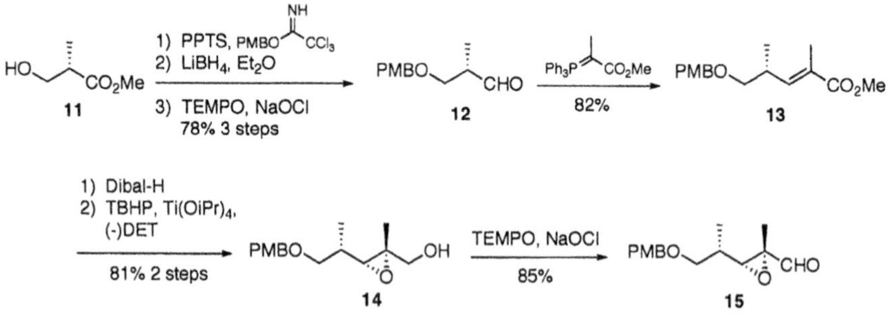

SCHEME 4

IV. C8-C9: The Best Model System is the Natural Substrate.

Incorporation of the C10-C11 epoxide early in our synthetic route provided an interesting opportunity to develop new methodology directed toward understanding stereoselective additions to α-methyl-α,β-epoxy aldehydes such as 15. We evaluated a number of transition metal promoted allylations for the formation of the C8-C9 myriaporone bond. Initially an indium allylation using 4-bromocrotonate, first reported by Loh,[9] was investigated as shown in Scheme 5. In model systems based on α-methyl-α,β-epoxy cinnamaldehyde **16**, this reaction proved to be quite selective; however, the undesired *anti, anti*-epoxy hydroxypropionate **17** was generated. We were disappointed to find that all attempts to reverse the facial selectivity of the addition were utterly unsuccessful.

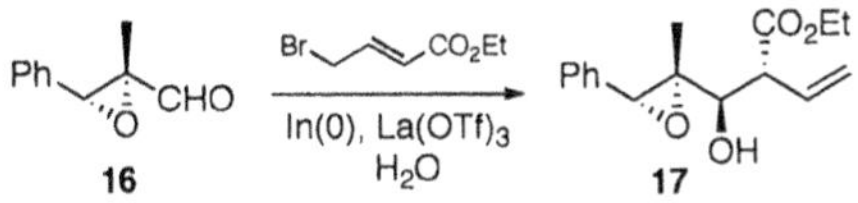

SCHEME 5

We next turned our attention to the Clark zirconium allylation for the synthesis of C8-C9 bond (Scheme 6). Clark had previously shown that *in situ*-generated allyl zirconium nucleophiles could selectively produce *anti*-hydroxypropionates of a variety of aliphatic and aromatic aldehydes.[10] In our lab, we found that reaction of aldehyde **15** with the same allyl zirconium nucleophile did selectively generate *anti*-hydroxypropionates. Unfortunately, the epoxide failed to control the facial selectivity of this addition; thus, a 1:1 mixture of separable diastereomers **18** and **19** was formed. This reaction represents our first successful, albeit limited, synthesis of the desired intermediate **18**. We next attempted to optimize the output of diastereomer **18** by modifying the Clark reaction conditions defined as THF or PhCH₃ solvent with no additives.

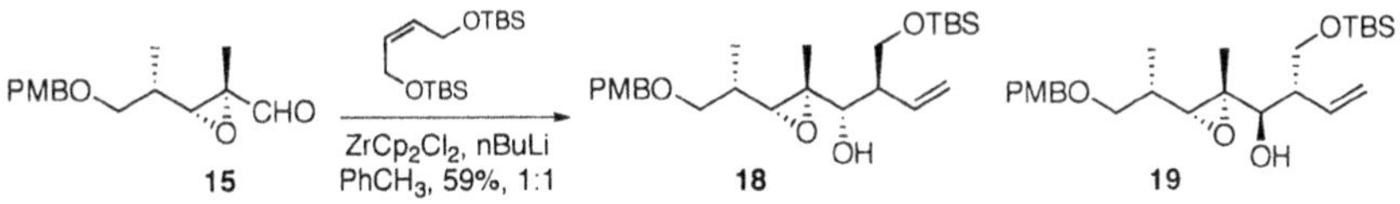

SCHEME 6

Because at that time we had access to only a limited supply of aldehyde **15**, we decided to optimize the Clark allylation using the cinnamaldehyde-based model system. We found that, by using external Lewis acids in nonpolar solvents (PhCH₃, Et₂O or CH₂Cl₂), the stereochemical outcome of the Clark allylation could be reversed to selectively produce *syn*-hydroxypropionates.[7c] Application of the optimized conditions (3 eq. $MgBr_2 \cdot OEt_2$ in $PhCH_3$) using epoxy aldehyde **16** generated *syn*-hydroxypropionate **20** as a single diastereomer in moderate yield. This interesting reversal of stereoselectivity was found to be a general effect as similar results were observed for a number aldehydes, both aliphatic and aromatic.[11]

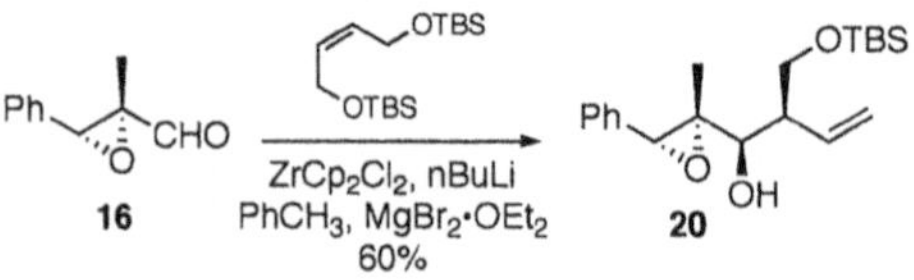

SCHEME 7

Although product **20** did not contain the stereochemistry found in the myriaporones, simple inversion of the C9 secondary alcohol could provide the same diastereomer required for the total synthesis of the myriaporones. As we soon learned, however, simple transformations of stereochemically complex substrates are rarely simple, and this was certainly no exception. Attempted inversion of a C9 mesylate with oxygen nucleophiles resulted exclusively in elimination. Oxidation-reduction sequences were only marginally more successful as mixtures of C9 epimers were obtained. Mitsunobu inversions also failed to generate the desired diastereomer. Finally, we decided to abandon the cinnamaldehyde model system and try our luck with the modified Clark conditions using the myriaporone aldehyde **15**. We had hoped that the distal C11-stereocenter and PMB ether might alter the diastereoselectivity of the allylation reaction. Unfortunately, a complex mixture resulted from the modified zirconium allylation of **15**.

After spending a significant amount of time trying to develop methodology for stereoselective additions to α-methyl-α,β-epoxy aldehydes, we had in fact made virtually no progress toward our primary goal of completing the total synthesis of the myriaporones. We had successfully uncovered very interesting reactivity with the zirconium allylation, but this chemistry could not be effectively incorporated into our total synthetic efforts. The cinnamaldehyde model system failed to accurately represent the reactivity of the myriaporone aldehyde **15**, and the epoxide did not effectively control the facial selectivity of zirconium allylation. We thus decided to utilize reagent-based stereocontrol for the formation of the C8-C9 bond.

V. Deja Vu all over again: Evans Aldol Approach

The Evans aldol reaction of chiral, crotonyl-substituted oxazolidinones had proven effective for the asymmetric synthesis of hydroxypropionates in 1986.[12] In fact, this approach was utilized by the Roush group as part of their own synthetic efforts toward the total synthesis of tedanolide.[2d] In our lab, reaction of imide **21** with myriaporone aldehyde **15** provided the desired diastereomer **22** in 78% yield. This reaction was highly selective and could be performed on multigram scale. Within a very short time of our decision to pursue the Evans aldol reaction for the formation of the C8-C9 bond, this previously problematic step was no longer an obstacle to progress toward the myriaporones.

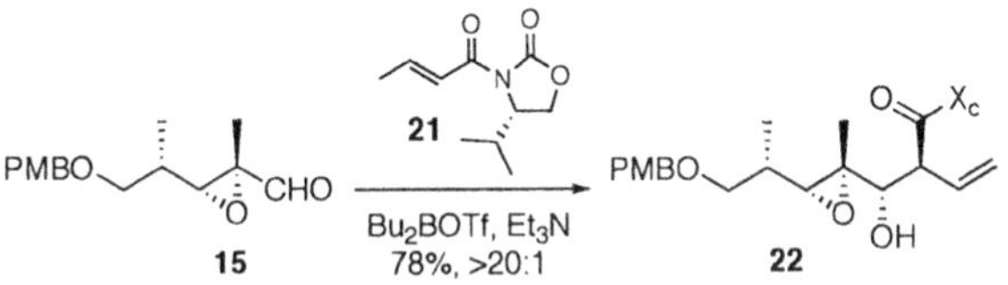

SCHEME 8

VI. Hindsight is 20:20: Myriaporone 1 or Myriaporones 3 and 4

After a series of functional group manipulations (Scheme 9), aldehyde **23** was generated in good yield, and this *common intermediate* could be utilized in the synthesis of either myriaporone 1 or 4 or even tedanolide. With one less unknown stereocenter, myriaporone 1 was the easy choice as an initial goal. Early attempts to create the C6-C7 bond involved modified Nozaki-Hiyama-Kishi couplings[13,14] with a functionally rich C1-C6 fragment but were unfortunately unsuccessful. However, a presentation at an NSF workshop by Professor Emmanuel Theodorakis (UCSD) provided a spark. He discussed an asymmetric homoallenylboration reaction his group had applied to the total synthesis of clerocidin.[15,16] This reaction produces a 2-substituted 1,3-diene, a potential intermediate in the synthesis of myriaporone 1 if selective functionalization of the monosubstituted olefin could be realized. In our lab, aldehyde **23** was treated with freshly prepared achiral boronate **24** to generate, after protection, bis-silyl ether **25**. Remarkably, only one C7 diastereomer was observed, and the identity of this stereocenter was tentatively assigned as indicated in **25**.

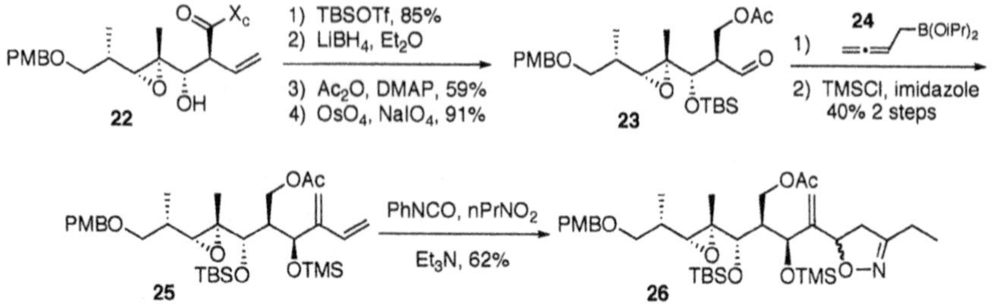

SCHEME 9

As an alternative to the aldol strategy for the formation of the C3-C5 β-hydroxy ketone, Rich suggested that we investigate a 1,3-dipolar nitrile oxide cycloaddition. Understandably, issues of chemo- and regioselectivity were of significant concern, but initial results were quite encouraging. Activation of 1-nitropropane with phenyl isocyanate generated the corresponding nitrile oxide that reacted with diene substrate **25**.[17] Much to our surprise (but not Rich's), this reaction generated isoxazoline **26** as a single chemo- and regioisomer in ~3:1 diastereoselectivity! Since the identity of the myriaporone 1 C5 stereocenter was unknown, an unselective cyclization was an ideal solution to this problem.

Although the diastereomers were inseparable at this stage, they were easily purified after installation of the C13-C14 Z-olefin (Scheme 10). Treatment of *bis*-silyl ether **27** with TBAF at 0°C resulted in deprotection of both silyl ethers, and the corresponding allylic alcohol was selectively oxidized using Dess-Martin periodinane.[18] At this point we were only one simple reductive hydrolysis removed from completing the total synthesis of myriaporone 1, but again we discovered that simple reactions of structurally complex substrates are rarely simple.

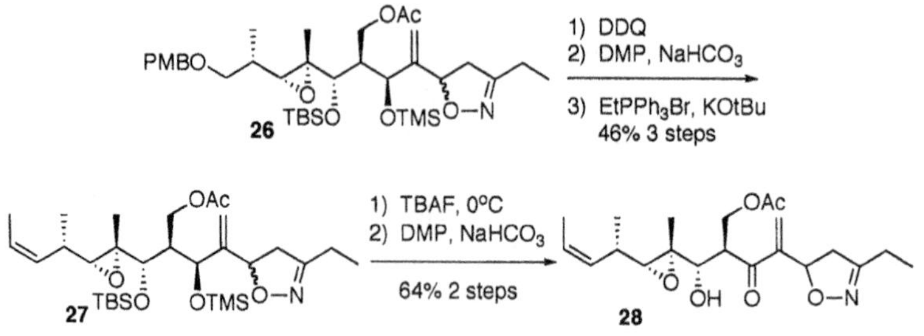

SCHEME 10

Hydrogenolysis of isoxazoline **28** with Raney nickel did successfully generate the C3-C5 β-hydroxy ketone; however, over-reduction of the C6 exo-olefin was also observed generating **29** in low yield. A variety of modified reaction conditions were screened, but over-reduction could not be avoided. We next attempted to protect the enone as the trimethylsilyl cyanohydrin, an approach that was successful in a model system, but this strategy was unsuccessful using myriaporone substrate **28**. Likewise, molybdenum hexacarbonyl failed to generate myriaporone 1. Because the late-stage reduction had proved difficult, we rearranged our end-game strategy to finish with a desilylation.

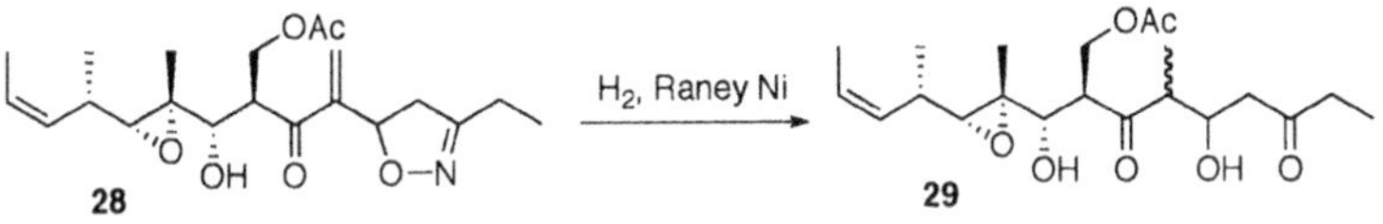

SCHEME 11

Since the C13-C14 olefin does not alter the reactivity of the C1-C9 portion of the myriaporone intermediates, subsequent efforts were explored on the C13-PMB ether (Scheme 12). Substrate **26** was successfully reduced with Raney nickel, and the trimethylsilyl ether was selectively cleaved with TBAF at –35°C. The C5 alcohol was selectively protected as the corresponding TBS ether, and C7 allylic alcohol was then oxidized under Dess-Martin conditions to provide bis-silyl ether **31**.

Again we found ourselves one simple reaction away from completing the total synthesis of myriaporone 1. A variety of desilylation conditions were screened, but none provided the desired product. Either no reaction was observed, or decomposition products, characterized by epoxide fragmentation and/or consumption of the enone, were recovered. After spending several months attempting final deprotection conditions (Scheme 12) and final reductions (Scheme 11), it was painfully clear that a new strategy would be required. At this point Brian and Jeff successfully defended their Ph.D. theses and Kristen took the lead. Her attempt to overcome the instability of our advanced synthetic intermediates would rely on a late-stage epoxidation, the stereochemical outcome of which was far from certain.

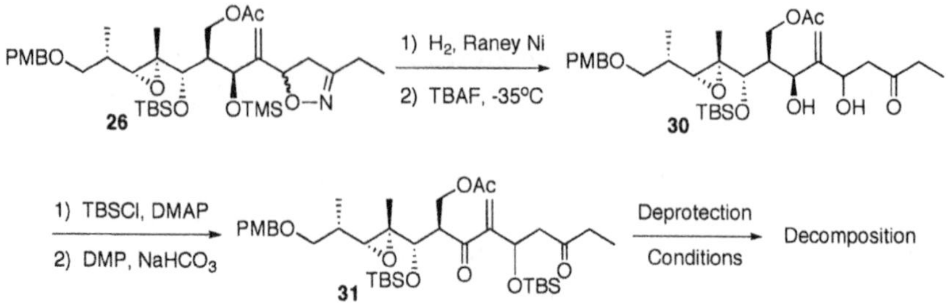

SCHEME 12

VII. A Critical Postponement: Late-Stage Epoxidation and the Targeting of Myriaporones 3 and 4

In 2001 and 2003, the Loh[2g] and Smith[4a,b] groups, respectively, whom chose to attempt late-stage epoxidations in their own tedanolide efforts, placed themselves at the forefront by obtaining unexpected but desired stereoselectivity. Much to the surprise of everyone, the desired diastereomer was formed exclusively in each case (Scheme 13). Even though the two substrates differ substantially, they both apparently adopt conformations favoring oxygen delivery to the same face as the hydroxyl group when m-CPBA is used. The yields in each case were moderate, but neither Loh nor Smith observed significant overoxidation.

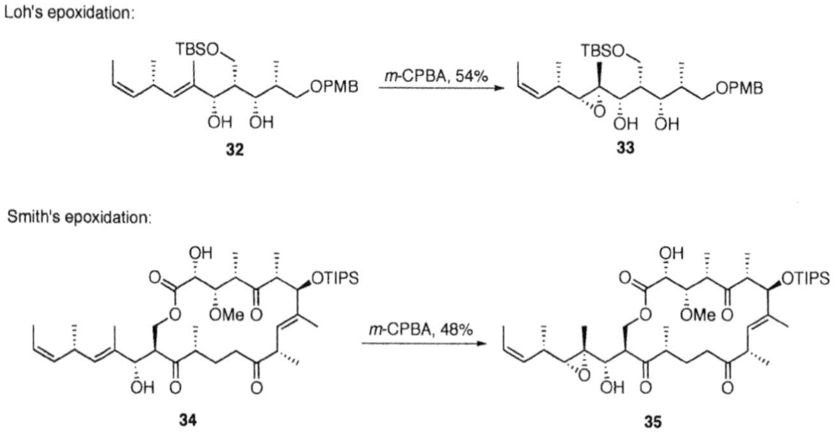

Loh's epoxidation:

Smith's epoxidation:

SCHEME 13

Prompted by the similarities between the Loh substrate and our own (and out of sheer frustration with the sensitivity of our advanced intermediates for myriaporone 1), we investigated our own late-stage epoxidation approach. Returning to the substrate previously used for the Sharpless asymmetric epoxidation, allylic alcohol **36** was oxidized to the corresponding aldehyde using IBX[19] (Scheme 14). This oxidant was chosen based on its ease of preparation and high yield. Following aldehyde formation, aldol employing chiral oxazolidinone **21** led to the formation of a single diastereomer in high yield. This product was concluded to be the desired diastereomer based on the high level of diastereoselectivity, an indication that the chiral oxazolidinone had exclusive control over the stereochemical outcome of the reaction. The aldol was followed by protection of the newly formed secondary hydroxyl as a TBS ether, reductive cleavage of the chiral auxiliary, and then protection with a second TBS group. Finally, selective dihydroxylation and oxidative cleavage of the diol provided a new common intermediate, aldehyde **39**.

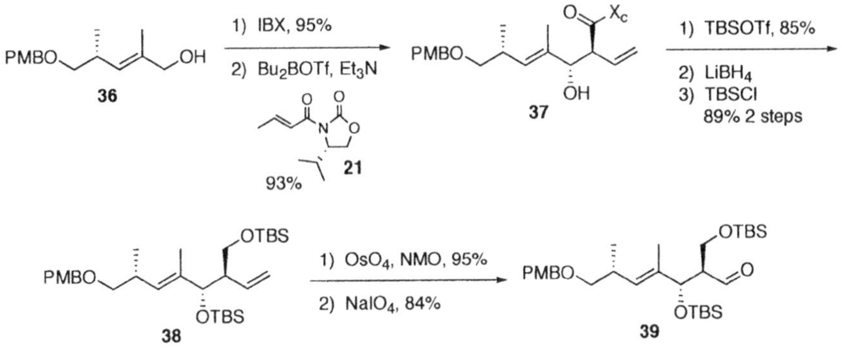

SCHEME 14

At this point, we shifted our target to myriaporones 3 and 4. With two unassigned stereocenters, C5 and C6, we simplified this problem by proposing that the C6 configuration was likely to be identical to that observed in tedanolide. Thus, the appropriate chiral auxiliary was chosen for a second aldol bond formation, and the C6-C7 bond was then formed in 99% yield using chiral oxazolidinone **40**, as shown in Scheme 15. Once again, the aldol led to formation of a single diastereomer, indicating the stereochemical outcome was controlled by the chiral oxazolidinone exclusively. The auxiliary was then cleaved via reduction with lithium borohydride, and the resulting primary hydroxyl protected with a third TBS group. Leaving the C7 hydroxyl group unprotected, a 1,3-dipolar cycloaddition using the nitrile oxide generated from 1-nitropropane and phenyl isocyanate generated isoxazoline **43** as a 1:1 mixture of inseparable diastereomers. Since the stereochemical orientation at C5 was not established, this center was intentionally formed as a mixture via the cycloaddition reaction as discussed previously. It was reasoned that, as long as the two diastereomers were separable at some point, they could aid in determination of the natural orientation at that site.

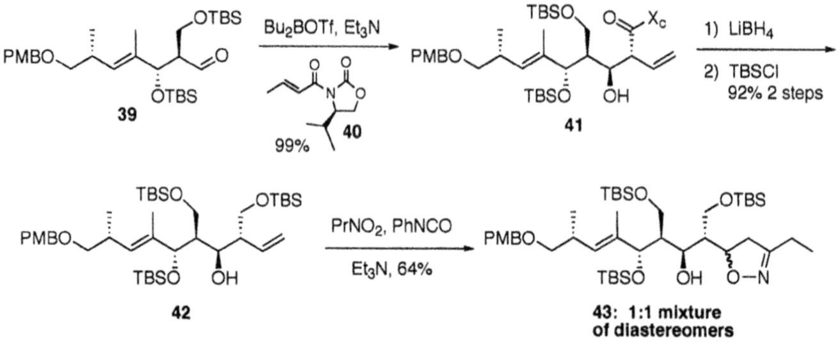

SCHEME 15

The mixture was taken on further, and the PMB ether removed to form the corresponding diol (Scheme 16). After selective oxidation of the primary hydroxyl with IBX, a Wittig olefination was used to form the C13-C14 bond, completing the carbon skeleton of myriaporone 4. Once the (Z)-olefin was installed, the two diastereomers **45a** and **45b** were now easily separated and each diastereomer was taken on independently.

At this point, the stereochemical identity of each diastereomer remained unknown, but for this treatise the less polar of the two is referred to as "**45a**" and the more polar referred to as "**45b**". Scheme 17 details chemistry for one of the two diastereomers, and an identical sequence was performed on the other without consequence. Dess-Martin periodinane was used to oxidize the C7 hydroxyl group to the corresponding ketone in very high yield. IBX proved to be quite slow in effecting this same transformation and generally the reaction did not go to

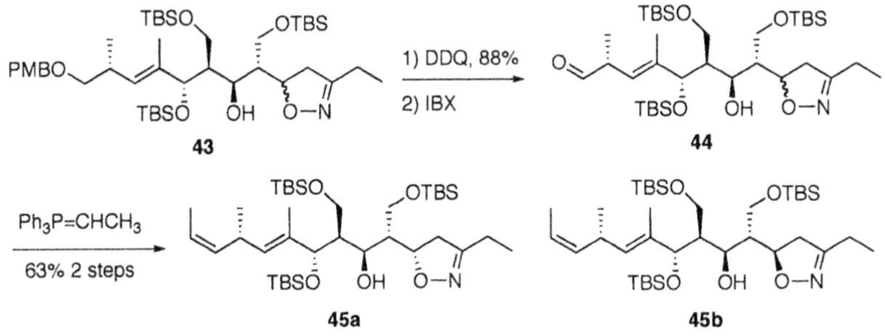

SCHEME 16

completion. At this point, the stereochemical identity of each diastereomer remained unknown, but for this treatise the less polar of the two is referred to as "**45a**" and the more polar referred to as "**45b**". Scheme 17 details chemistry for one of the two diastereomers, and an identical sequence was performed on the other without consequence. Dess-Martin periodinane was used to oxidize the C7 hydroxyl group to the corresponding ketone in very high yield. IBX proved to be quite slow in effecting this same transformation and generally did not go to completion. The next step, deprotection of the three TBS groups, proved to be somewhat challenging. The C9 TBS ether is quite hindered, and removal of this protecting group proved problematic in our previous efforts toward the completion of myriaporone 1. A number of conditions, including TBAF and aqueous HF, proved too harsh for this substrate and resulted in decomposition of the material. Although reaction with HF·Et₃N was slow, the desired triol was formed in good yield and was

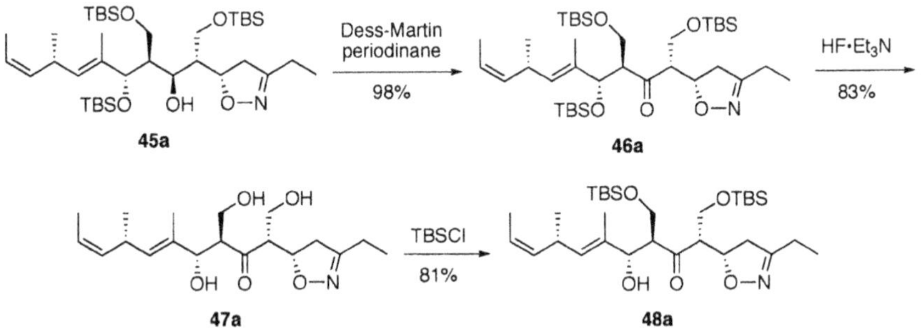

SCHEME 17

followed by reprotection of the two primary hydroxyl groups as TBS ethers to provide **48a**.

The next step was the reductive cleavage of the isoxazoline in order to unmask the β-hydroxy ketone. This transformation was previously accomplished in our efforts toward myriaporone 1 using hydrogenation conditions in the presence of Raney nickel. In fact, there are a number of known methods for effecting this conversion, and for this particular substrate, the cleanest, simplest, and highest yielding reaction was obtained using molybdenum hexacarbonyl,[20] as shown in Scheme 18. Although the yield is somewhat moderate, this can be attributed to the fact that the reaction tended not to go to completion. Fortunately, the unreacted starting material was easily recovered and then resubjected to the same reaction conditions.

Following formation of intermediate **49a**, the stage was set for the crucial epoxidation reaction. First employing the reaction conditions utilized by the Smith group, an excess of purified *m*-CPBA was used, and the reaction was run at a very dilute concentration. The *m*-CPBA was added at –30°C, and the reaction allowed to warm to room temperature and stir overnight. Unfortunately, when applied to our substrate, these conditions resulted in epoxidation of both olefins. Although the trisubstituted olefin would be expected to be more reactive in this type of reaction than the disubstituted olefin, both appeared to be equally reactive under these conditions. Recall that the Smith group observed no overoxidation in their epoxidation reaction (Scheme 13). The Loh group performed their reaction at a higher concentration with only a slight excess of *m*-CPBA and kept the reaction at –30°C for an extended period of time. When these reaction conditions were applied to substrate **49a**, only a small amount of the desired product was formed. However, the majority of the material consisted of the *bis*-epoxide product once again. When the reaction temperature was maintained at –78°C, no reaction took place. Finally, the *m*-CPBA was added at –78°C, and the reaction warmed to –50°C and maintained at this temperature overnight. This resulted in formation of mainly the desired product along with a small amount of the *bis*-epoxide product. Optimized conditions (0.9 equiv. *m*-CPBA and –50°C) resulted in exclusive formation of the desired product along with a small amount of unreacted starting material. The yield based on the recovered starting material was 81%!

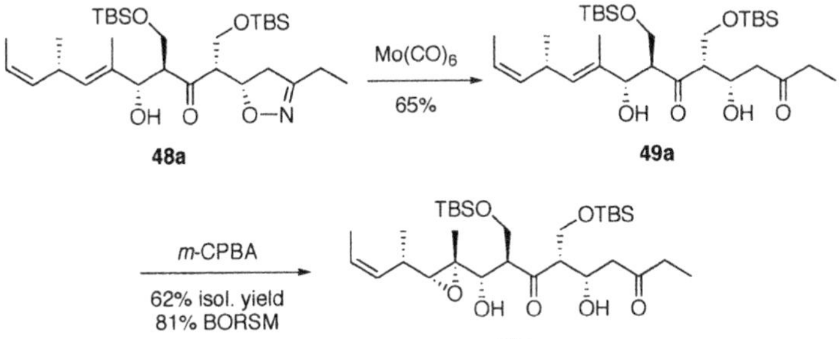

SCHEME 18

With the desired intermediate **50a** in hand, the final step required deprotection of the two primary TBS ethers. This deprotection is significantly different than the endgame of our initial attempts toward myriaporone 1. Our previous efforts called for deprotection of the very hindered C9 secondary hydroxyl with the epoxide present, while this new substrate simply requires deprotection of two primary hydroxyls. Also, very limited quantities of advanced intermediates toward myriaporone 1 were available due to two lower yielding reactions in the middle of the sequence. The current route toward myriaporones 3 and 4 proved to be highly efficient and provided plenty of material with which to examine the final deprotection step. A mild and neutral reagent, tris-(dimethyl-amino)sulfonium trimethylsilydifluoride (TASF),[21] was chosen for the final deprotection (Scheme 19) and in fact, reaction with **50a** at 0°C led to clean formation of an equilibrium mixture of myriaporones 3 and 4.

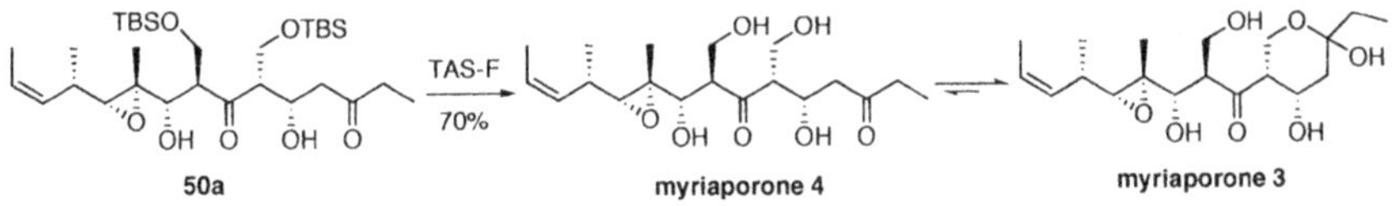

SCHEME 19

After comparing ^{1}H NMR spectra for the products obtained from the reactions of each of the two diastereomers with spectra corresponding to the natural products,[22] the product derived from diastereomer **50a** proved to be an exact match, while the product formed from **50b** had distinct differences. Figure 1 shows the spectrum for authentic myriaporones 3 and 4 compared to the product obtained from reaction of **50a** and TASF.

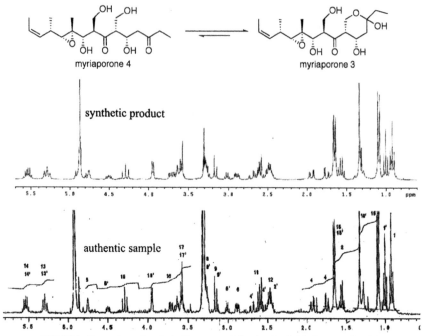

FIGURE 1. Comparison of synthetic diastereomer "a" and authentic Myriaporones 3 and 4.

The ¹H NMR spectra clearly show a match between the authentic and synthetic samples of myriaporones 3 and 4.[22] Moreover, optical rotation data collected on the synthetic sample defined the absolute configuration of the natural materials not previously determined as shown in this chapter.

At the time of isolation, 1.41 kilograms of *Myriapora truncata* yielded only 0.6 milligrams of the equilibrium mixture of myriaporones 3 and 4. Just as the two were inseparable at that time, it was impossible to separate them when formed synthetically. Additionally, myriaporone 3 was the dominant form in both instances. Without working on a particularly large scale, the route discussed led to the formation of 12 milligrams of the natural product.

VIII. Unexpected Myriaporone 1 Formation

During the completion of myriaporones 3 and 4, the final step proved somewhat problematic prior to using TASF for the deprotection.

Therefore, over the course of examining methods of effecting the desired transformation, the two primary hydroxyl groups were protected with acetates as shown in Scheme 20.[23] It was reasoned that these protecting groups should be even more readily removed under mild conditions than silyl protecting groups. As shown, triol **47a** was converted to the corresponding *bis*-acetate product **51a** in high yield. This was followed once again by reductive cleavage of the isoxazoline and then epoxidation to give the desired product **53a**. As in the case of the *bis*-TBS material, the Mo(CO)$_6$ and epoxidation reactions resulted in formation of the desired products as well as small amounts of recoverable unreacted starting materials.

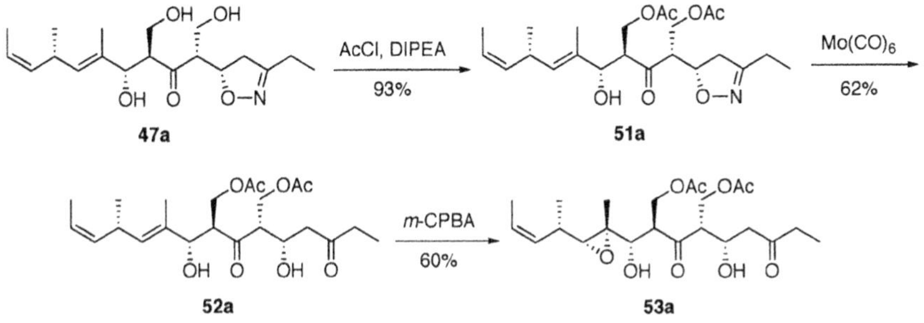

SCHEME 20

In order to deprotect the acetates, KCN in MeOH was chosen as a mild method to effect transesterification.[24] Since the NMR spectra corresponding to authentic samples of the natural products were obtained in CD$_3$OD and since methanol was the reaction solvent, the transformation was performed in an NMR tube and followed by NMR. Surprisingly, the transesterification reaction did not take place at all, and instead, the C6 acetoxy group eliminated selectively to provide myriaporone 1 exclusively. The identical reaction occurred when applied to both C5 diastereomers.

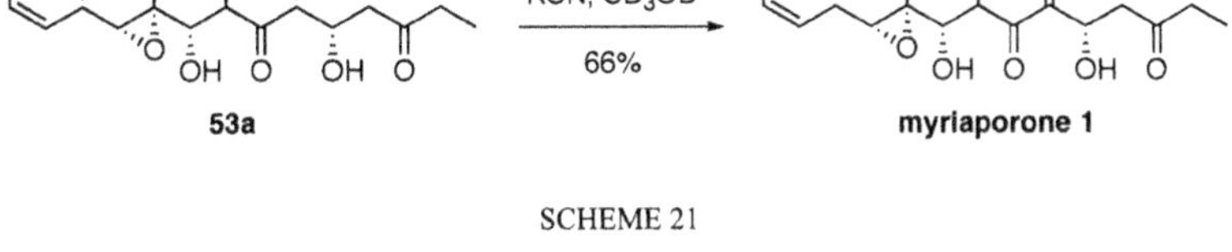

SCHEME 21

As in the case of myriaporones 3 and 4, ^{1}H NMR spectra corresponding to the two diastereomers "a" and "b" were compared to the spectrum for an authentic sample of myriaporone 1.[22] As Figure 2 shows, the product obtained from reaction of **53a** and KCN was an exact match. The product obtained from **53b**, however, showed distinct differences when compared to the authentic myriaporone 1 spectrum.

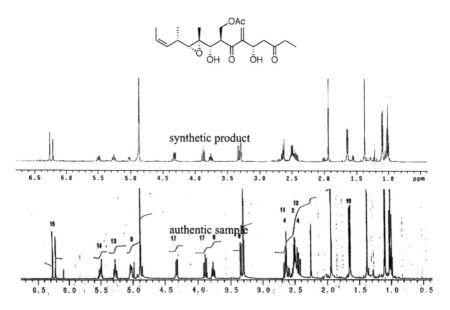

FIGURE 2. Comparison of synthetic diastereomer "a" and authentic Myriaporone 1.

One possible explanation for the selective elimination of the C6 acetoxy group is shown in Scheme 22. Following attack of the cyano anion on the C3 ketone, the resulting alkoxide ion could serve as an intramolecular base and deprotonate C6, leading to elimination of the acetoxy group. This theory is further supported by the fact that this deprotonation would occur via a six-centered transition state. Molecular modeling experiments will be used in the near future to predict the preferred conformations of the myriaporones in order to gain further insight into the cause of this selective elimination.

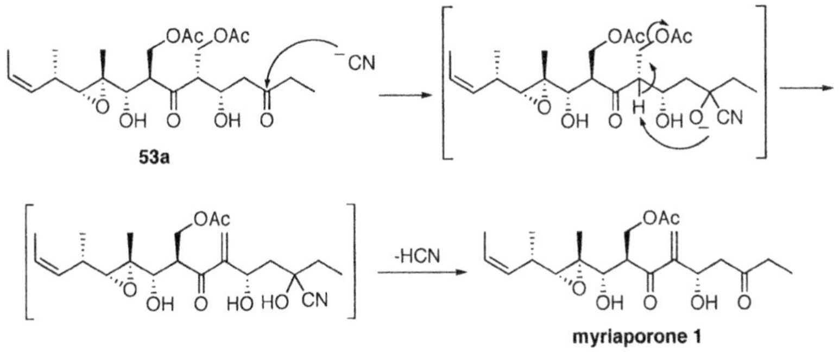

SCHEME 22

IX. Stereochemical Assignment of Myriaporones 1, 3, and 4

With the total synthesis of the myriaporones complete, the spectral data were used to determine the absolute configuration of this class of natural products. Although there was a strong possibility that the previously unassigned centers, C5 and C6, have configurations identical to the corresponding centers in tedanolide (C13, C14), irrefutable proof was needed to support this theory.

Before ascertaining the stereochemical configuration at C5, we compared coupling constants from the ^{1}H NMR spectra for the C8-C15 regions of authentic and synthetic myriaporones 1 and 3 to those for the side-chain of tedanolide in order to confirm their correlation, as shown in Table 1. This was also used to verify that our epoxidation had in fact produced the desired diastereomer. As the table indicates, not only do the coupling constants for this particular region correlate between the natural myriaporones and tedanolide, but our synthetic myriaporones match as well. This evidence supports the conclusion that the stereochemistry of the C8-C15 region and the geometry of the C13-C14 olefin of the natural myriaporones correspond identically to the side-chain of tedanolide and that our synthetic compounds possess the same configurations at these centers as well.

In order to assign the stereochemical configurations at C5, the ^{1}H NMR spectrum for myriaporone 3 was carefully examined. Figure 3 shows the C1-C7 portion of myriaporone 3 and the coupling constant assignments for each of the protons in this region. Although the proton at C5 appeared as a multuplet in the spectrum, the protons at C4 displayed vicinal coupling constants of 3.3 Hz for each. These values correspond to an axial/equatorial vicinal coupling and an equatorial/equatorial vicinal

TABLE 1

COMPARISON OF COUPLING CONSTANTS (IN Hz) FOR C8-C15 REGION OF MYRIAPORONES 1
AND 3 AND CORRESPONDING CENTERS IN TEDANOLIDE

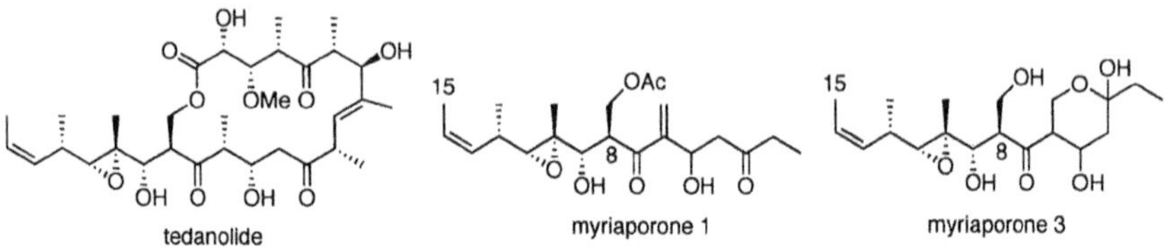

	H8/9	H11/12	H12/13	H13/14	H14/15	H13/15	H12/19
Tedanolide	9.5	9.4	10.8	10.8	7.4	1.7	6.5
Authentic Myr. 1	10.0	9.5	10.0	10.0	7.0	1.5	6.7
Synthetic Myr. 1	9.9	9.3	10.4	9.9	6.9	1.8	6.6
Authentic Myr. 3	9.7	9.2	10.0	10.0	7.0	1.5	6.8
Synthetic Myr. 3	10.0	9.5	9.6	11.0	6.5	1.5	6.5

coupling in a chair conformation. Thus, the proton at C5 must be equatorial. Likewise, the vicinal coupling constants for the proton at C6 are 12.0 Hz and 4.5 Hz. A high value for one of the coupling constants indicates an axial/axial vicinal orientation, meaning that the proton at C6

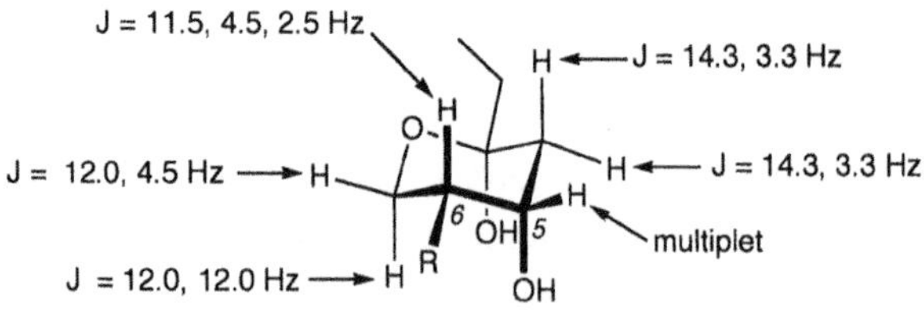

FIGURE 3

must be axial. Recall the absolute stereochemistry at C6 was defined by the choice of chiral auxiliary for the aldol reaction and the corresponding C14-stereogenic center of tedanolide. Combined with the optical rotation data, the stereochemistry of the myriaporone class of natural products has now been unambiguously determined.

X. Perspective

The successful synthetic effort is a result of the hard work and creativity of several exceptional graduate students. The total synthesis of the myriaporones represents only the first stage of a project designed to study this interesting class of natural products. Isolation of myriaporone's biological receptor, determination of the mode of action, and its relationship to tedanolide are currently being targeted by the next crop of coworkers.

References and Footnotes

1. (a) Schimtz, F. J., Gunasekera, S. P., Yalamanchili, G., Hossain, M. B., van der Helm, D., *J. Am. Chem. Soc.* **1984**, *106*, 7251. (b) O'Hagan, D., *Nat. Prod. Rep.* **1993**, 593.
2. (a) Matsushima, T., Horita, K., Nakajima, K., Yonemitsu, O., *Tetrahedron Lett.* **1996**, *37*, 385. (b) Liu, J.-F., Abiko, A., Pei, Z, Buske, D. C., Masamune, S., *Tetrahedron Lett.* **1998**, *39*, 1873. (c) Taylor, R. E., Ciavarri, J. P., Hearn, B. R., *Tetrahedron Lett.* **1998**, *39*, 9361. (d) Roush, W. R., Lane, G. C., *Org Lett.* **1999**, *1*, 95. (e) Jung, M. E., Marquez, R., *Org. Lett.* **2000**, *2*, 1669. (f) Matsushima, T., Nakajima, N., Zheng, B.-Z., Yonemitsu, O., *Chem. Pharm. Bull.* **2000**, *48*, 855. (g) Loh, T.-P., Feng, L.-C., *Tetrahedron Lett.* **2001**, *42*, 3223. (h) Matsui, K., Zheng, B. Z., Kusaka, S., Kuroda, M., Yoshimoto, K., Yamada, H., Yonemitsu, O., *Eur. J. Org. Chem.* **2001**, 3615. (i) Hassfeld, J., Kalesse, M., *Synlett* **2002**, 2007. (j) Roush, W. R., Newcom, J. S., *Org. Lett.* **2002**, *4*, 4739.
3. Fusetani, N., Sugawasara, T., Matsunaga, S., *J. Org. Chem.* **1991**, *56*, 4971.
4. (a) Smith III, A. B., Adams, C. M., Lodise, S, Barbosa, S. A., Degnan, A. P., *J. Am. Chem. Soc.* **2003**, *125*, 350. (b) Smith III, A. B., Adams, C. M., Barbosa, S. A., Degnan, A. P. *Proc. Natl. Acad. Sci.* **2004**, *101*, 12042.
5. Rinehart, K. L., Tachibana, K., *J. Nat. Prod.* **1995**, *58*, 344.
6. Rinehart, K. L., Cheng, J.-F., Lee, J.-S. US Patent 5, 514, 708 May 7, 1996.
7. (a) Taylor, R. E., Fleming, K. N., *Angew. Chem. Int. Ed.* **2004**, *43*, 1728. (b) Pérez, M., del Pozo, C., Reyes, F., Rodríguez, A., Francesch, A., Echavarren, A. M., Cuevas, C., *Angew. Chem. Int. Ed.* **2004**, *43*, 1724. (c) Taylor, R. E., Hearn, B. R., Ciavarri, J. P., *Org Lett.* **2002**, *4*, 2953. (d) Zheng, B.-Z., Yamauchi, M., Dei, H., Yonemitsu, O., *Chem. Pharm. Bull.* **2000**, *48*, 1761. (e) Taylor, R. E., Ciavarri, J. P., Hearn, B. R., *Tetrahedron Lett.* **1998**, *39*, 9361.
8. Rossiter, B.E., Verhoeven, T.R., Sharpless, K.B., *Tetrahedron Lett.* **1979**, *20*, 4733.
9. Diana, S.-C.H., Sim, K.-Y., Loh, T.-P., *Synlett*, **1996**, 263.
10. Clark, A.J., Kasujee, I., Peacock, J.L., *Tetrahedron Lett.* **1995**, *36*, 7137.
11. Hearn, B.R. "*A Zirconium-mediated Allylation Approach to the Total Synthesis of Tedanolide and the Myriaporones.*" Ph.D. Dissertation, University of Notre Dame: Notre Dame, IN, 2001.
12. Evans, D.A., Sjogren, E.B., Bartroli, J., Dow, R.L., *Tetrahedron Lett.* **1986**, *27*, 4957.
13. Taylor, R.E., Ciavarri, J.P., *Org. Lett.* **1999**, *1*, 467.

14. Ciavarri, J.P. *"Studies Toward the Total Synthesis of Myriaporone 1"* Ph.D. Dissertation, University of Notre Dame: Notre Dame, IN, 2001.
15. Soundararajan, R., Li, G., Brown, H.C., *J. Org. Chem.* **1996**, *61*, 100.
16. Xiang, A.X., Watson, D.A., Ling, T., Theodorakis, E.A., *J. Org. Chem.* **1998**, *63*, 6774.
17. Mukiyama, T, Hoshino, T., *J. Am. Chem. Soc.* **1960**, *82*, 5339.
18. Dess, D.B., Martin, J.C., *J. Am. Chem. Soc.* **1991**, *113*, 7277.
19. Frigerio, M., Santagostino, M., *Tetrahedron Lett.* **1994**, *35*, 8019.
20. Baraldi, P.G., Barco, A., Benetti, S., Manfredini, S., Simoni, D., *Synthesis* **1987**, *3*, 276.
21. Scheidt, K.A., Chen, H., Follows, B.C., Chemler, S.R., Coffey, D.S., Roush, W.R., *J. Org. Chem.* **1998**, *63*, 6436.
22. ^{1}H and ^{13}C NMR spectra for authentic samples of myriaporones 1-4 provided by Professor Kenneth L. Rinehart.
23. Ishihara, K., Kurihara, H., Yamamoto, H., *J. Org. Chem.* **1993**, *58*, 3791.
24. Mori, K., Tominaga, M., Takigawa, T., Matsui, M., *Synthesis* **1973**, *12*, 790.

ES AND TACTICS IN ORGANIC SYNTHESIS, VOL. 6
evier Ltd. All rights reserved.

er 9

NTURES IN NATURAL PRODUCT SYNTHESIS: M DEEP SEA SPONGE TO PILOT PLANT. THE GE SCALE TOTAL SYNTHESIS OF THE MARINE JRAL PRODUCT (+)-DISCODERMOLIDE

J. Mickel
J and Analytical Development
Pharma AG,
Basel, Switzerland

duction.

is chapter, I would like to share with you my experience with the scodermolide **1**, which is a novel polyketide natural product first olated in very small quantities from extracts of the marine sponge *iscodermia dissoluta* by researchers at Harbor Branch Oceanographic stitution (HBOI).[1]

![DISCODERMIA DISSOLUTA photograph]

DISCODERMIA DISSOLUTA

This is the most complex molecule whose total synthesis has ever been undertaken by Novartis process research. The synthetic route was not chosen or, better said, designed by my colleagues or me but the chemistry was scaled-up and optimized by us.[2] It caused us lots of stress, but we had lots of fun and in many ways it was an adventure, as you will read below.

1

But before I get into details, just what is it that a process research department does? It does two main things: firstly, process research has the

›e guaranteed throughout the clinical development process and
mes beyond, and the drug substances must be prepared under strict
conditions. What does cGMP mean? An adequate description is
l the scope of this chapter, but briefly, for the synthetic chemist it
es that the drug substance must be prepared in an absolutely
ucible quality and quantity. That is, each time the drug substance is
sized identical material must be produced each time regardless of
›cess or the equipment used. Secondly, in order to attain this goal
s research must develop robust and safe chemical processes and
tely introduce the process into chemical production for the
tion of market quantities of any given drug substance. As we will
:se points are crucial to any synthesis, and in this particular case
aused us many heart-stopping moments. This may be broadly
eted to mean that one must always expect the unexpected to occur
leed it did.

s story is not one of synthetic strategy; the synthesis was given. The
·y here deals with how to avoid scale-up problems, selecting
ts and reaction conditions to allow safe and efficient manufacture
lot plant to ultimately produce material of such a quality that it can
zed for human clinical trials.

›ments on the Structure

ıcturally, discodermolide consists of a linear polypropionate chain
ıing thirteen stereocenters, six of which are hydroxyl bearing, with
these esterified as a γ-lactone (C_5) and another as a carbamate
It also features seven methyl-bearing stereocenters and three Z-
ured alkenes, one of these being part of the terminal diene unit
r being a cis-trisubstituted system at $C_{13\text{-}14}$, which offers a synthetic
ıge in controlling the stereochemistry. Also present in the structure
mmon stereo triad (methyl, hydroxyl, and methyl) that is repeated
imes.

may be seen from the X-ray structure (Figure 1) and the solution
re[3,4] the molecule adopts a U-shaped conformation, where the
ıl (Z)-alkenes are functioning as conformational locks by
ızing 1,3-allylic strain between their respective substituents and
ntane interactions along the backbone are avoided, while the γ-
 is held in a boat-like conformation (Figure 1).
Schreiber group[5] has synthesized both antipodes, thus establishing
solute configuration of 1. Since the publication of Schreiber's

ynthesis, several total syntheses[6-9] and many syntheses of various
discodermolide fragments[10] have appeared in the literature. An excellent
review of the available synthetic approaches has recently been
published.[11]

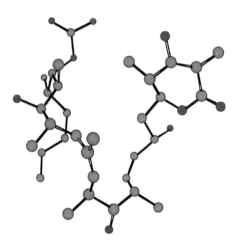

FIGURE 1: X-ray structure of discodermolide.

III. Project Background

A small, but structurally diverse collection of naturally occurring non-
taxane microtubule stabilizing agents (MTS) has been discovered over the
last decade. These include the epothilones (EPO), eleutherobin,
laulimalide, dicytostatin (Figure 2) and (+)-discodermolide.

Discodermolide, which stabilizes microtubules faster and more
potently than any of the other known MTS agents, is a potent inhibitor of
tumor cell growth *in vitro*, including paclitaxel- (PTX) and EPO-resistant
cells.[12] It also demonstrates significant human tumor growth inhibition in
hollow fiber and xenograft mouse models (including paclitaxel-resistant
tumors[13]). In 1998, Novartis licensed this compound from the HBOI and
also obtained some material prepared by Professor Smith[7c] and his group
for the initial preclinical biological evaluation.

The compound supply for development cannot be met through the
isolation and purification of discodermolide from *Discodermia sp.* (which
must be harvested using manned submersibles). Attempts to reproducibly
isolate a discodermolide-producing microorganism for fermentation have
not been successful to date. Therefore, all discodermolide used for late
preclinical research and development activities as well as for the ongoing

clinical trial has to be supplied by total synthesis. Discodermolide is currently undergoing phase 1 clinical trials.

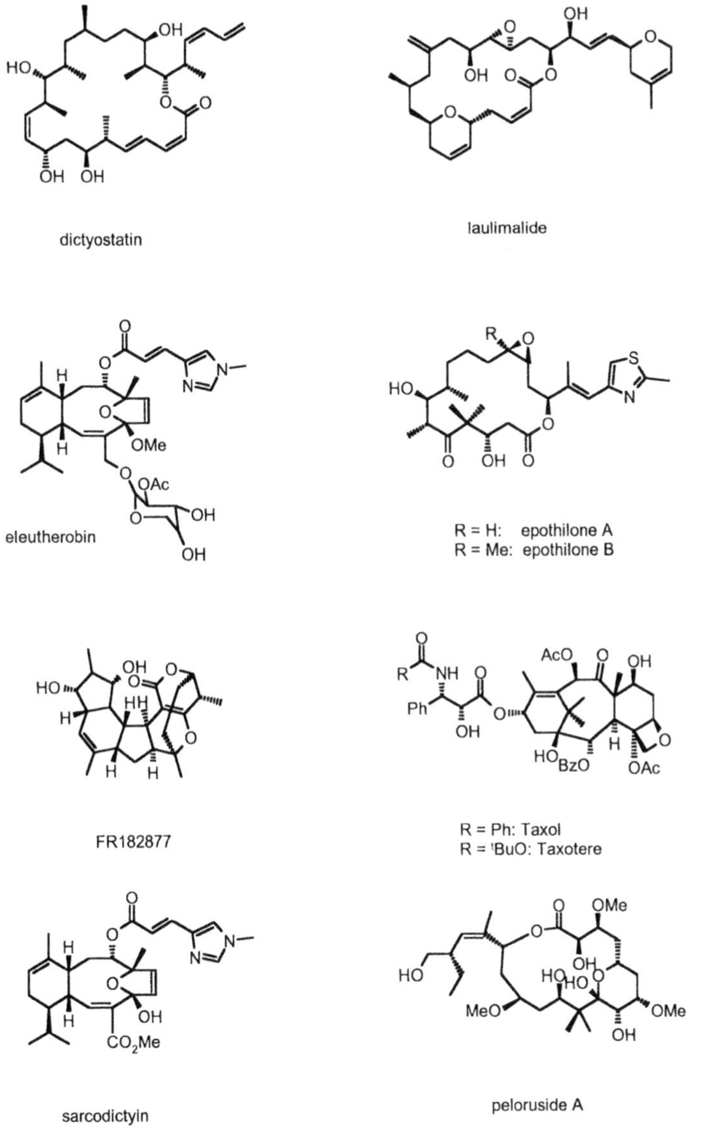

FIGURE 2. Microtubule stabilizing agents.

IV. Literature Evaluation and Route Selection

In late 1999, we established a working group that was given the task of evaluating the available literature routes (at that time around 5) and the fragment syntheses with a view to scale-up potential in order to deliver 100-gram quantities of discodermolide. To achieve an objective evaluation, a number of criteria were selected to aid this process including, amongst others, (a) any requirement for special equipment (e.g. low temperature), (b) scale-up problems, (c) the availability of reagents, (d) problematic reagents and steps (e.g. stereoselective reactions), (e) changes in oxidation state, (f) the total number of steps, (g) the number of steps and yield in longest sequence, (h) the number of crystalline intermediates, (i) the number of chromatographic purifications, (j) the yield of last steps, and (k) the final purification. After a discussion of concepts and these critical points two strategies were then selected to be the most promising for further investigation. Proposals for modifications of the favored routes were also discussed. The general conclusion was that the published syntheses couldn't be up scaled without major modifications. A few examples will suffice to illustrate these points.

Similar in all the reported routes was the total number of steps in the longest linear chain of the synthesis, around 30, with the total number of steps ranging from 35 to 50. The reported overall yields were all between 2–10%. Almost every step in the literature routes required chromatography with very few to no crystalline intermediates being reported. The crystallization of the final product was not defined in any of the publications. All the routes recognized the stereo-triad present in

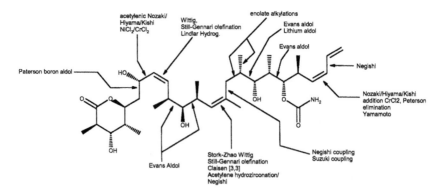

FIGURE 3. Strategic bond formation.

the structure. This allows a common precursor to be prepared that can, by suitable manipulation, be used to set the stereochemistry of various advanced fragments. These are ultimately assembled to produce (+)-discodermolide. While this approach makes the routes more efficient it does not make them any shorter.

Figure 3 shows a summary of the strategic bond-forming reactions employed to construct the various fragments of the discodermolide backbone. From this figure one can see that stereoselective processes form many of the bonds. Myles and co-workers stumbled across this exact point in their elegant approach to (-)[7a] and (+)-discodermolide[7b] (Figure 4).

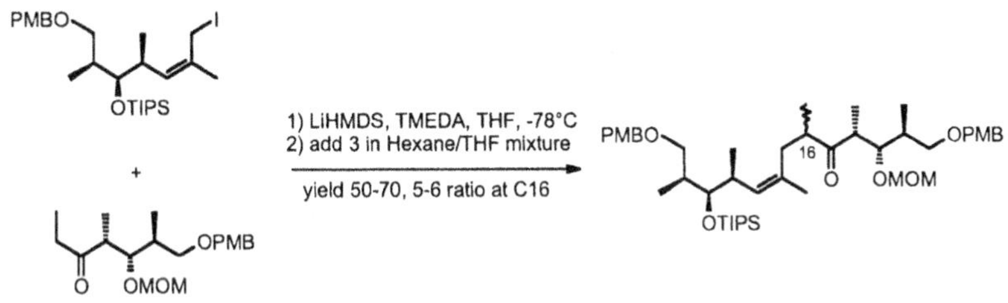

FIGURE 4. Myles' alkylation step.

The C_{16} (R):(S) ratio in the product varied from almost 1:1 to 6:1 in favor of the desired (R) configuration. Further study revealed how complex this alkylation procedure turned out to be.[7b] For example, appropriate control of the temperature, counterion, solvent system and concentration, amongst others, were surmised to be success factors in this reaction.

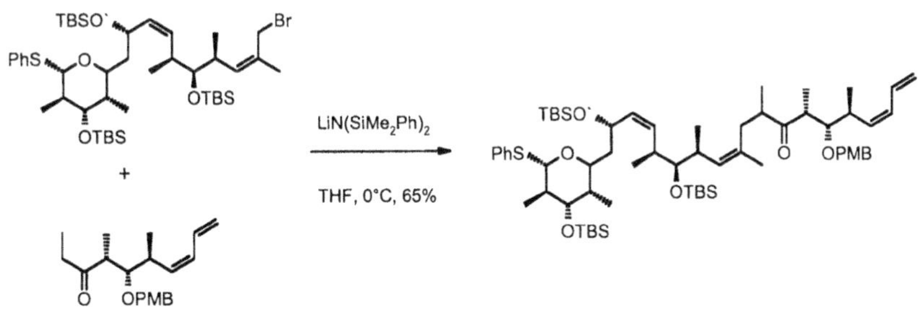

FIGURE 5. Schreiber's alkylation step.

Schreiber[5b] observed similar effects in his key alkylation processes for the construction of the same bond, although in this case the isomeric ratio was favored in the undesired direction (Figure 5).

This was partially overcome by fragment coupling employing a methyl ketone, followed by methylation to introduce the C_{16}-methyl group, (Figure 6). This produced a 3:1 mixture of isomers at C_{16} in the desired sense and the coupled product was produced using this two-step process.

For our requirements, this type of complex C-C bond construction was therefore considered very difficult to scale-up. Consequently, we decided to avoid this type of fragment coupling.

Special Equipment: quite a few of the available routes employ double bond ozonolysis to prepare small fragments. This was not an option for us, especially at the beginning of the synthesis where the quantities are expected to be rather large and scale-up problems are programmed in.

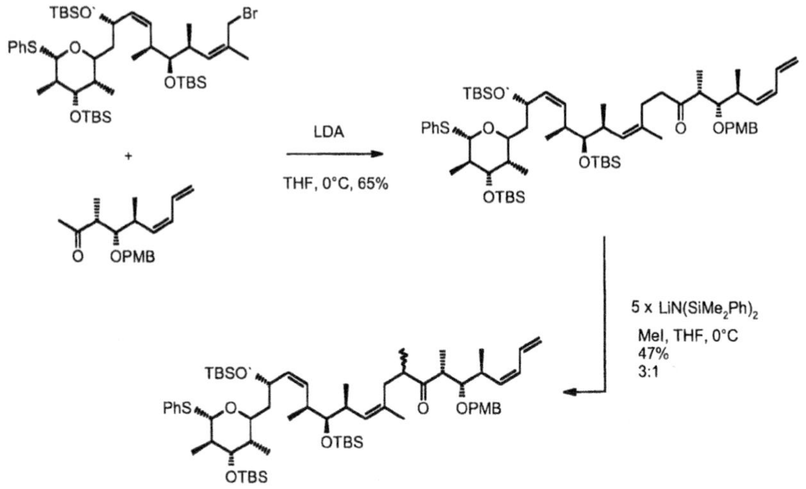

FIGURE 6. Schreiber's second attempt.

One sequence described by Smith[6c] employed a high-pressure reaction (12 kbar) to prepare a phosphonium salt for the construction of the C_{8-9} double bond (Figure 7). This type of process is not practicable on a 3–5 kilogram scale, at least not in our facilities. As a consequence, the following steps of the Smith route were unavailable to us.

This limitation was eventually overcome by a judicious choice of protecting groups so that the high-pressure reaction was avoided.[14] However, at the time we began the synthesis, this simple solution was not available to us.

The synthesis of the *cis*-trisubstituted C_{13-14} double bond is very difficult and several elegant methods have been presented for its construction. For example, Paterson employed an elegant route utilizing a Claisen rearrangement[9a] (Figure 8). This approach nicely sets the stereochemistry of the double bond via an 8-membered lactone ring. Subsequent ester hydrolysis reveals the desired stereochemical unit.

Another interesting approach is that described by Panek,[10c] who employed an acetylene hydrozirconation process followed by a Negishi coupling (Figure 9).

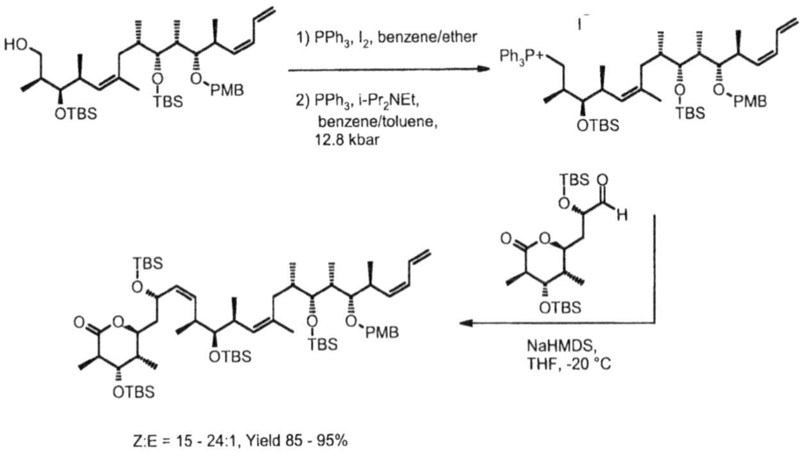

Z:E = 15 - 24:1, Yield 85 - 95%

FIGURE 7. High-pressure phosphonium salt formation.

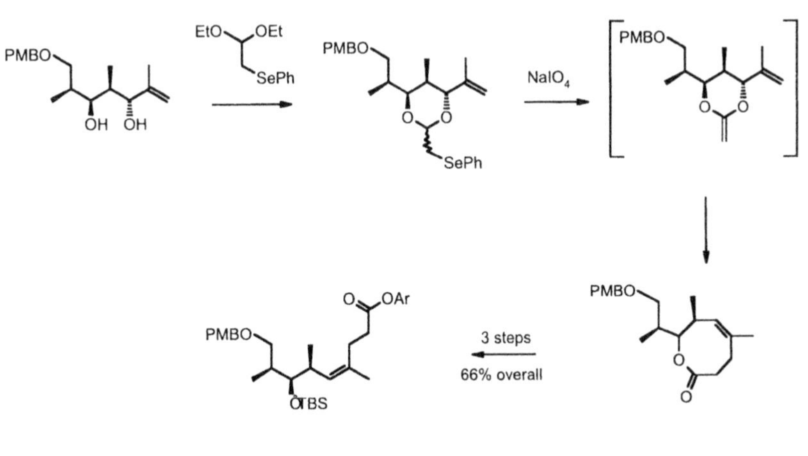

FIGURE 8. Paterson's Claisen approach to C_{13-14} subunit.

Unfortunately neither of these routes was an option for our proposed large scale approach and we decided to remain with the method, via a Stork/Zhao modification of the Wittig reaction, for this subunit as described by Smith,[6c] in spite of the reported yield (30 – 40%).

A large array of heavy metals was employed in the described syntheses and indeed the published routes appeared to be a walk through the periodic table. For example, Cr, Hg, Ni, Os, Pd, Sn, Zn, and Zr, to mention a few, were used at various points. Many of these reagents are

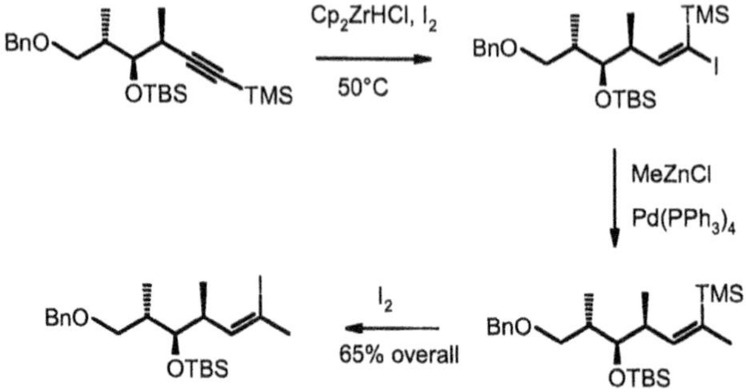

FIGURE 9. Panck's approach.

not commercially available and are difficult to prepare on the large scale, and when they are available the stability and quality of the reagents is suspect, as is their use in commercial synthesis (i.e., environmental concerns). However, they do have their place in modern organic chemistry and it remains a challenge for chemical development and chemical production to use them properly.

Schreiber utilized the attractive strategy of protecting the discodermolide lactone as a thioacetal (Figure 10). This was carried through the synthesis and reverted to the lactone at the end by treatment with mercuric chloride followed by oxidation.

Mercury is a complete taboo, especially at the end of the synthesis where extreme measures have to be taken to ensure its absence in the final drug product.

The use of chromium is almost unavoidable for the Nozaki-Hiyama-Kishi/Peterson elimination sequence. This was successfully employed by Paterson[9a] for the introduction of the cis-diene unit. (Figure 3)

The organic chemistry of boron also played a significant role in the total syntheses of discodermolide (Figure 3). For example, several Evans *syn* aldols are employed for C-C bond formation. Paterson utilized a boron aldol to complete a fragment coupling producing the complete carbon backbone of discodermolide (Figure 11).[9a]

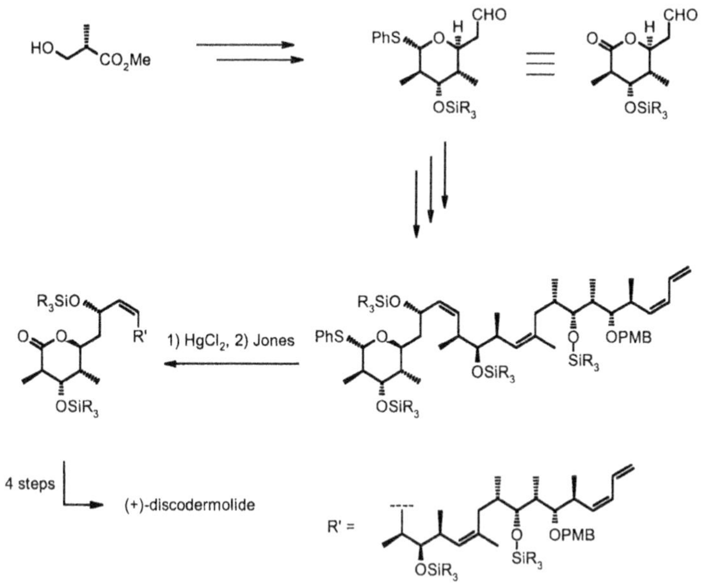

FIGURE 10. Schreiber's lactone strategy.

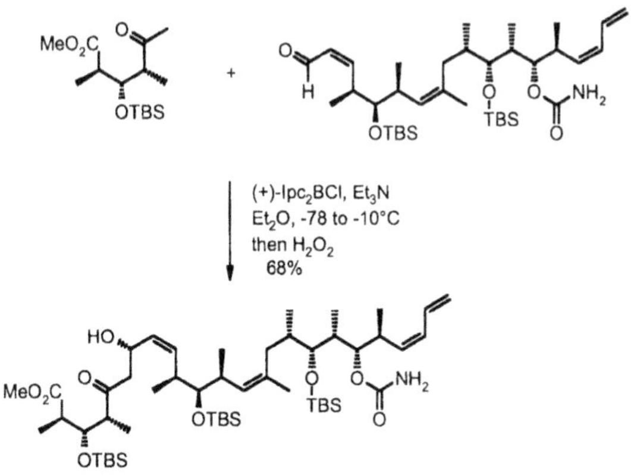

FIGURE 11. Paterson's endgame.

This reaction is mediated by (+)-diisopinocampheylboron chloride [(+)-DIP-Cl)] and produces the aldol product in good yield with reasonable selectivity. This is the first case in which this reagent system has been used successfully in a triple asymmetric induction situation to achieve the desired stereocontrol in a mismatched aldol coupling.

In all the literature syntheses, the final purification of the drug substance was not defined and this remained a large unknown at the start of this work.

The team agreed on a proposal to select the combination of the synthesis of the Smith and Paterson groups for further evaluation (NSP route). This combination employs the advantages of both routes, utilizing Professor Smith's route with a common intermediate up to the point just before the high pressure reaction then moving this intermediate into the Paterson end game. The retrosynthetic analysis is displayed in Figure 12 and the key intermediates of the chosen route are outlined in Scheme 1 (discodermolide numbering). While this route did not quite fit all of our criteria, we felt that this was a good compromise.

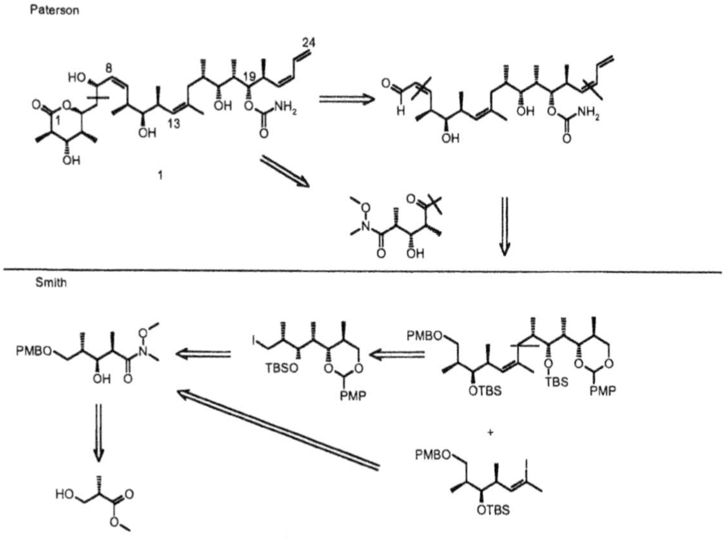

FIGURE 12. Retrosynthetic analysis of the NSP route to (+)-discodermolide.

The proposed route has a total of 27 steps in the longest synthetic linear sequence and takes advantage of the readily available ester **2**. Starting with this compound a common intermediate **3** is prepared in 6 steps via an Evans *syn* aldol. This amide contains the repeating stereo-triad that is present in discodermolide itself. Suitable manipulation

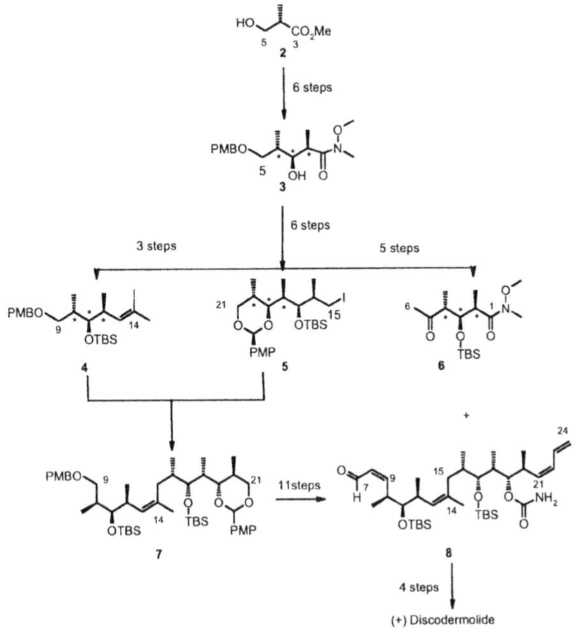

SCHEME 1. Intermediates in the NSP route. Discodermolide numbering, the "star" indicates the common stereochemistry.

then provides the three key advanced intermediates **4**, **5** and **6**.

The first fragment coupling occurs between compounds **4** and **5**. These two compounds are readily available in few steps from intermediate **3**. The fragment coupling can be achieved by either a Suzuki or Negishi-type coupling process. Further elaboration of the coupling product **7** then leads to the aldehyde **8**, which is coupled with **6** in a Paterson type aldol reaction. Three further steps (two chemical and one final purification step) would then deliver (+)-discodermolide. In total, there are 42 steps (including side chains).

We have now completed the NSP synthesis twice, producing 6 grams, 60 grams and at the time of writing we are currently two thirds of the way through a third campaign to produce 500 grams. In further discussion in this chapter, the problems and solutions will be related to these three synthesis campaigns. Initially, there were 18 chromatographic purifications involved. With increasing experience we were able to reduce this to 14. The number of crystalline compounds increased (currently 7) as did the number of reactions where we realized that, apart from a work-up, no further purification was necessary or desirable. The overall yield is low, around 0.05–0.6%, however this is not really relevant at this stage of development as one can easily compensate for this in the planning. Obviously, we would like to increase this overall yield, as it requires processing smaller amounts of material, especially at the beginning of the synthesis. For example, given this yield, to prepare 500 grams of discodermolide requires the manufacture of around 110 kg of the common intermediate **3**, which in turn necessitates the processing of 140 kg of ester **2**. The yield can certainly be significantly increased by careful optimization of the various reactions, but at this stage of development, this is not a top priority.

V. Execution

This turned out to be almost the correct choice of words! In this section I will select some of the steps in the synthesis and discuss in detail the problems we encountered and indicate solutions suitable for running the reactions in a pilot plant environment. The following discussion will be related to the synthetic schematic and will be divided into the sections indicated there (Scheme 1).

A. SYNTHESIS OF THE COMMON INTERMEDIATE 3

The sequence is presented in Scheme 2. Three reactions in this sequence deserve detailed comment: ester reduction, the aldol reaction to produce **12** and finally the formation of **3**.

1. *Reduction of ester 9*

This step displays a typical problem step for the development chemist. For the first (6 g) synthesis campaign, we employed $LiAlH_4$ for the reduction. This was fine on the scale of 5 kg of ester, but we noticed a problem in the second (60 g) campaign (50 kg ester). Large excesses of $LiAlH_4$ are employed for the reduction (4–5 equivalents). Consequently this generates significant quantities of aluminum salts after work-up. The only way we could remove them, in this case, was by filtration. The filtration was very slow and took 2–3 days to complete. For the third

tion run (500 g of **1**), we realized that this was unacceptable on the
of 200 kg of ester. We changed the reducing agent to lithium
dride. The yield remained the same and the work-up in this case is
le dilute acetic acid quench and extraction and solvent removal,
ig the work-up time down to 3 hours and generating around 70 m^3
ogen!

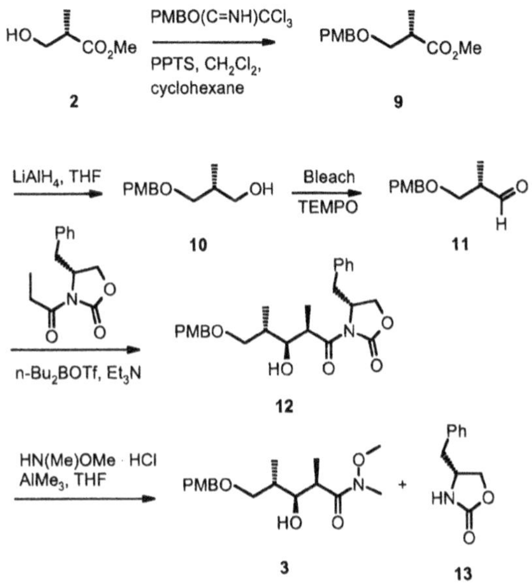

SCHEME 2. Sequence to the common intermediate **3**.

picture illustrates a typical set of pilot plant equipment used to
ut such reactions in a safe manner. It shows a 1000 L stainless steel
in a safety box.
heating and cooling system consists of an inert hydrocarbon
e. The height of the system is around 2.5 meters (to the top of the
and 5–6 meters in total. The metal-clad pipes on the right of the
are the condensers. The box is equipped with a direct venting
and a flame-retarding valve for safe removal of the hydrogen.

s Aldol (10→12)

reaction caused us a few more problems than the simple filtration
ed above. It is actually two processes in one, oxidation of **10** to the

ldehyde **11**, which is configurationally and chemically unstable and is therefore not purified, followed immediately by the aldol reaction at -78°C to deliver **12**. On a laboratory scale and in the first campaign the yield of this reaction was around 70%.

PILOT PLANT REACTOR

In the pilot plant, on a reaction scale of 24 kg of **11**, we were unable to reproduce this yield. The major problem with this chemistry is the quality of the dibutylboron triflate. Fresh samples behaved well, giving in the lab around 70% yield. If the triflate is "older" the yield drops drastically to 50% or less. Just what "older" means has not been defined. Certainly after about 3–4 weeks the yield starts to drop. Triflate from different sources (quality?) also had an effect, between no reaction and similar yields to our standard one. The process takes some 6 days of pilot plant time and as it is very dilute, due to the triflate being delivered as a 1.0M solution, we were unable to significantly increase the throughput. The triflate is pyrophoric at higher concentrations or in other solvents. Consequently, we did not want to prepare it nor distil it ourselves. These reagent quality problems along with an analytical method to measure the quality still need to be solved.

Initially, aldol **12** was obtained as an oil, however one laboratory batch eventually crystallized (after 12 months), resulting in the production of some seed crystals and the X-ray structure (Figure 13).

ce then it never stopped crystallizing! Unfortunately, the optimal
lization procedure we were able to work out proved to be
ely tedious, lasting some 5 days. Trying to speed this up caused the
und to appear as an oil.
procedure would have increased the pilot plant time to 11 days
ction! Thus, capacity problems in the plant were programmed into
cess and would have extended the time necessary for conversion of
0 (150 kg) to 12 to some 70 days.

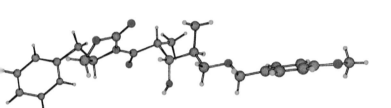

FIGURE 13. X-ray structure of aldol 12.

er returning to the "drawing board" further laboratory investigation
d that this was not necessary and the "crude quality" of 12 was
ent for the next steps.

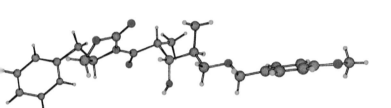

14

redeeming feature of this reaction is its complete stereoselectivity.
ver detected any of the other diastereoisomer, even when the whole
vas done at room temperature. One impurity, 14, was isolated. It
ites from small amounts of anisaldehyde being carried through the
s.

mon Intermediate 3

mination of the literature for the conversion of 12 to 3 reveals that
thod of choice is to employ a trimethylaluminum/Weinreb amine
ex. We attempted to replace trimethylaluminum by triisobutyl-
um in order to avoid the handling issues associated with the former

ompound (pyrophoricity). After a series of extensive laboratory investigations, we found that the only solvent producing an acceptable quality of **3** was tetrahydrofuran. All other solvents investigated produced varying quantities of the side product **15** resulting from non-regioselective amide formation.

We studied this reaction by calorimetry and found that the addition of triisobutylaluminum to a suspension of N, O-dimethyl-hydroxylamine hydrochloride in THF was highly exothermic. It was also noted that the resulting complex of triisobutylaluminum/N, O-dimethylhydroxylamine THF was thermally unstable. According to differential scanning calorimetry, this complex started to exothermically decompose at 30 °C and reached the maximum exotherm at 140 °C, resulting in the release of total of –406 kJ/kg of heat. This instability is presumably due to an aluminum-catalyzed polymerization reaction of tetrahydrofuran. Thus,

15

n case of a cooling failure in the plant, the chance of a thermal runaway ould be very high. The high risk in process safety made this process unsuitable for scale up. In an attempt to make this chemistry more amenable to the pilot plant, we investigated the following variations: inverse addition, extending the addition times, and lowering the temperature. In all cases, we observed the formation of significant amounts of by-product **15**, attributed to the opening of the oxazolidinone ring. We were unable to define conditions that could minimize this competitive ring-opening reaction. In view of these results, we decided to abandon the transamidation protocol and investigate other methods. The alternative approach is shown in Scheme 3.

Cleavage of the oxazolidinone was achieved by treatment with lithium hydroperoxide in methanol. Here, care must be taken to dilute the large quantities of oxygen, formed by the base-catalyzed decomposition of hydrogen peroxide, with nitrogen, so that the explosion limit is never attained. This route has the advantage that the acid can readily be isolated as its amine salt **16**, allowing a significant purification effect.

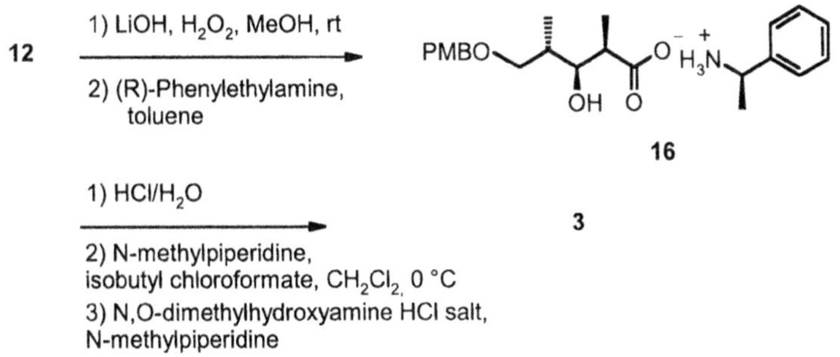

12

1) LiOH, H_2O_2, MeOH, rt
⟶
2) (R)-Phenylethylamine,
   toluene

16

1) HCl/H_2O
⟶   **3**
2) N-methylpiperidine,
isobutyl chloroformate, CH_2Cl_2, 0 °C
3) N,O-dimethylhydroxyamine HCl salt,
N-methylpiperidine

SCHEME 3. Alternative approach to **3**.

Oxazolidinone **13** is easily re-isolated in high yield without detriment to its optical activity and can be recycled. Conventional mixed anhydride amide formation ultimately produced the common intermediate **3** in reasonable yield and quality.

Table 1 indicates the overall yield and the absolute quantities of **3** for this modified 6-step sequence.

TABLE 1
Summary of the Three Campaigns to Produce **3**

Campaign	Overall yield	Absolute quantities of **3**
1st (6 g)	30%	3.5 kg
2nd (60 g)	35%	35.0 kg
3rd (500 g)	51%	110 kg

B. SYNTHESIS OF THE FIRST COUPLING PARTNERS

The route to **4** is outlined in Scheme 4. Two of the reactions here deserve further discussion. These are the reduction of **17** to form **18** and the preparation of the vinyl iodide **4**.

1. Reduction (17→18)

The success of converting amide **17** into aldehyde **18** was very dependent on the quality of silyl ether **17**. If **17** was not pure, the yield of aldehyde dropped to below 50%, so a chromatographic step was necessary here. The reaction time and temperature were also found to be critical. When the reaction was held too long at 0 °C, a competing desilylation reaction occurred leading to the formation of significant amounts of hydroxy aldehyde **21**.

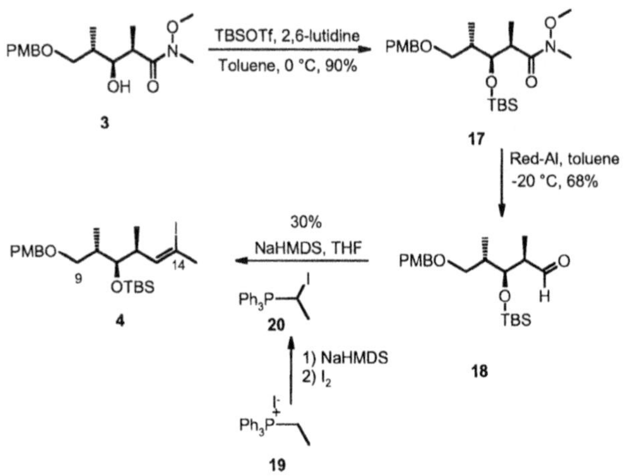

SCHEME 4. Synthesis of *cis*-vinyl iodide **4**.

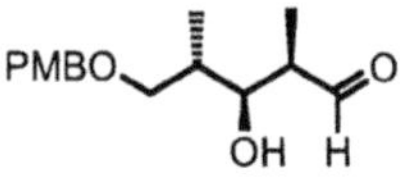

If necessary, aldehyde **21** could easily be isolated and subjected to the standard silylation conditions to regenerate **18**. By maintaining the reaction temperature between –5 to 0 °C, formation of **21** was minimized. If the temperature of the reaction mixture during the quench and work-up was not strictly held below 10 °C, significant quantities of **22** were isolated (currently we have around 2 kg of this compound). This is formed by β-elimination either of silyl alcohol from **18** or of water from **21**. We often observed such elimination of silyl alcohol or *para*-methoxybenzyl alcohol, especially from *beta* hydroxyl protected aldehydes. Attempts to convert all of **17** into **18** were unsuccessful. The final reaction conditions were a compromise to minimize by-product formation. Aldehyde **18** is stable if stored at < –10 °C.

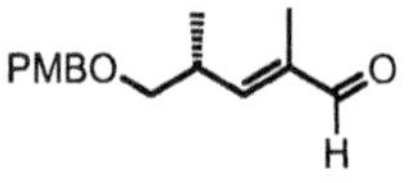

2. Synthesis of cis-Vinyl Iodide 4

The crucial step (**18** to **4**) here is the so-called Zhao olefination,[15] a variation of the Wittig reaction. This reaction is fraught with problems.

For the first (6 g) and the second (60 g) campaign, utilizing the Zhao olefination procedure, also used by Smith[7c] and Marshall[8], we obtained the desired *cis*-vinyl iodide **4** in 20 to 31% yield after chromatographic purification on silica gel. Only a small amount of the undesired *trans* isomer was detected (*cis: trans* = 10:1 to 15:1) and this could not be separated from the desired *cis* compound. This is fortunately not a problem as they can be separated after the next step in the sequence. We did not observe any *des*-iodo olefin **23**, suggesting that the formation of the iodo ylide from ethyltriphenylphosphonium iodide, *via* ylide iodination (Scheme 4), was complete before it was added to aldehyde **18**.

23

This olefination step was one of the most difficult reactions to scale-up. We consistently obtained variable and low yields on a maximum scale of 2 kg of aldehyde **18**. This lack of reproducibility indicated that we did not have the process under any sort of control. Complicated work-up procedures and the apparent instability of **4** to the work-up conditions most certainly contributed to the low yield.

Smith utilized iodine for the conversion of ethyltriphenylphosphonium iodide **19** into the iodo-ylide **20**. During process optimization work for the third campaign (500 g), we found that *N*-iodosuccinimide could be used to replace iodine without detriment. While this makes the reaction easier to handle and increased the reproducibility of the process, it did not contribute to an increase in yield.

During the first and second synthetic campaigns, we observed the formation of the methyl ketone **24**[16] during the work-up. For the third campaign, we changed this to a non-aqueous work-up.[2b] Consequently, this by-product and the aforementioned stability problems completely disappeared! This allowed a scale-up to 3 kg per reaction and the yield over 9 reactions on a scale of 3 kg per reaction of **18** was 31% ± 0.5%.

24   **25**

It was found, as reported,[8] that the reaction of 2-iodo ethyltriphenylphosphonium iodide **20** with **18** afforded epoxide **25** as a mixture of isomers in addition to the desired **4** in a 1:1 ratio. Alternative approaches were investigated in an attempt to minimize this major by-product, but they were unsuccessful. For example, employing a method described by Shen[17] (where the initially formed betaine intermediate was deprotonated with a second equivalent of base and then iodinated) produced des-iodo olefin **23**. Utilizing Hanessian's phosphonates[18] in this process also resulted in only des-iodo olefin **23**.

A mechanism has been proposed[16] for the formation of epoxide **25** in which intermediate betaine **A** plays an important role. This ring closes to the corresponding epoxyphosphonium salt **B** with the elimination of iodine (Figure 14). Phosphonium salt **B** can then eliminate triphenylphosphine oxide after an aqueous or methanolic work-up.

FIGURE 14. Proposed mechanism for the formation of epoxide **25**.

Apart from triphenylphosphine oxide, we found significant amounts of triphenylphosphine to be present and, interestingly, this was **not** observed during reaction monitoring. Triphenylphosphine was isolated after the chromatography on silica gel. This suggested the presence of an intermediate that decomposed during contact with silica gel, delivering the observed triphenylphosphine. If one accepts that the mechanism proposed by Smith is correct, the protonated form of intermediate **A** can also collapse to an unstable iodo epoxide with the elimination of triphenylphosphine. This may be especially relevant during a non-aqueous work-up.

The following anecdote illustrates the pitfalls into which one can stumble. During the scale-up of this process for the third (500 g) discodermolide campaign, the first reaction did not deliver the required yield of 30%. Only some 18% of **4** was isolated along with starting material. This result created some consternation until we realized that the sodium hexamethyldisilazide, used as a base in this sequence, is supplied as a 35% solution in THF. As we carried out the reaction in winter it was quite cold and somehow the THF solution of sodium hexamethyldisilazide apparently separated in the drum, but we assumed it was homogeneous! Thus, differing concentrations were present within the sample and we ended up simply not adding enough base for full

TABLE 2
Summary of the three campaigns to produce *cis*-vinyl iodide **4**

Campaign	Overall yield from 3	Absolute quantities of 4
1st (6 g)	22%	0.5 kg
2nd (60 g)	20-30%	3.4kg
3rd (500 g)	30.5%	11.8 kg

deprotonation. For the next reaction the drum was shaken vigorously! The result – a 32% yield! A label that stated, "shake well before use" was applied to the drum.

Table 2 indicates the overall yield and the absolute quantities of **3** for this modified 6-step sequence.

3. *Synthesis of Alkyl Iodide 5*

Within this sequence (Scheme 5) two of the reactions turned out to be really problematic, the Evans *syn* aldol and the reductive removal of the oxazolidinone to generate alcohol **30**.

a. Evans *syn* aldol

The comments made previously concerning the quality of dibutylboron triflate are equally valid in this case. Here, however, two further problems emerged. Firstly, product **28** is not configurationally stable for any length of time at 0°C or above. Complete epimerization occurs within one week via a retro-aldol/ aldol process (note that silyl ether **29** is configurationally stable at RT). The compound is stable indefinitely at storage below -10°C. Secondly, we noticed the formation of significant amounts of a side product **31** (together with trace amounts of **32**), which could not be separated (crystallization or chromatography) from the desired product.

The formation of both of these side products illustrates the lability of the *para*-methoxyphenyl acetal-protecting group. In the case of **31**, the excess dibutylboron triflate present catalyses the ring opening of the acetal and the resulting stabilized carbocation is trapped by the excess enolate[2c]. The tactic here is to establish conditions so that **31** can be eliminated. After some experimentation, we found that simply working up the reaction at temperatures between -10 and -5°C reduced its formation to <0.1% and we accepted the incomplete conversion of **27** (<0.5%).

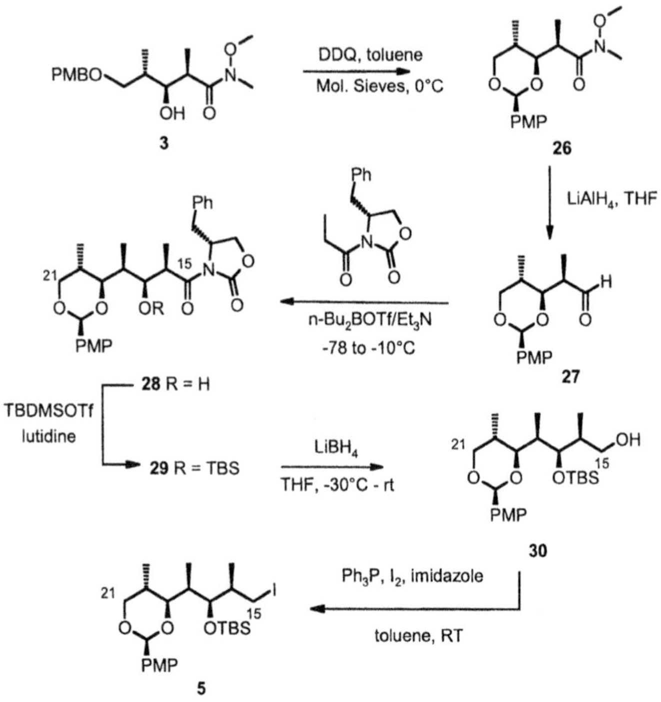

SCHEME 5. Route to alkyl iodide **5**.

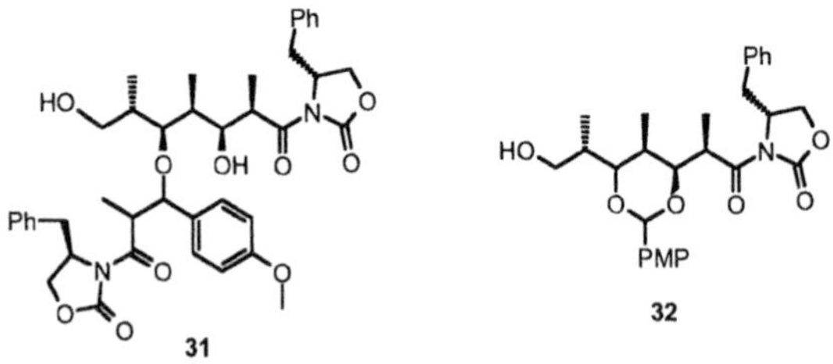

b. Reduction, forming **30**

The reductive removal of the chiral auxiliary proved to be troublesome. Here, after some 12 steps in the main chain, the first chromatographic purification is necessary.

The method of choice for the first two synthesis campaigns was to treat

29 with a solution of lithium borohydride in THF/EtOH.[19] This gave an average yield of 60% of the desired alcohol **30** after chromatography on silica gel. We were unable to reproduce the high yields (> 80%) reported in the literature. For the 500 g production run (32 kg of **29**) we substituted water for ethanol and increased the yield to 71%. However, the side product formation problem was not solved.

The problem with this reaction is caused by several competing reactions that produce at least four major by-products, **33–36**, which were isolated and identified. Compound **33** was the major by-product. It is obvious that they are generated by non-regioselective reduction of the carbonyl groups internal and external to the oxazolidinone ring. The exact pathway for the formation of **34** is unclear. The separation of **33** from the desired product is not trivial and requires very careful chromatography.

Some effort went into examining other conditions. For example reduction of **29** with sodium borohydride[20] was very slow and resulted in a very messy reaction. Attempts to reduce **29** to the corresponding aldehyde with the Schwartz reagent, according to a recently described procedure,[21] were also unsuccessful.

33

34

35

36

Another alternative, in which **29** was converted into a thioester[22] followed by reduction, did not improve the yield and chromatography was still required (Scheme 6).

In conclusion, fragment C$_{15-21}$ (**5**) was produced in 6 steps from the common precursor **3** with an overall yield of 24%. Although the yield was lower than that reported by Smith (56%), the ease of pilot plant operations made this route successful for the production of several kilograms of this

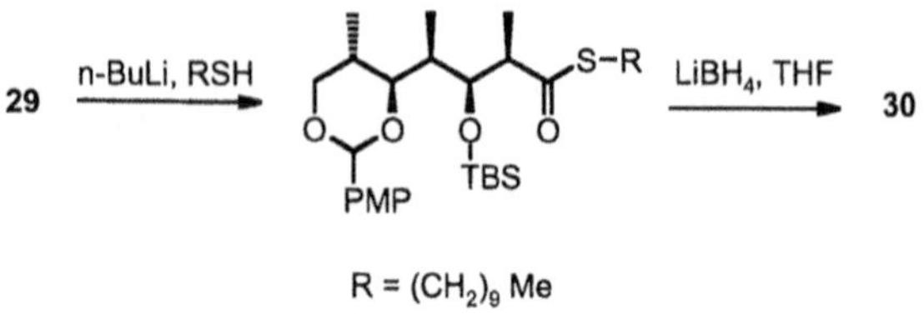

R = (CH$_2$)$_9$ Me

SCHEME 6. Thioester approach.

fragment with the high purity necessary for the difficult chemistry to come. Table 3 summarizes this portion of the work over three campaigns.

TABLE 3
Summary of the three campaigns to produce alkyl iodide 5

Campaign	Overall yield from 3	Absolute quantities of 5
1st (6 g)	25%	0.35 kg
2nd (60 g)	28%	5.7 kg
3rd (500 g)	40%	15.5 kg

C. FIRST FRAGMENT COUPLING

Moving down the main chain of the synthetic route (Scheme 1) we now arrived at the first fragment coupling to produce 7 (Scheme 7).

In preparation for the first two production runs, we initially examined a variation of the Negishi coupling as practiced by Smith.[6c] This process produced several side products, as indicated by the NMR spectrum of the crude reaction mixture after work-up, which were not separable from the desired product.

Marshall, in his approach to discodermolide, described[8] an alternative Suzuki type[23] cross-coupling step. Employing this procedure for our coupling reaction, [5/tert-BuLi/9-methoxy-9-borabicyclo[3.3.1]nonane added to 4/Cs$_2$CO$_3$/ Pd(dppf)Cl$_2$, resulted in a much cleaner reaction mixture. The only major by-product generated was the des-iodo compound 37. Some trans isomer of 7, carried over from the trans impurity in 4, was also observed. Pure 7 was easily obtained from the crude product in average yield (55%) by chromatography followed by crystallization from acetonitrile, the trans isomer 39 remaining in the mother liquors. The structure and absolute configuration of 7 was confirmed by single-crystal X-ray analysis of the corresponding alcohol 38 (Figure 15), obtained by removal of the p-methoxybenzyl protecting group in 7 with DDQ.

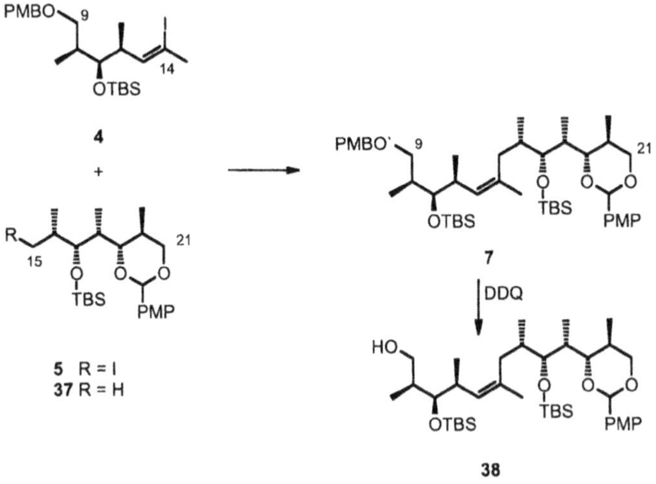

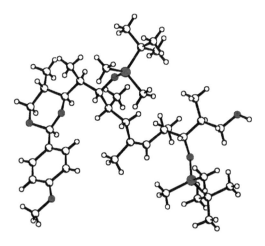

SCHEME 7. First fragment union.

the large scale (500 g) campaign, this Suzuki coupling reaction
examined with the objective of simplifying a very complex
ure and work-up. For example, the excesses of alkyl iodide **5,**

FIGURE 15. Single crystal X-ray structure of **38.**

must be employed in this coupling reaction (up to 2 equivalents)
be reduced in order to make the process more productive,
lly after such a long synthetic sequence to obtain **5.** The work-up

nd chromatography are extremely tedious and therefore direct isolation
of the product from the reaction mixture is also desirable.

By judicious choice of the base, we were able to reduce the number of
equivalents of **5** to 1.30. The quality of 9-methoxy BBN was also critical
for the success of the process; aged samples did not function well
although once again this has not been adequately defined. The product
was isolated by direct crystallization from acetonitrile after work-up of
the reaction mixture, thus eliminating a chromatographic purification
step. By employing these changes we were able to produce some 11.3 kg
of **7** corresponding to around 70% yield. None of the "wasted product" **37**
was observed.

Examination of the mother liquors from this crystallization revealed
the presence of two side products. The *trans* isomer **39** and a compound
tentatively identified as **40**, formed by dimerization of **4**. There is some
precedent for this "dimerization", at least in the case of the Negishi
coupling.[24]

With the first fragment union successfully completed, the transition
from the Smith approach to the Paterson route was now required in order
to arrive at the final C_{7-24} coupling partner. This necessitates the
elaboration of both termini to introduce the (Z)-enal and the terminal (Z)-
diene unit and introduction of the pendant carbamate moiety. Table 4
shows the results for the yields of **7** at various scales.

TABLE 4:
Summary of the 3 campaigns for 7

Campaign	Overall yield	Absolute quantities of 7
1st (6 g)	50%	0.15 kg
2nd (60 g)	53%	3.7 kg
3rd (500 g)	68%	11.4 kg

D. SYNTHESIS OF THE ENDGAME KEY INTERMEDIATES

Here we cross over from the Smith route into the Paterson chemistry in
preparation for the endgame. The reactions within this sequence, apart

from one, were scaled up with no tremendous problems! Attempted purification of aldehyde **42** (chromatography) produced a side product **50**. Closer examination of this elimination revealed that after 5 minutes exposure of **42** to silica gel, some 50% was converted to the olefin **50**.

Therefore, **42** was directly converted to *cis*-diene **44** by employing a combination of the Nozaki-Hiyama-Kishi reaction[25] followed by a Peterson elimination employing the conditions described by Paterson.[9a]

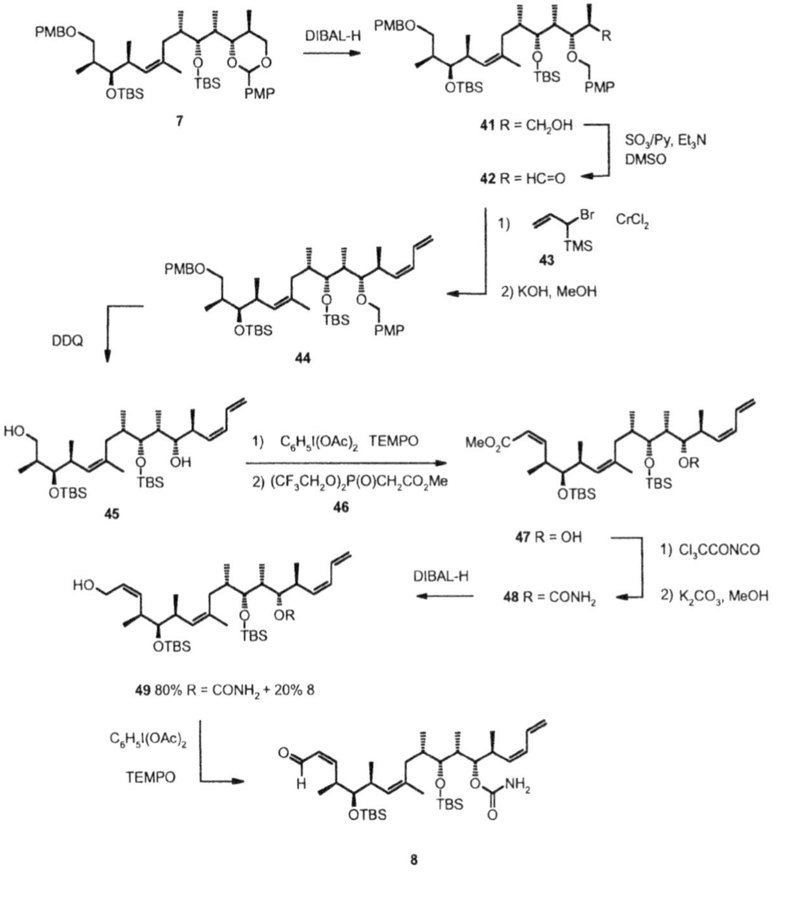

SCHEME 8. Route to final coupling fragment **8**.

Here the quality of chromous chloride was critical for the success of this reaction and once again this has not been adequately defined. There are suggestions that catalytic quantities of nickel (II) or palladium (II)

may be required for an efficient coupling process.[26] We did not examine
the nickel (II) content of our chromium, but this is a point to be filed for
future reference.

50

Indeed, this may be a reason for the extreme variability we observed
during the large scale (500 g) campaign, which was in direct contrast to
the second (60 g) campaign, where all the batches went to completion,
without problems, in 2 hours at room temperature. Each of the three batch
reactions behaved differently, the first was not complete after 2 hours and
required more (20%) CrCl$_2$ to be added with stirring for 18 hours at room
temperature before it was complete. Prior to the second batch, we made
sure that the reaction started by taking a small sample and running it in
the lab. It behaved as we expected. However, the large-scale reaction did
not budge one inch. We observed that the samples taken from the reactor
for reaction monitoring had all proceeded rapidly to completion. We
rationalized that this may have been an effect due to the contact of the
sample with air (oxygen). We could have taken 150,000 samples and let
them react but it was simpler to introduce air into the reaction vessel, a
procedure that is fraught with hazard. Lo and behold, the reaction went to
completion after a further 18 hour period. The third batch, however, when
run under the conditions used for the second batch did not move at all. As
a last resort it was warmed to 40°C and obliged us by going to completion
within 2 hours! I might add at this point that all the reagents and starting
materials were tested before starting the large-scale batches. No problems
were observed, all reactions going to completion within the 2 hour period.
The reason for this extreme variability remains a total mystery and will
need to be examined before any further scale-up is planned.

Furthermore, the chromium residues cannot be present in the
wastewater from the process and therefore they must be recycled for an
ecologically sound process, although at this stage of development we did
not investigate this.

The Still-Gennari reaction proceeded with very high selectivity for the
cis isomer, 30:1 to 40:1 to give **47**. The geometrical isomers were
separated by chromatography. This is advantageous as the *trans* isomer

51

52, R' = CO₂Me
53, R' = CH₂OH
54, R' = CHO

51 and the products derived from it in the synthetic sequence, (52-54) were useful analytical reference materials, especially in light of what was to come!

The DIBAL-H reduction of **48** produced a mixture of the alcohol **49** and aldehyde **8**. We were unable to halt the reduction completely at the aldehyde stage even by reducing the temperature to −100 °C. Fortunately, oxidation of the mixture proceeded without detriment to the integrity of the aldehyde component. Aldehyde **8** was purified by a tedious chromatography and was a stable compound, though unfortunately not crystalline. Several by-products were isolated (55–57). These are very probably formed from traces of the corresponding side products carried through the synthesis from the previous steps.

55

56

57

The key aldehyde **8** is the first of the two GMP (good manufacturing practices) starting materials. As such it is "subject to" special considerations. GMP definitions of API (active pharmaceutical ingredient) starting materials are as follows:

1. At least one chemical bond making or breaking away from the final product.
2. A significant part of the drug substance structure is present in the molecule (for example, stereochemistry).

3. The substance is stable and readily characterized.

Fortunately, this aldehyde fulfills all of these criteria. Table 5 summarized the numbers for **8** through the three campaigns.

The synthesis of the other API starting material, ketone **6**, (Scheme 1) will now be described. This 4-step sequence from the common precursor **3** (via **17**) is outlined in Scheme 9.

TABLE 5:
Summary of the 3 campaigns for **8**

Campaign	Overall yield from 7	Absolute quantities of **8**
1st (6 g)	50%	0.15 kg
2nd (60 g)	28%	0.52 kg
3rd (500 g)	30%	1.80 kg

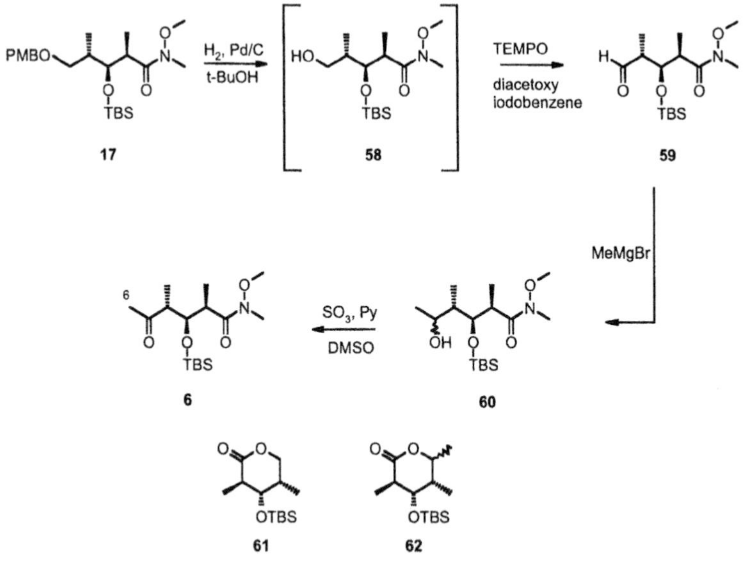

SCHEME 9. Synthesis of ketone **6**.

This sequence posed few problems. However, the hydrogenation step was always going to be problematic, for two reasons. First, the catalyst quality and second the propensity of alcohol **58** to cyclize to lactone **61**. In order to avoid this, the strategy was twofold; fast hydrogenation was required, with a minimum of manipulations, in order to minimize lactone formation. Further, the solvent must be compatible with the ensuing oxidation. Catalysts from a variety of manufactures were screened. One was found to promote the reaction very rapidly, (5 minutes to complete reaction). The solvent of choice was *tert*-butanol as this was unlikely to interfere in the oxidation step. Using this solvent with rapid hydrogenation led to complete formation of **58** with little or no **61** being formed. Indeed, only around 10% lactonization was observed when a solution of **58** in tert-butanol was left at room temperature for several days. So this turned out not to be a problem.

Oxidation to **59** in *tert*-butanol alone was found to be slow (>6 hours). However, it was observed that addition of *tert*-butylmethyl ether accelerated the reaction, which was then complete in around 2 hours at room temperature under these conditions.

The following observations illustrate the problems involved in carrying out oxidation reactions on a large scale: In some cases, the reaction appeared to stop at around 50% conversion! The reason for this was and still is unclear and the reaction had to be pushed to completion by the addition of more oxidant and TEMPO. Furthermore, some time after the addition of the oxidant was complete, we observed a sudden temperature rise of around 5 °C. This indicated that something "reactive" had accumulated in the reaction mixture that rapidly and exothermically reacted. This is a potentially dangerous situation and could lead to a runaway reaction when attempted on any significant scale. More work is required to sort this problem out.

Aldehyde **59** was purified by rapid chromatography at 10 °C due to its unwelcome propensity to epimerize on prolonged silica gel contact and, after removal of the solvent, it crystallized (mp 41-43 °C).

TABLE 6:
Summary of the 3 campaigns for **6**

Campaign	Overall yield from 3	Absolute quantities of 6
1st (6 g)	36%	0.5 kg
2nd (60 g)	36%	2.1 kg
3rd (500 g)	40%	8.1 kg

Reaction of **59** with methylmagnesium bromide produced secondary alcohol **60**, which also demonstrated lactonization tendencies to produce **62**. Interestingly, the epimer of **59** reacted more slowly with the Grignard reagent and traces of it were observed in the ketone (<1%). Direct oxidation under Doering/Parikh conditions (SO₃-pyridine complex, Et₃N, DMSO) in methylene chloride produced ketone **6** in good yield after chromatography. Ketone **6** is a stable oil that is readily characterized and as such fulfils the criteria given above for an API starting material. The results for its synthesis are shown in Table 6.

E. END GAME

After 18 months, 33 steps, and lots of sleepless nights, we now had the necessary fragments present in the appropriate quantities and commenced with the final few steps leading to the desired drug substance. Within this segment several "outside factors" started to make themselves felt. For example, to mention a few of them, current GMP regulations and the resulting requirements for quality and its assurance, toxicity and handling requirements including decontamination procedures, analysis, and final release of the drug substance for "human use". These aspects will be mentioned within the appropriate sections below.

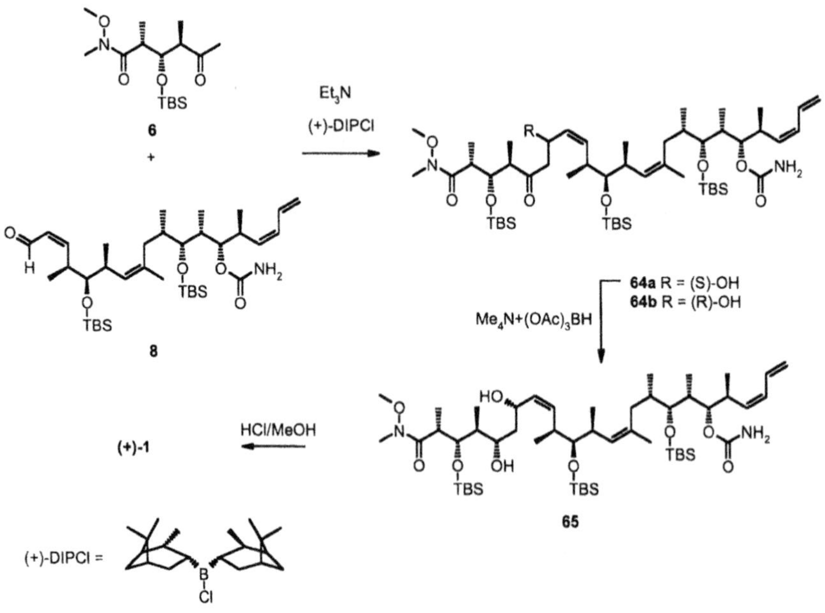

SCHEME 10. End game.

The laboratory experiments were carried out on a very small scale, something that is unusual for a process research department. This really precluded proper investigation of the reactions with the result that we nearly failed to produce the desired quantity when proceeding with the large scale reaction due to unknown factors. Two of the reactions in the end game (Scheme 10) will be described.

1. *Paterson aldol*

This complex aldol coupling requires the use of reagent control to reverse the intrinsic substrate selectivity, i.e., it is a mismatched reaction.[9a, 27]

Treatment of 6.6 equivalents of the corresponding boron enolate of **6**, [prepared by treatment of **6** with (+)-B-chlorodiisopinocampheylborane (DIP-Cl), **63**, and triethylamine in diethyl ether at 0 °C], followed by aldol addition with *cis*-α,β-unsaturated aldehyde **8** at -78 °C, led to alcohol **64a** in 50-55% yield after chromatography on reverse-phase silica gel, together with its epimer **64b** in a ratio of ~4:1 (Scheme 10).

This sounds like a simple process, however, in the event, this conversion of **8** to produce **64a** turned out to be one of the most difficult reactions that I have ever had to scale-up. The problems are manifold and complex. As of this writing, they still have not all been solved and the solutions presented here, while allowing the isolation of the product, are by no means optimal. Significant effort will have to be invested in order to make this a reasonable, well-behaved process!

During the first (6 g) campaign, we observed a considerable variation in the yield of this reaction, from 30% to 60%. While this was sufficient for the 6 g of discodermolide, such variation would not have allowed us to successfully prepare the larger quantities. The reason for this became clear after the first experiments during process research before the second campaign began. The quality of commercial (+)-DIP-Cl, **63**, was capricious.

We initially used commercially available solid (+)-DIP-Cl. This reagent is difficult to obtain and to handle in large quantities, as it is hygroscopic and inherently unstable. On storage it eliminates pinene, which reduces the quality of the reagent, producing undefined boron species. Obtaining a well-defined quality reagent on a large scale from a commercial supplier was problematic. Routine analytical methods are not really suitable for monitoring the quality of this boron reagent and the only method of testing its quality is to use it!

When the reaction was attempted with solid (+)-DIP-Cl, we did not obtain the desired **64a**, but the *trans*-aldol **66**, together with its epimer **67** in a 3:1 ratio together with isomers of aldehyde **54**.

66 R = (S)-OH
67 R = (R)-OH

Also obtained were significant amounts of allyl alcohol **49** and its *trans*-isomer **53**, resulting from the reduction of the aldehyde **54**. The structures were confirmed by comparison with authentic samples that were prepared from the *trans* Still-Gennari olefin **51**. The mechanism of this double bond isomerization is not clear. It occurs at -78 °C even before the aldol reaction takes place and may be the consequence of an addition/elimination process of chloride or triethylamine induced by boron coordination to the aldehyde oxygen atom, but this is speculative. However, I should note at this point that exposing aldehyde **8** separately to all the reagents used in the process produces no change apart from minimal reduction in the presence of (+)-DIP-Cl.

These problems were eliminated by employing a 70% solution of the reagent in hexane. This is commercially available and, according to the manufacturer, indefinitely stable! In this form it is easier to handle and is readily transferred into the reaction vessel from the cylinder it is supplied in. The aldol reaction now proceeded in a reproducible 50% yield on a small scale. Neither reduction nor isomerization was observed and the product **64a** was easily isolated by filtration through reverse phase silica gel.

The time had now come to attempt the first large-scale reaction, using 50 grams of aldehyde **8**. Utilizing the conditions we had worked out on the small scale gave 23% yield of the aldol **64a**! What happened? We obtained significant reduction of **8** back to **49** and this could not be reisolated. We surmised that the reason for this result was incomplete enolization of ketone **6** (2 hours, 0°C). We examined the enolization time on a small scale and extended it to 16 hours at 0°C. Once again on the 50 g scale 23% yield of aldol **64a** was obtained, this time with no reduction observed. Back to the drawing board! Here, we were in danger of not being able to produce the required quantities of drug substance.

We examined the fate of the aldol product at every stage of the process. Before work-up, the desired product was formed in 65% yield together with the epimer in 33% yield. That is, the aldol reaction proceeded almost quantitatively!

The reaction solvent was diethyl ether (no other solvent produced a viable result). Therefore, safety considerations demanded that before the standard oxidative work-up the solvent had to be changed due to the possible presence of ether peroxides. After an aqueous quench, solvent evaporation and re-dissolution in methylene chloride, examination of the fate of **64a** revealed that some 20% less was present in the mixture. Carrying out the oxidative work-up resulted in a further 20% loss. The product was not stable (in the reaction mixture) to the work-up conditions. To overcome this problem, we simply omitted these steps and after quenching the reaction mixture with water, it was diluted with the chromatography solvent and chromatographed directly on reverse phase silica gel. This led to a 60% yield of the desired alcohol **64a**, epimer free! The epimer **64b** was isolated by further elution and recycled as described below. The summarized column conditions are outlined in Table 7.

Combining and evaporating product fractions to 10% of their original volume followed by extraction with ethyl acetate and re-evaporation to dryness provided 150 grams of pure **64a**.

Prior to the third (500 g) campaign we reexamined the stoichiometry of the reaction with the aim of reducing the reagent excesses and increasing the product stability to the work-up conditions.[28] Several factors were examined:

i.   The effect of additives. Here the aim was to attempt to trap out any "reactive boron species" that may have been causing the product instability. The addition of various aldehydes, or dienes, boron specific ion exchange resins or other boron trapping agents just before work-up had no effect on the product stability.

ii. Solvent. As already mentioned, diethyl ether is not the solvent of choice for industrial applications. Dichloromethane and *tert*-butylmethyl ether were evaluated. Both solvent systems were detrimental to the ratio of **64a/64b**.

iii Alternative enolization method. Other methods of enolate generation could well be of use in reducing the reagent quantities. We examined firstly generating the lithium enolate with LDA and then carrying out a lithium–boron exchange. In a model system, this worked well, as demonstrated by following the reaction by FT-IR[28]. However, when applied to the real system, a 1/1 mixture of **64a/64b** was obtained.

iv Enolate excess: We examined the optimum stoichiometry of the reaction by systematically reducing the excesses of ketone **6**, (+) DIP-Cl and triethylamine (Table 7).

TABLE 7
Optimization of enolate excess

Entry	Equiv. of **6**	Equiv. DIP-Cl	Equiv. Et₃N	Yield of **64a** (%)	Ratio 64a/64b	Ratio 64a/66a	Ratio 64a/8
1	6.6	5.4	6.6	55	3.9/1	25/1	28/1
2	4.0	3.0	4.0	56	3.9/1	30/1	30/1
3	3.3	2.7	3.3	48	4/1	28/1	35/1
4	**3.3**	**2.5**	**3.3**	**68**	**3/1**	**22/1**	**22/1**
5	2.4	2.0	2.4	49	3.6/1	28/1	5/1
6	1.5	1.2	1.5	34	3.6/1	64/1	1/1
7	1.5	1.2	1.5	31	3.4/1	60/1	1/1
8	3.3	2.5	3.3	0	Trace of water added		

The observed yields in entries 2 and 3 are still a result of the product instability to reagent excesses and work-up conditions. Entries 5 and 6 indicate that when the enolate excess falls below a certain value, the reaction is very slow and incomplete conversion is observed. Entry 8 shows the remarkable effect of water. Here the (+)-DIP-Cl was exposed to atmospheric moisture for one to two seconds simply by removing the stopper of the weighing flask. No aldol product was formed and a 1/1 mixture of **8** and its isomer **54** was observed. The conditions of entry 4 are apparently optimal. However, the reaction is very intolerant of manipulative errors and very narrow limits between slow reaction and product stability are apparent. These conditions brought a benefit. The crude product was found to be stable indefinitely to the reagents and the work-up system. Stability tests of **64a** showed no loss in yield even when the mixture was kept at 40 °C.

This procedure is not optimal. For the 60 g campaign, we required some 20 m³ of solvent (Table 8) and at the time of writing, we still have to work out the process for the 500 g production run, but they will probably not be significantly different from those described above. For larger scale processes, i.e. > 2 kg of reaction mixture, one would require oceans of solvent, and indeed for any production quantities (>10 kg discodermolide) an entire planet full! Thus this process, while it works, still needs intensive optimization.

TABLE 8
Chromatography conditions.

700g Reaction mixture	dilute with 368 kg acetonitrile/t-butylmethyl ether/water 85/15/10
Apply to	120x30cm column packed with 20 kg RP-18 silica-gel
Elute	with 1060 kg acetonitrile/t-butylmethyl ether/water 85/15/10
Change eluent to	Acetonitrile/t-butylmethyl ether 1/1
Fractionate	20 kg fractions (8)

2. Protecting Group Cleavage

After the problems encountered with the final fragment coupling, we did not expect too many problems with the final step. As it turned out, the acidic hydrolysis of the 3 silyl ethers provided us with a few nasty surprises.

The reaction is quite complex and intensive investigation was required in order to maximize product and minimize side product formation. The literature is unclear on the best method for this transformation. Some authors use HF/pyridine, some HCl/MeOH, others employ HCl/THF, or *para*-toluenesulphonic acid. There is no obvious reason to be gleaned for utilizing differing acidic systems, especially within the same publication!

We settled on running 3 batches under carefully optimized HCl/MeOH conditions after a brief examination of the alternatives. The first reaction we attempted with **65** during the 60 g campaign produced a 40% yield of **1** after chromatography! The reason was simple although not obvious. The first reaction to occur is cyclization to the fully silylated discodermolide **68**. This compound oils out of the reaction mixture and distributes itself around the walls of the reactor. During the reaction monitoring this is not obvious and only becomes so after work-up. The trisilylated discodermolide **68** can be isolated and separately hydrolyzed (see below).

68

Continually washing the walls of the vessel with methanol in order to maintain **68** in solution and adding the HCl in portions over several hours avoided this problem. If this was done, (+)-**1** could be isolated in 70% yield after reverse-phase chromatography.

All the possible permutations of bis-silyl protected **1** could be observed in the HPLC of the reaction mixture. The slowest silyl group to cleave was that at the 3-position and we were able to isolate this compound (**69**) from the column chromatography of the reaction mixture. Forcing the reaction resulted in significant formation of side products. The isolation and formation of these will be discussed below.

69

3. *Isolation of (+)-Discodermolide*

The discodermolide isolated from the 6 g campaign proved to be a monohydrate. In agreement with our quality assurance department, we could only use the material from the 60 g campaign and eventually the 500 g campaign without repeating the toxicology if the following conditions were met: a) the synthetic route was identical, b) the material produced had the same side product profile c) the material was equally pure or better and d) the same crystal modification was produced (8 or 9 modifications are known) and e) the material was sterile.

The material isolated from the chromatography after the cleavage was anhydrous. So we now had to prepare the monohydrate and the required crystal modification and combine the three separate batches into one.[29] The anhydrous material turned out to be an equilibrium mixture of the lactone and the acid, around 9/1 (HPLC on top of Figure 16). This caused some consternation until we realized what was going on (Figure 16).

The solution was simple: Adjusting the pH to 4 kept the lactone ring closed (HPLC on bottom of Figure 16). Thus the material from the three columns was redissolved in acetonitrile/water 9/1 and the pH adjusted to 4 with HCl. After partial evaporation and crystallization (+)-discodermolide was isolated in 95% yield as the monohydrate with the desired crystal modification formed. The purity was 99.9% with loss on drying of 3.1%. The optical rotation was (+) 19° (0.5% in acetonitrile). It contained around 600 ppm of acetonitrile as residual solvent. This could not be removed by drying.

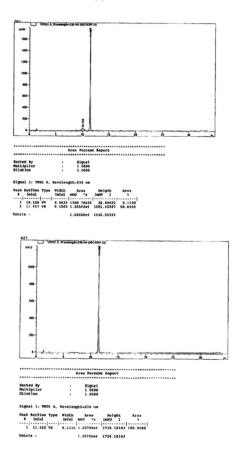

FIGURE 16. HPLCs of discodermolide.

isolation was carried out in a special laboratory in a laminar flow
ood in order to ensure sterility and provide personal protection. All
ts were rendered sterile by filtration through a 4-micron filter. The
r of spores was <20 cfu/g and the material was endotoxin free. A
structure is shown in Figure 17 and NMR data and the compound
re shown in Figure 18. A summary for the synthesis of **1** through
t two campaigns is shown in Table 9.

PRODUCTS

silyl derivative **68** was isolated from the chromatography of the
eavage reaction. This compound could be separately cleaved to **1**,
may then be isolated by crystallization. The mother liquors from

his crystallization contained several side products formed during the
leavage process. They were isolated by extensive chromatography on

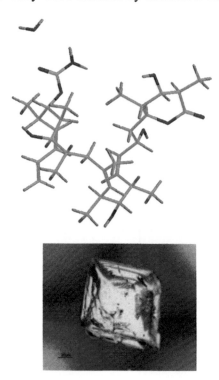

FIGURE 17. X-ray of (+)-discodermolide monohydrate and the crystal.

TABLE 9
Summary of the 3 campaigns for **1**

Campaign	Overall yield from **8**	Absolute quantities of **1**
1ˢᵗ (6 g)	30%	6 g
2ⁿᵈ (60 g)	20%	60 g
3ʳᵈ (500 g)	?%	? g

silica gel.[30] Their structures are shown in Figure 19. These compounds are
all formed under the acidic cleavage conditions in similar amounts. The
trick here is to push the desilylation to completion while at the same time
minimizing their formation. Their formation can be rationalized by the
mechanism shown in Scheme 11, in which the nucleophilicity of the

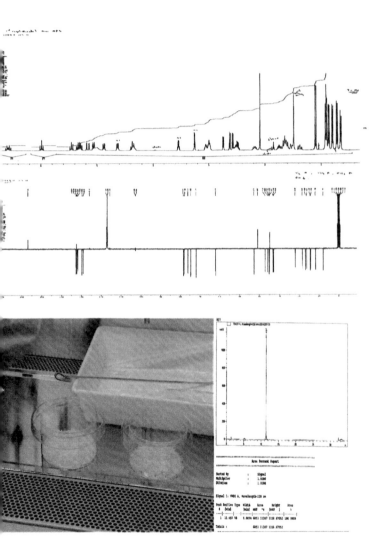

FIGURE 18. NMR data and final dry discodermolide product.

₄ double bond plays a significant role, certainly due to its proximity
cation (see X-ray structure). The allylic cation formed can either be
d by the double bond leading to **70** or the C_{11}-hydroxy group
ing **75**.
stereochemistry was determined by NMR experiments.[30]

Compounds 72–74 are presumably formed by hydration of the double bond. In **74**, the C16 methyl group apparently has the opposite configuration to that observed in discodermolide. This stereocenter is

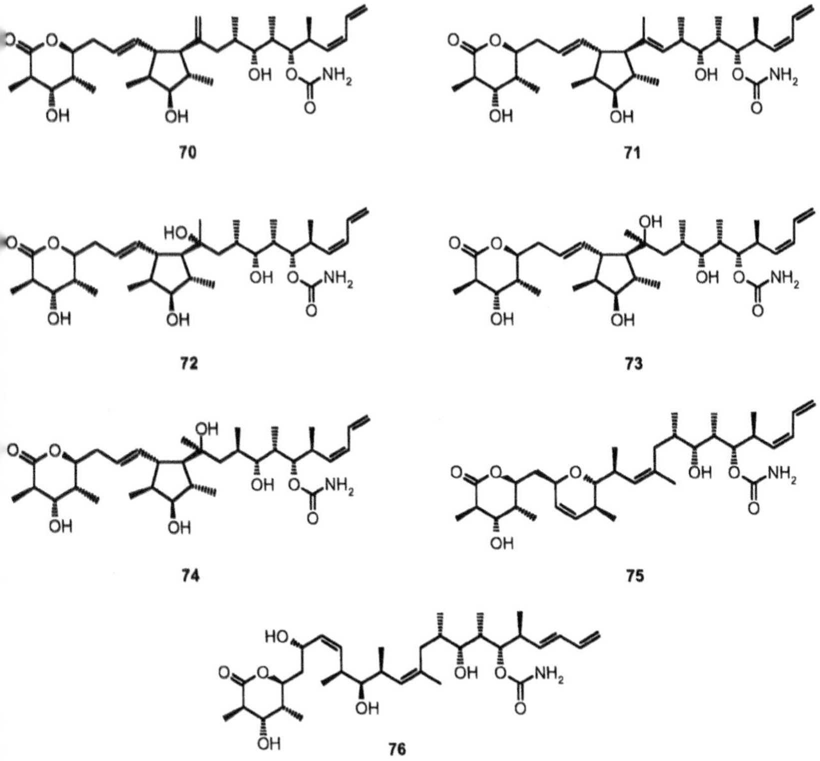

FIGURE 19. Side products.

formed in the aldol reaction described in Scheme 5 (compound **28**). Obviously, traces of the diastereoisomer have been formed and carried through the synthesis, although we did not observe the corresponding C_{16} (R) isomer of discodermolide.

One more side product was found, *trans*-diene **76**. This compound runs immediately after **1** on HPLC and unfortunately co-crystallizes with discodermolide. It is apparently formed in small amounts during the acid cleavage of the silyl groups, although its formation as a consequence of the Nozaki-Hiyama-Kishi reaction cannot be ruled out.

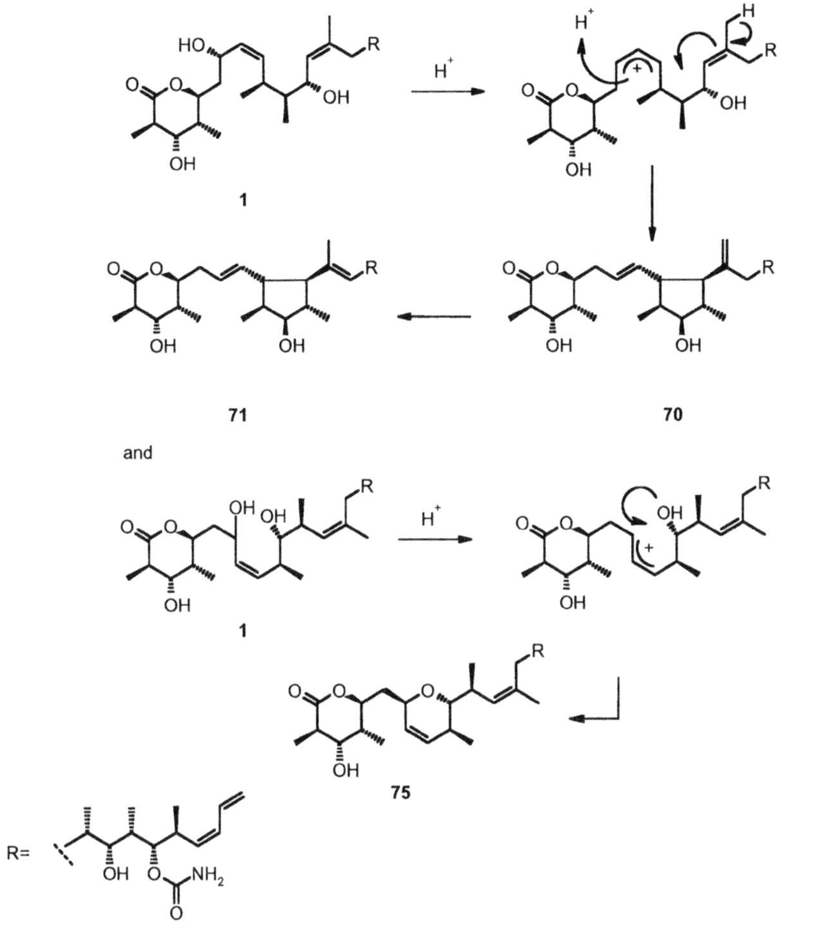

SCHEME 11. Proposed mechanism for formation of side products **70**, **71** and **75**.

VI. Statistics

A few figures relating to the second (60 g) campaign will help illustrate the magnitude of this synthetic effort. We began the synthesis on July 21, 2000 (first reaction started in pilot plant), and finished (compound released for human use) on February 1, 2002. This is 20 months, which equates to about two weeks per step! Here again we did not run the campaign in an overlapping mode, but rather progressed in a linear fashion. Therefore, there is a lot of room to obtain a quicker lead time. One could probably prepare the final product in around 12 months.

Around 100 persons were involved. This figure contains all laboratory and pilot plant personnel. The analytic effort is not included, but as an estimate around 10 more could be added.

The amount of discodermolide produced would equate to around 3000 kg of sponge, a quantity that probably does not exist! The total number of steps is 36 with an overall yield of 0.2% (main chain). This is on the low side, but it should be remembered that the yields for each step have not been optimized. The description of the various optimization processes in this article have only related to the various scale-up and reproducibility issues, not to obtaining the maximum possible yield, so there is plenty of scope for increasing the yield. There were, in this second campaign, some 18 chromatographic purifications. This has now been somewhat reduced to 14. The number of crystalline intermediates stands currently at 7.

VII. Outlook

The synthesis described here will probably never be the one of choice for chemical production. Two examples suffice to illustrate this: For the 60 g campaign, we used an estimated 300 tons of solvent per kg of **1** (Figure 20).

If one assumes a yearly supply requirement for 50 kg of drug substance, this would involve around 50,000 kg of the first Evans aldol product **12** to be produced! Thus, supply and logistical problems are programmed in. However, we do have extensive experience with this route and the problematic steps are easy to identify and optimize.

The next task is to design and scale-up a shorter more efficient alternative synthesis. Such a synthesis should eliminate all the problematic reagents and reactions along the lines described here. It should have the minimum number of chromatographic purifications and the maximum number of crystalline intermediates. High dilution reactions must be avoided in order to maximize reactor efficiency. Complex chemistry, for example aldol reactions, stereochemical introduction (alkylation, etc), should be placed at the beginning or in the middle of the synthesis; the final few steps leading to the final product need to be kept as simple as possible, for example, simple reductions or protecting group manipulations.

VIII. Conclusion and Summary

One major problem associated with a synthesis of this length is the proper laboratory examination of the later reactions in a sequence.

ly, there are no answers to these supply problems; one just has to
e small scale reaction and hope that on transfer to larger scale the

Solvents total = 33000Kg for 120g XAA296

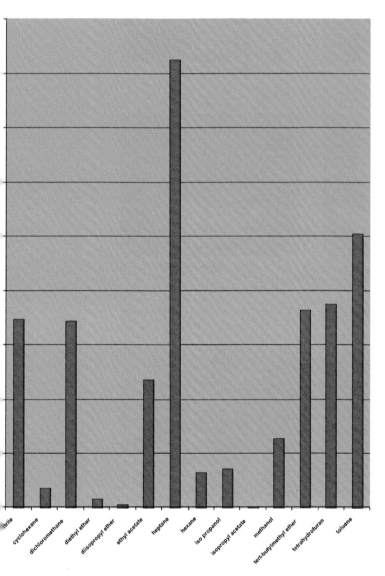

FIGURE 20. Solvents and quantities used for 60 g (+)-discodermolide.

large rotavap

sterile filtration

personal safety

final crystallization

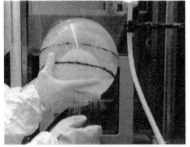

filtration

before drying

FIGURE 21. Scenes from the large scale synthesis of (+)-discodermolide.

reaction proceeds as expected. We certainly stumbled over this exact
point during the final aldol coupling. As a result, the project very nearly
ended in failure, or at least the required amount of discodermolide would
not have been able to be delivered.

The strategies and the tactics of the synthetic route should be "keep it simple, especially at the end", i.e. simple, easy chemistry, nothing like coupling reactions or aldol reactions, which are by their very nature complex processes.

In spite of all the problems, stress and sleepless nights spent wondering what would happen next, we all had great fun in the *disco*. I for one would love to see this compound make it to the marketplace, but there is a long way to go before that happens. Perhaps we will get another opportunity to attempt a project of this magnitude again, maybe with an even more complex substance.

The world of process chemistry is somewhat different from that experienced in many laboratories, particularly academic ones. To introduce the reader to some aspects of that world, Figure 21 includes some scenes from our synthetic adventure.

Acknowledgments

It would require a further 50 pages to properly acknowledge everyone involved in this project. I have written this article with them in mind and wish to thank everyone involved in the "disco" for their encouragement and dedication. Finally, Novartis itself is to be applauded for the courage to proceed with this project.

References and Footnotes

1. (a) Gunasekera, S. P., Gunasekera, M., Longley, R.E., Schulte, G.K. *J. Org. Chem.* **1990**, *55*, 4912, Correction *J. Org. Chem.* **1991**, *56*, 1346. (b) Gunasekera, S. P., Pomponi, S. A., Longley, R.E., U. S. Patent 5840750, November 24, **1998**. (c) Gunasekera, S. P., Paul, G. K., Longley, R.E., Isbrucker, R. A., Pomponi, S. A. *J. Nat. Prod.* **2002**, *65*, 1643.
2. (a) Mickel, S. J., Sedelmeier, G. H., Niederer, D., Daeffler, R., Osmani, A., Schreiner, K., Seeger-Weibel, M., Bérod, B., Schaer, K., Gamboni, R., Chen, S., Chen, W., Jagoe, C. T., Kinder, F. R. Jr., Loo, M., Prasad, K., Repic, O., Shieh, Wen-C., Wang, Run-M., Waykole, L., Xu, D., Xue, S., *Org. Process Res. Dev.* **2004**, *8*, 92. (b) Mickel, S. J., Sedelmeier, G. H., Niederer, D., Schuerch, F., Grimler, D., Koch, G., Daeffler, R., Osmani, A., Hirni, A., Schaer, K., Gamboni, R., Bach, A., Chaudhary, A., Chen, S., Chen, W. C., Hu, B., Jagoe, C. T., Kim, H. Y., Kinder, F. R. Jr., Liu, Y., Lu, Y., McKenna, J., Prashad, M.; Ramsey, T. M., Repic, O., Rogers, L., Shieh, Wen-C., Wang, Run-M., Waykole, L., *Org. Process Res. Dev.* **2004**, *8*, 101. (c) Mickel, S. J., Sedelmeier, G. H., Niederer, D., Schuerch, F., Koch, G., Kuesters, E., Daeffler, R., Osmani, A., Seeger-Weibel, M., Schmid, E., Hirni, A., Schaer, K., Gamboni, R., Bach, A., Chen, S., Chen, W. C., Geng, P., Jagoe, C. T., Kim, H. Y., Kinder, F. R. Jr., Lee, G. T., McKenna, J., Ramsey, T. M., Repic, O., Rogers, L., Shieh, Wen-C., Wang, Run-M., Waykole, L., *Org. Process Res. Dev.* **2004**, *8*, 107. (d) Mickel, S. J., Sedelmeier, G. H., Niederer, D., Schuerch, F., Seeger-Weibel, M., Schreiner, K., Daeffler, R., Osmani, A., Bixel, D.,

Loiseleur, O., Cercus, J., Stettler, H., Schaer, K., Gamboni, R., Bach, A., Chen, G. P., Chen, W. C., Geng, P., Lee, G. T., Loesser, E., McKenna, J., Kinder, F. R. Jr., Konigsberger, K., Prasad, K., Ramsey, T. M., Reel, N., Repic, O., Rogers, L., Shieh, Wen-C., Wang, Run-M., Waykole, L., Xue, S., Florence, G., Paterson, I., *Org. Process Res. Dev.* **2004**, *8*, 113. (e) Mickel, S. J., Niederer, D., Daeffler, R., Osmani, Kuesters, E., Schmid, E., H.; Schaer, K., Gamboni, R., Chen, W. C., Loesser, E., Kinder, F. R. Jr., Konigsberger, K., Prasad, K., Ramsey, T. M., Repic, O., Wang, Run-M., Florence, G., Lyothier, I., Paterson, I., *Org. Process Res. Dev.* **2004**, *8*, 122.

3. Smith, A. B. III, LaMarche, M. J., Falcone-Hindley, M., *Org. Lett.* **2001**, *3*, 695.
4. Monteagudo, E., Cicero, D. O., Cornett, B., Myles, D. C., Snyder, J. P., *J Am Chem Soc.* **2001**, *123*, 6929.
5. (a) Nerenberg, J. B., Hung, D. T., Sommers, P. K., Schreiber, S. L., *J. Amer. Chem. Soc.* **1993**, *115*, 12621. (b) Hung, D. T., Nerenber, J. B., Schreiber, S. L., *J. Amer. Chem. Soc.* **1996**, *118*, 11054.
6. (a) Smith, A. B., Qui, Y., Jones, D. R., Kobayashi K., *J. Amer. Chem. Soc.* **1995**, *117*, 12011. (b) Smith, A. B., Kaufmann, M. D., Beauchamp, T. J., LaMarche, M. J., Arimoto, H., *Org. Lett.* **1999**, *1*, 1823; Additions and corrections *Org. Lett.* **2000**, *2*, 1983. (c) Smith, A. B., Beauchamp, T. J., LaMarche, M. J., Kaufmann, M. D., Qui, Y., Arimoto, H., Jones, D. R., Kobayashi K., *J. Amer. Chem. Soc.* **2000**, *112*, 8654.
7. (a) Harried, S. S., Yang, G., Strawn, M. A., Myles, D. C. *J. Org. Chem.* **1997**, *62*, 6098. (b) Harried, S. S., Yang, G., Lee, T. I. H., Myles, D. C., *J. Org. Chem.* **2003**, *68*, 6646.
8. Marshall, J. A., Johns, B. A., *J. Org. Chem.* **1998**, *63*, 7885.
9. (a) Paterson, I., Florence, G. J., Gerlach, K., Scott, J. P., *Angew. Chem. Int. Ed.* **2000**, *39*, 377. (b) Paterson, I., Florence, G. J., *Tetrahedron Lett.* **2000**, *41*, 6935. (c) Paterson, I., Florence, G. J., Gerlach, K., Scott, J. P., Sereinig, N., *J. Amer. Chem. Soc.* **2001**, 123, 9535. (d) Paterson, I., Delgado, O., Florence, G. L., Lyothier, I., Scott, J. P., Sereinig, N., *Org. Lett.* **2003**, *5*, 35.
10. (a) Bazain-Tejeda, B., Georgy, M., Campagne, J-M., *Synlett* **2004**, 720. (b) Francavilla, C., Chen, W., Kinder, F. R., Jr., *Organic Lett.* **2003**, *5*, 1233. (c) Arefolov, A., Panek, J. S. *Org. Lett.* **2002**, *4*, 2397. (d) Shahid, K. A., Mursheda, J., Okazaki, M., Shuto, Y., Goto, F., Kiyooka, S., *Tetrahedron Lett.* **2002**, *43*, 6377. (e) Shahid, K. A., Li, Y. N., Okazaki, M., Shuto, Y., Goto, F., Kiyooka, S., *Tetrahedron Lett.* **2002**, *43*, 6373. (f) Yadav, J. S., Abraham, S., Reddy, M. M., Sabitha, G., Sankar, A. R., Kunwar, A. C., *Tetrahedron Lett.* **2001**, *42*, 4713. (g) Miyazawa, M., Oonuma, S., Maruyama, K., Miyashita, M., *Chem. Lett.* **1997**, 1191. (h) Miyazawa, M., Oonuma, S., Maruyama, K., Miyashita, M., *Chem. Let.,* **1997**, 1193. (i) BouzBouz, S., Cossy, J., *Org. Lett* **2001**, *3*, 3995. (j) Golec, J. M. C., Jones, S., *Tetrahedron Lett.* **1993**, *34*, 8159. (k) Evans, P. L., Golec, J. M. C., Gillespie, R. J., *Tetrahedron Lett.* **1993**, *34*, 8163. (l) Golec, J. M. C., Gillespie, R. J., *Tetrahedron Lett.* **1993**, *34*, 8167. (m) Marshall, J. A., Lu, Z. H., Johns, B. A., *J. Org. Chem.* **1998**, *63*, 817. (n) Yang, G., Myles, D. C. *Tetrahedron Lett.* **1994**, *35*, 1313. (o) Yang, G., Myles, D. C. *Tetrahedron Lett.* **1994**, *35*, 2503. (p) Paterson, I., Schlapbach, A., *Synlett*, **1995**, 498.
11. Paterson, I., Florence, G. J., *Eur. J.Org. Chem.* **2003**, *12*, 2193.

12. (a) Jordan, M. A., *Curr. Med. Chem.: Anti-Cancer Agents* **2002**, *2*, 1. (b) Altmann, K. H., *Curr. Opin. Chem. Chem. Biol.* **2001**, *5*, 424. (c) He, L. F., Orr, G. A., Horwitz, S. B., *Drug Discovery Today* **2001**, *6*, 1153. (d) He, L., Chia-Ping, H. Y., Horwitz, S. B., *Mol. Cancer Ther.*, **2001**, *1*, 3. (e) Kowalsky, R. J., Giannakakou, P., Gunasekera, S. P., Longley, R. E., Day, B. W., Hamel, E., *Mol. Pharmacol.* **1997**, *52*, 613. (f) Kalesse, M. *ChemBiochem.* **2000**, *1*, 171. (g) Longley, R. E., Caddigan, D., Harmody, D., Gunasekra, M., Gunasekra, S. P., *Transplantation* **1991**, *52*, 650. (h) Longley, R. E., Caddigan, D., Harmody, D., Gunasekra, M., Gunasekra, S. P., *Transplantation* **1991**, *52*, 656. (i) Martello, L. A., LaMarche, M. J., Lifeng He,, Beauchamp, T. J., Smith, A. B., Horwitz, S. B. *Chem. Biol.* **2001**, *8*, 843.

13. *Synthesis and antitumor activity of analogues of the novel microtubule stabilizing agent discodermolide.* Kinder, Jr., F. R., Bair, K. W., Chen, W., Florence, G., Francavilla, C., Geng, P., Gunasekera, S., Guo, Q., Lassota, P. T., Longley, R. E., Palermo, M. G., Paterson, I., Pomponi, S., Ramsey, T.M., Rogers, L., Sabio, M., Sereinig, N., Sorensen, E., Wang, R.M., Wright, A., *Abstracts of Papers*, 224th American Chemical Society National Meeting, Boston, MA, August 18-22, 2002; MEDI-236.

14. Smith, A. B. III, Freeze, B. S, Brouard, I., Hirose, T.; *Org Lett.* **2003** *5* 4405.

15. Chen, J., Zhao, K., *Tetrahedron Lett.* **1994**, *35*, 2827.

16. Arimoto, H., Kaufmann, M. D., Kobayashi, K., Qiu, Y., Smith, A. B., *Synlett.* **1998**, 765.

17. Shen, Y., Gao, S., *J. Chem. Soc. Perkin Trans. I*, **1995**, 1331.

18. (a) Stowell, M. H. B., Ueland, J. M., McClard, R. W., *Tetrahedron Lett.* **1990**, *31*, 3261. (b) Patois, C., Savignac, P., *Tetrahedron Lett.* **1991**, *32*, 1317. (c) Hanessian, S., Bennani, Y. L., Delorne, D., *Tetrahedron Lett.* **1990**, *31*, 6461. (d) Hanessian, S., Bennani, Y. L., *Tetrahedron Lett.* **1990**, *31*, 6465.

19. Penning, T. D., Djuric, S. W., Haack, R. A., Kalish, V. J., Miyashiro, J. M., Rowell, B. W., Yu, S. S., *Synth. Commun.* **1990**, *20*, 307.

20. Prashad, M., Har, D., Kim, Hong-Yong, Repi, O., *Tetrahedron Lett.* **1998**, *39*, 7067.

21. White, J. M., Ashok, R T., Georg, G. I., *J. Amer. Chem. Soc.* **2000**, *122*, 11995.

22. Evans, D. A., Wu, L. D., Wiener, J. J. M., Johnson, J. S., Ripin, D. H. B., Tedrow, J. S., *J. Org. Chem.* **1999**, *64*, 6411.

23. Miyaura, N., Suzuki, A., *Chem. Rev.* **1995**, *95*, 2457.

24. Panek, J. S., Hu, T., *J. Org. Chem.*, **1997**, *62*, 4912.

25. (a) Okude, Y., Hirano, S., Hiyama, T., Nozaki, H., *J. Amer. Chem. Soc.*, **1977**, *99*, 3179. (b) Hiyama, T., Kimura, K., Nozaki, H., *Tetrahedron Lett.*, **1981**, *22*, 1937. (c) Hiyama, T., Okude, Y., Kimura, K., Nozaki, H., *Bull. Chem. Soc. Jpn.*, **1982**, *55*, 561. (d) Taki, K., Kimura, K., Kuroda, T., Hiyama, T., Nozaki, H., *Tetrahedron Lett.*, **1983**, *24*, 5281.

26. (a) Jin, H., Uenishi, J., Christ, W. J., Kishi, Y., *J. Amer. Chem. Soc.*, **1986**, *108*, 5644. (b) Kress, M. H., Ruel, R., Miller, W. H., Kishi, Y., *Tetrahedron Lett*, **1993**, *34*, 6003. (c) Chen, C., Tagami, K., Kishi, Y., *J. Org. Chem.*,**1995**, *60*, 5386. (d) Stamos, D. P., Sheng, C., Chen, S. S., Kishi, Y., *Tetrahedron Lett.* **1997**, *38*, 6355. (e) Wan, Z. K., Choi, H. W., Kang, F. A., Nakajima, K., Demeke, D., Kishi, Y., *Org. Lett.*, **2002**, *4*, 4431. (e) Choi, H. W., Demeke, D., Kang, F. A., Kishi, Y., Nakajima, K., Nowak, P., Wan, Z. K., Xie, C., *Pure Appl. Chem.*, **2003**, *75*, 1.

27. Cowden, C. J., Paterson, I., *Org. React.* **1997**, 51, 1.

STRATEGIES AND TACTICS IN ORGANIC SYNTHESIS, VOL. 6

Chapter 10

SYNTHESIS OF APREPITANT

Todd D. Nelson
Merck Research Laboratories Inc.
Department of Process Research
Merck & Co.
Wayne, Pennsylvania   19087

I. Introduction

This chapter chronicles the synthetic development of aprepitant [EMEND®] from its initial entry into preclinical development through the design and implementation of a manufacturing process. In addition to the traditional synthetic odyssey, this story is peppered with pragmatic and anecdotal vignettes related to synthesis and process development that are not only useful on a multi-kilogram scale, but are also useful in a general academic laboratory setting.

II. Original Medicinal Chemistry Route and Process Research Evaluation

In December 1993, aprepitant entered the preclinical development pipeline at Merck. At this time, the Process Research group was

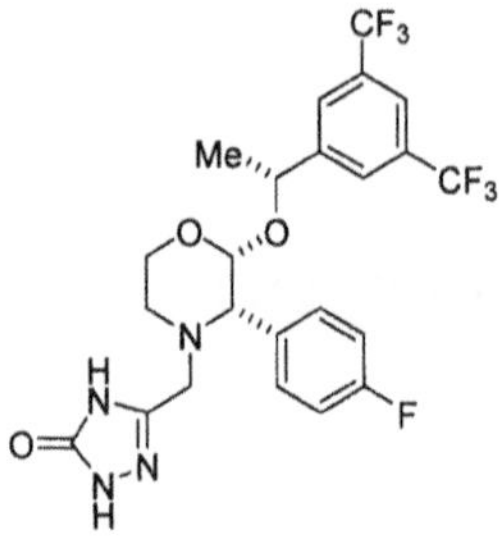

aprepitant [Emend®]

provided with the original synthetic scheme and a set of experimental procedures that the Medicinal Chemistry team had used to synthesize this drug candidate.[1]

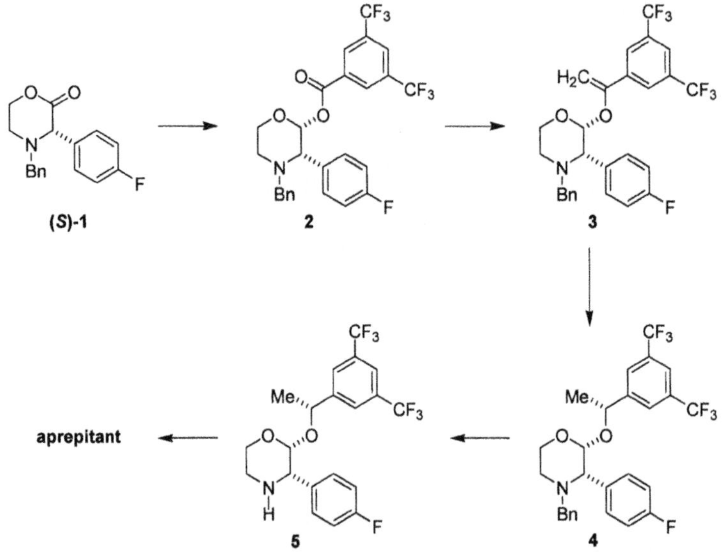

SCHEME 1

A key feature of this molecule is the *cis* vicinally-substituted morpholine core. This skeleton also contains a mixed acetal juncture and a third stereocenter that is adjacent to the oxygen ether linkage. The route that the medicinal chemistry group designed contained many attractive features that were initially envisioned to be the cornerstone of a longer term synthesis.   Central to this was the reduction and acylation of oxazinone **1**, olefination of ester **2**, and diastereoselective hydrogenation of vinyl ether **3** (Scheme 1). The initial reduction/acylation sequence was used to relay the stereochemical information of a pre-existing stereocenter during the construction of the second stereocenter. Thus, an enantiomerically pure oxazinone could be converted into vicinally substituted morpholine acetal **3**. In a similar fashion, the stereochemical relay of this substrate was used to establish the third stereogenic center of the carbon backbone.

Intermediate **(S)-1** is simply *N*-benzyl-4-fluorophenylglycine that has been capped with an ethylene unit. The original synthesis in which 4-fluorophenylacetic acid was transformed to the corresponding chiral oxazolidinone **6** is depicted in Scheme 2.[1]   Masked α-azido acid **7** was formed diastereoselectively from this intermediate.[2] Hydrolysis and azide reduction afforded enantiomerically pure (*S*)-4-fluorophenyl glycine (**8**). Reductive amination with benzaldehyde introduced the *N*-benzyl unit and subsequent *N,O*-dialkylation with ethylene dibromide provided chiral oxazinone **1**.

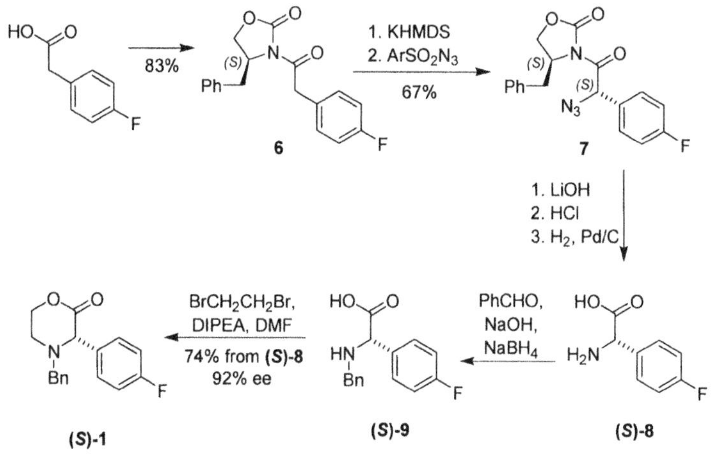

SCHEME 2

Prior to the initiation of work on this drug candidate, which ultimately would result in bulk drug for toxicology studies, a process research feasibility analysis was performed.   The conclusion was that while the overall synthetic strategy that the medicinal chemistry group used to set the latter two stereocenters was attractive for scale-up, the entry into precursory oxazinone **1** needed to be modified.

III. Oxazinone Synthesis

Multiple routes that by-passed the oxazolidone and the azide-forming step were immediately designed. The common key intermediate that was targeted in these approaches was 4-fluorophenylglycine *(S)*-**8**. One such concise approach was by the enzymatically-mediated transamination of 4-fluorophenylglyoxylic acid (**10**) (Scheme 3).[3] This method resulted in a high level of asymmetric induction, but the assay yield was moderate and the dilute conditions required for rapid scale-up prohibited this from being translated to a preparative scale.

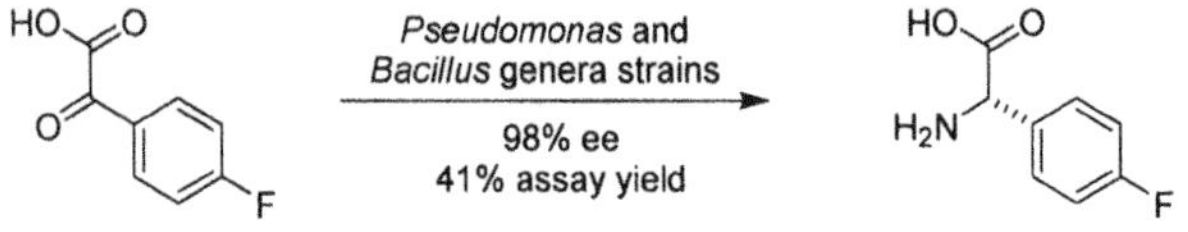

SCHEME 3

In parallel to the above effort, an alternative straightforward approach was also being developed, which utilized a classical entry into amino acids: the Strecker reaction. As a result, inexpensive and readily available 4-fluorobenzaldehyde was condensed with NaCN and $BnNH_2$ to afford amino nitrile **11**,[4] which was hydrolyzed and esterified to afford crystalline amino ester hydrochloride **13** (Scheme 4). Hydrogenolysis and resolution with (-)-dibenzoyl tartaric (DBT) acid resulted in diastereomerically enriched (95% de) amino ester *(S)*-**14·DBT**.[5] After freebasing the amine, a modified ethylene installation was utilized to access the enantiomerically pure oxazinone that was crystallized as the hydrochloride salt [*(S)*-**1·HCl**].

While this second generation approach offered a safe, robust and inexpensive route to enantiomerically pure oxazinone *(S)*-**1·HCl**, it was also lengthy and suffered from the inherent 50% theoretical yield incumbent to the classical resolution. An additional problem that arose was the epimerization of the benzylic hydrogen during alkylation with dibromoethane. Ultimately, conditions were developed that did not compromise the stereochemical integrity of the phenylglycine backbone.

However, this facile epimerization phenomenon was viewed as an opportunity, rather than a liability, to dramatically refine the oxazinone route. Within our process research group, it is commonplace to consider a crystallization-driven racemization resolution protocol whenever an epimerizable stereocenter exists in a molecule. Such an approach towards diastereomerically pure substrates simply relies on differences in the solubility of the diastereomeric products and on the ability of the undefined stereocenter to epimerize under the conditions used in the crystallization. In fact, these processes can effectively rival catalytic asymmetric synthesis and the literature is replete with recent examples of resolution/racemization approaches to chiral molecules.[6]

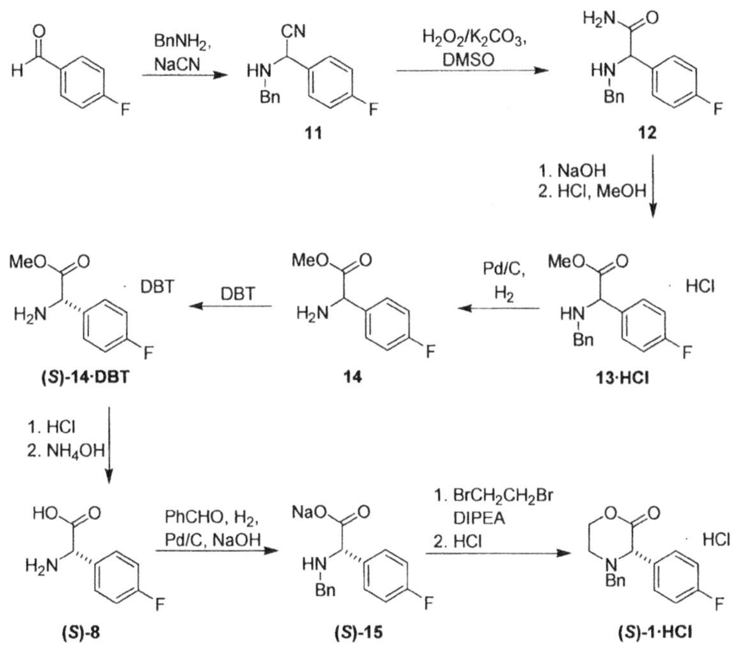

SCHEME 4

The first criterion was to identify a crystalline diastereomeric salt of oxazinone (±)-1. Many chiral acids were screened and [(1S)-(endo,anti)]-(-)-3-bromocamphor-8-sulphonic acid (BCSA) afforded a 27% yield of the diastereomeric enriched (88% de) BCSA salt (Scheme 5).[7] This was upgraded to 99% de with an 88% recovery after recrystallization from DMF/iso-propyl acetate (IPAC). A necessary physical property of crystalline oxazinone **(S)-1·BCSA** was low solubility when compared to the undesired diastereomeric salt. The second requisite for this process required that epimerization occur under the reaction conditions, and since the resolving agent was a sulfonic acid, the epimerization needed to rely on acid-catalyzed tautomerization of the oxazinone.

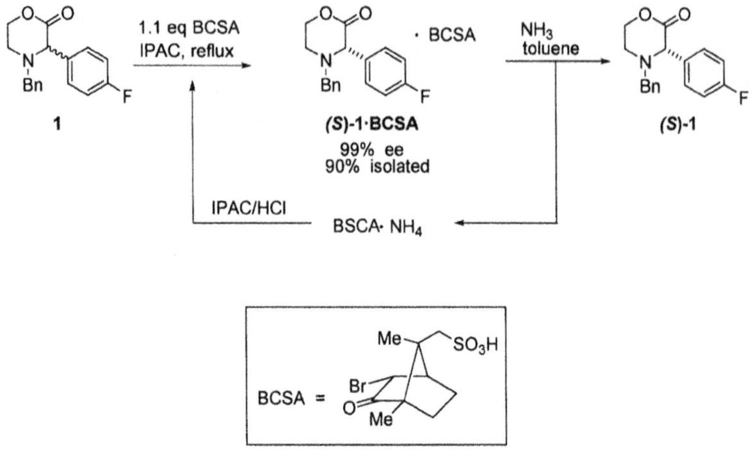

SCHEME 5

Initial experiments to induce racemization of oxazinone **(R)-1** under acidic conditions (HOAc) were successful. However, when these conditions were applied to a system in which the desired salt **(S)-1·BCSA** had low solubility (HOAc/IPAC, 1:4 at 50 °C), a low yield (25%) was obtained. A system of TFA/IPAC (3:97) improved the yield to 90%, but the conversion time (7 days) was prohibitive. Ultimately, a slight excess of BCSA (1.15 eq) was found to catalyze the racemization. Thus, the initial step in the crystallization-induced resolution procedure was to heat a solution of racemic oxazinone **1** at reflux in IPAC followed by the slow addition of an IPAC solution of BCSA.

A commonly used protocol used in controlling crystallization processes is to establish a seed bed prior to the bulk crystallization process. This significantly aids in the reproducibility of the crystallization. This is effected by determining the saturation point of the system and then adding authentic product prior to reaching this point. Once the seed bed is maintained, the system by default has been saturated with the desired product and by either addition of anti-solvent or with controlled cooling, both nucleation and crystal growth can be controlled.

This mitigates the risk of the system becoming supersaturated, and as a result, poised for uncontrolled crystallization.

In this specific case, seeding was crucial to the success of the process. If the system was not seeded, the initially crystallized BCSA·oxazinone was 90% de. The major factor in the length of reaction time was the slow turnover of the undesired diastereomer that had crystallized. A three hour addition of BCSA solution to a solution of racemic oxazinone 1 that contained 10% seed of BCSA oxazinone avoided supersaturation problems and high diastereopurity of the crystallized solids was maintained throughout the process. With this modified procedure, since very little of the undesired diastereomer was formed early in the process, the time cycle was reduced from 7 days to 2 days

Thus, the refined process established a seed bed of desired oxazinone salt **(S)-1·BCSA** with 5% authentic product in the refluxing IPAC solution prior to the addition of the BCSA solution. This resulted in a racemic mixture of oxazinone being converted into diastereomerically pure (99% de) crystalline **(S)-1·BCSA** in a 90% isolated yield.

The next step was to perform a salt break of the oxazinone **(S)-1·BCSA**. Initially, this was accomplished by partitioning the oxazinone with the two phase system of aqueous ammonium hydroxide and toluene. This was convenient since the next step required that oxazinone **(S)-1** be in toluene; however, recycling the aqueous stream of BCSA was problematic. Ultimately, the method of choice for the generation of oxazinone freebase was to simply add NH_3 (g) to a slurry of oxazinone **(S)-1·BCSA** in toluene. Filtration of the resulting BCSA·NH_4 salt provided the desired toluene stream of oxazinone that was used in the subsequent reduction/acylation step. The advantage of this free-basing technique was that it provided a high recovery of crystalline BCSA·NH_4, which could be recycled through another crystallization-induced resolution procedure.

For the second part of the recycle process, an IPAC stream of BCSA was initially generated from BCSA·NH_4 by lengthy recycle procedure that involved ion exchange chromatography followed by a cumbersome solvent switch from water to IPA and then to IPAC. An improved procedure was developed in which HCl (g) was added directly to an IPAC slurry of BCSA·NH_4. Filtration removed ammonium chloride and distillation provided an anhydrous IPAC stream of BCSA that was devoid of HCl, which was directly used in the oxazinone racemization-resolution procedure. Regardless of the number of recycle sequences, degradation of the enantiopurity or effectiveness of BCSA was not observed.

With this efficient crystallization-induced resolution in place, a more concise preparation of racemic oxazinone **1** was desired. A modified variant of Scheme 4 was rapidly instituted (Scheme 6).

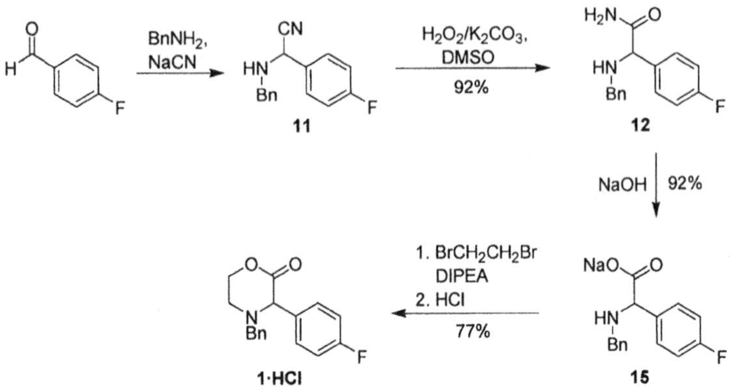

SCHEME 6

While this route worked well on multigram scale, significant issues were identified during the scale-up process. First, reproducibility of the alkylation was problematic. Sometimes acid **8** (Scheme 4) would form a gel when dissolved in DMF. Therefore, the intermediate sodium salt **15** was isolated and used as the alkylation precursor. This was done with NaOH/IPA; however, tedious azetropic drying of the IPA solution was required to obtain reasonable yields. Also, product isolation was difficult due to the hygroscopic nature of sodium salt **15**. In order to successfully perform the subsequent alkylation, sodium salt **15** was dissolved in DMF, which also needed to be azeotroped again in order to provide a dry stream for the alkylation.

In general, on multigram scale, these types of drying operations are simply performed by drying the organic stream over a desiccant; however, on a multi-kilogram scale it is more desirable to perform drying operations by distillation. Solvents that form good azeotropes with water (e.g. toluene, IPAC) are often used for this purpose. The progress of the operation is quantitatively monitored by measuring the water content of aliquots by Karl Fisher titration.

After the azeotropic drying of the DMF steam of sodium salt **15**, oxazinone hydrochloride **1·HCl** was formed in a typical 77% yield. Overall, both the tediousness of this process and the hygroscopicity of using the sodium salt **15** en route to oxazinone **1** needed to be addressed.

At this time, an even more streamlined approached to oxazinone **1** was investigated. A modified Strecker reaction between 4-fluoro-benzaldehyde and *N*-benzylethanolamine (NBEA) afforded aminonitrile **16**, which was extracted into IPAC (Scheme 7). The subsequent Pinner reaction was effected by addition of HCl (g) to this solution. Imidate hydrochloride **17·HCl** was formed and residual water caused spontaneous hydrolysis to lactone hydrochloride **1·HCl**. This slurry could be filtered to afford the crystalline lactone hydrochloride (containing ca. 15 wt% NH_4Cl) or alternatively and preferably it could be treated with aqueous $KHCO_3$ to freebase the amine. The resulting IPAC stream fed directly into the crystallization-induced resolution procedure that had been established.

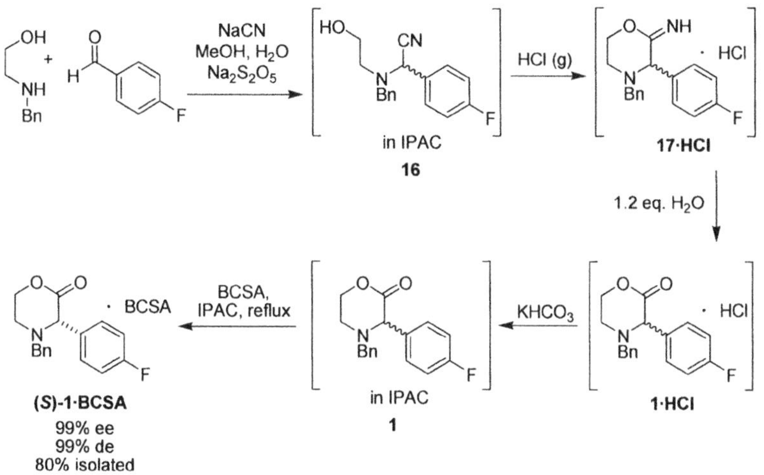

SCHEME 7

Two issues with the route, however, needed attention. In the first transformation, NaCN is charged to the system, and is followed by the addition of HCl gas. A considerable amount of time was devoted to monitoring and minimizing the amount of cyanide that tracked through the first step of this process. This insured an extremely high level of safety, understanding, and accountability for cyanide mass balance.

The second issue with this scheme was that elevated amounts of adventitious water present in the hydrolysis of imidate **17·HCl** would further hydrolyze resulting lactone **1** into the corresponding hydroxy acid. Although the hydroxy acid could potentially be lactonized, we preferred not to pursue this route, and exclusively focused on minimizing this side reaction. One solution was to dry the IPAC stream of aminonitrile **16** and subsequently charge 1.1 equivalents of water to the system. A much simpler solution was to wash the IPAC solution of aminonitrile **16** with 15 wt% NaCl prior to HCl addition. This brine wash reproducibly adjusted the water content of the IPAC stream to contain 1.1-1.2 equivalents of water.

Overall, this improved oxazinone synthesis now allowed for diastereomerically pure BCSA oxazinone salt **(S)-1·BCSA** to be formed in a single vessel without the necessity of isolating intermediates. The overall isolated yield for this process was 89% (>98% de).

IV. Diastereoselective Acetal Formation and Olefin Hydrogenation

After the salt break of **(S)-1·BCSA** was complete, the oxazinone solution was cooled to -70 °C and 1 equivalent of lithium tri-*sec*-butyl borohydride was added. The resulting alkoxide was acylated with 3,5-bis trifluoromethylbenzoyl chloride (**22**) to provide 80% yield of enantiomerically pure *cis*-ester **2** (Scheme 8).[13] Strict adherence to the reaction parameters for this reduction/acylation sequence was required. If at any time prior to the completion of acylation the temperature rose above -60 °C, a significant degradation in diastereoselectivity occurred. A series of experiments demonstrated that the intermediate was the kinetically-formed alkoxy anion. If the internal temperature for either of these steps was allowed to exceed -60 °C, then elevated levels of *trans*-ester **21** would result. In fact, if the temperature was allowed to reach ambient temperature at any point, then *trans*-ester **21** would predominate (99:1, *trans:cis*). If, however, the temperature was maintained <-60 °C, then the desired *cis* diastereomer would preferentially form. Reaction workup of the kinetically-controlled system followed by a direct crystallization resulted in an 80% isolated yield of *cis*-ester **2** (>99% de).

Numerous other reducing agents were investigated and none were found to give higher *cis/trans* ratios than lithium tri-*sec*-butyl borohydride. In addition, a series of experiments that investigated the *cis/trans* ester ratio as it related to the age time of alkoxide prior to acylation were performed. These experiments indicated that high *cis/trans* ratios could be obtained at room temperature if the alkoxide was rapidly acylated. These results were in turn substantiated by the predominant formation of *cis*-ester **2** with a flow cell apparatus (i.e. streams of oxazinone **1** and reducing agent were simultaneously added at room temperature and acylated after a fixed residence time). Although this latter method could potentially remove cryogenics from the process, the overall yield of cis-ester 2 was still lower than the simple reduction at -60 °C.

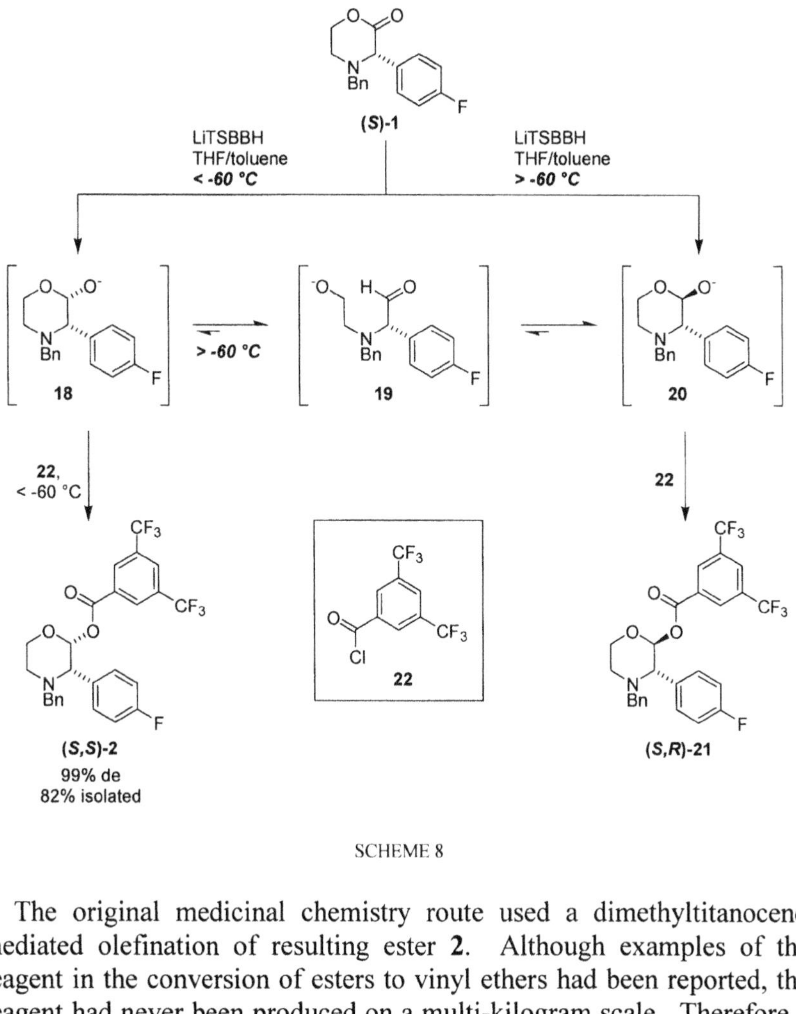

SCHEME 8

The original medicinal chemistry route used a dimethyltitanocene-mediated olefination of resulting ester 2. Although examples of this reagent in the conversion of esters to vinyl ethers had been reported, this reagent had never been produced on a multi-kilogram scale. Therefore, a significant amount of fundamental research was performed that resulted in an improved preparation,[8] mechanistic elucidation,[9] titration method for its quantitative determination,[10] and process optimization for the particular ester substrate.[11]

The process optimization centered on developing a safe, robust, and economical olefination. An initial modification utilized MeMgCl in THF, instead of MeLi in ether, to generate dimethyl titanocene **24**. In the solid state, dry **24** decomposed within minutes by turning black with gas evolution. Therefore, a process was developed that avoided isolation of intermediate **24** and resulted in the direct formation of dimethyl titanium carbene **25** stream, which converted ester **2** to desired vinyl ether **3**. One additional facet was that the reaction needed to be stopped quickly once

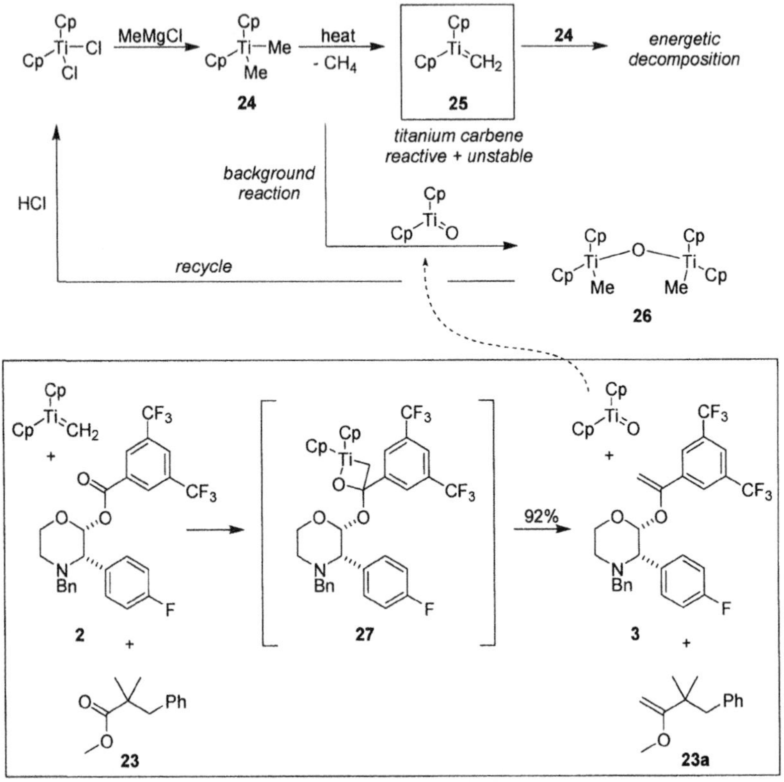

SCHEME 9

ester **2** was consumed, otherwise excess titanium carbene would further react with vinyl ether **3**. If, however, a precise stoichiometric dimethyl titanocene charge was used, the reaction would not proceed rapidly or to completion. It was discovered that a sacrificial ester could be charged to the reaction mixture in order to react with excess reagent.[11] If this ester were to be olefinated with surplus dimethyl titanocene **24**, but at a relative rate that was much less than substrate **2**, then this sacrificial ester could even be co-charged at the beginning of the reaction. A screen of esters revealed that acetates of tertiary alcohols had these properties. The inexpensive ester 1,1-dimethyl-2-phenylethyl acetate (**23**) was the optimal co-additive for this reaction. The oxo-bridged titanocene dimer **26** was the predominant titanium byproduct from this reaction. This was recovered by filtration at the end of the reaction and recycled to crystalline titanocene dichloride in 94% by treatment with HCl in toluene or THF. Thus, this recycling procedure reduced waste and helped to minimize costs by generating the starting reagent for this process. A detailed experimental procedure for the step has been reported and has resulted in the conversion of ester **2** to vinyl ether **3** in 91% yield.[11]

The next steps were to diastereoselectively hydrogenate vinyl ether **3** and replace the *N*-benzyl group of **4** with the correct appendage. A quick screen of heterogeneous catalysts in 1:1 EtOAc/EtOH showed that better diastereoselectivity (86:14 ratio, **4:28**) was observed with Pd/Al$_2$O$_3$/H$_2$ rather than with the originally used Pd/C.

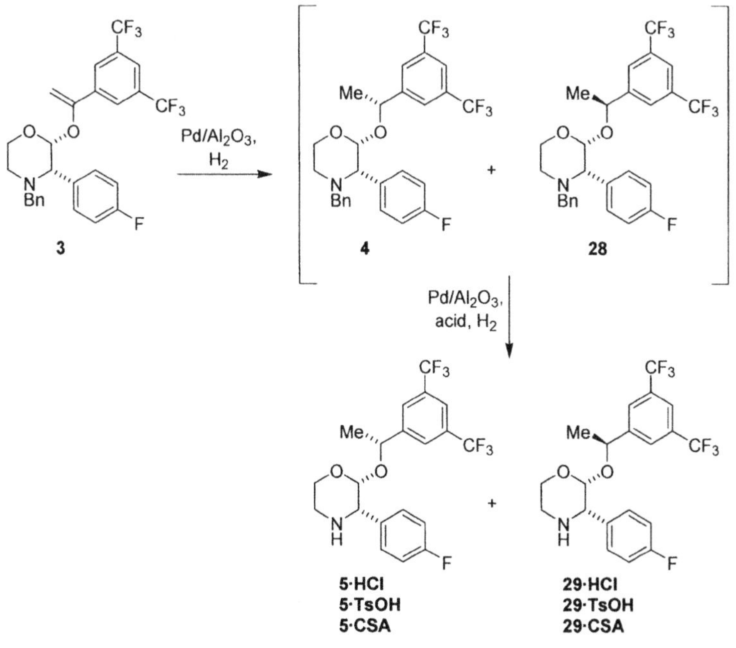

SCHEME 10

During early work on the debenzylation step, a small amount of an *N*-Et by-product was observed. This was presumably formed from the reductive alkylation of acetaldehyde, which in turn was most likely formed from dehydrogenation of the solvent EtOH. To avoid this by-product formation, the debenzylation was performed in MTBE or acetone, which resulted in a 91:9 ratio of diastereomers (**4:28**). Another advantage of MTBE or acetone was that a direct crystallization of the *sec*-amine TsOH salt **29·TsOH** or the corresponding (*R*)-camphorsulphonic acid salt **29·CSA** could be accomplished without having to perform a solvent turnover. These conditions resulted in a 91:9 diastereoselectivity and 86% isolated yield of acetal **29·TsOH** or **29·CSA** (99% de) after crystallization.

V. End Game and Triazolinone Synthesis

Secondary amine **5** was alkylated with *N*-methyl-carboxyl-2-chloroacetamidrazone (**30**) (prepared from chloroacetonitrile and methyl hydrazonocarboxylate in 70% yield). Initially, DMF, powdered K_2CO_3 and 1.1 equivalents of the alkyl halide **30** were used. The time cycle for this reaction was important – the reaction would darken after prolonged reaction times. An even better reaction profile was observed using 1.05 equivalents of the alkylating agent in DMSO.

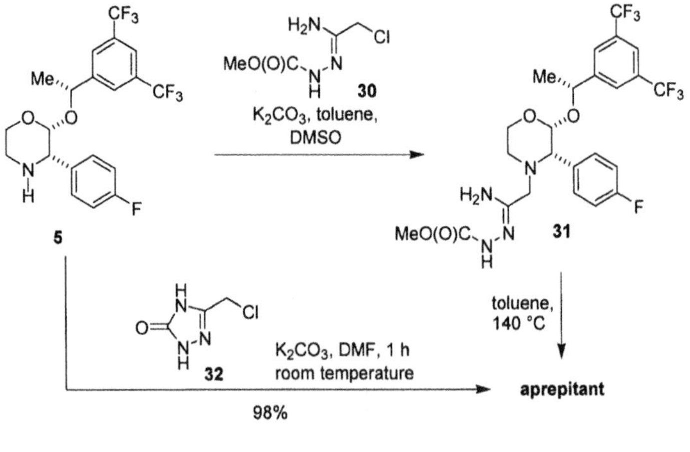

SCHEME 11

The subsequent thermal cyclization that afforded aprepitant was originally performed in refluxing xylenes with DIPEA. Alternative systems (acidic or strongly basic) did not offer an advantage. Thus, the alkylation of amine **5** was conducted in $DMSO/K_2CO_3$, partitioned between MTBE and water, and the organic layer solvent switched to xylene/DIPEA at 135 °C to afford crystalline aprepitant in 85% yield. In addition to the extractive work-up, since color was still formed in this reaction, a carbon treatment was necessary. Overall, aprepitant was synthesized from *cis* ester **2** in 57% yield.

It was envisioned that the triazoline-5-one **32** could be made off-line and introduced by an S_N2 displacement. A method to rapidly synthesize 3-chloromethyl-1,2,4-triazolin-5-one [CMT (**32**)] and alkylate secondary amine **5** was developed.[12]   The optimized procedure condensed semicarbazide hydrochloride with the orthoester of chloroacetic acid in 90% yield. Alkylation of secondary amine **5** (K_2CO_3, DMF, 1 h, room temperature) cleanly and rapidly produced aprepitant that was devoid of color in 98% yield (Scheme 11). This route was more convergent and provided a better overall yield of aprepitant from secondary amine **5** than the initial route that went through intermediate **31**.

VI. Synthesis Revision

The chemistry described above was used to in multiple pilot campaigns (1995-1999) to prepare hundreds of kilograms of aprepitant that were used in clinical studies. The overall yield to aprepitant was 48%. During the course of these campaigns, three key factors drastically affected the future course of research and development of this molecule within our Process Research Group: 1) hNK1 receptor occupancy data for this antagonist indicated that higher doses of the drug might be needed 2) positive clinical trials for the lead indication [chemotherapy-induced nausea and vomiting (CINV)] and 3) the possibility that this drug candidate was clinically efficacious for another indication. All of these factors led to an aggressive undertaking to design and demonstrate a more efficient method for producing aprepitant.

Critical evaluation of the existing process revealed three key areas for improvement.

1. BCSA: The synthesis of enantiomerically pure oxazinone **1** used an efficient crystallization-induced resolution process; however, the only chiral acid that could be used in this protocol was BCSA. Numerous attempts to find an alternative acid for the crystallization driven resolution were unsuccessful. As mentioned early, the recycling of BCSA was imperative. However, even with the recycle, enantiomerically pure BCSA

was a still a very expensive reagent and the unreacted BCSA was lost in the mother liquor of the resolution-racemization procedure.

2. Lithium tri-*sec*-butyl borohydride: This reagent came off patent protection during this development period. Accordingly, new potential vendors of this expensive reagent could have had positive economic impact on this step. Nevertheless, long-term projected costs still indicated that this reagent would remain a cost driver in the bulk drug synthesis. Equally important was the inherent temperature restrictions of the chemistry. This kinetically controlled step required strict adherence to a cryogenic operating window.

3. Dimethyl titanocene: Once again, an expensive reagent with an imperative recycle was required.

One critical factor that the team needed to be cognizant of was the impurity profile of penultimate intermediate and API (advance pharmaceutical intermediate) that would be derived from a new synthesis. A significant number of clinical trials had already been conducted with kilograms of material that had already been synthesized. A new batch of API that contained an elevated level of an impurity or that contained a new impurity could not immediately be used for clinical trials without a time consuming qualification process. Hence, the task at hand was the discovery and development of a more efficient synthesis with the caveat that no new impurities could be present in the final bulk drug.

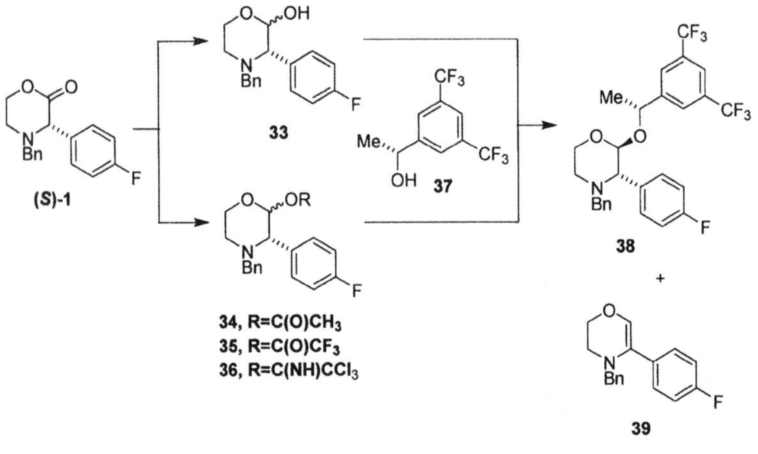

SCHEME 12

One attractive disconnection was at the acetal center of the morpholine core. Early in the program, the acetalization between lactols **33-36** and 3,5-bis(trifluoro)-*sec*-phenethyl alcohol (**37**) was investigated (Scheme 12). Addition to the β-face of the intermediate oxonium ion, which provided *trans* ester **38**, or elimination that afforded olefin **39**, occurred. The reciprocal approach was to conduct an S_N2 displacement of an activated 3,5-bis(trifluoro)-*sec*-phenethyl alcohol such as triflate **40** and morpholine alkoxide **18** (Scheme 13). The problem in this case was not the morpholine substrate; indeed, this was the same alkoxide that was acylated on multi-kilogram scale (Scheme 8). The issue with this approach resided with the electrophile. When triflate **40** was treated with alkoxide **18**, extensive elimination occurred, although the corresponding des-Me triflate cleanly underwent displacement with a similar alkoxide from an earlier generation analog synthesis.[13]

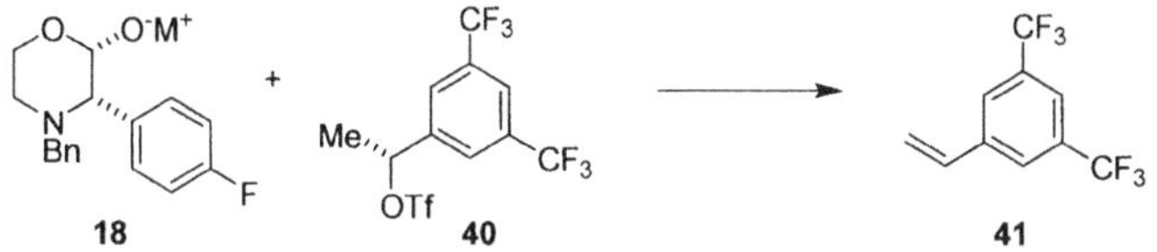

SCHEME 13

VII. 2nd Total Synthesis: Hofmann Elimination

An alternative approach to vinyl ether **3** was designed, which relied on a quaternary ammonium bicyclic acetal undergoing a Hofmann elimination.[14] A requisite synthon for this process was aminodiol **44** (Scheme 14). An asymmetric dihydroxylation of styrene **41** afforded the corresponding enantiomerically pure diol **42**.[15] Activation and ethanolamine addition afforded crystalline **44**. The single stereocenter contained in this substrate was the stereocontrol element for subsequent asymmetric induction.

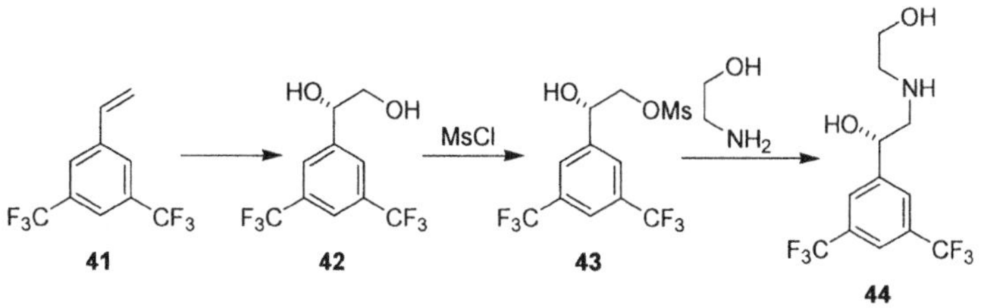

SCHEME 14

The morpholine skeleton was assembled by coupling aminodiol **44** with glyoxal and 4-fluorophenyl boronic acid (**45**) (Scheme 15). This produced a diastereomeric and regioisomeric mixture of five isomers. When the reaction was performed with racemic amino alcohol **44**, desired racemic *trans* lactol **47** was crystallized directly from the reaction mixture. However, when enantiomerically pure amino alcohol **44** was used, direct crystallization of **47** did not occur. Only diastereomerically pure *cis* lactol **46** could be coaxed into selectively crystallizing from the reaction mixture under conditions that allowed for the interconversion of the isomeric morpholines in the supernatant. However, the isomer needed for the assembly of bicyclic acetal **48** was *trans* **47**.

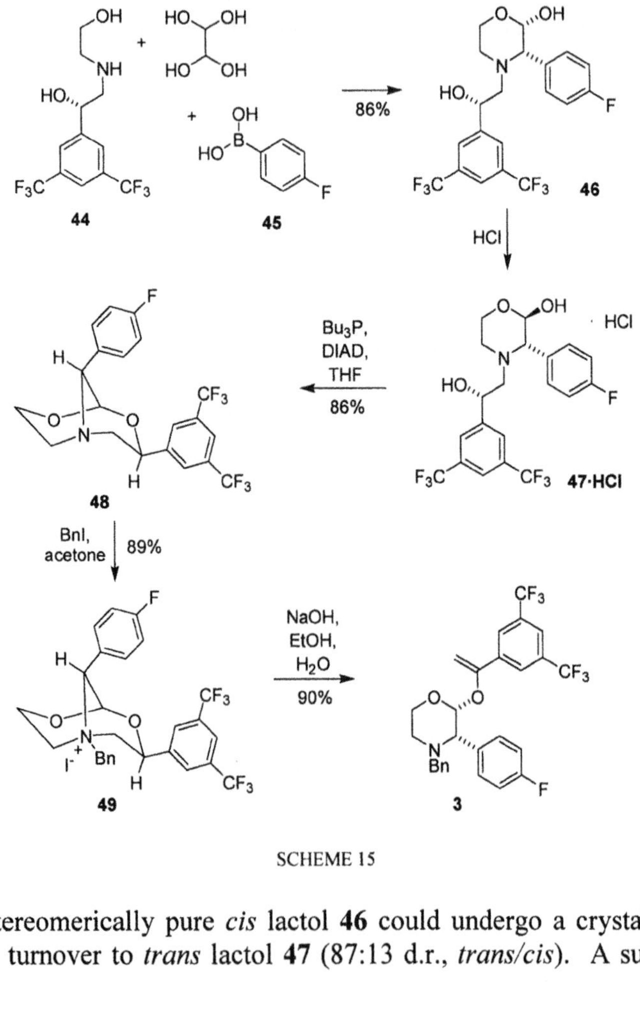

SCHEME 15

Diastereomerically pure *cis* lactol **46** could undergo a crystallization-induced turnover to *trans* lactol **47** (87:13 d.r., *trans/cis*). A subsequent

Mitsunobu reaction afforded crystalline bicyclic acetal **48** in 86% isolated yield from diastereomerically pure **47·HCl**.   The key Hofmann elimination precursor was prepared by quaternization of bicyclic acetal **48** with BnI in 89% yield. Treatment of the resulting quaternary ammonium salt **49** with 1.1 eq of NaOH cleanly provided desired vinyl ether **3** in 90% yield.

This process intercepted the vinyl ether intermediate from the existing bulk route in 24% overall yield and is highlighted by a crystallization driven diastereoselective condensation, an intramolecular cyclization that controlled the selectivity of the glycosidation, and by a selective Hoffman elimination. It also addressed the main objectives: BSCA was removed from the process, the cryogenic step that utilized lithium tri-*sec*-butylborohydride had been circumvented, and a dimethyl titanium carbene olefination was not required. However, this was not viewed as the ultimate manufacturing route. A limitation was that the sequence was lengthy and it incorporated a large number of operations.

VIII. 3rd Total Synthesis: Trans Acetalization Reaction

The next total synthesis relied on setting the vicinal stereochemistry of the morpholine ring by a stereoselective hydrogenation. The goal was to use the acetal stereocenter to control the stereochemical induction at the adjacent carbon. A proof of concept was needed to validate this route.

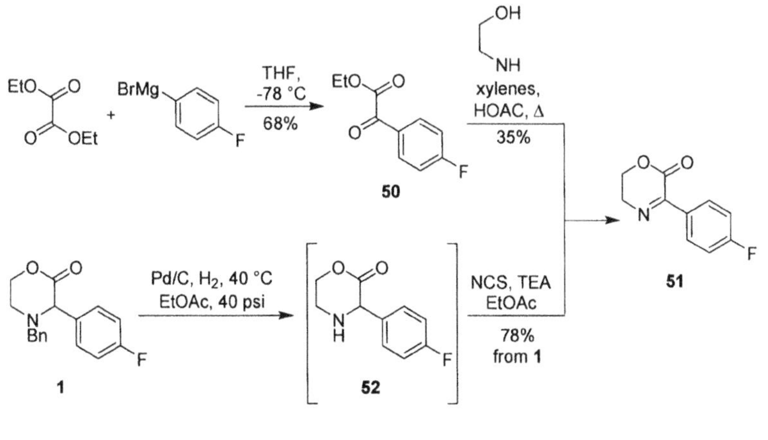

SCHEME 16

The first approach to imine **51** was accomplished in a two step fashion by the formation of glyoxylate **50**, followed by the condensation with ethanolamine (Scheme 16). The yield for this sequence was poor. Therefore, an alternative strategy to access imine **51** was used. Oxazinone **1**, an intermediate from the pilot plant process, was abundant. This was hydrogenolyzed to secondary amine **52**, which was then halogenated with NCS and eliminated to form ketoimine intermediate **51** in 78% overall yield.

Selective reduction of lactone **51** afforded intermediate imine lactol **53**, which was esterified with 3,5-bis(trifluromethyl)benzoic acid (**54**) to give imine ester **55** (Scheme 17). Alternatively, the intermediate alkoxide could be directly acetylated with 3,5-bis(trifluromethyl)benzoyl chloride (**22**). It was not necessary for this intermediate to be enantiomerically pure: at this stage we were only concerned with the relative stereochemical outcome from the reduction of imine **55**.

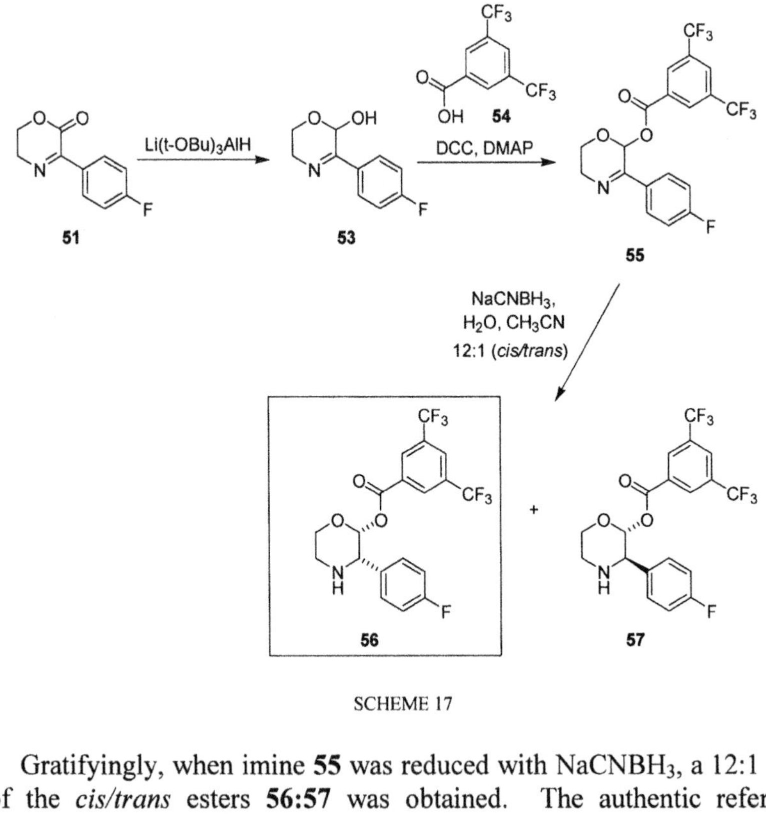

SCHEME 17

Gratifyingly, when imine **55** was reduced with NaCNBH$_3$, a 12:1 ratio of the *cis/trans* esters **56:57** was obtained. The authentic reference products had been synthesized by debenzylation of known *cis* ester **2** and *trans* ester **21**. This result indicated that the stereochemical information at the acetal juncture could indeed be used to establish the adjacent stereocenter adjacent to the nitrogen. This concept, a substrate controlled diastereoselective imine reduction to establish the requisite *cis*-disubstituted morpholine motif, was successfully used in the 3rd generation synthesis and was found to be operative in the ultimate manufacturing route.

The next key milestone was to synthesize the analogous enantiomerically pure substrate that contained the chiral *sec*-phenylethyl ether linkage and investigate the corresponding diastereoselective imine reduction. The starting point for this route was enantiomerically pure oxazinone **61**.[16] This was similar to the enantiomerically pure oxazinone that was currently being used in the pilot plant via the BCSA-mediated crystallization-induced resolution process – except the *N*-Bn group had been replaced with an *N*-α-MeBn group (Scheme 18). This chiral appendage elegantly allowed for a dynamic crystallization-induced resolution of oxazinone **61**. An advantage of this oxazinone was that it obviated the external and expensive chiral control element (BCSA) with an internal source of chirality that could be derived from inexpensive α-methylbenyzyl amine. The two key starting materials, amino alcohol **58** and glyoxal hydrate **59**, were synthesized in three steps and one step, respectively. Amino alcohol **58** was prepared from ethyl oxalyl chloride and α-methylbenzyl amine in 81% yield. Glyoxal hydrate **59** was synthesized by oxidation (HBr/DMSO, 80%) of the corresponding acetophenone.

The desired oxazinone was created by the condensation of 4-fluorophenylglyoxal hydrate **59** with amino alcohol **58**. The initial diastereoselectivity of this condensation was 2:1 (**60:61**). A number of crystallization-induced resolution conditions were examined. The 2:1 mixture could be treated with HCl in IPAC at 70 °C to selectively crystallize the desired diastereomeric hydrochloride salt **61·HCl** and concomitantly epimerize the undesired oxazinone diastereomer **60**. Overall, the process afforded a 90% yield of oxazinone hydrochloride **61·HCl** (98% de).

Reduction of oxazinone **61** with DIBAL-H afforded the corresponding lactol, which was converted to the trichloroacetimidate **62**. As mentioned earlier with the *N*-Bn case (Scheme 12), all attempts at installing a *cis*-acetal juncture resulted in elimination or predominantly *trans*-acetalization. It was envisioned that this phenomenon of selective *trans*-acetalization could be successfully utilized. The plan was to start with the enantiomerically pure antipode (at the benzylic position) and perform a selective *trans* acetalization. The correct absolute stereochemistry would be established at the anomeric center. Subsequent, imine formation would destroy the original morpholine ring stereocenter, as well as excise the α-methylbenzylamine stereocenter, and imine reduction would create the *cis*-vicinally morpholine core with the correct absolute stereochemistry.

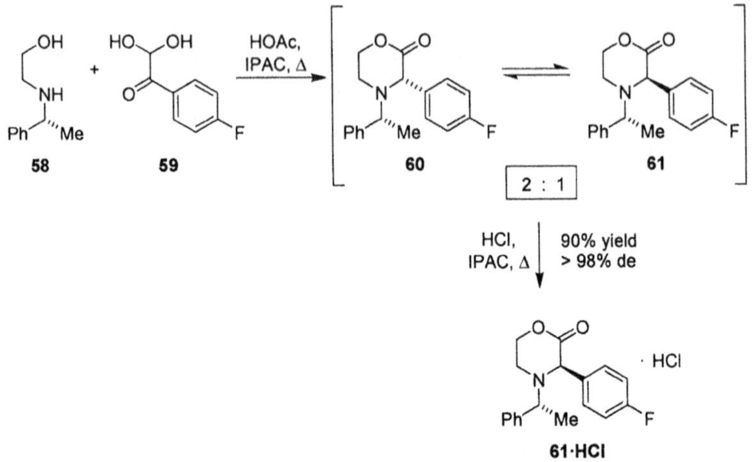

SCHEME 18

Lewis-acid catalyzed acetalization with enantiomerically pure alcohol **37** afford crystalline *trans*-acetal **63** in high yield (85% from oxazinone **61**) and diastereoselectivity (96:4 *trans/cis*). Hydrogenolysis gave secondary amine **64** that was cleanly converted to imine ether **65** by a chlorination/elimination procedure. The diastereoselectivity of the subsequent imine hydrogenation was >99:1 (*cis/trans*). The resulting penultimate amine **5** was isolated as either the HCl or TsOH salt in 81% yield (Scheme 19).

This route also successfully addressed most of the objectives that were desired in a manufacturing process. It used inexpensive reagents, contained no cryogenic steps and it was extremely high yielding. It was, however, an operationally lengthy synthesis. The fourth and final route that was being investigated on a parallel time frame was also inexpensive and high yielding; however, a significant advantage of this alternative route was the concise assembly of the carbon backbone.

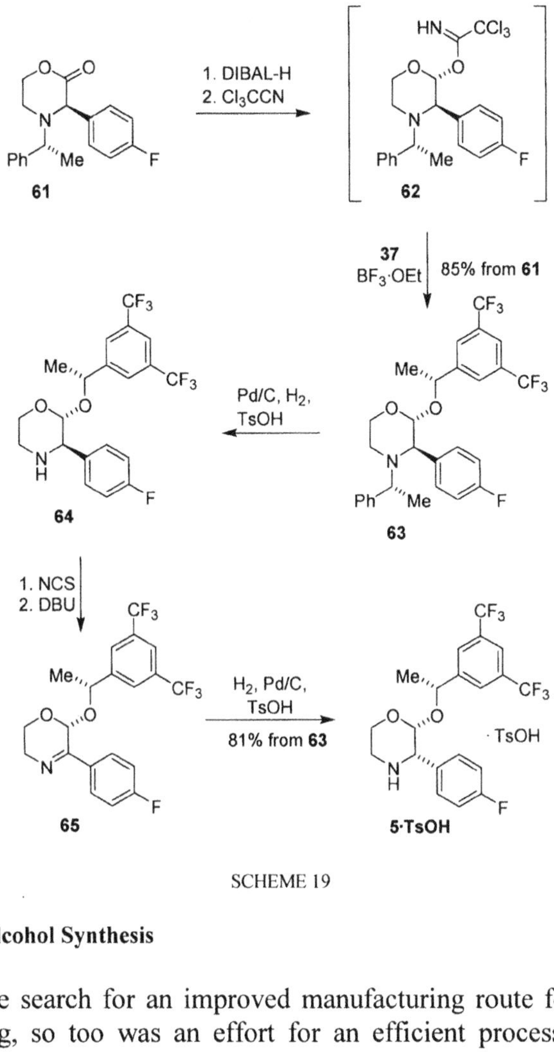

SCHEME 19

IX. Chiral Alcohol Synthesis

While the search for an improved manufacturing route for aprepitant was ongoing, so too was an effort for an efficient process to (R)-3,5-bistrifluoromethylphenyl ethanol (37), which was a critical chiral building block to both the 3rd and 4th generation total syntheses of aprepitant. To this end, a robust catalytic asymmetric transfer hydrogenation of the corresponding ketone 66 was discovered (Scheme 20).[17] The preferred ligand for the transfer hydrogenation of ketone 66 was (1S,2R)-cis-1-aminoindan-2-ol (67) and the preferred metal source for the reduction was dichloro(p-cymene)Ru(II)dimer (68).

A problem with the isolation of enantiomerically pure alcohol 37 was its high solubility in most organic solvents [52 mg/g heptane (mp 52 °C)] and the ease at which it sublimed. The situation was complicated by the lower solubility and high melting point of racemic **37** [21 mg/g heptane (mp 72 °C)]. This issue was side-stepped by forming a 2:1 crystalline inclusion complex of chiral alcohol **37** with DABCO. The level of

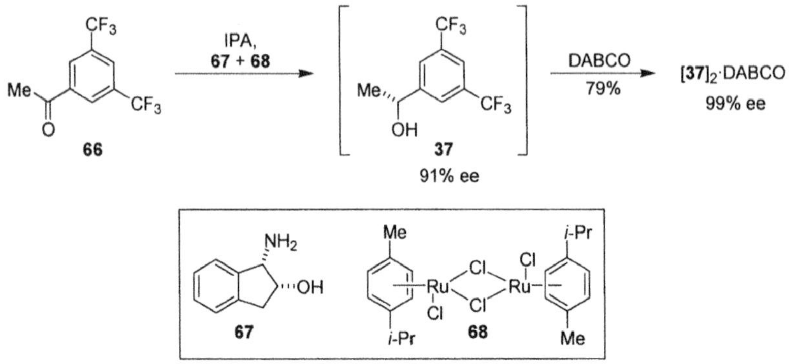

SCHEME 20

asymmetric induction in the crude reduction stream equated to 91% ee and an assay yield of 87%. Isolation of this as the DABCO inclusion complex resulted in enantiomeric upgrade to 99% ee and good recovery (79% isolated yield). This work was complimented by an improved preparation of 3,5-bis(trifluoromethyl)acetophenone (**66**).[18] This included an efficient bromination of 3,5-bis(trifluoromethyl)benzene [6.5:1 (v/v) H_2SO_4:HOAc, 0.54 eq 1,3-dibromo-5,5-dimethylhydantoin, 99% isolated yield] and a safe and reliable conversion to ketone **66** via the Grignard reagent and acetic anhydride (88% isolated yield).

X. 4th Total Synthesis: The Manufacturing Process – Discovery

The final route, which has ultimately become the manufacturing process for aprepitant, utilized many of the same concepts as the previously discussed routes, but there was a significant strategic difference. The novel aspect of the plan, from a retrosynthetic analysis perspective, was that the stereocontrol element was chiral alcohol **37** and not one of the two stereocenters contained in the morpholine backbone. The strategic themes that were conserved from some of the other routes included an acetalization and resolution at an epimerizable center.[19]

Activation of lactam lactol **69** as trichloracetimidate **70** and subsequent Lewis acid-catalyzed acetalization reaction afforded acetals **71** and **72** as a

1:1 diastereomeric mixture (Scheme 21). Another key component was the successful 1,2-addition of 4-fluorophenylmagnesium bromide to the

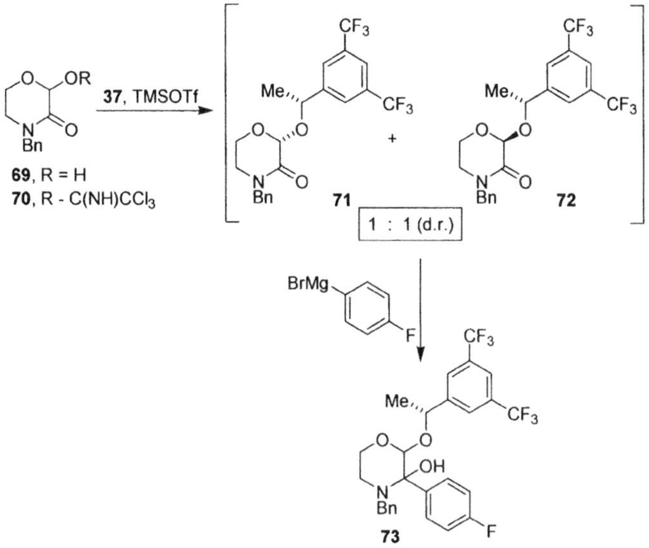

SCHEME 21

lactam carbonyl. These two reactions were very exciting. Initial attempts to epimerize the diastereomeric acetal mixture that contained **71** as well as to hydrogenate adduct **73** were unsuccessful. However, even with this tempering, this was a route that needed to be aggressively investigated.

XI.   4th Total Synthesis: The Manufacturing Process – Optimization

Three synthetic priorities were set for this route 1) development of an efficient and economical way to access lactam lactol **69**; 2) optimization of the acetalization, effecting a crystallization-induced resolution, and isolation the diastereomerically pure desired acetal **71**; 3) investigation of the aryl Grignard addition and demonstration of the envisioned debenzylation/reduction step. In addition, a significant chore of ensuring that secondary amine **5** was sufficiently pure was of paramount importance. Assiduous analysis of secondary amine **5** impurity profiles was necessary to assure that this penultimate intermediate, as well as the final drug substance that would result from this process, was identical (or better) compared with previous batches. The final step in the synthesis, the attachment of the triazolinone ring via 3-chloromethyl-1,2,4-triazolin-5-one (**32**), was to remain unchanged.

A. 2-HYDROXYMORPHOLIN-3-ONE

The intermediate lactol was to be subjected to a diastereoselective acetalization with (*R*)-3,5-bistrifluoromethylphenyl ethanol. In order to probe this diastereoselective coupling, a variety of chiral *N*-R lactols **76** was desired. In addition, since the ultimate success to aprepitant via this route depended on crystalline intermediates, numerous *N*-substituted lactols **76** needed to be prepared in order to investigate the crystalline properties of downstream intermediates. Accordingly, a versatile procedure to access a variety of *N*-substituted morpholine-2,3-diones **75** and 2-hydroxymorpholin-3-ones **76** was required. This was accomplished by the condensation of amino alcohols **74** with diethyl oxalate to afford diones **75**, which could be selectively reduced to the corresponding lactols **76** (Scheme 22).[20] A host of these substrates were synthesized; however, none offered any significant advantage over the simple *N*-Bn lactol **69**. Thus, the objective became the development of a concise synthesis for this substrate.

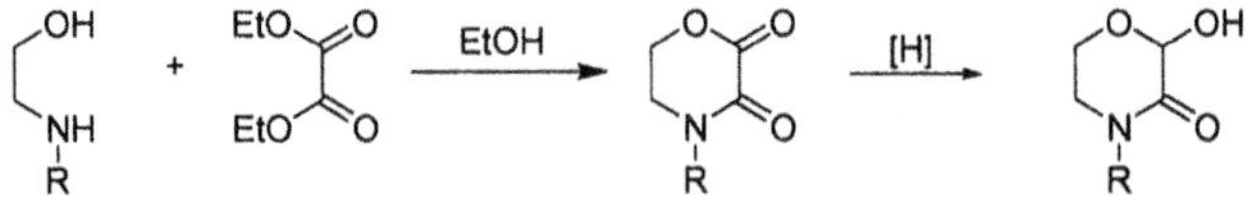

SCHEME 22

After investigating routes that required superfluous chemical transformations and other routes in which the oxidation state of the molecule was correct but required multiple steps, a condensation between *N*-benzylethanolamine (NBEA) and aqueous glyoxylic acid that afforded *N*-Bn lactol **69** in 75% yield was discovered (Scheme 23). Rapid development led to a robust process that was quickly scaled to produce hundreds of kilograms of this intermediate. The optimized conditions for the formation of lactol **69** were to slowly add neat NBEA to a preheated solution of THF and 50 wt% glyoxylic acid (2.2:1; v/v). Prewarming the glyoxylic acid to a point below reflux drove off dissolved CO_2 that could potentially react with NBEA. After heating for 18 h, the THF was distilled and lactol **69** was crystallized from water (73% isolated).[19] It was important to stop the distillation when the solvent composition was ca. 3-8% (v/v) of THF/H$_2$O.   If there was too much THF present, increased mother liquor losses occurred.   If there was <3% (v/v) of THF/H$_2$O, then impurities were not completely rejected.

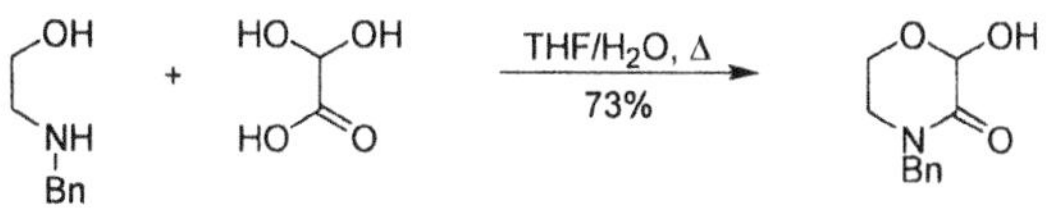

SCHEME 23

B. ACETAL

A few rapid observations were made in the acetal system.   The trichloroacetonitrile activation did not scale well (Scheme 24).   A variety of leaving groups were screened and the activation of lactol **69** was ultimately performed with trifluoroacetic anhydride.   Crucial to the success of this route, it was found that the diastereomeric mixture of acetals **78** and **79** could be equilibrated with *t*-BuOK in *t*-BuOH.   In addition, the desired diastereomer could be selectively crystallized from heptane.

The stage was set for yet another crystallization-induced transformation.   Capitalizing on these lead results, however, was not straightforward: *t*-BuOK was not soluble in heptane and other equilibrating conditions were inferior to tertiary potassium alkoxides in tertiary alcohols.   The potassium salt of the long chain fatty alcohol 3,7-dimethyl-3-octanol was found to be soluble in heptane.   When this was applied to the resolution/racemization of acetals **78** and **79** the result was spectacular.   An initial 55:45 mixture of **78** and **79** from the crude acetalization reaction was dissolved in heptane and 0.9 eq of 3,7-dimethyl-3-octanol was added, cooled to -5 °C, 0.3 eq of the potassium salt was added and the system seeded with authentic **78**. Without the addition of 3,7-dimethyl-3-octanol, which was soluble in heptane, a competing 1,2-Wittig rearrangement occurred.[19] After 5 hours, the 55:45 mixture had been transformed to 96:4 ratio of **78:79**.   The desired diastereomeric acetal could be crystallized from the reaction mixture in an overall 84% (99% de) from starting lactol **69**.

C. SECONDARY AMINE

4-Fluorophenylmagnesium bromide cleanly added to the carbonyl of lactam **71** (Scheme 25). The diastereoselectivity of this step was inconsequential, since the hope was to destroy this stereocenter upon in situ formation of imine **65**. After workup, the crude mixture was hydrogenated with Pd(OH)$_2$. Although a complex mixture of products

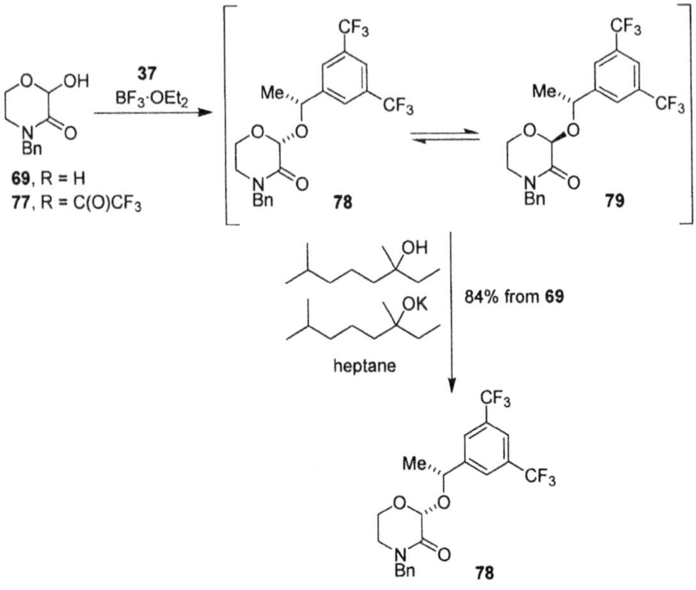

SCHEME 24

was produced, desired amine **5** was crystallized and isolated as the tosylate salt **5·TsOH** in 64% from lactam **71**. A significant observation regarding the stability of *N,O*-hemiacetal **80** was noted. Upon workup, significant levels of two byproducts that could be attributed to a [1,3]-sigmatropic rearrangement were isolated and characterized. To avoid these side reactions, it was necessary to use the crude Grignard stream directly in the hydrogenation. The first attempt at this through process was to quench the crude Grignard reaction stream into MeOH and hydrogenate this directly; however, this disappointingly led to a low yield of desired *sec*-amine **5**. Optimization of this reaction revealed that the addition of TsOH to the mixture prior to hydrogenation was crucial to the success of the reaction. A significant amount of work was conducted to better understand the mechanism and hydrogenation kinetics of this process.[21]   After adding 1.3 equivalents of 4-fluorophenylmagnesium bromide to lactam **71**, the reaction was quenched into MeOH followed by the addition of 1.8-2.2 equivalents of TsOH.   Hydrogenation of this mixture led to a cis/trans selectivity of >300:1.   After workup, hydrochloride salt **5·HCl** was formed and isolated in a 91% overall yield from lactam **71**.[19] The impurity profile of amine hydrochloride **5·HCl** generated by this new route compared favorably to amine tosylate **5·TsOH** that had already been synthesized.   The final attachment of triazolinone **32** as previously described (Scheme 11) afforded aprepitant. This fourth generation synthesis was immediately scaled to produce hundreds of kilograms and has become the preferred method to synthesize aprepitant.

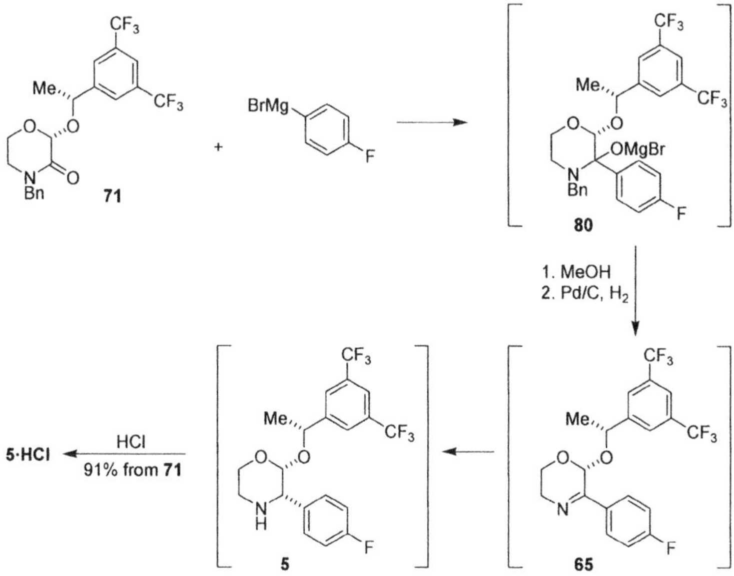

SCHEME 25

XII. Conclusion

In summary, this final route was concise, high yielding, operationally simple, avoided cryogenics, and utilized inexpensive and readily available raw materials. The overall yield to aprepitant was 55%. On March 25,

2003, just over 9 years after our Process Research group initiated work on this drug candidate, aprepitant received FDA approval for prevention of acute and delayed nausea associated with highly emetogenic chemotherapy. This nine year development period resulted in numerous synthetic organic discoveries and the demonstration of multiple routes to this opulently substituted morpholine target molecule. A recurring theme throughout this time frame was the power of crystallization-induced resolutions. These types of reactions can indeed rival catalytic asymmetric synthesis. During this work on aprepitant, such crystalline-induced diastereomeric transformations were used early in the program, highlighted in alternative approaches, and became the foundation of the manufacturing process.

Overall, this final route to an architecturally complex target molecule is robust and provided a high level of atom economy. The outcome was an inexpensive and concise manufacturing route to aprepitant.

Acknowledgments

The synthesis and drug development process is time consuming and requires the efforts of a multitude of scientists. Clearly, although the author of this review, I played a much smaller role in the overall process. The multiple synthetic iterations of this complex molecule are a testament to the creative scientists with whom I have worked. I greatly acknowledge these chemists and they are listed as authors on the Merck publications in the following references.

References and Footnotes

1. Hale, J. J., Mills, S. G., MacCoss, M., Finke, P., Casieri, M. A., Sadowski, S., Ber, E., Chicchi, G. G., Kurtz, M., Metzger, J., Eiermann, G., Tsou, N. N., Tattersall, F. D., Rupniak, N. M. J., Williams, A. R., Rycroft, W., Hargreaves, R., MacIntyre, D. E., *J. Med Chem.* **1998**, *41*, 4607.
2. Evans, D. A., Britton, T. C., Ellman, J. A., Dorow, R. L., *J. Am. Chem. Soc.* **1990**, *112*, 4011.
3. Cameron, M., Cohen, D., Cottrell, I. F., Kennedy, D. J., Roberge, C., Chartrain, M., *J. Molec. Catal. B: Ezym.* **2001**, *14*, 1.
4. Davies, A. J., Ashwood, M. S., Cottrell, I. F., *Synth. Commun.* **2000**, *30*, 1095.
5. Moseley, J. D., Williams, B. J., Owen, S. N., Verrier, H. M, *Tetrahedron: Asymm.* **1996**, *7*, 3351.
6. For a review, see: Ebbers, E. J., Ariaan, G. J. A., Houbiers, J. P. M., Bruggink, A., Zwanenburg, B., *Tetrahedron* **1997**, *53*, 9417.
7. Alabaster, R. J., Gibson, A. W., Johnson, S. A., Edwards, J. S., Cottrell, I. F., *Tetrahedron: Asymm.* **1997**, *8*, 447.
8. (a) Payack, J. F., Hughes, D. L., Cai, D., Cottrell, I. F., Verhoeven, T. R., *Org.*

Prep. Proced. Int. **1995**, *27*, 707.   (b) Payack, J. F., Hughes, D. L., Cai, D., Cottrell, I. F., Verhoeven, T. R., *Org. Synth.* **2002**, *79*, 19.

9.   Hughes, D. L., Payack, J. F., Cai, D., Verhoeven, T. R., Reider, P. J., *Organometallics* **1996**, *15*, 663.

10. Valilaya, A., Wang, T. Chen, Y., Huffman, M. A., *J. Pharm. Biomed. Anal.* **2001**, *25*, 577.

11. Payack, J. F., Huffman, M. A., Cai, D., Hughes, D. L., Collins, P. C., Johnson, B. K., Cottrell, I. F., Tuma, L. D., *Org. Proc. Res. Develop.* **2004**, *8*, 256.

12. Cowden, C. J., Wilson, R. D., Bishop, B. C., Cottrell, I. F., Davies, A. J., Dolling, U. H., *Tetrahedron Lett.* **2000**, *41*, 8661.

13. Ashwood, M. S., Cottrell, I. F., Davies, A. J., *Tetrahedron: Asymm.* **1997**, *8*, 957.

14. Pye, P. J., Rossen, K., Weissman, S. A., Moliakal, A., Reamer, R. A., Ball, R., Tsou, N., Volante, R. P., Reider, P. J., *Chem. Eur. J.* **2002**, *8*, 1372.

15. Weissman, S. A., Rossen, K., Reider, P. J., *Org. Lett.* **2001**, *3*, 2513.

16. Zhao, M. M., McNamara, J. M., Ho, G.-J., Emerson, K. M., Song, Z. J., Tschaen, D. M., Brands, K. M. J., Dolling, U. H., Grabowski, E. J. J., Reider, P. J., Cottrell, I. F., Ashwood, M. S., Bishop, B. C., *J. Org. Chem.* **2002**, *67*, 6743.

17. Hansen, K. B., Chilenski, J. R., Desmond, R., Devine, P. N., Grabowski, E. J. J., Heid, R., Kubryk, M., Mathre, D. J., Varsolona, R., *Tetrahedron: Asymm.* **2003**, *14*, 3581.

18. Leazer Jr., J. L, Cvetovich, R., Tsay, F.-R., Dolling, U., Vickery, T., Bachert, D., *J. Org. Chem.* **2003**, *68*, 3695.

19. Brands, K. M. J., Payack, J. F., Rosen, J. D., Nelson, T. D., Candelario, A., Huffman, M. A., Zhao, M. M., Li, J., Craig, B., Song, Z. J., Tschaen, D. M., Hansen, K., Devine, P. N., Pye, P. J., Rossen, K., Dormer, P. G., Reamer, R. A., Welch, C. J., Mathre, D. J., Tsou, N. N., McNamara, J. M., Reider, P. J., *J. Am. Chem. Soc.* **2003**, *125*, 2129.

20. Nelson, T. D., Rosen, J. D., Brands, K. M. J., Craig, B., Huffman, M. A., McNamara, J. M., *Tetrahedron Lett.* **2004**, *45*, 8917.

21. Brands, K. M. J., Krska, S. W., Rosner, T., Conrad, K. M., Corley, E. G., Hartner, F. W., Kaba, M., Larsen, R. D., Reamer, R. A., Sun, Y., Tsay, R.-R., *J. Org. Chem.* manuscript submitted.

STRATEGIES AND TACTICS IN ORGANIC SYNTHESIS, VOL. 6

Chapter 11

TOTAL SYNTHESIS AND MECHANISM OF ACTION STUDIES ON THE ANTITUMOR ALKALOID, (-)-AGELASTATIN A

Karl J. Hale and Mathias M. Domostoj
The UCL Centre for Chemical Genomics
The Christopher Ingold Laboratories
The Department of Chemistry, University College London
20 Gordon Street, London WC1H 0AJ, United Kingdom

and

Mohamed El-Tanani, F. Charles Campbell, and Charlene K. Mason
The Department of Surgery
Queen's University of Belfast
Grosvenor Road, Belfast, Northern Ireland
BT12 6BJ, United Kingdom

I. Introduction: The Challenges Faced by Scientists Involved in 21st Century Genomics Research, and the Place of Synthetic Organic Chemistry in the Modern-Day Genomics Era

As a result of the recent sequencing of the human genome, it is now estimated that there are between 30,000 and 35,000 different genes within the human body encoding for up to 450,000 different proteins.[1] Identifying the multiple functions of all these genes and proteins, elucidating their three-dimensional molecular architectures, and delineating their mutual interconnectivities within cells, are going to be some of the greatest challenges facing scientists in the 21st Century; challenges that hopefully will be met, head on, by many who are active in the chemistry and molecular medicine fields.

Deconvoluting the inner workings of the human genome will almost certainly require large multidisciplinary teams of researchers coming together from the currently separate disciplines of chemistry, biology and medicine. Not only will their collaborative efforts enable many biological phenomena to be properly understood at molecular level, they will also greatly improve our prospects for creating the future drug treatments needed to combat human disease.

The currently used "gene-silencing" techniques of mutational analysis,[2] gene knockout,[3] gene reporter/transcription factor reporter assaying, and RNA interference (RNAi)[4] will continue to play a dominant role in human genomics and proteomics over the coming years. However, because these methods do not allow the *multiple* functions of many proteins to be readily dissected, and because they do not produce the small-molecule leads needed to fuel drug-discovery programmes, many are predicting that they will eventually be superceded by the more sophisticated techniques of chemical genomics and chemical proteomics; protocols that are heavily underpinned by the methods of modern-day synthetic organic chemistry.

In chemical genomics and proteomics, tiny low molecular weight probe molecules are used to perturb the functions of individual genes and proteins.[5] The functional consequences of that perturbation are then examined and analyzed. By using small, yet relatively complex, molecular structures to disrupt cellular events and behavior, biologists can often pick apart some of the most complex and intricate signaling processes that are operational within cells, and thus provide a biomolecular panoptic that would otherwise be unattainable by purely biological means.[5]

Small molecule modulators of gene and protein function have the special advantage that they can often be designed to either "switch on" or "switch off" the actions of particular genes and proteins within cells, in a highly controlled fashion. By way of contrast, the currently used techniques of gene knockout,[3] RNAi,[4] and mutational analysis[2] can usually only "switch off" the functions of individual proteins and genes, which is one of their great drawbacks.

Low molecular weight inquisitors of gene and protein function can further be designed to selectively inhibit *one* particular function of a targeted protein or gene, whilst leaving other essential actions intact. This is often not possible with gene-knockout methods, which usually oblate all the actions of a particular gene under study.

Small molecule genomics tools can additionally serve as highly useful lead structures for low molecular weight drug design, once a novel biological target has been shown to be modulated. This is a particularly valuable facet of the chemical genomics approach.

With careful planning and appropriate structural modification, many chemical genomics tools can be anchored onto latex microsphere or sepharose supports and used as "capture substrates" for the separation and identification of novel protein or gene targets that are present within the nuclear or cytoplasmic lysates of cells.[6] An excellent example of such work has been provided by the Takahashi group,[7] who recently attached a truncated FR225659 derivative onto a latex microsphere support, exposed it to a cytoplasmic fraction from rat hepatocytes, and isolated three candidate protein receptors for the inhibition of glucagon-induced gluconeogenesis from the lysate. The three proteins in question had apparent molecular weights of 36, 37, and 60 kDa respectively,[7a] and they were later shown to actually be subunits of the serine/threonine type 1 (PP1) and type 2A (PP2A) protein phosphatases.[7b] The roles of these two enzymes in the onset and development of diabetes is currently being investigated.

Complexes formed between small molecule probes and their target gene/protein receptors can also often give important structural insights into how important biomolecules operate and interact at molecular level, particularly if these complexes are studied by NMR[8] or ultrafast laser spectroscopic[9] or X-ray crystallographic[10] techniques.

For all these reasons, there is massive biological interest in the infant scientific disciplines of chemical genomics and proteomics.

At present, there are three main methods used by chemical genomicists to access the small molecule tools needed to do this kind of work. These are: a) Diversity Oriented Synthesis (DOS);[11] b) Bioassay-Guided Natural Product Isolation (BGNPI),[12] and; c) Target-Oriented Synthesis (TOS) of moderately complex natural products and their analogues.[13]

Natural products (and their analogues) are generally the molecules of choice for many chemical genomics and chemical proteomics applications, as their molecular frameworks have already undergone substantial evolutionary selection for binding to particular gene and protein domains within living organisms, and many naturally-occurring molecules have also been produced with a particular survival purpose in mind.[1] According to Waldmann,[1] small molecule natural products generally have a 2-3 times higher "hit-rate" than "man-made structures" when it comes to screening against therapeutically relevant drug targets.

For all these reasons, chemists and biologists alike continue to remain enthralled by the prospect of gaining access to novel natural products and their analogue structures, particularly if the molecules in question can potentially be used to identify or probe a new biological target whose function has yet to be properly understood.

In the present chapter, we will attempt to give an overview of one very important chemical genomics tool that we have recently synthesized in our laboratories; it is the antitumor alkaloid, (-)-agelastatin A. Its synthesis was undertaken solely for the purpose of helping us answer some important questions we had concerning the workings of the Wnt/GSK-3β/β-catenin/E-cadherin cell signaling pathways[14] and the downstream genes and proteins they control.

II. The Discovery of Agelastatin A and its Postulated Biogenesis

In 1993, Pietra and coworkers at the University of Trento in Italy reported their isolation of the powerful antitumor alkaloid, (-)-agelastatin A, from aqueous extracts of *Agelas dendromorpha*, an axinellid sponge indigenous to waters of the Coral Sea off New Caledonia (Scheme 1).[15] Its absolute stereochemistry was deduced by a range of spectroscopic methods, including a new application of the exciton spitting technique.

Since its initial isolation as a 4:1 mixture alongside (-)-agelastatin B, two other family members, agelastatins C and D, have also been isolated and characterized. The latter were discovered by Molinski and coworkers[16] in a collection of the Indian Ocean sponge *Cymbastela* sp. (Axinellida) gathered off the Western Australia coastline. During this study, the latter researchers also reisolated (-)-agelastatin A but, on this occasion, as a single entity.

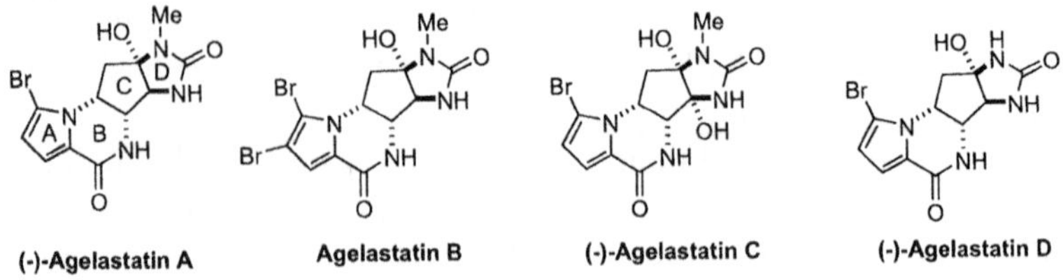

(-)-Agelastatin A   Agelastatin B   (-)-Agelastatin C   (-)-Agelastatin D

SCHEME 1. The agelastatin family of antitumor alkaloids.

From a taxonomic perspective, the agelastatins belong to the Oroidin family of alkaloids, and two chemically distinct pathways have been suggested for their biogenesis by the groups of Pietra and Potier. In Pietra's proposal (Scheme 2),[15a] (-)-agelastatin A derives from an enzyme mediated nucleophilic addition of the C(7)-C(8) alkene in **1** onto one of the tethered imidazolinone carbonyl groups.

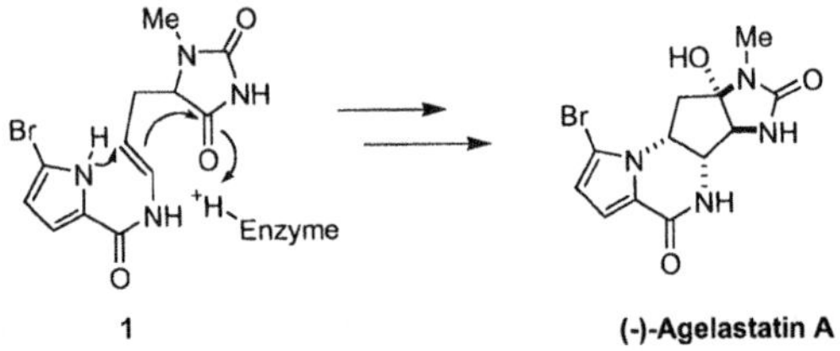

1   (-)-Agelastatin A

SCHEME 2. Pietra's suggested biogenesis of (-)-agelastatin A.

This is then followed by a tandem attack of the pyrrole-*N* on C(7), a subsequent dehydration, and a site-selective hydroxylation to give the imidazolone hemiaminal; the latter event completes assembly of the natural product.

A somewhat similar biosynthetic scheme was advanced by Potier and Al Mourabit in 2001.[17] In their proposal (Scheme 3), the C(7)-C(8)-alkene of **3** attacks a pendant enzyme-coordinated cyclic iminium ion prior to pyrrole conjugate addition. Subsequent protonation, regioselective *N*-methylation, and iminium ion hydration are then suggested to afford (-)-agelastatin A.

III. Initial Biological Screening of (-)-Agelastatin A

The initial *in vitro* antitumor screening of (-)-agelastatin A revealed that it could powerfully arrest the growth of a murine L1210 leukemia cell line (IC_{50} = 0.033 µg/ml) and a doxorubicin-resistant L1210/DX subline (IC_{50} = 0.469 µg/ml). (-)-Agelastatin A could also potently inhibit

owth of a human KB nasopharyngeal cancer cell line (IC_{50} = 0.075
).[15]

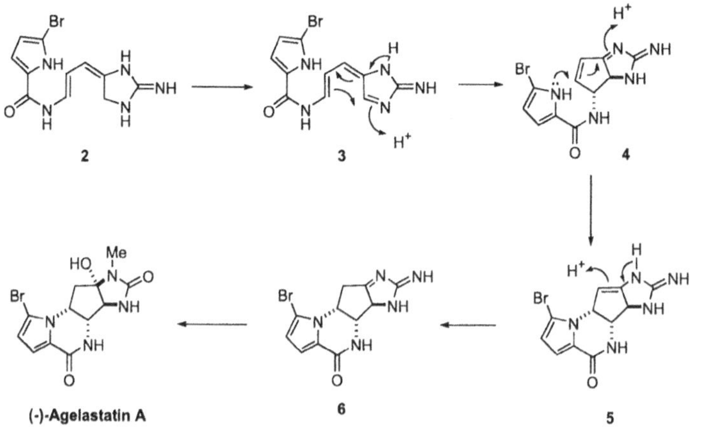

SCHEME 3. The Potier-Al Mourabit biosynthetic proposal for (-)-agelastatin A.

light of this, Pietra and coworkers submitted (-)-agelastatin A to
n's Oncological Research Laboratories in Italy for detailed
nor evaluation in mice. There, the alkaloid was found to confer a 63
-extension on mice that had been xenografted with murine L1210
nia, when it was repeatedly administered intraperitoneally at the
)sage of 2.6 mg/kg.[15] However, notwithstanding these promising *in*
ndings, a decision was eventually made not to move (-)-agelastatin
/ard into full preclinical development, based on the observation that
wed low antitumor efficacy in mice when it was administered
enously. Unfortunately, neither the toxicity profile nor the
acokinetic parameters of (-)-agelastatin A were ever reported in
)apers.[15]

lowing this preliminary evaluation in 1996, four more years elapsed
any further biological testing was done on (-)-agelastatin A. Then,
year 2000, the groups of Laurent Meijer and George Pettit
atively screened (-)-agelastatin A against a range of neurological
that included casein kinase 1, CDK1/cyclin B, CDK5/p25 and
β (glycogen synthase kinase-3β), and significantly, they observed
inhibited the latter with an IC_{50} of 12 μM.[18]

K-3β is a ubiquitous serine/threonine kinase[19] that plays a pivotal
regulating β-catenin levels within cells through its involvement in

ιe Wnt/β-catenin/TCF4 cell-signaling pathway (Scheme 4).[14] The latter
ontrols the expression of a range of genes and proteins that are critically
ivolved in regulating cell growth and proliferation.

Ordinarily, the APC/Axin/GSK-3β multimeric protein complex
ιaintains low levels of cytoplasmic β-catenin within cells by binding to
ιnd phosphorylating free cytoplasmic β-catenin, which causes it to be
legraded by the ubiquitin proteasome. However, when active Wnt
ignaling occurs, the APC/Axin/GSK-3β protein complex is prevented
rom binding to and phosphorylating cytosolic β-catenin, and this causes

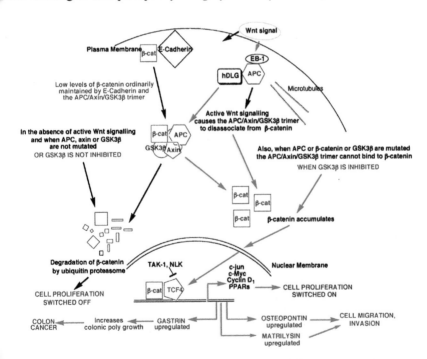

CHEME 4. The Wnt/β-catenin/GSK-3β/APC/E-cadherin cell signalling pathways and the downstream genes
nd proteins they control, and their potential roles in the onset of cancer and metastasis when deregulated.[14i]

t to accumulate in the cytosol and thereafter undergo translocation to the
ιucleus, where it subsequently drives transcription from the TCF4
ιromoter. Target genes for β-catenin/TCF4 mediated transcription include
-myc,[20] cyclin D1,[21] matrilysin (MMP7),[22] TCF1,[23] the multidrug
esistance 1 (MDR1) gene,[24] c-jun, fra-1, the urokinase plasminogen
ιctivator receptor,[25] gastrin,[26] osteopontin,[27] EPHB2/3,[26a] and the
ιeroxisome-proliferator-activated receptor (PPAR) δ-gene.[28]

Significantly, functionally inactivating mutations to the APC, GSK-3β, or β-catenin proteins, as well as upregulated Wnt signaling, are all thought to play a major role in the onset of many tumors.

IV. Past Total Synthesis Studies on Agelastatin A

The first total synthesis of (±)-agelastatin A was achieved by Weinreb and coworkers in 1999.[29] Their approach (Scheme 5) revolved around the use of a hetero-Diels-Alder /[2,3]-sigmatropic rearrangement sequence to

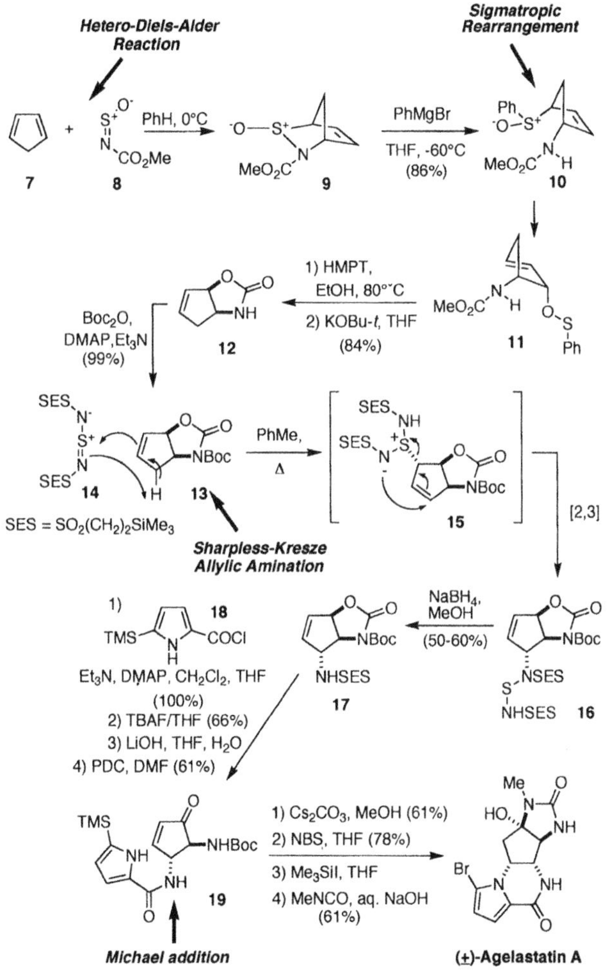

SCHEME 5. Weinreb's racemic route to agelastatin A (1999).

introduce the *O*- and *N*-functionality present at C(8a) and C(5b) in the target structure. After *N*-protection of the product **12**, the bicyclic cyclopentene **13** was subjected to a Sharpless-Kresze allylic amination reaction[30] with the SES sulfodiimide **14**, and this was followed by a reductive N-S bond cleavage with NaBH$_4$ to anchor the requisite *N*-substituent needed at C(5a) for further elaboration into agelastatin A. Importantly, their recourse to this rarely used amination tactic elegantly allowed them to forge the correct *trans*-arrangement needed for the *N*-substituents at C(5a) and C(5b) in **17**.

The next four steps in their route involved *N*-acylation of the SESNH in **17** with acid chloride **18**, selective cleavage of the oxazolidinone ring with aqueous LiOH, SES group removal with *n*-Bu$_4$NF, and allylic alcohol   oxidation with PDC (pyridinium dichromate) to yield the pyrrolocyclopentenone **19**, which was subjected to an internal conjugate addition reaction to complete assembly of the ABC tricycle. A chemoselective bromination was next effected at the pyrrole carbon bearing the trimethylsilyl substituent. Finally, the Boc group was detached from the brominated α-amidoketone with iodotrimethylsilane, and the trimethylsilylurethane intermediate hydrolyzed and *N*-carbamoylated with *N*-methyl isocyanate in a two step, one-pot, process that concluded this 16 step route to ($\pm$)-agelastatin A.

The first asymmetric synthesis of (-)-agelastatin A was achieved by Feldman and Saunders in July 2002,[31] whilst we were in the middle of our own synthetic campaign on this compound.

From a sheer planning perspective, the Feldman route is brilliantly conceived, boldly creating the enantiomerically pure natural product in a mere 16 steps (Scheme 6), in a highly elegant way. It too featured an internal heteroconjugate addition reaction of a tethered pyrrole nitrogen onto a highly functionalized cyclopentenone for ABC-tricycle assembly, and significantly, it avoided the use of *N*-methyl isocyanate for the final D-ring construction.

Feldman sculpted his intermediary cyclopentanone **29** by applying a sequential hydrolysis/oxidation/β-elimination sequence to **28**. The latter was forged through a novel conjugate addition reaction between **26** and *o*-nitrobenzylamine, and an *N*-acylation with the pyrrole acid chloride **27**. The key bicyclic cyclopentenyl phenylsulfone **26** was built up through an even more intriguing process that involved stereospecific vinylcarbene insertion. The necessary vinylcarbene was generated from the alkynyliodonium salt **24** by treatment with NaSO$_2$Ph. Unfortunately, the key vinylidene insertion step produced **26** in a rather low 34% yield, the major reaction product actually being the rearranged alkynylsulfone **25** (formed in 41% yield). The synthesis of **24** commenced from the chiral epoxide **20** which itself was accessed through a Jacobsen kinetic resolution.[32]

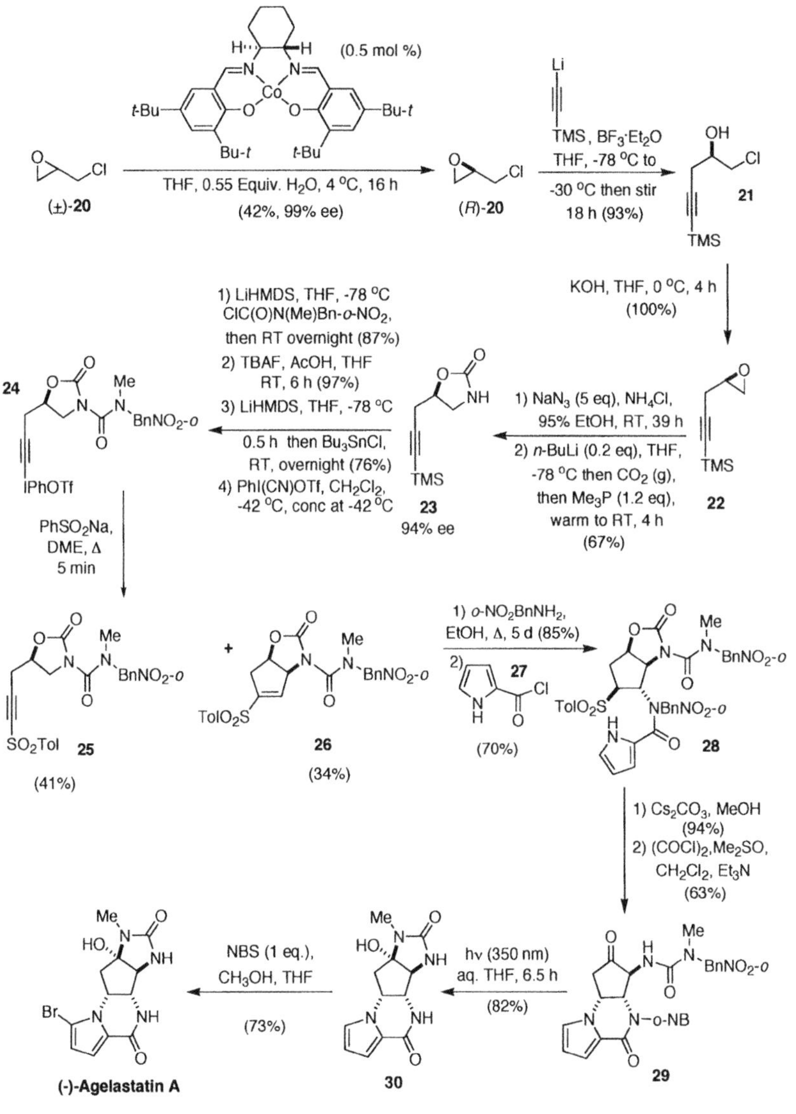

SCHEME 6. Feldman's asymmetric synthesis of (-)-agelastatin A (2002).

Although O'Brien and coworkers[33] have not yet completed a total synthesis of (-)-agelastatin A, they have nevertheless reported a very interesting method for arriving at the differentially protected racemic diaminocyclopentenone **37** (Scheme 7); an intermediate that could prove of value for a future total synthesis. Their route to **37** commenced with the highly selective aziridination of cyclopentadiene first reported by Knight and Muldowney in 1995.[34] Racemic **31** was then ring-opened at the more activated aziridine carbon with ammonia, and the product amine capped with a Boc group to obtain the diaminated cyclopentene **32**; the latter was then epoxidized with *m*-CPBA. The newly installed epoxide emerged *syn* to the NHBoc group, almost certainly because the oxidant coordinated to the carbamate grouping. Epoxide **33** was then subjected to an elimination reaction with 4 eq. of the racemic amide base **34**. The latter process furnished an inseparable mixture of **35** and **33** in an 85:15 ratio. To obtain **35** in a pure state it proved necessary to O-acetylate the mixture, separate the O-acetate **36** by SiO$_2$ flash chromatography, and then cleave the OAc group. A PDC oxidation was then implemented in DMF to access **37**. Future reports on the successful elaboration of **37** into (±)-agelastatin A are awaited with interest.

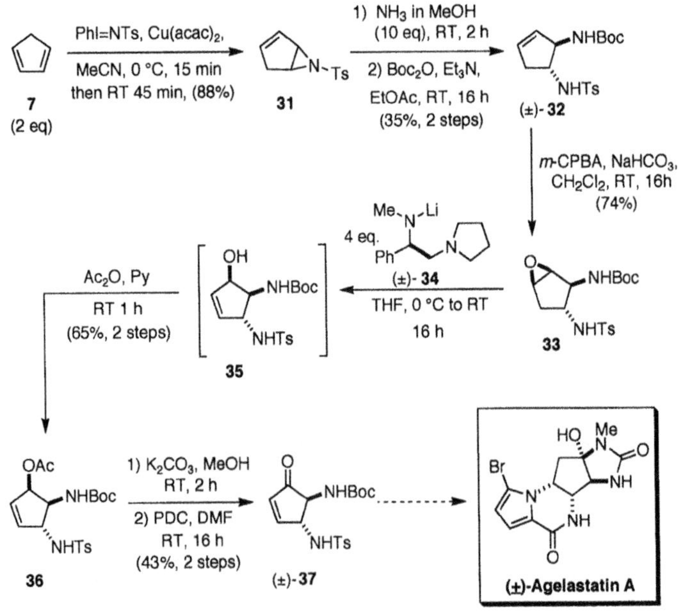

SCHEME 7. O'Brien's Synthetic Approach to Racemic Agelastatin A (2002)

V. Our Reasons for Developing a New Total Synthesis of (-)-Agelastatin A

The observation that (-)-agelastatin A could selectively inhibit GSK-3β suggested to us that molecules of this class might potentially be useful chemical genomics tools for probing the intimate workings of the Wnt/β-catenin/TCF4/E-Cadherin cell signaling pathways; networks whose functioning continues to be poorly understood, and which we have been attempting to deconvolute with the aid of small molecules for quite some time. Given the great scarcity of the agelastatins in Nature, it was our belief that the only way we could gain access to such probes would be if we developed our own "in-house" total synthesis of (-)-agelastatin A, and with this in mind, we began work on this problem back in September 2001.

It was our hope that our own total synthesis of (-)-agelastatin A might yield enough material to allow the growth inhibitory effects of the natural product to be determined against a much greater range of human tumor cell lines than had originally been screened by the Pietra group.[15] We also hoped that it would provide enough material to test the molecule (for the very first time) against xenografted solid human tumors in mice, and allow its toxicological and pharmacokinetic parameters to be precisely determined in these animals.

With this as background, we will now discuss our total synthesis of (-)-agelastatin A in more detail, and thereafter present details of our first chemical genomics studies on the synthetic (-)-agelastatin A so obtained.

VI. Our Initial Retrosynthetic Plan

In Weinreb's total synthesis,[29] he successfully converted the cyclopentenone **19** into (±)-agelastatin A by a four-step sequence that featured an intramolecular conjugate addition to assemble the ABC-tricycle (Scheme 5). Given that our primary motive for embarking on the total synthesis (-)-agelastatin A was for chemical genomics and medicinal chemistry purposes, it seemed logical for us to capitalize on his past synthetic achievements in this area, and intersect with this advanced intermediate. However, in our proposed synthesis of **19**, we would aim to prepare it in optically pure form, as this would enable the natural product and its analogues to be obtained as single enantiomers (Scheme 8).

We envisioned generating the scalemic enone **19** from the allylic alcohol **38** by a multiple N-manipulation/oxidation sequence. A ring-closing metathesis (RCM) reaction[35] was considered strategic for constructing its cyclopentene ring-system, and **39** would in turn potentially emanate from aldehyde **40** by a Wittig process. The fact that aldehyde **40** would almost certainly exist as mixture of hemiacetal anomers **(40α,β)** was not of special concern to us at this stage, as compounds of this sort had previously been converted into open-chain alkenes by the Wittig reaction on many occasions.

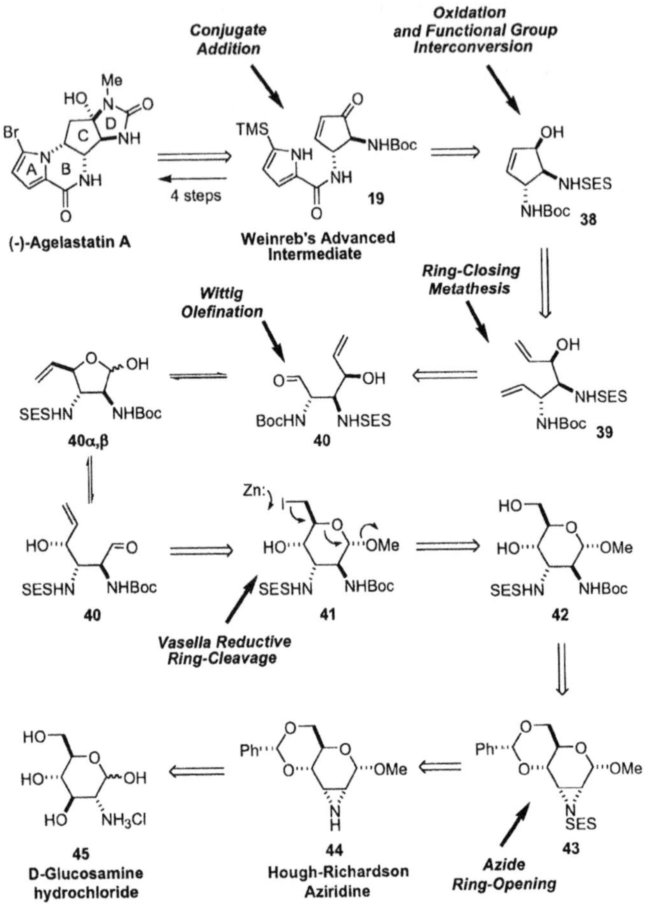

SCHEME 8. Our initial retrosynthetic strategy for (-)-agelastatin A.

A logical precursor of aldehyde **40** was deemed to be the 6-iodo-6-deoxy-D-altropyranoside **41**, as a Vasella reductive ring cleavage with Zn dust could be expected to implement the forward process. Iodide **41** would be available from the diol **42** by a selective iodination.   A regioselective *trans*-diaxial ring-opening on aziridine **43** with sodium azide was considered appropriate for setting up the vicinal diamido functionality present in **42**. A SES (β-trimethylsilylethanesulfonyl) protecting group[37] was selected for blockade of this particular nitrogen since, as the route progressed, acid and base mediated reactions would have to be implemented, and a long-term *N*-protecting group was required that could withstand these conditions, and yet still be removable with reasonable facility at a later stage in the synthesis. Epimine **43** was considered accessible from the well-known Hough-Richardson aziridine **44**[38] by *N*-sulfonylation. As **44** had already been prepared from readily available D-glucosamine hydrochloride, we considered **44** to be an ideal choice of starting material for the new route.

VII. Attempted Implementation of Our Initial Strategic Planning for (-)-Agelastatin A

One of the early synthetic objectives in our route was the synthesis of aziridine **44** in copious amounts. Happily, we were able to prepare this compound fairly readily on large scale (0.5 kg) using a slightly modified version of the Hough and Richardson five-step procedure that was first published in *Carbohydrate Research* in 1965 (Scheme 9).[38] In their route to **44**, the cheap and readily available monosaccharide, D-glucosamine hydrochloride, is chemoselectively *N*-acylated with benzoic anhydride in methanolic sodium methoxide to obtain **46**, and a Fischer glycosidation is then performed to access the α-methylpyranoside **47**. An *O*-benzylidenation is subsequently implemented on **48** with benzaldehyde and zinc chloride (although in our case we preferred to use benzaldehyde dimethylacetal and a catalytic amount of *p*-toluenesulfonic acid in DMF at 55 °C for this purpose).

Compound **48** is *O*-mesylated at C(3) with methanesulfonyl chloride in pyridine to obtain **49**, and this is then treated with sodium hydroxide in hot 2-methoxyethanol to instigate ring closure to the epimine **44**. Because *N*-acyl aziridines are base-labile, the *N*-deprotected aziridine **44** is the final product that actually emerges from this reaction, which was highly desirable for the purposes we had in mind.

One of the great hallmarks of the Hough-Richardson route to **44** is the remarkable crystallinity of all the intermediates that are formed *en route*. This means that all of the purifications can be readily performed on large scale by simple recrystallisation, and as a consequence, it is possible to routinely access aziridine **44** in 0.5 kg lots without much fuss or struggle.

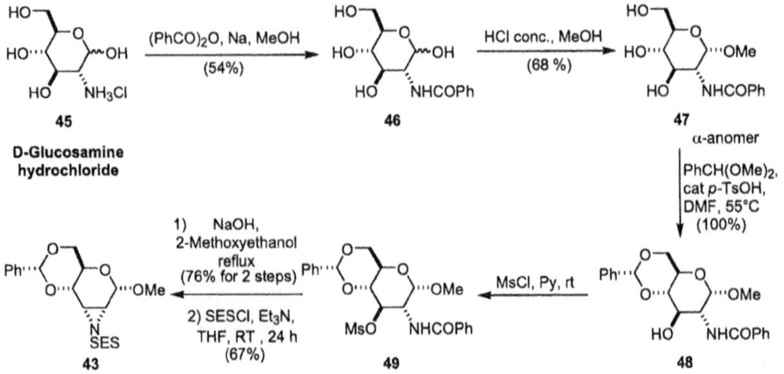

SCHEME 9. The synthetic route used to obtain the SES-protected aziridine **43**.

With aziridine **44** in hand, we reacted it with 1.2 equivalents of SES chloride and excess triethylamine in THF to obtain the *N*-SES protected aziridine **43** in 67% yield (Scheme 9); this nicely set the stage for the subsequent Furst-Plattner ring-opening with azide ion.

During their seminal work on the nucleophilic ring-opening of 2,3-epimino-glycosides, the Hough and Richardson group,[38,39] as well as that of Guthrie,[40] both independently observed that 4,6-*O*-benzylidenated *N*-acyl aziridines such as **50**, **52**, and **54** all react with azide ion (and other nucleophiles) to give the *trans*-diaxial ring-opening products with high regioselectivity.

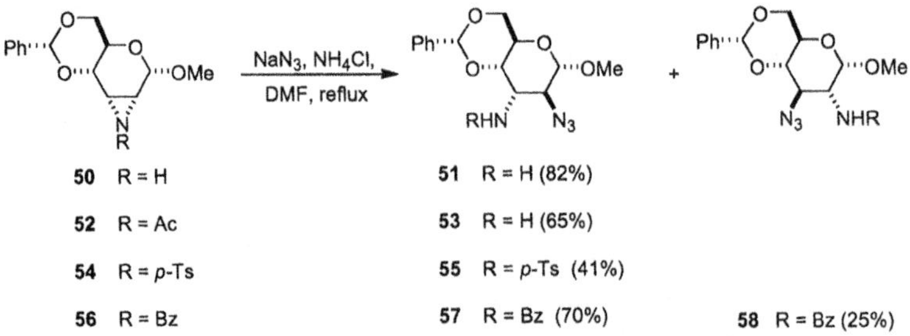

SCHEME 10. *Trans*-diaxial ring-openings in monosaccharide aziridines.

Nevertheless, one apparent exception to this rule was presented by Guthrie and Murphy in their 1965 paper. They stated that the reaction of *N*-benzoylaziridine **56** with NaN₃ and NH₄Cl in DMF at 140 °C furnished the D-glucopyranoside **58** *exclusively* in 70% yield. However, a subsequent reinvestigation of their work by Chaby and coworkers[41] in

1998 demonstrated that this early claim was in fact inaccurate. The latter researchers observed that the major component of this reaction mixture was in actual fact the *trans*-diaxial ring-opened product **57**, and they found that it was formed in 70% yield, alongside a 25% yield of the highly crystalline D-glucopyranoside **58**.

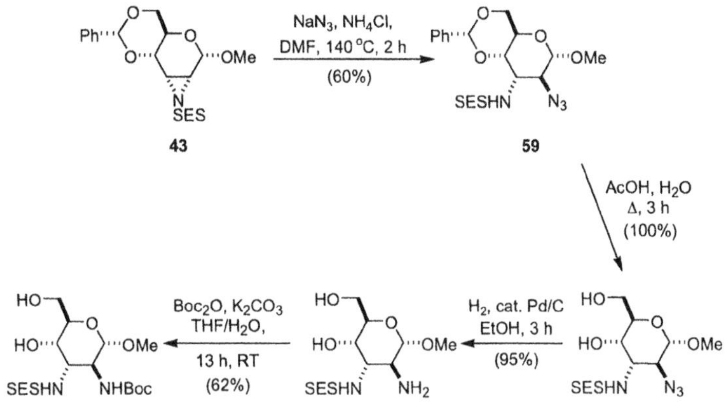

SCHEME 11. Regioselective *trans*-diaxial ring-opening of **43** *en route* to **42**.

Given this raft of successful past precedents, we were reasonably confident that the SES[37] (β-trimethylsilylethanesulfonyl)-protected aziridine **43** would react with NaN$_3$ and NH$_4$Cl in DMF to give **59** as the primary reaction product (Scheme 11). In the event, the ring-opening of **43** afforded the expected D-altropyranoside **59** exclusively in 60% isolated yield. The *trans*-diaxial relationship between the N-substituents at C(2) and C(3) was confirmed by the 500 MHz ^{1}H-NMR spectrum of **59**, which revealed a fairly small coupling constant ($J = 2.7$ Hz) between the equatorial C(2) and C(3) protons, which was consistent with the expected structure.

Cleavage of the benzylidene acetal from **59** was next accomplished by heating it in an 8:2 mixture of acetic acid/water. Catalytic hydrogenolysis of the azido group in **60** was thereafter effected in ethanol, in the presence of 10% wet palladium on carbon. The newly liberated amine was subsequently N-acylated under standard biphasic conditions to give **42** in 62% yield.

Our next goal was to prepare the iodide **41** (Scheme 12). Although the carbohydrate literature abounds with examples of alkyl glycosides and oligosaccharides having been selectively halogenated at C(6),[42] most of the standard halogenation methods failed to give the desired 6-deoxy-6-halogeno-D-altroside in good yield when they were applied to **42**. Some of the protocols examined are shown in Scheme 12 along with the reaction outcome in each instance.

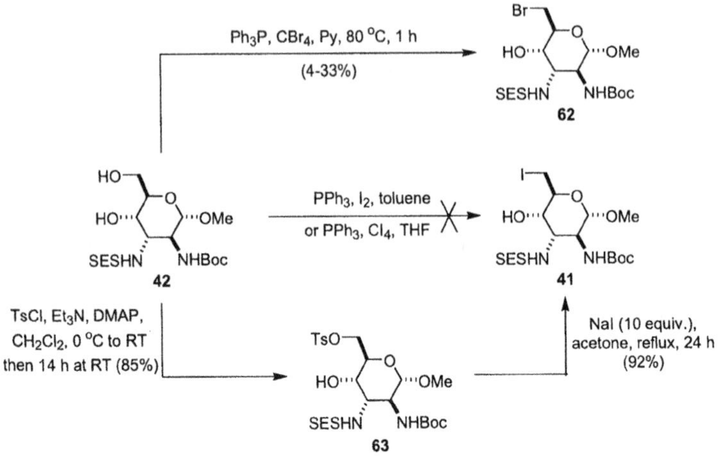

SCHEME 12. Selective Halogenation Studies on Diol **42**

Fortunately, we were able to regioselectively tosylate the primary hydroxyl in **42** with *p*-toluenesulfonyl chloride, triethylamine, and DMAP in dichloromethane at room temperature over 14 h; tosylate **63** was isolated in 85% yield. The subsequent nucleophilic displacement with iodide also proceeded satisfactorily delivering the 6-deoxy-6-iodopyranoside **41** in 92% yield. With the desired 6-iodo sugar successfully prepared, we examined its Vasella reductive ring cleavage[36] with Zn dust (size <10 μm) in a 4:1 THF / water mixture at reflux (Scheme 13). A single product was formed according to TLC analysis. However, our ^{1}H-NMR examination of this product revealed that it was in fact a 1:1 mixture of the two anomeric hemiacetals **40α** and **40β** formed in 80% yield.

Our next objective was to methylenate the hemiacetal **40α,β** so as to obtain diene **39** (Scheme 14). Standard Wittig olefination[43] procedures on **40** with methyl triphenylphosphoranylidene, generated *in situ* from methyl triphenylphosphonium bromide and *n*-BuLi, or with KOBu-*t* in THF (Entries 1 and 2, Scheme 14) simply furnished unreacted starting material. Tebbe olefinations[44] were also unsuccessful; a complex mixture of products arising under the conditions investigated in Entry 3 (Scheme

14). Significantly, none of these corresponded to **39**. Takai's recently described $PbCl_2$/diiodomethylzinc protocol[45] was also screened (Entry 4, Scheme 14), but again no apparent reaction took place. Kocienski's

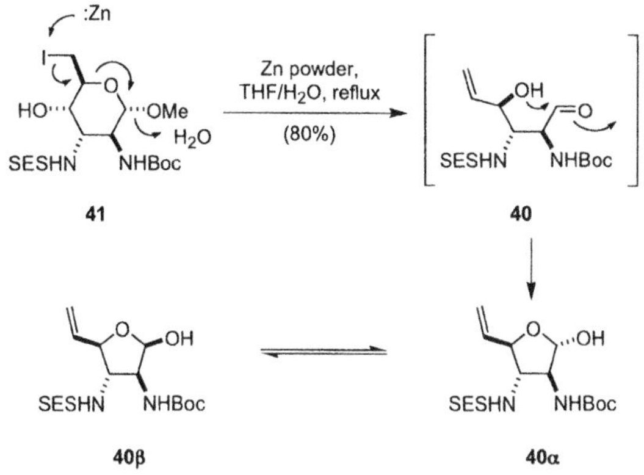

SCHEME 13. The Vasella reductive ring cleavage of iodopyranoside **41**.

olefination method,[46,47] using the carbanion derived from 5-methanesulfonyl-1-phenyl-1*H*-tetrazole, also left the **40α,β** mixture untouched (Entries 5 and 6, Scheme 14). In fact, all our efforts at converting the hemiacetal **40α,β** into diene **39** failed uniformly.

It had now become clear that if we wished to successfully methylenate the C(1)-carbonyl of **40**, we would have to prevent cyclic hemiacetal formation from occurring. A synthetic retreat was therefore beaten, and a triethylsilyl (TES) group was positioned on the C(4)-OH of **63** (Scheme 15). Nucleophilic displacement of the tosyloxy unit now afforded the iodopyranoside **65**, and Vasella reductive ring cleavage[36] delivered the desired aldehyde **66** in 85% yield.

Notwithstanding us having side-stepped the internal hemiacetalisation problem, difficulties were once again encountered when we attempted to methylenate **66** under Wittig or Tebbe conditions. In both cases, starting aldehyde was always recovered untouched from either reaction. Fortunately, when **66** was subjected to Kocienski/Julia olefination with 5-methanesulfonyl-1-phenyl-1*H*-tetrazole and potassium hexamethyl-disilazide at low temperature, diene **67** was formed in 77% yield.[46,47]

With the pathway to diene **67** secure, our attention was now directed towards the ensuing ring closing metathesis (RCM)[35] reaction to form cyclopentene **68** (Scheme 16). The first-generation Grubbs catalyst **69** was our first port of call, as this was the cheapest of the commercially

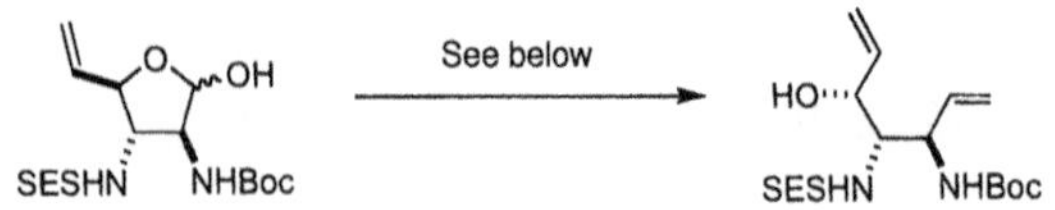

Entry	Conditions	Outcome
1	n-BuLi (5 eq.), Ph$_3$PCH$_3$Br (5 eq.) THF, RT to Δ	No Reaction
2	KOBu-t (3.2 eq.), Ph$_3$PCH$_3$Br (2.2 eq.) THF, RT to Δ	No Reaction
3	Cp$_2$TiCH$_2$AlCl(Me)$_2$ (1 eq.), THF/PhMe, 0 °C to RT	Complex Mixture
4	CH$_2$I$_2$ (3 eq.), Zn (10 eq.), Me$_3$SiCl (3 eq.) cat. PbCl$_2$ (0.1 eq.), THF, RT	No Reaction
5	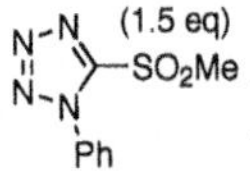 (1.5 eq) LiN(SiMe$_3$)$_2$ (3.7 eq.) THF, -78 °C, 2 h, then -30 °C 14 h; add LiN(SiMe$_3$)$_2$ (3.7 eq.), stir 1h at -78 °C	No Reaction
6	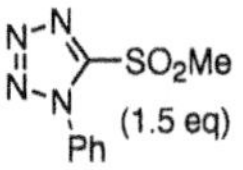 (1.5 eq) KN(SiMe$_3$)$_2$ (4.7 eq.) THF, -78 °C, 2 h	No Reaction

SCHEME 14. Some of the unsuccessful attempts at converting **40α,β** into **39**.

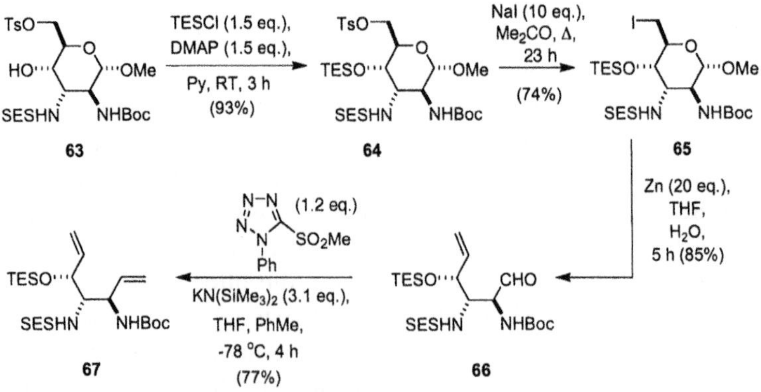

SCHEME 15. Synthesis of the new RCM precursor **67**.

le catalysts, and it was known to be tolerant of a wide range of
nality. Unfortunately, the desired RCM process did not take place
ne 16, Entries 1 and 2). The second-generation Grubbs catalyst **70**
erefore evaluated, and it performed much more effectively in this
ty, it giving rise to **68** in 53% yield (Scheme 16, Entry 3).
heless, starting diene always remained, and the reaction could
be driven to completion, suggesting that the catalyst was being
vated or destroyed in some way. This encouraging result prompted
y an even more active ruthenium alkylidene catalyst.

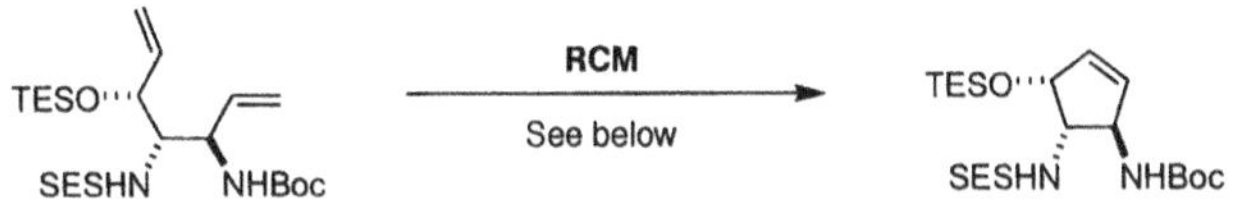

Entry	Catalyst (Mol%)	Solvent	Time	Yield
1	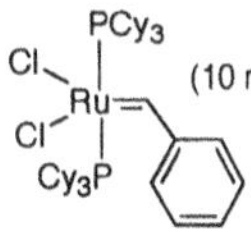 **69** (10 mol %)	CH_2Cl_2	8 h	0 %
2	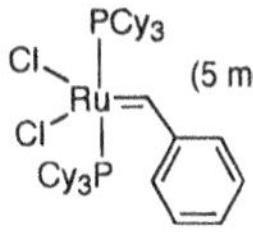 **69** (5 mol %)	PhMe	28 h	0 %
3	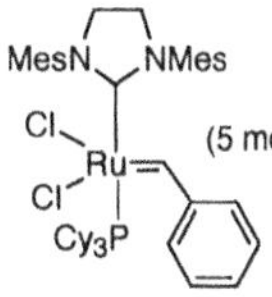 **70** (5 mol %)	PhMe	8 h	53 %
4	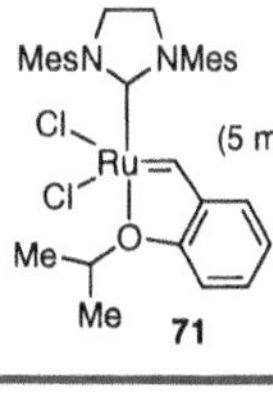 **71** (5 mol %)	PhMe	15.5 h	92 %

SCHEME 16. The ring-closing metathesis reaction of diene **67**.

In this regard, the Hoveyda catalyst **71**[48] was next examined. To our great delight, it afforded the desired cyclopentene **68** in an excellent 92% yield after overnight heating in toluene at 110 °C (Entry 4). Importantly, the reaction could be successfully performed, without any special precautions, in non-distilled commercial solvent in a reaction vessel that was open to the air.

Our good fortune in being able to prepare **68** with negligible material loss now allowed us to potentially channel into the final stages of the Weinreb synthesis. We had originally envisioned detaching the two silicon protecting groups from **68** with fluoride ion (Scheme 17) and converting **69** into the amine **70** by oxazolidinone formation with phenyl chloroformate, and cleavage of the Boc group with trifluoroacetic acid. Acylation of **70** with acid chloride **18** and protection of the oxazolidinone nitrogen with a Boc group were then expected to lead to the Weinreb advanced intermediate **71** except now, of course, we would have secured it in optically pure form.

Disappointingly, all our attempts at removing the SES group from **68** with a variety of fluoride sources failed miserably (Scheme 17). CsF in DMF at 95°C, and TBAF in THF at 50 °C, were just two sets of conditions that we examined but, in both instances, decomposition was the sole result. On one occasion, a single product was isolated in 76% yield from the reaction of **68** with 5 eq. of TBAF in acetonitrile at 80 °C. However, its structure was later shown to be that of **72** by detailed mass spectroscopic and ^{1}H-NMR analysis. It is likely that the hydroxide ion that is invariably present in TBAF solutions was responsible for the observed Boc group cleavage in this instance; amazingly, the SES group survived these deprotection conditions.

A closer examination of the rather sparse literature that exists on SES group removal very quickly revealed that Ward and coworkers[49] had experienced similar problems during their attempted removal of a SES group from a primary amine in their synthetic studies on bactobolin. In that instance, a range of fluoride sources were examined unsuccessfully, including TBAF in THF at reflux, and CsF in DMF at 95 °C.

Given that Campbell and Hart[50] had reported that SES groups can be cleaved much more readily from amide systems, we sought to prepare oxazolidinone **74** with a view to attempting the cleavage of its SES group and ultimately accessing **71**. Accordingly, sulfonamide **73** was treated with sodium hydride and phenyl chloroformate to form **74** in an unoptimised 31% yield. The latter compound was then exposed to TBAF in THF but the desired oxazolidinone **75** was not formed; instead an unidentified slower-moving major product was produced according to TLC analysis. Despite our isolating and purifying this material, we were never able to positively identity it by spectroscopic means.

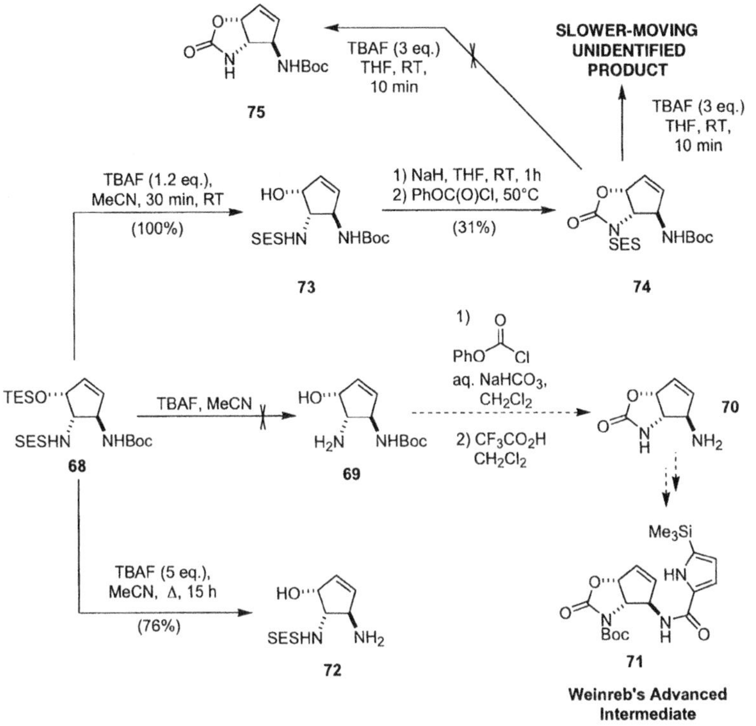

SCHEME 17. Our failed attempts at intersecting with Weinreb's advanced intermediate 71.

VIII. A Second-Strike Strategy for (-)-Agelastatin A

It had now become clear that our original plan for (-)-agelastatin A could go no further, and that a major synthetic redesign was now needed. Notwithstanding the demise of our originally conceived approach, we had learned much from it. For example, we had discovered how to successfully olefinate alkenyl aldehydes such as **66**. We had also managed to identify conditions that could effect the RCM reaction[35] on a diene of the sort that would be needed to eventually reach the agelastatins. In light of this, we decided not to totally abandon the idea of arriving at (-)-agelastatin A through a monosaccharide-based RCM strategy. Rather, we sought to dramatically modify our protecting group strategy so as to comply with the Weinreb end game at a much earlier stage. Weinreb's oxazolidinone **77**[29] (Scheme 18) was now selected as the new intersection point for our revised enantiospecific formal total synthesis. Weinreb had previously converted **77** into the enone **19** by a three step protocol that involved *N*-acylation with the pyrrole acid chloride **18**, oxazolidinone hydrolysis with aqueous LiOH, and oxidation of the resulting alcohol with pyridinium dichromate (PDC).

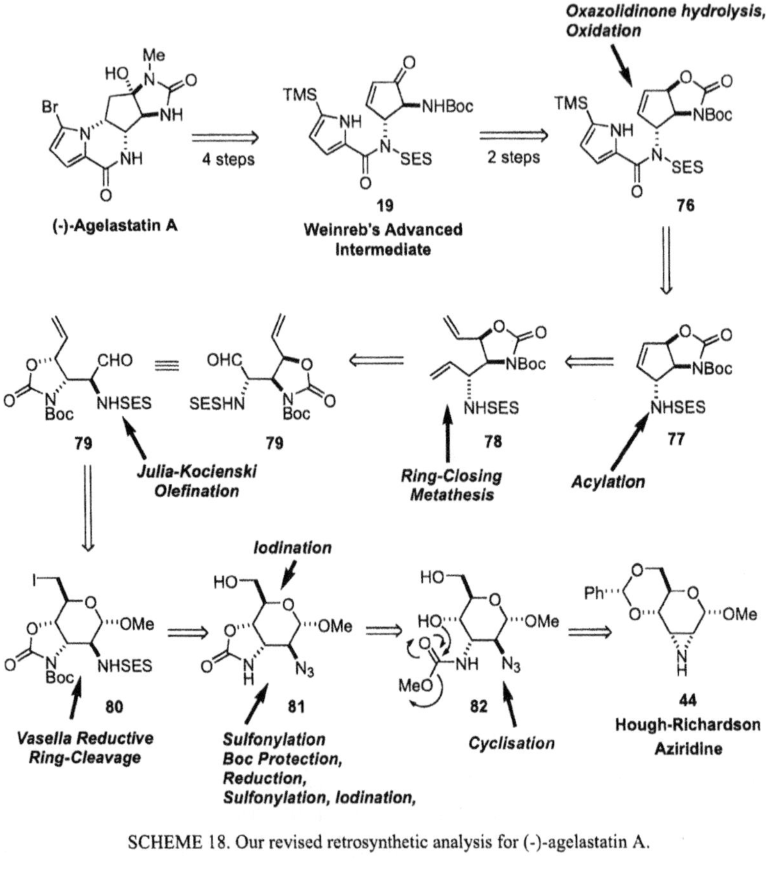

SCHEME 18. Our revised retrosynthetic analysis for (-)-agelastatin A.

By applying a RCM retrosynthetic transform[35] directly to cyclopentene **77**, the new diene **78** was now identified as a key intermediate for the modified route. We had deliberately left the oxazolidinone ring in place in our retrosynthetic planning, as we believed that it might facilitate the desired ring-closure process as a result of it imparting turn-structure on the diene. Yet again, a Julia-Kocienski methylenation[46,47] would be used to prepare **78** from the alkenyl aldehyde **79** and, as previously, a Vasella reductive ring cleavage[36] would potentially secure aldehyde **79** from the iodide **80**. In principle, the latter could be derived from the azido-alcohol **81**. We envisaged fashioning the oxazolidinone ring in **81** through a cyclization of the diol **82** as shown in Scheme 18. Finally, we considered that carbamate **82** might potentially be attainable from an azide ring-opening of the methyl carbamate obtained from **44**.

Accordingly, the Hough-Richardson aziridine **44** was N-acylated with methyl chloroformate in pyridine and dichloromethane, and carbamate **83** was isolated as a crystalline solid in 86% yield (Scheme 19). Although the azide ring-opening of carbamates derived from **44** had never previously been studied in the literature, we were gratified to find that aziridine **83** ring-opened readily with 4 equivalents sodium azide in DMF at 140 °C to give the desired D-altropyranoside **84** as the sole product in excellent yield. The O-benzylidene acetal was removed from **84** by treating it with methanolic HCl, generated *in situ* by the addition of acetyl chloride to anhydrous methanol at 0°C; the 4,6-diol was liberated in 94% yield.

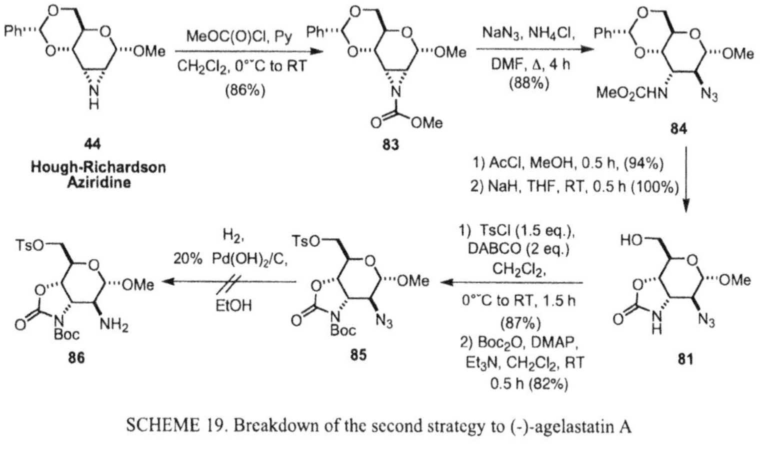

SCHEME 19. Breakdown of the second strategy to (-)-agelastatin A

It had been hoped that the acidic conditions of this deprotection would induce *in situ* oxazolidinone formation and give compound **81** directly. However, the tandem cyclization process did not occur. To bring about ring closure, **82** had to be exposed to 1.5 eq. of sodium hydride in THF at room temperature for 30 minutes, whereupon the desired oxazolidinone **81** was formed in essentially quantitative yield.

Selective tosylation of the primary alcohol in **81** was now attempted in the presence of the highly nucleophilic nitrogen of the oxazolidinone. We had counted on the much less hindered alcohol in **81** being sulfonylated preferentially to this nitrogen. However, when we applied a standard set of tosylation conditions to **81** (TsCl 1.1 eq, triethylamine, DMAP, CH$_2$Cl$_2$), our worst fears were soon confirmed, inasmuch as a 1:1 mixture of the mono-*O*-tosylated and the *N,O*-ditosylated oxazolidinones was formed. Poor results were also obtained when pyridine was used as the reaction solvent. After much experimentation, we eventually found that by employing 2 eq. of the quinuclidine base, DABCO, and 1.5 equivalents of TsCl in CH$_2$Cl$_2$,[51] we could create a hindered *p*-toluenesulfonylammonium salt that selectively *O*-tosylated the primary alcohol in **81** in excellent yield (87%), without producing any of the ditosylated product according to careful TLC analysis.

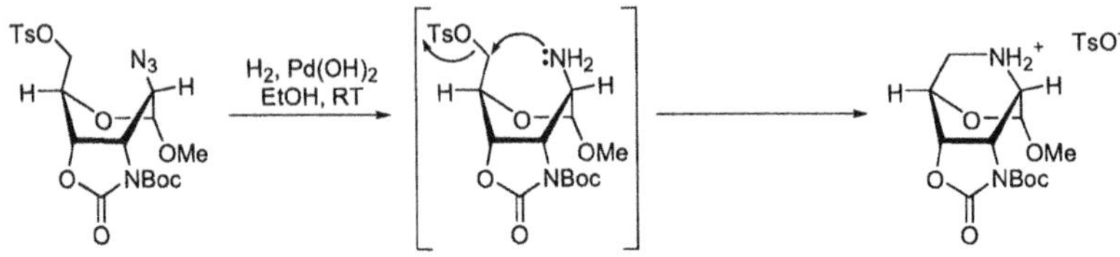

SCHEME 20. A possible but unproven fate for **85** after hydrogenolysis.

With this obstacle surmounted, all that now remained was for us to anchor a Boc group on the oxazolidinone nitrogen of the product, and then attempt reduction of the azido group. Stirring of an ethanolic solution of **85** under an atmosphere of hydrogen with 20% Pd(OH)$_2$ on C for 2 hours caused complete disappearance of the starting azide according to TLC analysis (Scheme 19). However, it proved exceedingly difficult to isolate any products from the reaction mixture after work-up. We suspect that the reason for this was that the transiently generated amino group in **86** effected an internal nucleophilic displacement of the tosylate to generate the highly polar salt **87**, which was difficult to liberate from the catalyst even after extensive washing with polar solvents such as MeOH (Scheme 20). Molecular models of **86** suggest that its oxazolidinone ring could actually favor attainment of the boatlike transition state needed for cyclisation to occur. Another explanation of our inability to isolate any products could simply be due to a polymerisation process occurring.

Whatever the actual source of our problems, in our third-strike strategy to (-)-agelastatin A, we resolved to delay oxazolidinone formation until after the cyclopentene ring had been formed and, accordingly, we accommodated this thinking in our new retrosynthetic analysis of the target.

IX. A Third-Strike Strategy for (-)-Agelastatin A: A Formal Total Synthesis

In our new antithetic plan (Scheme 21), Weinreb's oxazolidinone 77^{29} would remain a key synthetic objective. However, the Hoveyda-Grubbs reaction^{35,48} would now be applied to the diene **88**. The new route would also block O(4) of the vinyl aldehyde precursor with a TES ether to prevent hemiacetal formation from occurring. The crucial Vasella

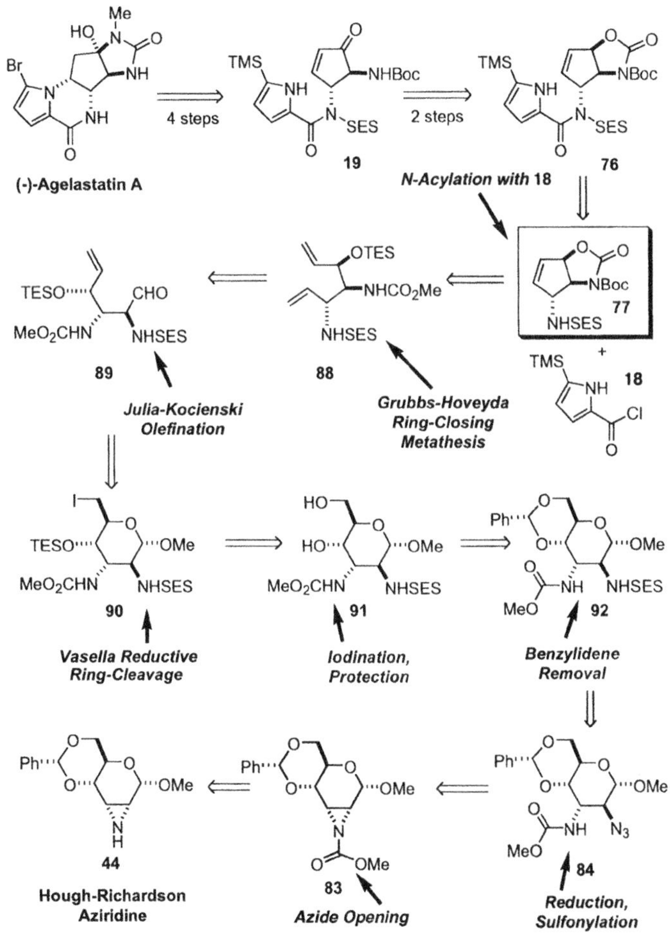

SCHEME 21. A third-strike retrosynthetic strategy for (-)-agelastatin A.

reaction[36] would now be performed on the iodide **90** and the latter would in turn be fashioned from **91** and **44**, respectively.

With this in mind, the hydrogenation of **84** was attempted with Pearlman's catalyst and found to give amine **93** in essentially pure condition (Scheme 22); this allowed the sulfonylation step to be prosecuted directly. The traditional method for introducing a SES group, which entails reacting the amine with SESCl (1.5 eq) and Et$_3$N in DMF, gave poor results with amine **93**. The desired sulfonamide **92** was never isolated in more than 23% yield even after 11 days, and a number of unidentified products were also formed. Employing 1.5 equivalents of

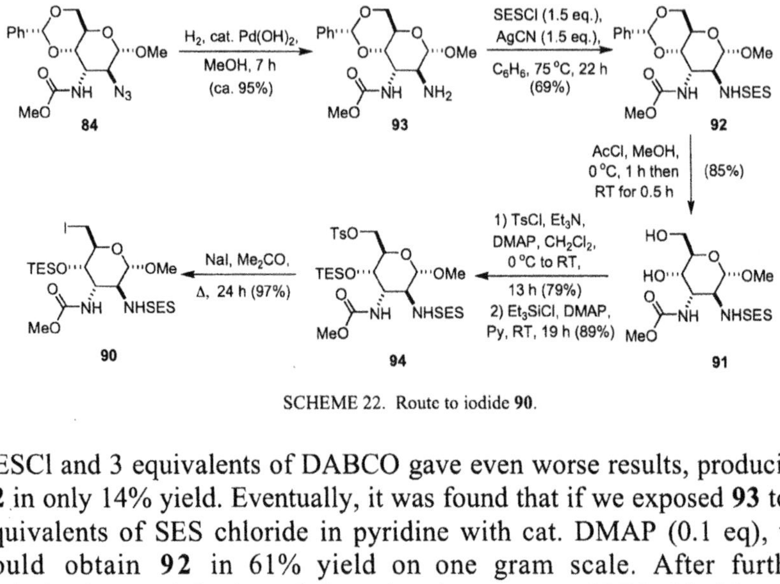

SCHEME 22. Route to iodide **90**.

SESCl and 3 equivalents of DABCO gave even worse results, producing **92** in only 14% yield. Eventually, it was found that if we exposed **93** to 3 equivalents of SES chloride in pyridine with cat. DMAP (0.1 eq), we could obtain **92** in 61% yield on one gram scale. After further optimization, mainly through lowering the amount of SES chloride used (1.5 eq), and decreasing the substrate concentration to 0.2 M, an 80% yield of **92** was eventually obtained on ten gram scale. However, this protocol proved considerably less successful on larger scale, the yields of **92** typically ranging from 50 to 54%.

In light of this, a new silver cyanide mediated procedure was investigated for the *N*-sulfonylation of **93** with SESCl. Significantly the new protocol proved much more dependable than the base-mediated methods when used on large scale, typically furnishing **92** in 69% yield. Originally developed in the late 1980's and early 1990s for carrying out amidations on unreactive and base-sensitive amino-acid chlorides,[52] this is the first time that the AgCN procedure has been applied for the sulfonylation of a primary amine.

Cleavage of the benzylidene acetal from **92** was best achieved with anhydrous HCl in methanol; diol **91** was produced in 85% yield. Regioselective *O*-tosylation of **91** could also be efficiently accomplished (79% yield) by the portionwise addition of 1.1 eq. of TsCl, over several hours, to a mixture of the diol, DMAP and Et$_3$N at 0 °C. *O*-Silylation of this product with triethylsilyl chloride and DMAP in pyridine again occurred, readily delivering **94** in 89% yield. The nucleophilic displacement of tosylate **94** with sodium iodide in hot acetone thereafter furnished **90** in nearly quantitative yield.

The stage was now set for the final push forward towards oxazolidinone **77** (Scheme 23). This commenced with our implementation of a Vasella reductive ring-opening[36] on **90** with zinc dust in aqueous THF. Aldehyde **89** was then olefinated under Kocienski's conditions[46,47] with the carbanion derived from 5-methanesulfonyl-1-phenyl-1*H*-tetrazole. It was not usually possible to remove all of the oxytetrazole by-product from diene **88** on large scale. However, this impurity did not adversely affect the Grubbs-Hoveyda ring-closing metathesis reaction[35,48]

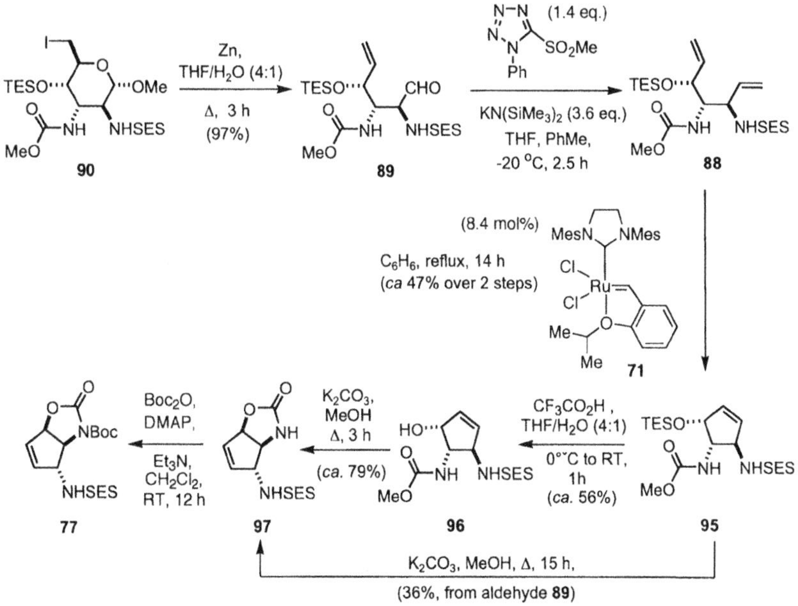

SCHEME 23. Completion of the synthesis of Weinreb's advanced intermediate (-)-**77** for (-)-agelastatin A.

that followed, which efficiently transformed **88** into the desired cyclopentene **95** in a very clean manner. Yet again, it still proved difficult to fully remove all of the tetrazole byproduct from **95** by SiO_2 flash chromatography, and so slightly impure material was used directly for the next step.

Even though the O-desilylation of **95** with TBAF turned out to be problematical, the TES ether could be cleaved in 56% yield with aqueous CF_3CO_2H in THF. Compound **96** was thereafter converted to the oxazolidinone **97** by treatment with potassium carbonate in hot methanol. Generally, however, it was found far more convenient to directly convert **95** into the oxazolidinone **97** by heating the former with K_2CO_3 in MeOH at reflux for 3 hours. Using this procedure, oxazolidinone **97** was isolated pure in 36% overall yield for the three steps from aldehyde **89**.

We were now faced with the challenging issue of having to chemoselectively N-acylate the oxazolidinone nitrogen of **97**, in the presence of the sulfonamido functionality, to obtain (-)-**77**. Fortunately, this could be achieved reasonably cleanly and efficiently (in *ca.* 63% yield), by the slow addition of a solution of Boc_2O in CH_2Cl_2 to a solution of **97**, DMAP, and Et_3N in CH_2Cl_2. To our delight, the ^1H and ^{13}C NMR spectra of our optically pure (-)-**77** matched those recorded by Weinreb and coworkers[29] for ($\pm$)-**77** in the same solvent. Our enantiopure **77** also had a large negative $[\alpha]_D$ (−88 ° at *c* 0.22 in CH_2Cl_2); its relative and absolute stereostructure were further confirmed by X-ray crystallography.

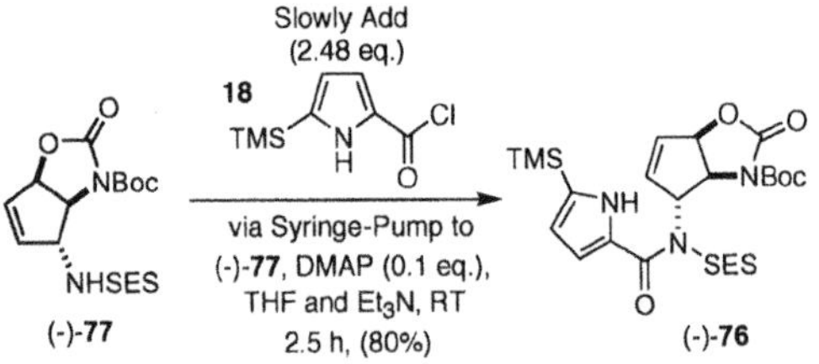

SCHEME 24. Enantiospecific synthesis of another Weinreb intermediate (-)-**76**.

To confirm our formal total synthesis of (-)-agelastatin A, we went on to reproduce the N-acylation reaction that was performed by Weinreb on ($\pm$)-**77** with acid chloride **18**. It proceeded in 80% yield, but required a slow syringe pump addition of the acid chloride in CH_2Cl_2 to amide (-)-**77** and DMAP for success. As previously, the spectra of (-)-**76** matched those reported by Weinreb and coworkers[29] for their racemic version of the same intermediate.

X. A New End Game and Total Synthesis of (-)-Agelastatin A

With an enantiospecific pathway to the natural product now secure, rather than simply repeating the Weinreb endgame, we sought a new way of converting the chiral oxazolidinone (-)-**97** into (-)-agelastatin A, safe in the knowledge that we would have a fall-back position for arriving at (-)-agelastatin A in the event of failure. We were particularly keen to devise a more "user friendly" end game that would avoid the use of iodotrimethylsilane (which reacts violently with water) for Boc-deprotection, and N-methyl isocyanate (which is highly toxic, dangerous, expensive, and nowadays, hard to obtain) for imidazolone-hemiaminal elaboration.[29]

Given that Feldman and Saunders[31] had successfully deployed an N-(methyl)benzylcarbamate group as a precursor of the 1,3-imidazolone hemiaminal system in their first-generation racemic agelastatin A synthesis, we opted to position the same grouping on the oxazolidinone ring nitrogen of (-)-**97**. We believed that this might ultimately allow us to create the tricycle **101**, and totally avoid use of the aforementioned reagents (Scheme 25).

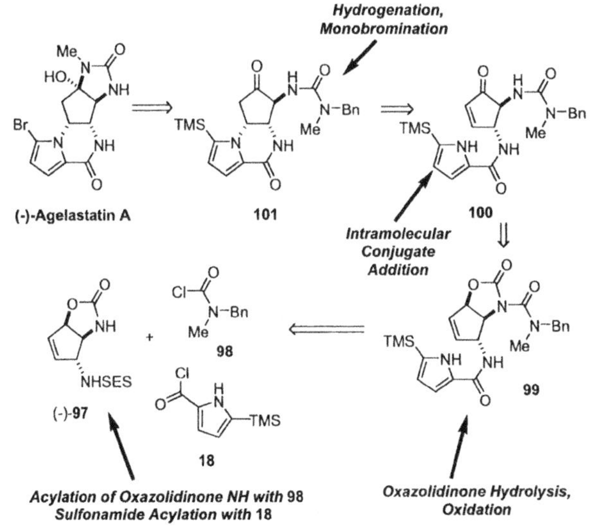

SCHEME 25. A new retrosynthetic end game for (-) agelastatin A from the enantiopure oxazolidinone (-)-**97**.

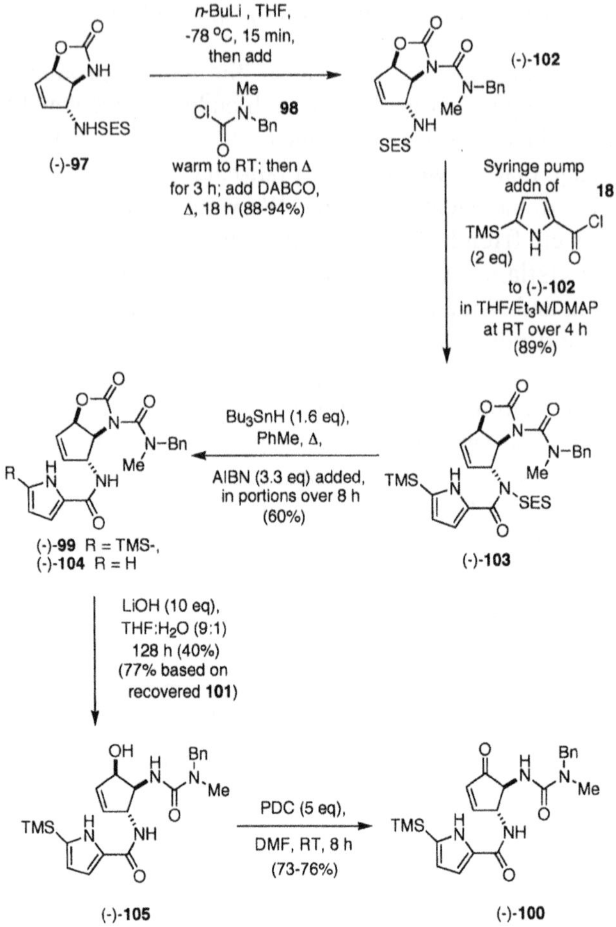

SCHEME 26. Synthesis of Cyclopentenone (-)-**100**

With the above plan in mind, we examined the regioselective *N*-carbamoylation of oxazolidinone (-)-**97** with carbamoyl chloride **98**[53] and a combination of *n*-BuLi and DABCO as bases. In this reaction (-)-**97** was treated with *n*-BuLi in THF at low temperature, and the carbamoyl chloride **98** was added. The mixture was then heated at reflux for 3-4 h, and DABCO (3 eq) was introduced, and reactants were thereafter stirred at reflux for a further 18 hours. The highly crystalline *N*-carbamoyl oxazolidinone (-)-**102** was the end-result; it was isolated in 88% yield. Its structure was confirmed by spectroscopic means, and subsequently verified by single crystal X-ray analysis. A second *N*-acylation with the pyrrole acid chloride **18**,[29] triethylamine and DMAP now ensued to afford the pyrrolocarboxamide **103** in 89% yield. As with **77**, a slow syringe-pump addition of a solution of the acid chloride in CH_2Cl_2 to a solution of sulfonamide **102** in THF was essential for this acylation to proceed cleanly (Weinreb's procedure).

Given our past difficulties with SES group cleavage, we were not at all surprised to find that fluoride-induced removal of the SES group from **103** with 1 eq. of TBAF in THF at 0°C gave disappointing results, it producing **99** in a rather meager 34% yield. A much more satisfactory way of effecting this deprotection treated **103** with 1.6 eq. of Bu_3SnH and 3.3 eq. of AIBN in PhMe at reflux, adding the AIBN in small portions over 8 h. Via this new SES deprotection protocol for amides, the free pyrroloamide was typically isolated in around 60% yield in a *highly reproducible* fashion. This procedure was originally developed by Amos Smith and coworkers for the desulfonylation of multifunctional β-keto-phenylsulfones under neutral conditions,[54] and was subsequently applied to the deprotection of arylsulfonylated amides by Andrew Parsons and his colleagues at the University of York.[55] A small amount of the *C*-desilylated product **104** (up to 7%) was also sometimes encountered in this Bu_3SnH reduction depending on its overall duration.

Our next objective was to hydrolyze the oxazolidinone ring in **99** whilst leaving the urethane unit intact. Although this reaction could be accomplished cleanly with aqueous LiOH in THF at room temperature, it did require long reaction times (128 h) to deliver workable yields of product. In this regard, the desired allylic alcohol **105** was usually isolated in 40% yield along with 48% of recovered **99**, which was then recycled. Based on the quantity of **99** that was typically recovered, the yield of **105** was calculated to be 77%. Although the oxidation of alcohol **105** with TPAP/NMO[56] in acetonitrile produced the enone **100** in 66% yield, pyridinium dichromate proved to be the more effective oxidant of the two for this system, affording the enone **100** in a reproducible 73% yield.

The base-mediated ring-closure of enone **100** was next investigated with a view to securing ketone **101** (Scheme 27). Triethylamine in MeOH, or 10 mol% Bu_3P in THF,[57] were both completely ineffective at mediating this cyclisation, with enone **100** being recovered untouched in either case. When enone **100** was treated with 2 eq. of DBU in THF at room temperature, the rearranged enone **106** was produced as essentially a single product in 70% yield![58] Sodium hydride was also investigated for effecting the desired ring-closure, but this also failed to deliver the cyclised ketone **101**. Instead, a complex mixture of compounds arose in

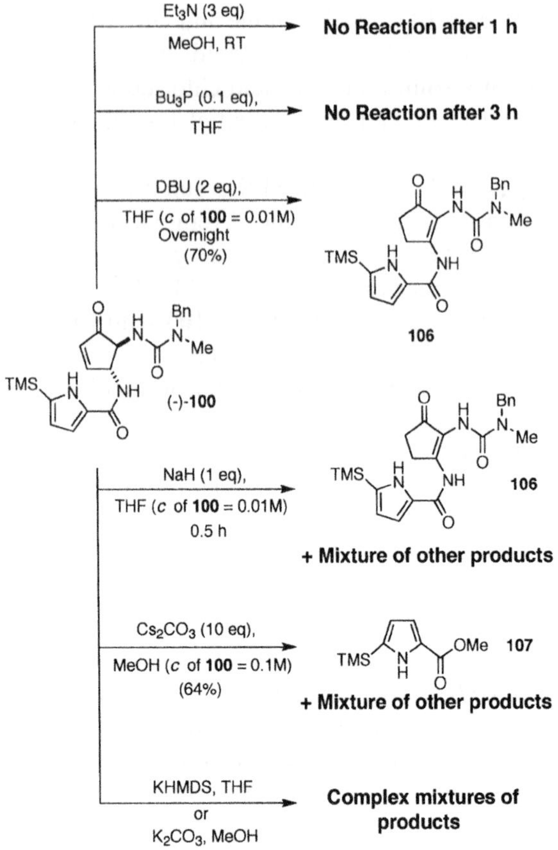

SCHEME 27 Attempts at cyclizing (-)-**100** with various bases.

which the dreaded cyclopentenone **106** was present. Potassium hexamethyldisilazide in THF likewise gave rise to complicated reaction mixtures, as did K₂CO₃ in MeOH. Surprisingly, Cs_2CO_3 in MeOH afforded **107** as the only readily isolable reaction product in 64% yield, along with other decomposition products.

The cyclopentenone isomerization seen here is analogous to some of the well known base-mediated rearrangements seen with certain prostaglandins (e.g. PGA1 to PGB1).[58] Presumably **106** arises from a γ-deprotonation occurring in **100** to create the cyclopentadienol **108**, which then undergoes further facile deprotonation to give the aromatic 6π-anion **109**, prior to reprotonation in the manner shown (Scheme 28).

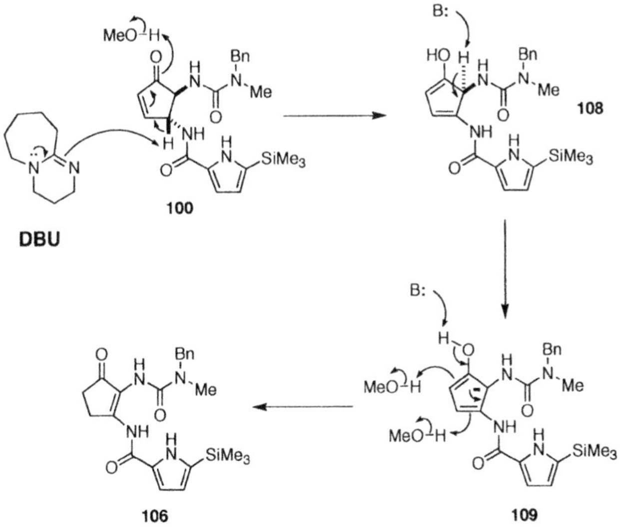

SCHEME 28. Base-mediated isomerization of cyclopentenone **100**.

In light of all these failures, a range of Brønsted and Lewis acids were surveyed for their ability to instigate the desired cyclisation. PPTS (5 eq) in THF and MeOH was initially screened in this role. Unfortunately, this led to an *ipso*-substitution of the pyrrole TMS group in **100** by hydrogen. The latter product was also formed when TMSOTf (1.1 eq) was used to activate the enone **100** in methanol; it was coproduced with **106**.

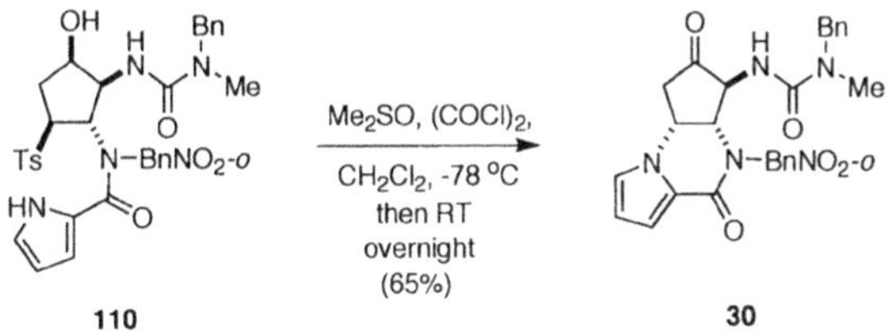

SCHEME 29. Feldman's Tandem Swern Oxidation/Elimination/Cyclization
Conditions Applied on **110**

Given this farrago of adverse results, we once again stepped backwards, and attempted the Swern oxidation and *in situ* cyclisation of **105** under conditions analogous to those reported by Feldman and co-workers[31] for the preparation of **30** (see Scheme 29). On our substrate, a repetition of their conditions resulted in a 16% isolated yield of **106**, a 20% yield of the enone **100**, and a 62% recovery of **105**.

Because Et$_3$N in MeOH had already been shown to leave enone **100** intact (after several hours exposure), we postulated that we might be able to bring about the desired cyclization with a hindered trialkylamine in MeOH, if we could find a way of somehow lowering the pKa of the pyrrole nitrogen in **100**.

After giving the problem considerable thought, we eventually concluded that we might be able to bring about the desired change in pKa by selectively dibrominating the pyrrole ring in **100**. With this in mind, **100** was reacted with 2 eq. of *N*-bromosuccinimide in THF for 3 h, in the expectation that a single 2,3-dibromopyrrole would form, whose pKa would be significantly lower (Scheme 30). To our surprise, a complex mixture of mono-, di-, and tri-bromopyrroles arose according to TLC analysis, and none of the starting enone remained. In light of this, we decided to add Hunig's base (5 eq) directly to the reaction mixture, and observed that the desired internal conjugate addition proceeded smoothly without the offending enone rearrangement snaring us.

Notwithstanding our success in this key cyclization reaction, we still had to deal with an unwieldy mixture of cyclized bromopyrroles. Therefore, rather than attempting to fractionate the mixture at this stage, we extractively worked it up, and submitted the crude concentrated residue directly to hydrogenolysis with a catalytic amount of 10% Pd/C in MeOH in the presence of NaOAc (3 eq) for 2 h. By following this protocol, the desired ketone **113** could be isolated in 35% yield for these 3 steps (after SiO$_2$ flash chromatography).

Inclusion of NaOAc in the aforementioned reaction mixture was absolutely essential for a correct outcome. When omitted, the 5b,8a-di-*epi*-debromoagelastatin A **115** was formed as the primary reaction product in a 13% yield (unoptimised) over the 3 steps (Scheme 31). The structure of **115** was assigned by 2D NOESY NMR spectroscopy, which revealed a strong nuclear Overhauser effect (NOE) between H(5a) and H(5b). The ^{1}H NMR spectrum of **115** also showed a large ^3J coupling constant of 7.3 Hz between H(5a) and H(5b). The unwanted epimerisation was most likely caused by enolisation and reprotonation occurring as the catalytic debromination took its course, induced by the HBr that was being generated.

Initially, we ran the hydrogenolytic *N*-debenzylation of **113** in methanol in the presence of Pearlman's catalyst (20% Pd(OH)$_2$/C) (Scheme 30). Unfortunately, this led to the methyl hemiaminal **114** being formed in 36% yield (unoptimized). By conducting this reaction in THF, however, in accord with Feldman and Saunders' originally published protocol, we could totally avoid this problem, and (-)-debromoagelastatin A **31** was obtained in 66% yield. Our spectra for **31** matched perfectly those reported by Feldman and Saunders[31] for the very same compound. There was also a perfect spectral correlation between our product and theirs for ketone **113**. Regioselective monobromination of **31** with NBS in a 1:2 mixture of MeOH/THF as described by Feldman *et al.*[31] thereafter provided (-)-agelastatin A in 84% yield as a single compound.

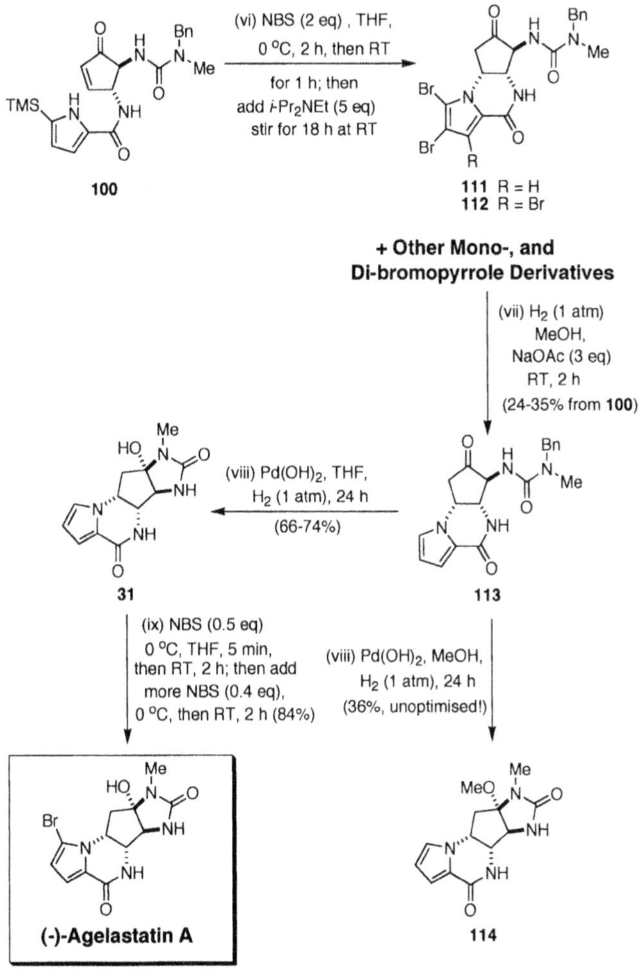

SCHEME 30. Completion of the total synthesis of (-)-agelastatin A.

At last, we had achieved our goal of completing a full enantiospecific total synthesis of (-)-agelastatin A, and with our new route we then went on to prepare 223 mg of the pure natural product to allow us to commence the biological studies we will now describe.

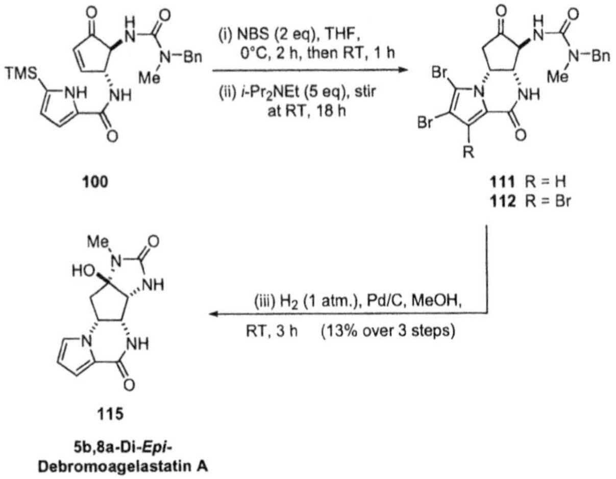

100

111 R = H
112 R = Br

(i) NBS (2 eq), THF, 0°C, 2 h, then RT, 1 h

(ii) i-Pr₂NEt (5 eq), stir at RT, 18 h

(iii) H₂ (1 atm.), Pd/C, MeOH, RT, 3 h   (13% over 3 steps)

115

5b,8a-Di-*Epi*-
Debromoagelastatin A

SCHEME 31. The outcome when NaOAc is omitted from the hydrogenolysis reaction .

XI. A Preliminary Comparison of the Antitumor Properties of (-)-Agelastatin A and Cisplatin Against Various Human Tumor Cell Lines

In Scheme 32 we show the comparative growth inhibition data obtained by Quintiles, in Scotland, for our synthetic (-)-agelastatin A and cisplatin against a range of human tumor cell lines. It can be seen that (-)-agelastatin A is considerably more potent than cisplatin at inhibiting the growth of identical sets of human cancer cell lines.

XII. Toxicological Studies on (-)-Agelastatin A

An independent preliminary toxicological evaluation of our synthetic (-)-agelastatin A was also undertaken by Quintiles. Specifically, (-)-agelastatin A was formulated for dosing by suspending it in a minimal volume of DMSO, and sonicating the sample while gradually adding 0.9% w/v sterile saline containing 0.1% w/v Tween 80. The final dosing solution had vehicle component concentrations that were 5% v/v DMSO, 0.855% w/v sterile saline, and 0.095% w/v Tween 80. It was found that at dose levels of 10 mg/kg, (-)-agelastatin A was toxic to female nude mice. However, when it was administered daily at 2.5 mg/kg for 4 successive

) such mice, there were no signs of toxicity. It has thus been
ded that future studies of the antitumor effects of (-)-agelastatin A
transplanted with human tumors should be conducted at doses of
/kg/day, which is in line with the original observations made by the
group.[15]

	(-)-Agelastatin A	Cisplatin
Human Tumour Cell Line	Mean IC_{50} (μM)	Mean IC_{50} (μM)
NCI-H460 (Lung)	0.194	0.832
LoVo (Colon)	0.975	3.315
DLD-1 (Colon)	0.398	2.362
HCT116 (Colon)	0.344	2.249
HT-29 (Colon)	0.670	4.736
ACHN (Renal)	2.133	1.057
MDAM B435s (Breast)	0.497	5.627
MES-SA (Uterine)	0.162	0.285
MES-SA/Dx5 (Uterine)	8.420	5.826
HepG2 (Hepatic)	0.846	3.300
DU145 (Prostate)	0.701	1.051
BxPC-3 (Pancreatic)	0.414	2.409
AsPC-1 (Pancreatic)	0.642	6.151
RT112/84 (Bladder)	0.234	3.793
SK-MEL-5 (Melanoma)	0.485	6.393

HEME 32. Comparative antitumor screening of (-)-agelastatin A and cisplatin against various human
tumor cell lines.

eliminary Mechanism of Action Studies on (-)-Agelastatin A

h the synthetic (-)-agelastatin A that we prepared by the above
we have been able to preliminarily investigate its effects on the
PC/β-catenin cell signalling pathway. In this regard, we have found
10 nM concentrations, (-)-agelastatin A dramatically decreases the
of cellular β-catenin within a metastatic Rama 37 C9-Met DNA
ary epithelial cell line.[27,59,60] We have also observed that it
/ely down-regulates expression of the metastasis-inducing
hosphoprotein, osteopontin (Opn) in such cells at this low

oncentration, and that it additionally down-regulates Tcf-1. At 10 nM oncentrations, (-)-agelastatin A also potently up-regulates Tcf-4 in such ells. Our preliminary mechanistic work has thus revealed that one of the rimary antitumor targets of (-)-agelastatin A is the Wnt/APC/β-catenin ell signaling pathway, and given the fact that it modulates all these individual components of the pathway at much lower drug concentrations han it inhibits GSK-3β, this suggests that the latter enzyme is not the rimary arbiter of its effects.

In light of the remarkable down-regulatory effects of (-)-agelastatin A n Opn levels within metastatic human breast cancer cells (metastatic Rama 37 C9-Met DNA mammary epithelial), we conducted a preliminary in vitro biological evaluation of its ability to repress the metastasis of uch cells in Matrigel. Our results have indicated that at 10 nM oncentration, (-)-agelastatin A can significantly reduce colony formation, nd that it can attenuate the invasive, the adhesive, and the migratory otential of these highly invasive cancer cells. We have thus concluded hat (-)-agelastatin A might potentially be useful as an antimetastatic drug. We have also shown that the Wnt pathway might potentially provide a useful new set of drug targets for the development of future antimetastatic herapies.

XIV. Conclusions

As a consequence of our completing an enantiospecific total synthesis of (-)-agelastatin A[61] that could deliver meaningful quantities of this highly scarce alkaloid for biological tests, we were able to prove that the Wnt signalling pathway is one of the primary antitumor targets of this drug. Specifically, we were able to show that (-)-agelastatin A can significantly down-regulate β-catenin, Opn and Tcf-1 levels within metastatic human breast cancer cells at concentrations as low as 10 nM. We also demonstrated that it can simultaneously upregulate Tcf-4 expression and considerably reduce the metastatic potential of such cells. Our total synthesis and chemical genomics studies have thus validated the Wnt signalling pathway as a viable and tractable therapeutic target for future antimetastatic drug discovery.

Acknowledgements

We would like to thank Ultrafine for their very generous financial support of this work, and Quintiles for supplying us with some of the biological data reported.

References and Footnotes

1. Breinbauer, R., Vetter, I.R., Waldmann, H., *Angew. Chem. Int. Ed.* **2002**, *41*, 2878.
2. For a leading review on the uses of ethylnitrosourea (ENU)-induced mutagenesis for genome analysis, see: Balling, R., *Annu. Rev. Genomics Hum. Genet.* **2001**, *2*, 463.
3. For an excellent article on the first rats to have their breast cancer suppressor genes *Brca1* and *Brca2* knocked out see: Zan, Y., Haag, J.D., Chen, K.-S., Shepel, L.A., Wigington, D., Wang, Y.-R., Hu, R., Lopez- Guajardo, C.C., Brose, H.L., Porter, K.I., Leonard, R.A., Hitt, A.A., Schommer, S.L., Elegbede, A.F., Gould, M.N., *Nat. Biotech.* **2003**, *21*, 645.
4. For some excellent reviews on RNA interference see: (a) Dykxhoorn, D.M., Novina, C.D., Sharp, P.A., *Nature Rev. Molecular Cell Biology* **2003**, *4*, 457. (b) Hannon, G.J., *Nature* **2002**, *418*, 244. (c) Brummelkamp, T.R., Bernards, R., *Nat. Rev. Cancer* **2003**, *3*, 781.
5. For discussions of the utility of small molecules for dissecting cell signaling pathways see: (a) Mitchison, T.J., *Chem. Biol.*, **1994**, *1*, 3. (b) Schreiber, S.L., *Biorg. Med. Chem.* **1998**, *6*, 1127. (c) Hinterding, K., Alonso-Diaz, D., Waldmann, H., *Angew. Chem. Int. Ed.* **1998**, *37*, 688. (d) Crews, C.M., Splittgerber, U. *TIBS*, **1999**, *24*, 317. (e) Stockwell, B.R., *Trends Biotechnol.* **2000**, *18*, 449. (f) Stockwell, B.R., *Nat. Rev. Genetics* **2000**, *1*, 116. (g) Alaimo, P.J., Shogren-Knaak, M.A., Shokat, K.M., *Curr. Opin. Chem. Biol.* **2001**, *5*, 360. (h) Schreiber, S.L., *Chem. Eng. News* **2003**, March 3, 51.
6. For examples of the use of latex microspheres in the affinity chromatographic purification of drug receptors, see: (a) Shimizu, N., Sugimoto, K., Tang, J., Nishi, T., Sata, I., Hiramoto, M., Aizawa, S., Hatakeyama, M., Ohba, R., Hatori, H., Yoshikawa, T., Suzuki, F., Oomori, A., Tanaka, H., Kawaguchi, H., Watanabe, H., Handa, H., *Nat. Biotechnol.* **2000**, *18*, 877; (b) For a recent successful example of the immobilization of the antibiotic, moenomycin A, on sepharose, and its use for the purification of the penicillin binding protein 1b (PBP 1b), see: Stembera, K., Buchynskyy, Vogel, S., Knoll, D., Osman, A.W., Ayala, J.A., Welzel, P., *ChemBioChem* **2002**, *3*, 332.
7. (a) Zenkoh, T., Hatori, H., Tanaka, H., Hasegawa, M., Hatakeyama, M., Kabe, Y., Setoi, H., Kawaguchi, H., Handa, H., Takahashi T., *Org. Lett.* **2004**, *6*, 2477; (b) Hatori, H., Zenkoh, T., Kobayashi, M., Ohtsu, Y., Shigematsu, N., Setoi, H., Hino, M., Handa, H., *J. Antibiotics* **2004**, *57*, 456.
8. Hale, K.J., Frigerio, M., Manaviazar, S., Hummersone, M.G., Roberts, G.C.K., Fillingham, I., Barsukov, I., Gescher, A., *Org. Lett.* **2003**, *5*, 499.
9. Ge, N.-H., Hochstrasser, R.M., *PhysChemComm* **2002**, *5*, 17.
10. (a) Mittl, P.R.E., Gritter, M.G., *Curr. Opin. Chem. Biol.* **2001**, *5*, 402; (b) Stevens, R.C., *Curr. Opin. Struct. Biol.* **2000**, *10*, 558.
11. Burke, M.D., Schreiber, S.L., *Angew. Chem. Int. Ed.* **2004**, *43*, 46.
12. See for example: (a) Blunt, J.W., Copp, B.R., Munro, M.H.G., Nothcote, P.T. Prinsep, M.R., *Nat. Prod. Rep.* **2004**, *21*, 1. (b) Blunt, J.W., Copp, B.R., Munro, M.H.G., Nothcote, P.T. Prinsep, M.R., *Nat. Prod. Rep.* **2003**, *20*, 1. (c) Faulkner, D.J., *Nat. Prod. Rep.* **2002**, *19*, 1. (d) Faulkner, D.J., *Nat. Prod. Rep.* **2001**, *18*, 1.
13. For a recent monograph on the total synthesis of complex natural products see: "*The Chemical Synthesis of Natural Products*", Ed. Hale, K.J., Blackwell: Oxford, 2000.

14. For some reviews and papers that discuss the Wnt/APC/β-Catenin Cell Signalling Pathway, see: (a) Macleod, K., *Curr. Opin. Gen. Dev.* **2000**, *10*, 81. (b) Watson, S.A. *Lancet* **2001**, *357*, 572. (c) van Noort, M., Clevers, H., *Dev. Biol.* **2002**, *244*, 1. (d) Polakis, P., *Genes Dev.* **2000**, *14*, 1837. (e) Takahashi, M. Tsunoda, T., Seiki, M., Nakamura, Y., Furukawa, Y., *Oncogene*, **2002**, *21*, 5861. (f) van Es, J.H., Barker, N., Clevers, H. *Curr. Opin. Gen. Dev.* **2003**, *13*, 28. (g) He, X. *Dev. Cell* **2003**, *4*, 791. (h) Nelson, W.J., Nusse, R., *Science* **2004**, *303*, 1483. (i) Chung, D.C. *Gastroenterology* **2000**, *119*, 854.

15. (a) D'Ambrosio, M., Guerriero, A., Debitus, C., Ribes, O., Pusset, J., Leroy, S., Pietra, F., *J. Chem. Soc. Chem. Commun.* **1993**, 1305. (b) D'Ambrosio, M., Guerriero, A., Chiasera, G., Pietra, F., *Helv. Chim. Acta* **1994**, *77*, 1895. (c) D'Ambrosio, M., Guerriero, A., Ripamonti, M., Debitus, C., Waikedre, J., Pietra, F., *Helv. Chim. Acta* **1996**, *79*, 727.

16. Hong, T.W., Jimenez, D.R., Molinski, T.F., *J. Nat. Prod.* **1998**, *61*, 158.

17. Al Mourabit, A., Potier, P., *Eur. J. Org. Chem.* **2001**, 237.

18. Meijer, L., Thunnissen, A.M., White, A.W., Garnier, M., Nikolic, M., Tsai, L.H., Walter, J., Cleverley, K.E., Salinas, P.C., Wu, Y.Z., Biernat, J., Mandelkow, E.-M., Kim, S.-H., Pettit, G.R., *Chem. Biol.* **2000**, *7*, 51.

19. For some further information on GSK-3β see: (a) Lovestone, S., Reynolds, C.H., Latimer, D.; Davis, D.R., Anderton, B.H., Gallo, J.-M., Hanger, D., Mulot, S., Marquardt, B., *Curr. Biol.*, **1994**, *4*, 1077; (b) Harwood, A.J., *Cell*, **2001**, *105*, 821; (c) Ali, A., Hoeflich, K.P., Woodgett, J.R., *Chem. Rev.* **2001**, 101, 2527.

20. He, T.C., Sparks, A.B., Rago, C., Hermeking, H., Zawel, L., da Costa, L.T., Morin, P.J., Vogelstein, B., Kinzler, K.W., *Science* **1998**, *281*, 1509.

21. (a) Tetsu, O.; McCormick, F., *Nature*, **1999**, *398*, 422. (b) Shutman, M., Zhurinsky, J., Simcha, I., Albanes, C., D'Amico, M., Pestell, R., Ben-Ze'ev, A., *Proc. Nat. Acad. Sci, USA* **1999**, *96*, 5522.

22. (a) Crawford, H.C., Fungleton, B.M., Rudolph-Owen, L.A., Goss, K.J., Rubinfeld, B., Polakis, P., Matrisian, L.M., *Oncogene* **1999**, *18*, 2883; (b) Brabletz, T., Jung, T., Dag, S., Hlubek, F., Kirchner, T., *Am. J. Pathol.* **1999**, *155*, 1033.

23. Roose, J., Huls, G., van Beest, M., Moerer, P., van der Horn, K., Goldschmeding, R., Lotgenberg, T., Clevers, H., *Science* **1999**, *85*, 1923.

24. Yamada, T., Takaoka, A.S., Naishiro, Y., Hayashi, R., Maruyama, K., Maesawa, C., Ochiai, A., Hirohashi, S., *Cancer Res.* **2000**, *60*, 4761.

25. Mann, B., Gelos, M., Siedow, A., Hanski, M.L., Gratchev, A., Ilyas, M., Bodmer, W.F., Moyer, M.P., Rieken, E.O., Buhr, H.J., Hanski, C., *Proc. Natl. Acad. Sci USA* **1999**, *96*, 1603.

26. (a) Lei, S., Dubevkovskiy, A., Chakladar, A., Wojtukiewicz, L., Wang, T.C. *J. Biol. Chem.* **2004**, *279*, 42492. (b) Koh, T.J., Bukitta, C.J., Fleming, J.V., Dockray, G.J., Varro, A., Wang, T.C., *J. Clin. Invest.* **2000**, *106*, 533.

27. El-Tanani, M.K., Barraclough, R., Wilkinson, M.C. and Rudland, P.S., *Oncogene* **2001**, *20*, 1793.

28. He, T.C., Chan, T.A., Vogelstein B., Kinzler K.W., *Cell* **1999**, *99*, 335.

29. (a) Stein, D., Anderson, G.T., Chase, C.E., Koh, Y-h., Weinreb, S.M., *J. Am. Chem. Soc.* **1999**, *121*, 9574. (b) Anderson, G.T., Chase, C.E., Koh, Y-h., Stein, D., Weinreb, S.M., *J. Org. Chem.*, **1998**, *63*, 7594.

30. (a) Sharpless, K.B., Hori, T., *J. Org. Chem.* **1976**, *41*, 176. (b) Bussas, R., Kresze, G., *Liebigs Ann. Chem.* **1980**, 629.

31. (a) Feldman, K.S., Saunders, J.C., *J. Am. Chem. Soc.* **2002**, *124*, 9060. (b) Feldman, K.S., Saunders, J.C., Laci Wrobleski, M., *J. Org. Chem.* **2002**, *67*, 7096. (c) Feldman, K.S., *Arkivoc*, **2003**, 179.

32. Furrow, M.E., Schaus, S.E., Jacobsen, E.N., *J. Org. Chem.* **1998**, *63*, 6776.

33. Baron, E., O'Brien, P., Towers, T.D., *Tetrahedron Lett.* **2002**, *43*, 723.

34. Knight, J.G., Muldowney, M.P., *Synlett* **1995**, 949.

35. (a) Grubbs, R.H.; Chang, S., *Tetrahedron*, **1998**, *54*, 4413. (b) Furstner, A., *Angew. Chem. Int. Ed. Engl.*, **2000**, *39*, 3012.

36. (a) Bernet, B., Vasella, A., *Helv. Chim. Acta.* **1979**, *62*, 1990. (b) Bernet, B., Vasella, A., *Helv. Chim. Acta.* **1979**, *62*, 2400.

37. We have found that SESCl is most conveniently prepared by the following literature method: (a) Huang, J., Widlanski, T.S., *Tetrahedron Lett.* **1992**, 33, 2657. Other protocols for SESCl preparation include: (b) Weinreb, S.M.; Demko, D.M., Lessen, T.A., Demers, J.P., *Tetrahedron Lett.* **1986**, *27*, 2099. (c) Weinreb, S.M., Chase, C.E., Wipf, P. Venkatraman, S., *Org. Synth.* **1997**, *75*, 161.

38. (a) Gibbs, C.F.; Hough, L.; Richardson, A.C. *Carbohydrate Res.* **1965**, *1*, 290. (b) Ali, Y.; Richardson, A.C.; Gibbs, C.F.; Hough, L. *Carbohydrate Res.* **1968**, *7*, 255.

39. Buss, D.H.; Hough, L.; Richardson, A.C., *J. Chem. Soc.*, **1965**, 2736.

40. Guthrie, R.D.; Murphy, D., *J. Chem. Soc.* **1965**, 3828.

41. Charon, D., Montage, M., Pons, J-F., Le Bray, K., Chaby, R., *Biorg. Med. Chem.* **1998**, *6*, 755.

42. For a detailed discussion of the methods for the selective halogenation of carbohydrates, see: Boons, G.J., Hale, K.J. *"Organic Synthesis with Carbohydrates"* Blackwell Science: Oxford, 2000, Chapter 3, 56.

43. Edmonds, M., Abell, A. *"The Wittig Reaction"* in *"Modern Carbonyl Olefination"* Ed. Takeda, T., Wiley-VCH: Weinheim, 2004, Chapter 1, p 1-17.

44. Takeda, T., Tsubouchi, A. *"Carbonyl Olefination Utilizing Metal Carbene Complexes* in *"Modern Carbonyl Olefination"* Ed. Takeda, T., Wiley-VCH: Weinheim, 2004, Chapter 4, p 151-199.

45. Takai, K., Kakiuchi, T., Kataoka, Y., Utimoto, K., *J. Org. Chem.* **1994**, 59, 2668.

46. (a) Blakemore, P.R., Cole, W.J., Kocienski, P.J., Morley, A., *Synlett* **1998**, 26. (b) Blakemore, P.R., Kocienski, P.J., Morley, A., Muir, K., *J. Chem. Soc. Perkin Trans.* *1* **1999**, 955. (c) Baudin, J.B., Hareau, G., Julia, S.A., Ruel, O., *Tetrahedron Lett.* **1991**, *32*, 1175.

47. For a leading and up-to-date review on the Julia reaction, see: Dumeunier, R., Marko, I.E. *"The Julia Reaction"* in *"Modern Carbonyl Olefination"* Ed. Takeda, T., Wiley-VCH: Weinheim, 2004, Chapter 4, p104-150.

48. (a) Kingsbury, J.S., Harrity, J.P.A., Bonitatebus, Jr., P.J., Hoveyda, A.H., *J. Am. Chem. Soc.* **1999**, *121*, 791. (b) Garber, S.B., Kingsbury, J.S., Gray, B.L., Hoveyda, A.H., *J. Am. Chem. Soc.* **2000**, *122*, 8168.

49. Ward, D.E., Gai, Y., Kaller, B.F. *J. Org. Chem.* **1996**, *61*, 5498.

50. Campbell, J.A., Hart, D.J. *J. Org. Chem.* **1993**, *58*, 2900.

51. Hartung, J., Hunig, S., Kneuer, R., Schwarz, M., Wenner, H., *Synthesis*, **1997**, 1433.

52. For some recent examples of the AgCN mediated amidation reaction, see: (a) Durette, P.L., Baker, F., Barker, P.L., Boger, J., Bondy, S.S., Hammond, M.L., Lanza, T.J., Pessolano, A.A., Caldwell, C.G., *Tetrahedron Lett.*, **1990**, *31*, 1237. (b) Hale, K.J., Cai, J. *J. Chem. Soc. Chem. Comm.* **1997**, 2319. (c) Hale, K.J.,

Lazarides, L., Cai, J., *Org. Lett.* **2001**, *3*, 2927; (d) Hale, K.J., Lazarides, L., *Org. Lett.* **2002**, *4*, 1903.

53. Yoakim, C.; Ogilvie, W.W.; Cameron, D.R.; Chabot, C.; Guse, I.; Hache, B.; Naud, J.; O'Meara, J.A.; Plante, R.; Deziel, R., *J. Med. Chem.* **1998**, *41*, 2882.

54. Smith, III, A.B., Hale, K.J., McCauley, Jr. J., *Tetrahedron Lett.* **1998**, *30*, 5579.

55. Parsons, A.F., Pettifer, R.M., *Tetrahedron Lett.* **1996**, *38*, 1667.

56. Ley, S.V., Norman, J., Griffith, W.P., Marsden, S.P., *Synthesis*, **1994**, 639.

57. Stewart, I.C., Bergman, R.G., Toste, F.D., *J. Am. Chem. Soc.* **2003**, *125*, 8696.

58. The cyclopentenone isomerisation seen here is analogous to the well-known base-mediated prostaglandin A_1 to B_1 rearrangement reported by Corey in the late 1960s. See: (a) Corey, E.J., Andersen, N.H., Carlson, R.M., Paust, J., Vedejs, E., Vlattas, I., Winter, R.E.K., *J. Am. Chem. Soc.* **1968**, *90*, 3245; (b) Newton, R.F., Roberts, S.M. *"Prostaglandins and Thromboxanes"* Eds. Roberts, S.M., Newton, R.F., Butterworths, 1982; Chapter 6, p 84.

59. (a) El-Tanani, M., Barraclough, R., Wilkins, M.C., Rudland, P.S., *Cancer Res.* **2001**, *61*, 5619-29. (b) Oates, A.J., Barraclough, R., Rudland, P.S., *Oncogene* **1996**, *13*, 97-104.

60. El-Tanani, M., Mason, C.K., Campbell, F.C., Domostoj, M.M., Hale, K.J. *Unpublished Results*.

61. For the preliminary communications on our total synthesis of (-)-agelastatin A, see: (a) Hale, K.J.; Domostoj, M.M.; Tocher, D.A., Irving, E., Scheinmann, F., *Org. Lett.* **2003**, *5*, 2927. (b) Domostoj, M.M., Irving, E., Scheinmann, F., Hale, K.J., *Org. Lett.* **2004**, *6*, 2615.

STRATEGIES AND TACTICS IN ORGANIC SYNTHESIS, VOL. 6

Chapter 12

DESIGN AND SYNTHESIS OF COOPERATIVE "PINWHEEL" FLUORESCENT SENSORS

Joseph Raker, Kristen Secor, Ricardo Moran, Ellen Feuster, John Hanley, and Timothy Glass

Department of Chemistry
University of Missouri
Columbia, Missouri 65211

and

Department of Chemistry
Pennsylvania State University
University Park, Pennsylvania 16802

I. Introduction to Cooperative Recognition and Chemical Sensing

Fluorescent probes have been extremely useful in elucidating biochemical mechanisms and processes inside of living cells via fluorescent microscopy. This technique is particularly valuable because it is non-destructive and the probes can be observed in real time over the course of cellular events. Fluorescent probes fall into two main classes: chemosensors and biosensors. Biosensors[1] are fluorescently labelled proteins, most often antibodies. These types of probes have the disadvantage of poor cell permeability, but can be generated with specificity for any macromolecule against which an antibody can be raised. Chemical sensors[2] are typically based on synthetic compounds and have been used in cells mainly to quantify the concentration of certain inorganic ions (Na^+, K^+, Ca^{+2}, Mg^{+2}, Zn^{+2}, and Cl^-). Despite significant advances in molecular recognition and chemical sensing, chemical sensors have enjoyed limited success in practical intracellular applications for sensing analytes beyond these ions and cellular conditions such as pH and pO_2. There are currently very few intracellular fluorescent sensors for small organic compounds because they are not easily targeted by antibodies and chemical sensors for organic analytes are not yet useful in a practical sense.

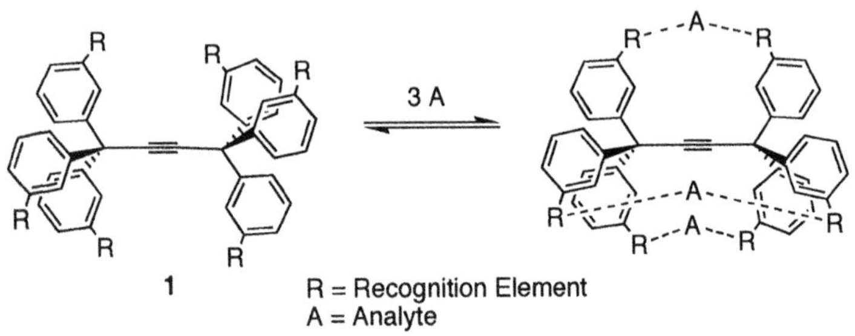

R = Recognition Element
A = Analyte

SCHEME 1

As part of an ongoing project geared toward creating more effective chemical sensors, we designed a novel class of cooperative receptors based on bistrityl acetylene compounds, termed pinwheel receptors (Scheme 1).[3] Each phenyl group of the receptor was appended with recognition elements in the *meta* position wherein a pair would bind one analyte across the receptor framework. This arrangement of recognition elements was designed to give cooperative binding of three analytes in three identical binding pockets. Cooperative, or allosteric, recognition is seen when a receptor has two or more binding sites which are linked. For compound 1, in the absence of analyte, the trityl groups will freely rotate about the acetylene axis.[4] Binding of the first analyte will force the receptor into an eclipsed rotamer. The loss of rotational freedom as well as the build up of steric interactions will make the first binding event a weak one. However, based on the symmetry of the molecule, upon binding of the first analyte, the remaining binding sites will be well aligned for binding the second and third analytes. Thus, a weak first binding event will be reinforced by stronger subsequent binding events. There are a number of such allosteric receptors in the literature.[5] This mechanism of cooperativity is conceptually similar to the restricted rotational freedom employed by Rebek's cooperative biphenyl bis-crown ether[6] and Shinkai's cooperative porphyrin sandwich complexes.[7] Up to that time, cooperative recognition had not been applied toward chemical sensor applications to any large extent.[8] From a design perspective, this method of preorganizing an analyte binding pocket was very attractive. Most receptors and sensors used multiple recognition elements to bind an analyte. Strong binding was most often observed when the recognition elements are held rigidly apart such that there is little entropic loss to the receptor upon binding the guest.[9] The challenge of receptor design became one of creating the optimal spacing and orientation of the recognition elements in a rigid fashion.[10] For receptors of the class depicted in Scheme 1, there would be a range of distances which the recognition elements can span owing to free rotation about the acetylenic axis. Binding of the first analyte rigidly would set the correct distance between the remaining pairs of binding groups. Thus, the entropic loss upon binding the analyte would be effectively averaged over three consecutive binding events, diminishing the overall entropic penalty to binding. Adjusting the distance between the trityl groups by adjusting the spacer, the rough distance between recognition elements could be varied. Upon binding of the first guest, the receptor would then be fully organized to match the size of the guest.

The one downside to cooperative recognition was that cooperativity is typically associated with a sharp transition between the unbound and bound state of the receptor. This implies that the sensor goes rapidly from "OFF" to "ON" and this limits the range over which the sensor can actually sense the analyte. Nevertheless, in many sensor applications, sensitivity is really the limiting factor, so the smaller range of operation is an acceptable trade-off.

II. Synthesis of a Symmetrical Bis-trityl Mono-acetylene Sensor

As a simple proof of concept system, we designed the metal ion receptor **1a** which used ethylenediamine recognition elements for the tetrahedral coordination of three metal ions. For applications to fluorescent sensing, the fluorescent version **1b** was also desired. To achieve the appropriate geometry for metal chelation, the R groups of compound **1** were necessarily close in space and presented an interesting synthetic challenge.

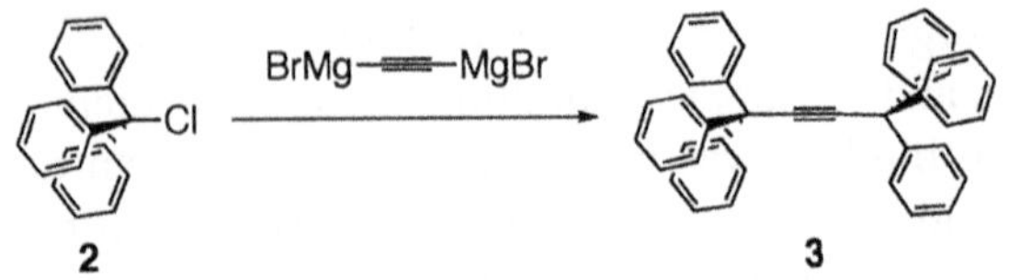

SCHEME 2

At the outset of this work, only the parent compound **3** was known,[11] having been synthesized from the corresponding trityl chloride and acetylene bis-Grignard. However, the formation of trityl acetylenes from trityl alcohols was generally well established.[12] Thus, it appeared that this class of compounds would be easily accessible.

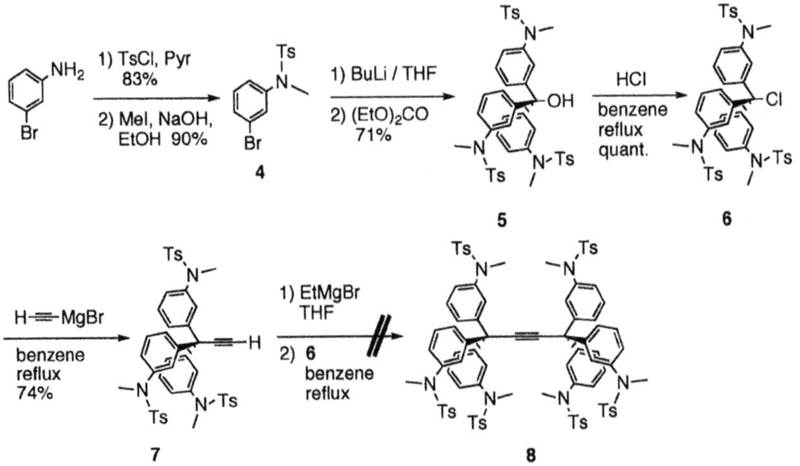

SCHEME 3

The initial synthesis of **1a** was attempted as shown in Scheme 3. The protected trityl alcohol **5** was conveniently prepared from 3-bromoaniline. Typical conditions for formation of trityl acetylenes from the corresponding alcohols entail chlorination under harsh acidic conditions followed by treatment with the acetylene Grignard reagent at elevated temperature. The tosyl protecting group was used because it was stable to strong acid and strong base conditions and could be readily removed under reductive conditions.[13] Indeed, chlorination of trityl alcohol **5** and acetylene addition proceeded smoothly in refluxing benzene to afford the trityl acetylene **7**. Unfortunately, deprotonation of compound **7** to form the alkynyl Grignard and addition of a further equivalent of chloride **6** returned only starting materials. No trace of compound **8** was detected under a variety of conditions, even at elevated temperatures. Given that nearly identical conditions yielded the parent compound (**2** to **3**), it appeared that the *meta*-sulfonanilide groups were simply too sterically demanding to permit addition of the alkyne to the second trityl group. Other less sterically demanding nitrogen protecting groups were ineffective for the formation of the mono-tritylacetylene analogous to **7**; therefore, an alternate means of introducing the recognition element was required.

In principle, the amine binding group could be introduced via amine substitution of an aryl halide. A halogen was sufficiently small to permit formation of the bistrityl acetylene skeleton. Several elegant procedures for amination of aryl halides using palladium cross coupling have been developed.[14] Unfortunately, these sterically sensitive reactions were found to be ineffective for this system. Direct nucleophilic substitution was a promising alternative. However, many aryl halide substitutions proceed via the intermediacy of a benzyne, which would raise regiochemical concerns here. An $S_N Ar$ mechanism could be favored by strong activating groups (e.g., nitro groups). Unfortunately, formation of trityl acetylenes with electron withdrawing substituents was extremely inefficient due to the instability of the intermediate carbocation.[15] Indeed, electron donating groups were preferred. Along this line, Wynberg has reported that *ortho*-fluoroanisole can be substituted by secondary lithium amides at room temperature to give the $S_N Ar$ product as the major product over the benzyne mediated pathway.[16] Thus, it seemed that an *ortho*-fluoroanisole substitution pattern might be a reasonable solution to the problem.

This strategy was pursued as shown in Scheme 4. The trifluoro-trimethoxytritylacetylene **10** was prepared in the typical fashion. As anticipated, the electron donating *para*-methoxy substituents facilitated both the chlorination and Grignard displacement reactions. Acetyl chloride at room temperature quantitatively converted the alcohol **9** to the chloride. Addition of acetylene Grignard at room temperature produced the trityl acetylene **10** in good yield. Formation of the Grignard from **10** by deprotonation with ethyl magnesium bromide followed by addition of the trityl chloride derived from compound **9** gave the bis-tritylacetylene **11** in 61% yield. Addition of an excess of the lithium amide derived from trimethylethylene diamine resulted in the clean replacement of six fluorines to give compound **12** which is analogous to **1a** except for the

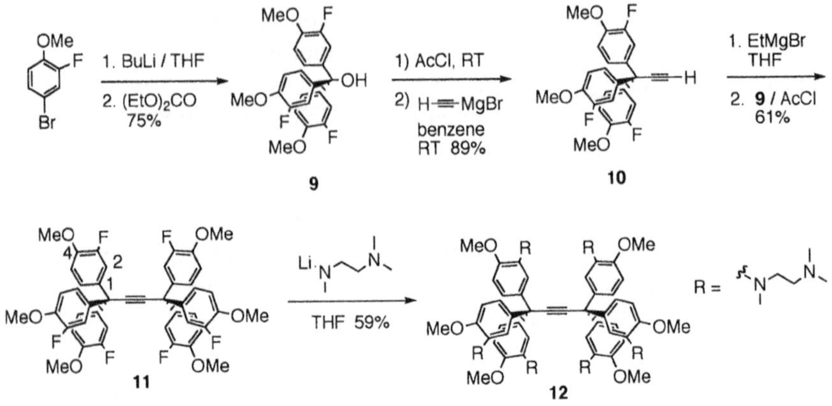

SCHEME 4

ortho-methoxy groups. The regiochemistry of the addition was established by 2D NMR (NOESY) and no other regiochemical isomers were detected. The methoxy substituents on **11** served to block benzyne formation from the 4-position of the aromatic ring and the 2-position is simply too sterically hindered to react. Thus, only direct displacement was possible.

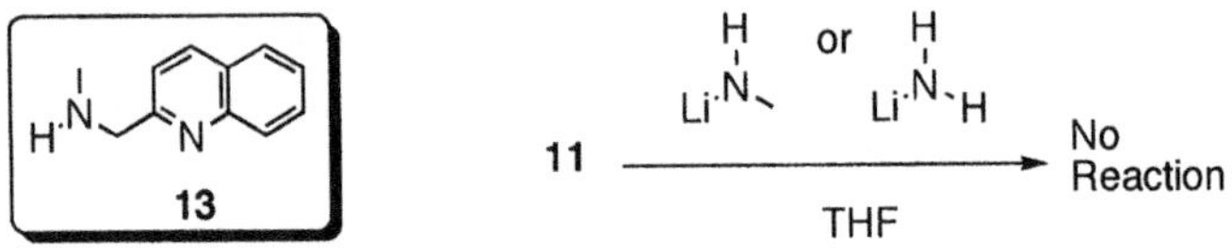

SCHEME 5

In order to append a fluorescent recognition element, the N-methyl quinaldine amine **13** (Scheme 5) would have to be installed on the trityl framework. However, the lithium amide of this compound was inaccessible due to the high acidity of the alpha methylene protons. An alternate approach would involve substitution of the fluoroanisole **11** with a primary lithium amide followed by alkylation of the formed anisidine with an appropriate alkylating agent. However, neither lithium amide or methyl lithium amide reacted with compound **11** under conditions identical to those in which the secondary lithium amide reacted smoothly. A screen of solvents, conditions, and substituents on the primary amide returned only unreacted **11**. This lack of reactivity may be related to the state of aggregation of the primary lithium amides or their inherently lower nucleophilicity compared to secondary amides.

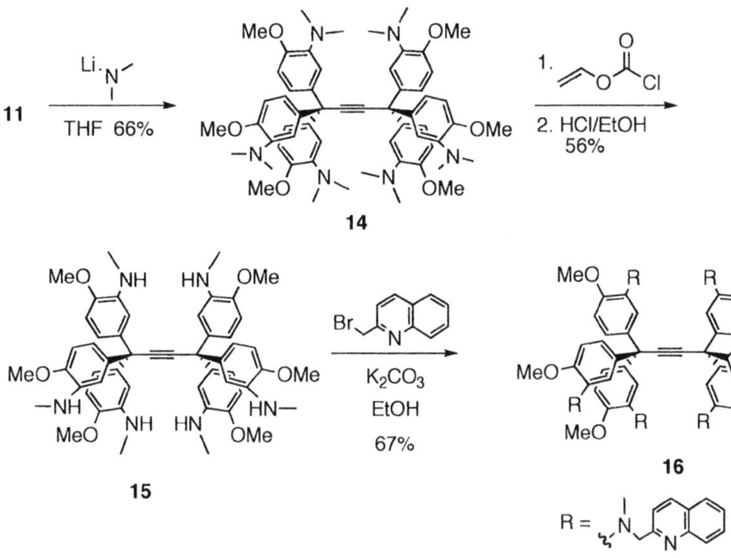

SCHEME 6

This problem was circumvented by a three-step procedure (Scheme 6). As expected, fluoroanisole **11** could be cleanly substituted with lithium dimethylamide to produce the hexa-aniline **14**. Each nitrogen of compound **14** was mono-demethylated using a modification of a dealkylation procedure developed by Olofson[17] giving compound **15** in good overall yield considering six dealkylations. Interestingly, this dealkylation is typically employed only with aliphatic (basic) amines. Its success in the case of **14** is attributed to the lack of conjugation between the aryl ring and the nitrogen due to steric buttressing of the *ortho*-methoxy substituent. This effect is most notable when looking at the polarity of compound **15** compared to **14**. Despite the six N-H groups on **15**, it is much less polar than **14** by thin layer chromatography. The aniline nitrogen of **15** is planar and more like an aromatic amine and the nitrogen of **14** is non-planar with the aromatic ring and more like an aliphatic amine. Finally, alkylation of **15** with an excess of bromomethylquinoline[18] gave the fluorescent receptor **16** which is analogous to **1b**.

The binding properties of compound **16** were thoroughly explored and the advantages of cooperativity were demonstrated compared to a similar non-cooperative sensor.[3] With this encouraging data, similar sensors for organic compounds were then pursued.

III. Design of a Second Generation Fluorescent Pinwheel Sensor

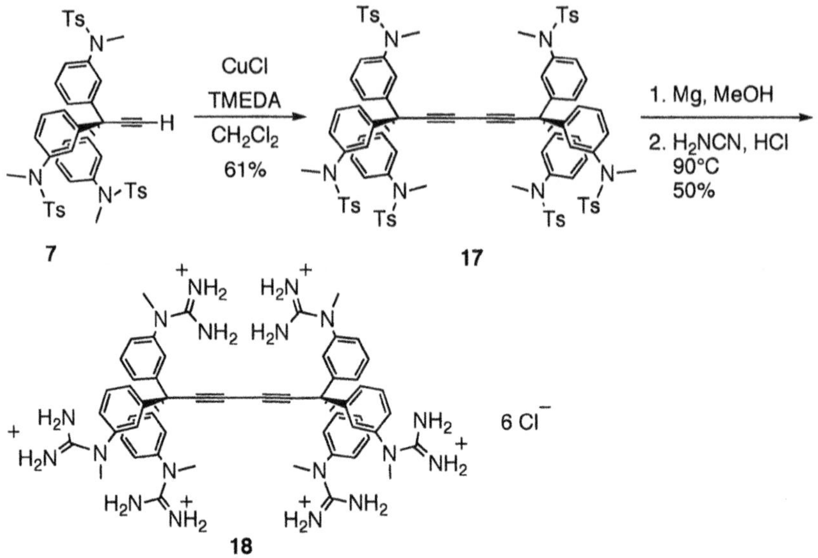

SCHEME 7

In order to prepare sensors for organic analytes a larger scaffold was required. Thus, the tris-tosyl sulfonamide **7** was dimerized by Glaser coupling to produce the bis-trityl bis-acetylene **17**. Here, steric constraints were less of an issue than for the synthesis of the bis-trityl mono-acetylene sensor series, though a great deal of effort was expended in optimizing the Glaser coupling conditions. Deprotection of the tosyl groups followed by acid treatment with cyanamide produced the hexa-guandinium compound **18** as the chloride salt. The larger spacer of **18** would be more amenable to recognition of larger (organic) analytes. Compound **18** appeared to bind dicarboxlates cooperatively in methanol, however studies were hampered by the need to follow binding by small changes in the UV.[19] Compound **18** was not fluorescent nor was it possible to make it fluorescent by simple addition of a fluorescent reporter group. Thus, a more general strategy for introducing fluorescence was required.

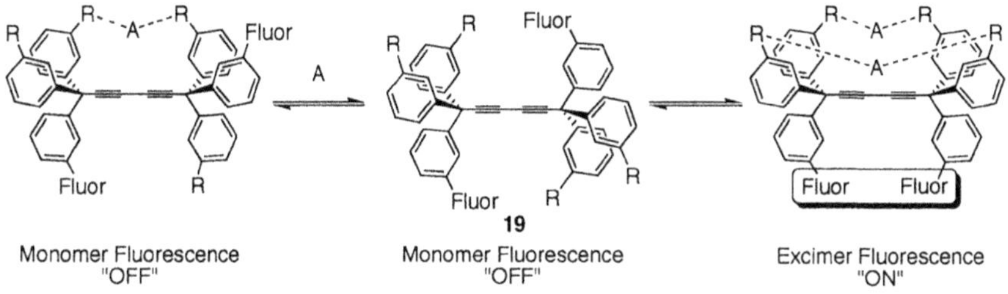

| Monomer Fluorescence "OFF" | Monomer Fluorescence "OFF" | Excimer Fluorescence "ON" |

19

SCHEME 8

To address the issue of fluorescent read-out, we sought a solution that would not depend on the particular recognition element used or the analyte targeted. Thus, a second generation sensor was designed as shown in Scheme 8. On each trityl group was appended two recognition elements (R) and one fluorophore (Fluor). The recognition elements were designed to be spatially convergent such that a pair could bind an analyte across the butadiyne axis. The fluorophores were similarly designed to interact across the butadiyne axis. In the unbound state of the sensor, the trityl groups would have free rotation relative to each other that would result in little interaction between the fluorescent groups (center in Figure 8). The fluorescence in this state would be dominated by the emission of the monomeric fluorophore, which would be read as the "OFF" state. Binding of the first analyte would restrict the natural rotation of the trityl groups and align the second pair of recognition elements for stronger binding of the second analyte. This alignment would create the cooperative response (right in Figure 8). The loss of rotational freedom would similarly force the two fluorescent groups into spatial proximity that could produce an excimer[20] between the fluorophores. The excimer emission would be read as the "ON" state. It should be noted that an alternate binding mode (left in Scheme 8) could be accessed which would not produce a fluorescent response. However, positive cooperative recognition implies that the second binding event would be stronger (higher K_a) than the first binding event. Consequently, a singly bound sensor would be disfavored relative to the doubly bound sensor such that the major fluorescent species in solution would be the unbound and the doubly bound sensor.

This design utilized the same mechanism of cooperativity as the original sensor. The primary advantage of this design was that the cooperative recognition domain was entirely separated from the read-out domain. Based on this mechanism, it was anticipated that any recognition

event that restricted the rotational freedom of the sensor framework
would elicit a fluorescent response. Therefore, once the appropriate
fluorescent groups were identified, little or no modification of the
fluorescence system would be required regardless of the chosen
recognition elements or target analyte.

V. General Synthetic Strategy for Fluorescent Bis-trityl Sensors

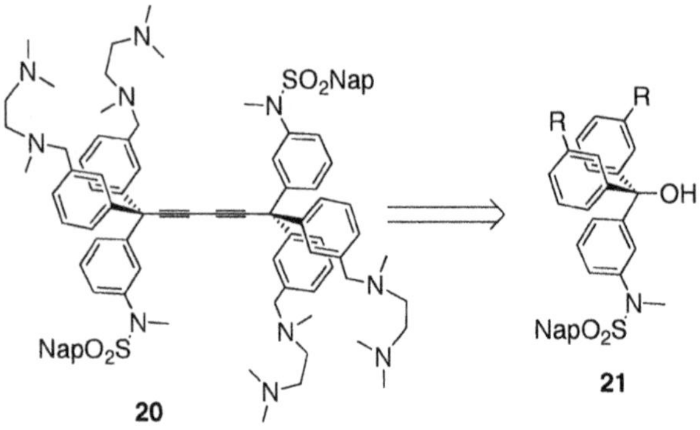

SCHEME 9

As a proof of concept for the second generation sensor system,
compound **20** was initially prepared. The ethylene diamine groups of **20**
served as appropriate non-fluorescent metal-binding elements and the
naphthalene sulfonanilides were used as fluorophores. The naphthyl
sulfonanilides were particularly useful because they could be deprotected
under reductive conditions in high yields compared to the analogous tosyl
group. Once removed, the aniline nitrogen which results could then be
used as a handle to attach different fluorophores. In principle, sensor **20**
could be derived from trityl alcohol **21**. The only remaining issue was
finding a suitable functionality (R), which could be converted into the
ethylene diamine group of **20**.

The simplest approach to sensor **20** was to start with the diamine
recognition elements in place.[21] Thus, trityl alcohol **24a** was prepared in
three steps from ethyl aminobenzoate as shown in Scheme 10. Conversion
of the trityl alcohol to the trityl acetylene was attempted using the typical
procedure. Unfortunately, the diamine recognition elements did not
survive these conditions, yielding only decomposition, though a variety of
permutations were explored. A similar fate was met with several potential
precursors to the desired diamine appended trityl acetylene (e.g., **24b** and

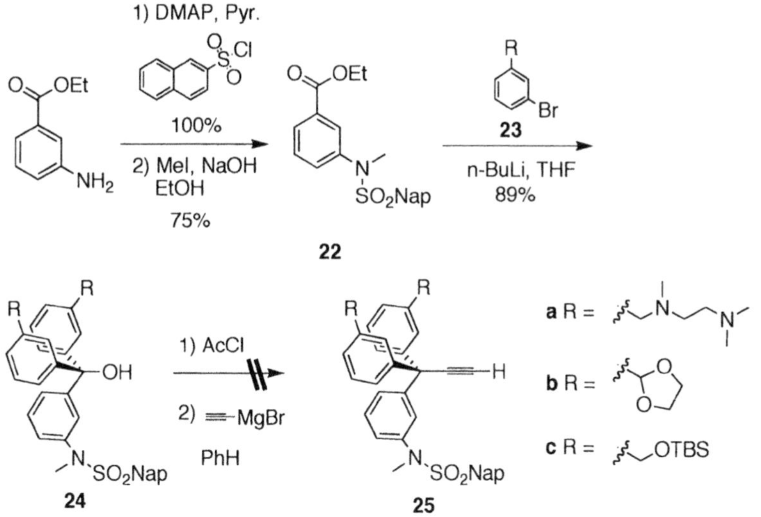

SCHEME 10

24c). Thus, it became immediately apparent that the range of functionality tolerated in this system was quite limited.

As it became clear that benzylic functionality was not tolerated in this system, simple methyl substituents were used, anticipating that free

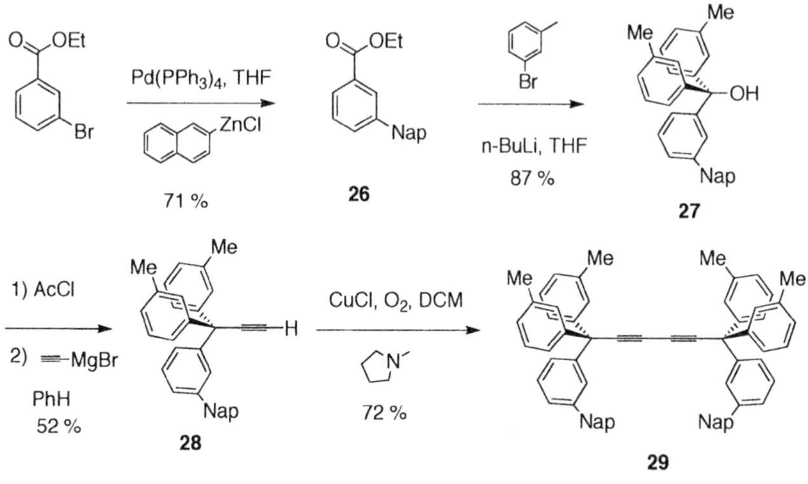

SCHEME 11

radical halogenation followed by alkylation would produce the desired recognition elements (Scheme 11). The preparation of the relatively unfunctionalized bis-trityl bis-acetylene (**29**) was straightforward following the methods outlined above. In this case, a rather simple fluorophore was employed to eliminate alternate sites of reactivity in the halogenation. Unfortunately, several sets of free radical halogenation conditions yielded mostly alkene products from compound **29**. These products presumably arose from attack of halogen radical on the butadiyne core. We had reasoned that the butadiyne was sterically protected by the capping trityl units, therefore we were surprised that the butadiyne reacted under these conditions. In fact, this was the only evidence of reactivity of the butadiyne spacer that we have observed (vide infra).

Frustrated by the low functional group tolerance of the synthetic procedure, we explored alternative methods of introducing the required functionality at a later stage. We designed intermediate **31** that contained the original naphthalene sulfonanilide fluorophore and four anisole groups (Figure 12). A modified sensor **30** could be derived from **31** by appending the recognition elements via electrophilic aromatic substitution. The C3 positions of the anisole groups are the most reactive toward electrophiles since the sulfonanilide deactivates the remaining aromatic moieties. Furthermore, the butadiyne spacer should be unreactive toward electrophiles, as it is sterically encumbered by the trityl groups. Importantly, all of the sensitive functionality was installed after construction of the sensor core. The methodology was particularly attractive because the wide range of electrophilic aromatic substitution

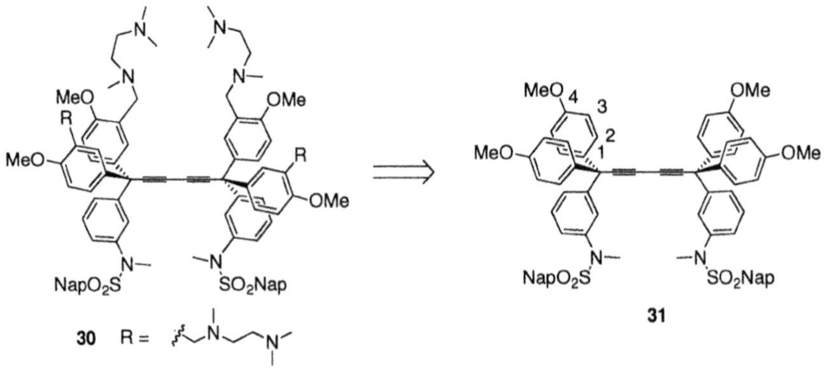

SCHEME 12

chemistry would lend itself to a general strategy for incorporation of recognition elements. Indeed, by removal of the sulfonamides at the appropriate point in the synthesis, a variety of fluorophores could be attached. Thus, the synthesis is rather divergent – by generating a large quantity of intermediate **31**, one could make and test a variety of recognition elements and fluorescent reporters in a very efficient manner.

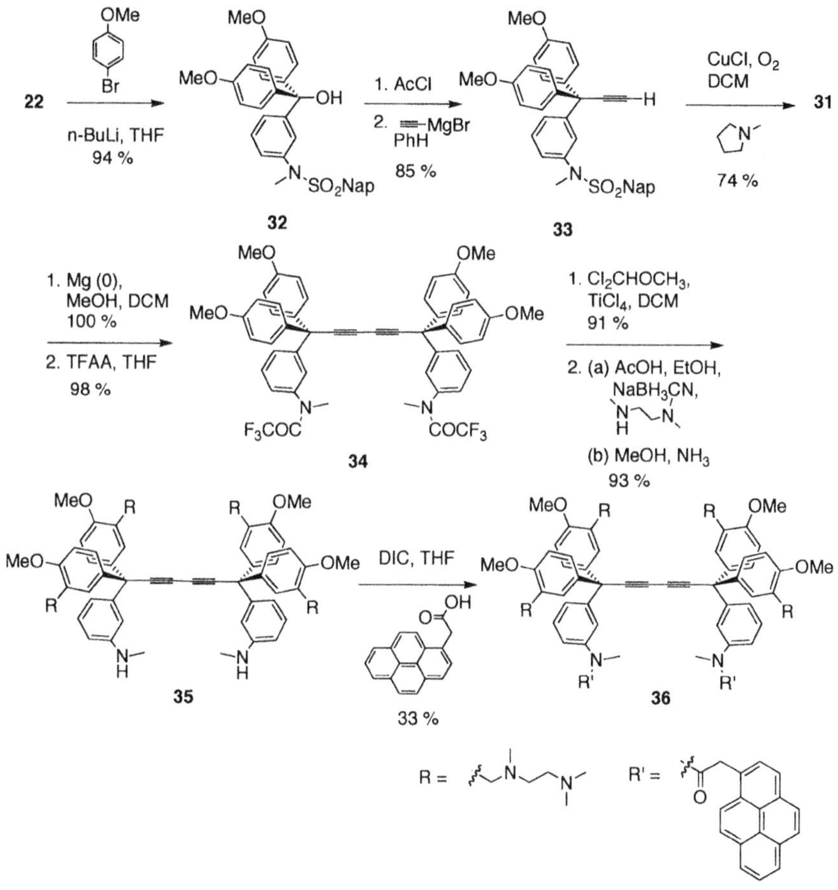

SCHEME 13

Intermediate **31** was synthesized as shown in Scheme 13. Trityl alcohol **32** was prepared from ester **22**[22] and was smoothly converted to trityl acetylene **33**. Glaser coupling of compound **33** gave the tetra-anisole intermediate **31** in good yield. We found that the sulfonanilide group could not be removed at a late stage, so compound **31** was deprotected under mild reductive conditions. The resulting anilines were reprotected as the trifluoroacetanilides to produce compound **34** in good overall yield. Reprotection was necessary in order to electronically deactivate the aniline rings prior to the electrophilic substitution that follows. The recognition elements were installed by formylation and reductive amination with trimethyl ethylene diamine. Compound **35** was produced following deprotection of the trifluoroacetanilides. The high yield of the tetraformylation validated our assumptions about the relative reactivity of the anisole rings versus the amide functionalized rings and the bis-acetyelene. At this stage, any number of amine reactive fluorescent groups could be appended.[23] For example, pyrene acetic acid was coupled to compound **35** in order to yield sensor **36**.

Several sensors with alternate fluorophores were prepared and tested using this method, allowing optimization of the read-out.[24] In fact, good fluorescence properties as well as cooperative binding of metal ions were observed in these systems. Therefore, this sensor design was utilized to prepare sensors for other analytes.

VI. Applications to a Dicarboxylate Sensor

Given the success of the fluorescent read-out in the second generation cooperative sensor, attention was again turned toward organic analytes. Thus, a cooperative dicarboxylate sensor (**37**, Scheme 14) was designed that contained four guanidinium groups such that one dicarboxylate would be bound between each pair of guanidinium groups. Two anthracene sulfonanilide groups would serve as the fluorescent read-out. The bis-acetylene spacer of our previous system was not sufficiently long to accommodate dicarboxylate guests, so the phenyldiyne was necessary for this application.[25]

This system was used to demonstrate that organic analytes could be targeted in the same way as the metal ions. Furthermore, this system was used to demonstrate that cooperativity contributed to selectivity as well as affinity. Sensors such as **37** are expected to be quite selective for dicarboxylates over monocarboxylates. Monocarboxylates that bind to only one guanidinium of the sensor (e.g. acetate) would not produce a fluorescent response since the response is coupled to the cooperative binding event rather than to binding to one of the recognition elements. Therefore, to the extent that monocarboxylates bind to the sensor, they should not turn the sensor "ON." Furthermore, guests that do not bridge

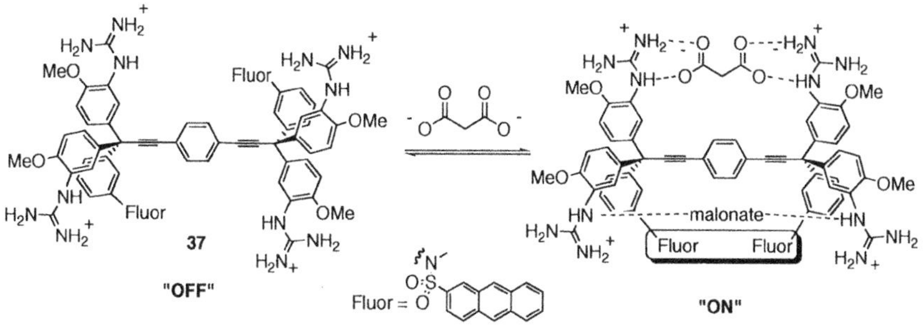

SCHEME 14

a pair of guanidinium groups would not enjoy the cooperative binding interaction since they cannot freeze the rotational freedom of the trityl system. Thus, cooperatively bound guests would have a higher affinity than the non-cooperatively bound guests. Because of this effect, high background concentrations of monocarboxylates should not interfere with the ability of the sensor to recognize the intended guest, a dicarboxylate. For these reasons it was anticipated that sensors based on this design would be very selective for their target analytes over other potential (monofunctional) contaminants.

The synthesis of compound **37** is shown in Scheme 15. The naphthalene sulfonanilide of compound **33** was exchanged with a trifluoroacetanilide to give compound **38**. This trityl acetylene was coupled to 1,4-diiodobenzene under Sonogashira conditions[26] to give the complete skeleton **39** in excellent yield. Nitration of **39** with ammonium nitrate and trifluoroacetic anhydride gave the desired tetranitro compound with no trace of undesired regioisomers. Reduction of the nitro groups with Sn(II) gave tetra-amine **40**. The guanidine recognition elements were then installed by reaction with bis-BOC thiourea in the presence of mercury (II) chloride.[27] Removal of the trifluoroacetanilides gave the protected tetra-guanidine **41**. Addition of the anthracenesulfonyl chloride appended the fluorophores and deprotection of the BOC groups gave the final sensor **37**. The nitration/reduction sequence used here to install the guanidinium groups demonstrated the versatility and generality of the synthetic method. Even the phenyl group of the spacer was unreactive toward the electrophilic substitution conditions. These results indicated that a wide variety of recognition elements (and therefore sensors) could be constructed in this fashion from intermediates common to the synthesis already described.

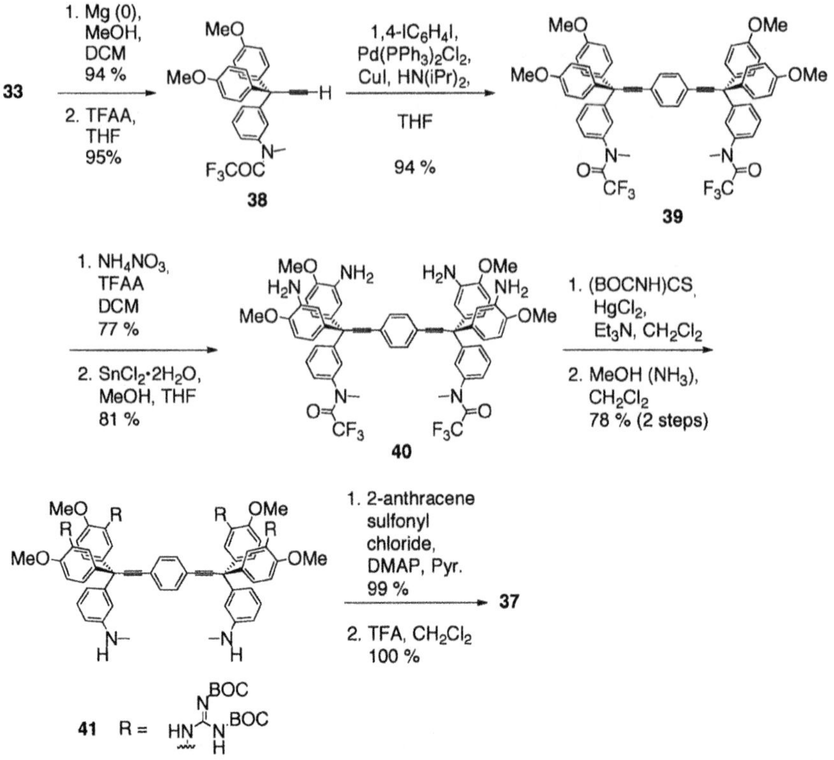

SCHEME 15

This sensor also performed well, giving excellent binding constants for the recognition of several dicarboxylate guests. Furthermore, the sensor gave no response to the presence of monocarboxylates guests to the extent that the solution could be buffered with acetate without consequence.

VI. Applications to a Carbohydrate Sensor

One important area of current research is on the recognition of carbohydrates.[28] In particular, cell surface carbohydrates have implications in various disease states. In some forms of cancer, closely packed cell surface carbohydrates effect cancer progression and invasiveness.[29] Many of these carbohydrates have terminal sialic acid groups. The pinwheel receptor 42 was designed as a divalent receptor for anionic sugars including sialic acid with a view toward making sensors for cancer markers. Sensor 42 used a boronic acid to bind a diol on the

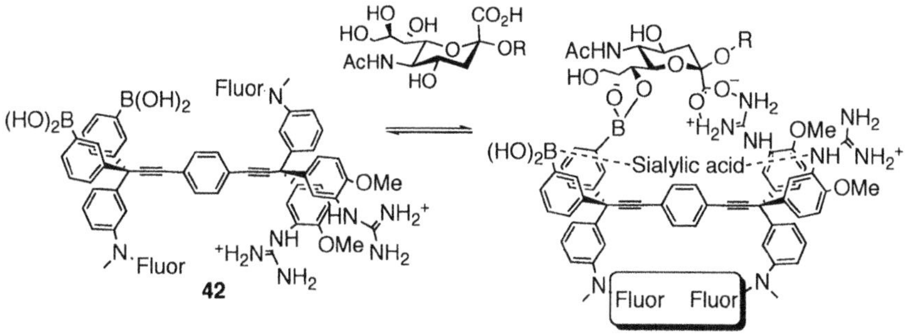

SCHEME 16

sugar and a guanidinium group for the carboxylate (Scheme 16). We anticipated that this receptor would exhibit cooperative binding of two sialic acids. Moreover, **42** would also take advantage of the multivalent effect. Because **42** was designed with two binding pockets, it could bind two sugars nearby on the cell surface. Since the target carbohydrates were known to be present on the cell surface in high concentration, a divalent receptor such as **42** should bind very tightly. Therefore, it was anticipated that the effects of cooperative recognition and multivalent recognition would combine to make sensors like **42** effective reagents for recognition of cell surface carbohydrates.[30]

The synthesis of **42** was complicated by the boronic acid binding groups. The guanidinium and fluorophore substitution followed the established procedures. However, boronic acids and their esters were known to be somewhat sensitive functionalities and would be best installed at the end of the synthesis. The typical method for preparing a phenyl boronic acid was to generate an phenyl anion and quench with a trialkyl boronate. However, strong anion chemistry was not possible in the late stages of the sensor synthesis. Alternatively, palladium catalyzed coupling of an aryl iodide with a diboron reagent would produce a phenylboronic ester. However, the aryl iodide could not be incorporated early on without complicating the Sonogashira chemistry. So, to circumvent issues of functional group compatibility, the trimethylsilyl group was used as a stable iodine precursor.

Scheme 17 depicts the synthesis of the first generation carbohydrate sensor. The trimethylsilyl appended trityl alcohol **43** was prepared by the standard method. Conversion of **43** to the trityl acetylene and mono-Sonogashira coupling gave the first half of the receptor (**45**). Trityl acetylene **47** was constructed in the standard fashion from trityl alcohol

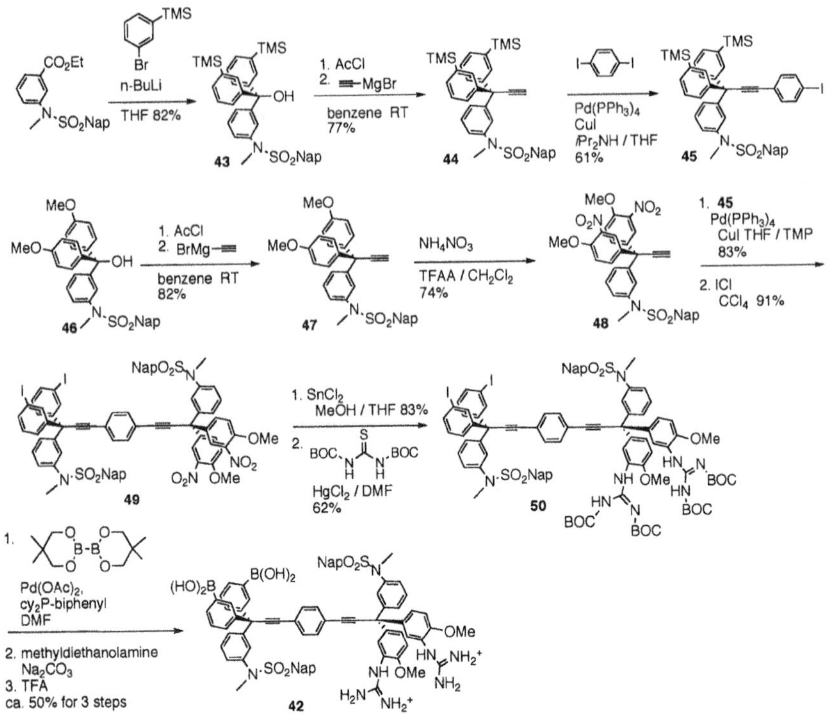

SCHEME 17

46. Nitration gave dinitroacetylene **48**, which was coupled to iodophenyl **45** followed by ipso iodination of the TMS groups to give the full skeleton **49**. It is interesting to note that the order of steps here is essential. Nitration cannot be performed in the presence of the TMS groups, and conversion of the TMS to the iodide cannot be performed in the presence of the aniline functionality. These restrictions required a chemoselective reduction of the nitro groups in the presence the iodides. This reduction, followed by formation of the protected guanidines, gave **50.** Palladium-mediated installation of the boronic esters followed by deprotection of the neo-pentyl glycols[31] and the BOC groups gave guanidinium boronic acid **42**, which was purified by reverse phase HPLC. By exchanging the sulfonamide fluorophores with trifluoroacetamides in the early stages, a number of fluorophores could be installed using this methodology as well. Studies of sensor **42** and related structures are underway in our laboratories.

VI. Conclusions

The pinwheel sensor system has proven to be an effective way of generating selective, high affinity fluorescent sensors for a variety of analytes. In fact, the main challenge with these systems was their synthesis. We have developed a general sensor framework that gives strong, selective, cooperative recognition that is coupled to a fluorescent read-out. We have developed a divergent synthesis of this class of compounds in which a number of different sensors with different recognition elements and fluorophores can be generated from common intermediates. The method is suitable to fully non-symmetrical sensors for complex analytes.

Acknowledgments

This work was supported by the National Institutes of Health.

References and Footnotes

1. Hall, E. E. H. *Biosensors*; Prentice-Hall: Englewood Cliffs, 1991.
2. (a) *Chemosensors of Ion and Molecule Recognition.* Desvergne, J. P.; Czarnik, A. W., Eds.; NATO ASI Series C: 492; Kluwer: New York, 1997. (b) de Silva, A. P., Gunaratne, H. Q. N., Gunnlaugsson T., Huxley, A. J. M., McCoy, C. P., Radermacher, J. T., Rice, T. E. *Chem. Rev.* **1997**, *97*, 1515. (c) Czarnik, A. W. *Chem. Biol.* **1995**, *2*, 423. (d) *Fluorescent Chemosensors for Ion and Molecule Recognition.* Czarnik, A. W., Ed.; ACS Symp. Ser. 538; ACS: Washington DC, 1993.
3. Glass, T. E., *J. Am. Chem. Soc.* **2000**, *122*, 4522-4523.
4. Kelly, T. R., Bowyer, M. C., Bhaskar, K. V., Bebbington, D., Garcia, A., Lang, F., Kim, M. H., Jette, M. P., *J. Am. Chem. Soc.* **1994**, *116*, 3657.
5. (a) Wipf, P., Venkatraman, S., Miller, C. P., Geib, S. J. *Angew. Chem. Int. Ed. Engl.* **1994**, *33*, 1516. (b) Blanc, S., Yakirevitch, P., Leize, E., Meyer, M., Libman, J., Van Dorsselaer, A., Albrecht-Gary, A.-M., Shanzer, A., *J. Am. Chem. Soc.* **1997**, *119*, 4934. (c) Takeuchi, M., Shioya, T., Swager, T. M., *Angew. Chem. Int. Ed.* **2001**, *40*, 3372.
6. Rebek, J., Costello, T., Marshall, L., Wattley, R., Gadwood, R. C., Onan, K. *J. Am. Chem. Soc.* **1985**, *107*, 7481.
7. Shinkai S, Ikeda M, Sugasaki A, Takeuchi M., *Acc. Chem. Res.* **2001**, *34*, 494and Takeuchi, M., Ikeda, M., Sugasaki, A., Shinkai, S., *Acc. Chem. Res.* **2001**, *34*, 865.
8. (a) Marquis, K., Desvergne, J.-P., Bouas-Laurent, H., *J. Org. Chem.* **1995**, *60*, 7984. (b) Chen., C.-T., Huang, W.-P., *J. Am. Chem. Soc.* **2002**, *124*, 6246.
9. (a) Williams, D. H., Westwell, M. S., *Chem. Soc. Rev.* **1998**, 57. (b) Williams, D. H., Searle, M., Mackay, J. P., Gerhard, U., Maplestone, R. A., *Proc. Natl. Acad. Sci. USA* **1993**, *90*, 1172.
10. Yang, W., He, H., Drueckhammer, D. G., *Angew. Chem. Int. Ed. Engl.* **2001**, *40*,

1714.

11. Wieland, H., Kloss, H., *Justus Liebigs Ann. Chem.* **1929**, *470*, 202.

12. Shi, M., Okamoto, Y., Takamuku, S., *J. Chem. Soc., Perkin Trans. 1* **1991**, 2391.

13. Brettle, R.; Hilton, N. A.; Shibi, S. M., *J. Chem. Res. Miniprint* **1984**, *12*, 3712.

14. (a) Guram, A. S.; Rennels, R. A.; Buchwald, S. L., *Angew. Chem., Int. Ed. Engl.* **1995**, *34*, 1348. (b) Driver, M. S.; Hartwig, J. F., *J. Am. Chem. Soc.* **1996**, *118*, 7217.

15. Oyler, R. E.; Ketz, B. E.; Glass, T. E., *Tetrahedron Lett.* **2000**, *41*, 8247.

16. ten Hoeve, W., Druse, C. G., Luteyn, J. M., Thiecke, J., Wynberg, H., *J. Org. Chem.* **1993**, *58*, 5101.

17. (a) Acosta, K.; Cessac, J. W.; Rao, P. N.; Kim, H. K., *J. Chem. Soc., Chem. Commun.* **1994**, 1985. (b) Olofson, R. A.; Schnur, R. C.; Bunes, L.; Pepe, J. P., *Tetrahedron Lett.* **1977**, *18*, 1567.

18. Rieger, B., Abu-Surrah, A. S., Fawzi, R., Steiman, M., *J. Organomet. Chem.* **1995**, *497*, 73.

19. Moran, R., unpublished results.

20. Birks, J. B. *Photophysics of Aromatic Molecules*; John Wiley: New York, 1970.

21. Raker, J., Glass, T. E., *Tetrahedron* **2001**, *57*, 6505.

22. Ramage, R., Wahl, F. O., *Tetrahedron. Lett.* **1993**, *34*, 7133.

23. Haugland, R. P. *Handbook of Fluorescent Probes and Research Chemicals*, 6th ed.; Molecular Probes, Inc.: Eugene, OR, 1996.

24. Raker, J., Glass, T. E., *J. Org. Chem.* **2001**, *66*, 6505.

25. Dominguez, Z., Dang, H., Strouse, M. J., Garcia-Garibay, M. A., *J. Am. Chem. Soc.* **2002**, *124*, 2398.

26. Sonogashira, K., Tohda, Y., Hagihara, N., *Tetrahedron Lett.* **1975**, *50*, 4467,.

27. Kim, K. S., Quian, L. *Tetrahedron Lett.* **1993**, 34, 7677.

28. (a) Davis, A. P., Wareham, R. S. *Angew. Chem. Int. Ed.* **1999**, *38*, 2978. (b) James, T. D., Sandanayake, S., Shinkai, S., *Angew. Chem. Int. Ed.* **1996**, *35*, 1910.

29. (a) Hakomori, S., *Adv. Can. Res.* **1989**, *52*, 257. (b) Hounsell, E. F., Young, M., Davies, M. J. *Clinical Science* **1997**, *93*, 287. (c) Hakomori, S. *Cancer Res.* **1996**, *56*, 5309.

30. (a) Yang, W., Gao, S., Gao, X., Karnati, V., Ni, W., Wang, B., Hooks, B., Carson, J., Weston, B., *Bioorg. Med. Chem. Lett.* **2002**, *12*, 2175. (b) Yang, W., Fan, H., Gao, X., Gao, S., Karnati, V. V. R., Ni, W., Hooks, W. B., Carson, J., Weston, B., Wang, B. *Chem. Biol.* **2004**, *11*, 439.

31. Takeuchi, M., Imado, T, Shinkai, S., *Bull. Chem. Soc. Jpn.* **1998**, *71*, 1117.

STRATEGIES AND TACTICS IN ORGANIC SYNTHESIS, VOL. 6

Chapter 13

FUNCTIONALIZATION OF PYRIDINES AND THIAZOLES VIA THE HALOGEN-DANCE REACTION, APPLICATION TO THE TOTAL SYNTHESIS OF CAERULOMYCIN C AND WS75624 B

Tarek Sammakia, Eric L. Stangeland, and Mark C. Whitcomb
Department of Chemistry
Universityof Colorado
Boulder, Colorado 80309

I. Introduction

This article describes strategies for the use of the halogen-dance reaction in the functionalization of aromatic heterocycles, and the synthesis of two pyridine-derived natural products, caerulomycin C and WS75624 B (Figure 1).[1,2] The origins of this research lie in the

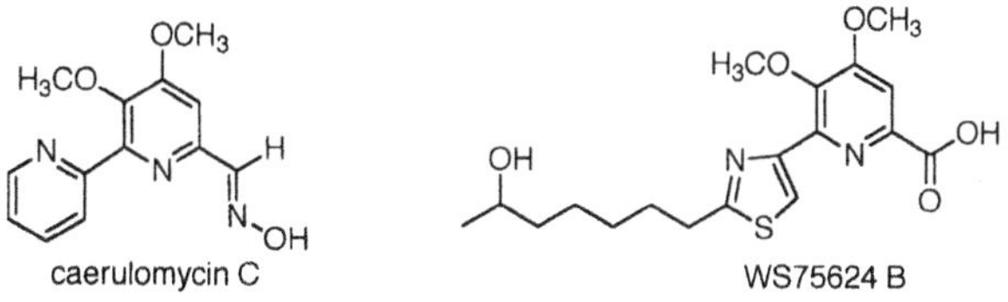

FIGURE 1. The structures of caerulomycin C and WS75624 B.

observation shown in Scheme 1. While a graduate student in the laboratories of Professor Stuart Schreiber at Yale, one of the authors attempted the reduction of β-keto ester **1** with sodium borohydride in 95% ethanol. As expected, the ketone was reduced to the alcohol, but what was not expected was that the ester underwent hydrolysis to the carboxylic acid (**2**).[3] This result, while unwelcomed, was nevertheless

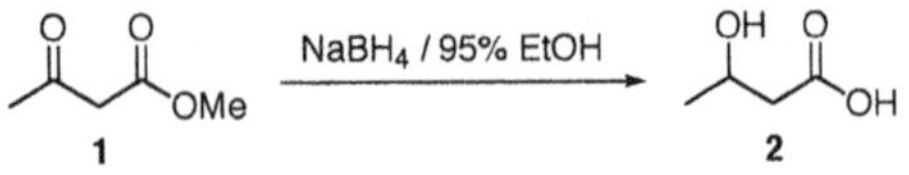

SCHEME 1

interesting, and a search of the literature revealed that ester hydrolysis can be promoted by borate salts.[4] Thus, Okuyama and coworkers have shown that α-hydroxythioesters (e.g., **3**, Scheme 2) undergo borate-mediated hydrolysis, and detailed mechanistic studies suggest that the reaction is directed by the α-hydroxyl group, such that the borate moiety delivers hydroxide via intermediate **4**.

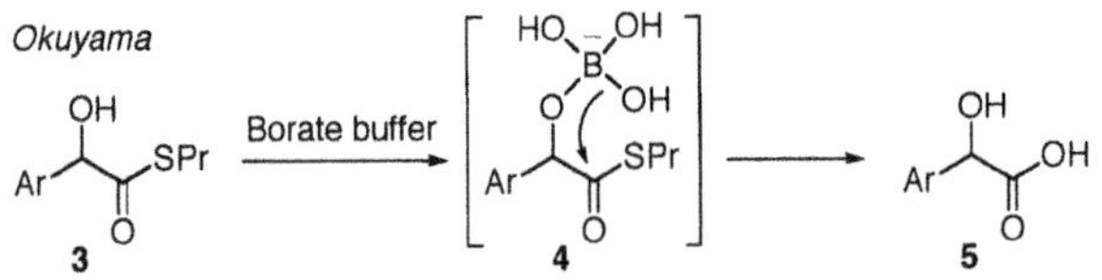

SCHEME 2

Subsequently, at the University of Colorado, our group wished to design catalytic systems that took advantage of this type of reactivity. A graduate student, T. Brian Hurley, prepared complexes bearing borate esters in proximity to amines, and studied their use as hydroxyl-directed acyl transfer catalysts. The amine was to serve two roles: it was to facilitate the exchange of the hydroxyl group of the substrate onto the boron,[5] and act as a general base for the deprotonation of any water bound to the boron, and thereby promote attack of hydroxide onto the carbonyl. Unfortunately, these species did not have the desired catalytic activity. However, a related class of compounds, the 2-acyl-4-amino pyridines, did (**8**, Scheme 3). These compounds display remarkable selectivity for the methanolysis of α- and β-hydroxy esters over the corresponding alkoxy esters (e.g., **6** vs. **7**) with rate differences of over 1,700:1.[6]

R	Relative rates
HO	1,700
MeO	<1*

*No methanolysis detected
after 1 week after

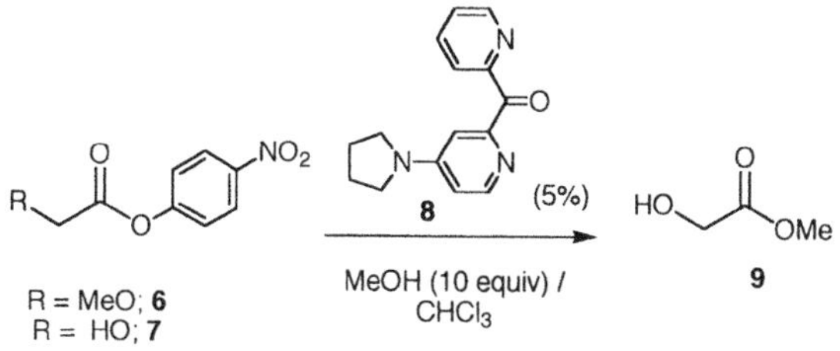

R = MeO; **6**
R = HO; **7**

(5%)

MeOH (10 equiv) /
CHCl$_3$

8

9

SCHEME 3

We have extensively studied the mechanism of this reaction, and have good evidence that the reaction proceeds by the mechanism shown in Scheme 4, wherein the carbonyl serves as a binding site for the hydroxyl group of the substrate, and provides a latent nucleophile in the form of a hemiacetal hydroxyl group (vide infra).[7] Thus, addition of the hydroxyl group of the substrate to the ketone of the catalyst provide a hemiacetal (**11**), the hydroxyl group of which adds to the active ester in a step promoted by the proximal basic nitrogen to provide the unusual dioxalanone **12**. Methanolysis of **12**, again with assistance from the proximal basic nitrogen, provides the product bound to the catalyst as a hemiacetal (**13**). In the final step, the product (**9**) is released and the catalyst (**10**) is regenerated.

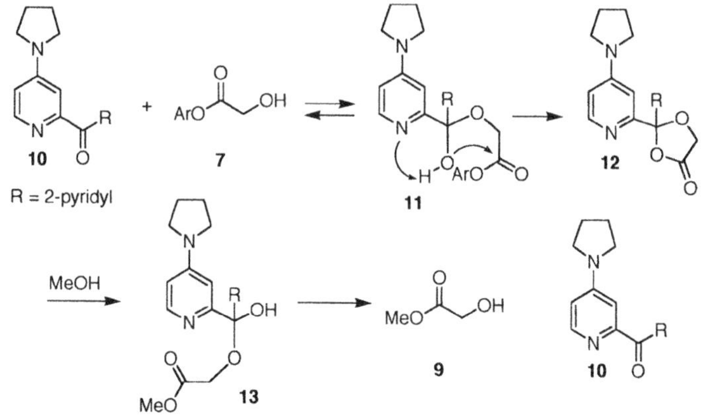

10   +   **7**   **11**   **12**

R = 2-pyridyl

MeOH

13   **9**   **10**

SCHEME 4

II. The Functionalization of Pyridines – The Halogen-Dance Reaction

In the course of this research, we prepared over a dozen 2-acyl-4-

aminopyridine catalysts (such as **10**), and we required reliable methods for functionalization of the various sites around the pyridine ring. We found that an interesting reaction, the halogen-dance reaction,[8] serves this purpose well, and we set out to study its use in the synthesis of the pyridine-derived natural products, caerulomycin C[9] and WS75624 B.[10,11]

Pyrdines bear an obvious resemblance to benzenes, but they pose very different problems in their functionalization. While both classes of compounds are aromatic, the π-system of pyridines is significantly less electron rich, and electrophilic aromatic substitution reactions are of limited utility on pyridine derivatives. This is not only due to the electronics of the pyridine ring, it is also because the presence of the nucleophilic / basic nitrogen can interfere with the typical reagents used to activate electrophiles for electrophilic aromatic substitution. Thus, strong Brönsted acids will protonate the nitrogen, and Lewis acidic additives will either directly interact with the nitrogen and attenuate the reactivity of both reagents, or produce an activated complex that will interact with the nitrogen in preference to the π-system of the pyridine (Scheme 5). These interactions further deactivate the system, and in general, electrophilic aromatic substitution on pyridines requires harsh conditions.[12]

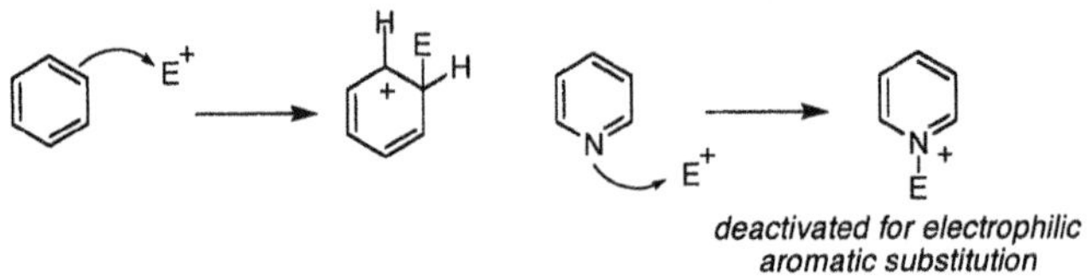

deactivated for electrophilic
aromatic substitution

SCHEME 5

In contrast, pyridine rings are well suited for functionalization by base-promoted processes. Their electron deficient nature renders them intrinsically more acidic than similar benzene ring systems, and in the presence of even moderate directed metalation groups (DMG's), pyridine C-H bonds can be deprotonated by lithium amide bases.[13] This facile deprotonation is the basis for the halogen-dance reaction. This reaction was first observed by Wotiz and Huba in 1959 in the attempted formation of benzyne **15** by treatment of **14** with NaNH$_2$ (Scheme 6).[14] In addition to the expected benzyne-derived product **16**, the side product **17**, in which the halogen at C-1 has migrated by one carbon, was produced in 33% yield. Since this initial report, others, including Bunnett and Quéguiner, have studied the mechanism and synthetic utility of this reaction, and it has been developed into a powerful and reliable process with suitably designed systems.[8]

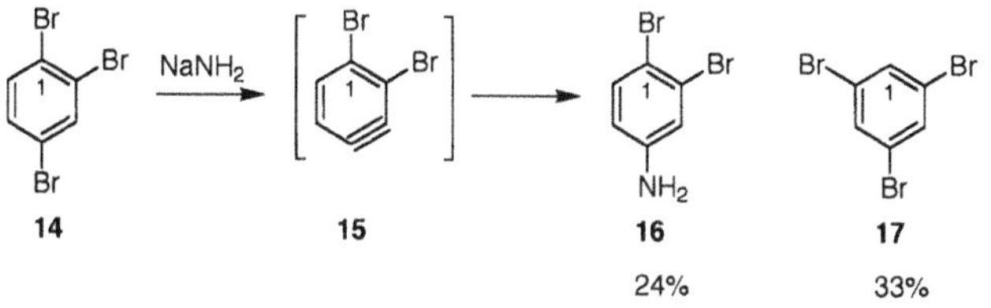

SCHEME 6

A typical halogen-dance reaction from the work of Quéguiner is shown in Scheme 7. In this example, treatment of 2-fluoro-3-iodopyridine (**18**) with LDA and methyl iodide provides pyridine **19**, in which the iodine has migrated to the 4-position and a methyl group has been incorporated at the 3-position.   The   mechanism   of   related   halogen-dance

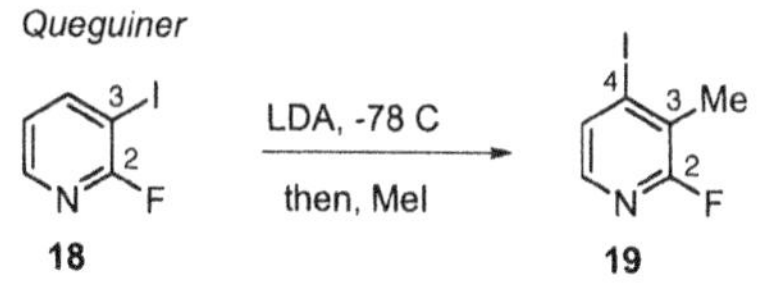

SCHEME 7

reactions has been studied by Bunnett and by Quéguiner and a positive halogen transfer mechanism via the polar chain process shown in Scheme 8 has been proposed by both groups.[15] The first step of this mechanism involves a halogen-directed lithiation of the 4-position of the substrate (**20**) to provide **21**. In an initiation step, **21** undergoes a metal-halogen exchange with **20** to provide the dihalo species **22** and the 3-lithio species **23**, which does not react further. The reaction is then propagated by a halogen-metal exchange between compounds **22** and **21**, which provides **24** and regenerates **22**. Compound **22** thereby acts as the carrier in this chain process. Treatment of **24** with an electrophile, provides compound **25** in which the halogen has migrated by one carbon (a 1,2-shift) and the electrophile is in the position originally occupied by the halogen. The driving force for this reaction is the greater stability of the anion in compound **24**, which is stabilized by two DMG's as opposed to only one in compound **21**.

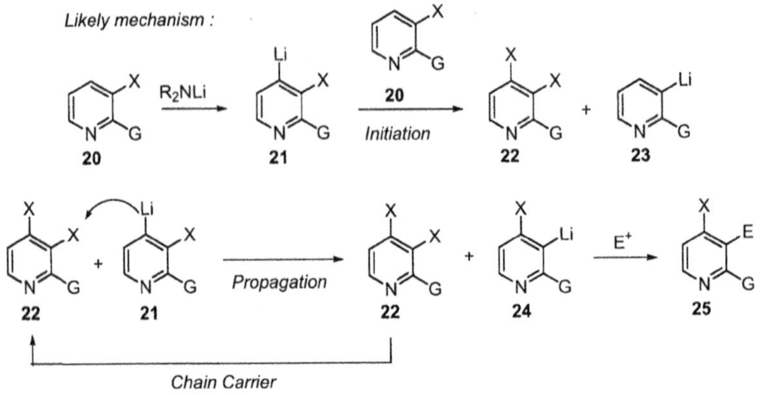

Likely mechanism :

SCHEME 8

This reaction involves a series of lithium-halogen exchange steps, and as such, it is not intrinsically limited to 1,2-migrations. Thus, if the initial lithiation were to take place at a site more distant to the halogen, a 1,3- or 1,4- migration could occur (Scheme 9).[16] Because the reaction is driven by the greater stability of the anion in the product than in the initial metalation, the site where the anion ultimately resides must either be intrinsically more acidic than other sites on the ring, or be rendered more

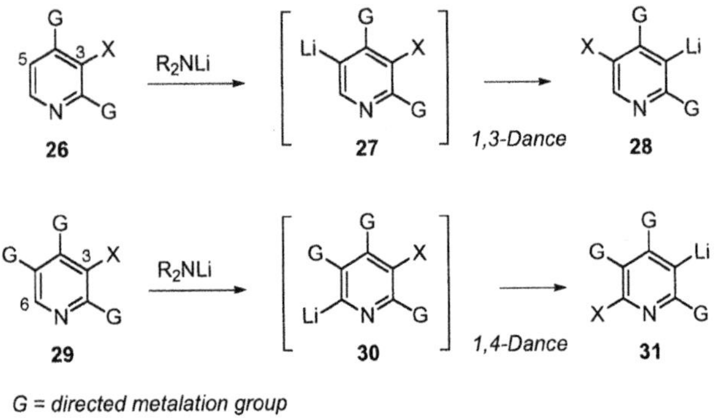

G = directed metalation group

SCHEME 9

acidic by the presence of anion stabilizing groups. In the latter case, these groups must either be incorporated into the molecule, or be part of the target structure. Thus, molecules which have differing sites of acidity are potential substrates for the halogen-dance reaction. Strategically, this can be considered a two-step process in which a halogen is first introduced into a molecule at an acidic and easily accessed site, and then rearranged to a less acidic and less accessible site. However, in order to initiate the reaction, the less accessible site must still be acidic enough to undergo deprotonation with a lithium amide base (alkyl lithium reagents cannot be used as they undergo lithium-halogen exchange either competitively with, or in preference to, deprotonation of the pyridine). In the prototypical examples shown in Scheme 9, the 3-position of compounds **26** and **29** is flanked by two directed metalation groups. In contrast, 5- and 6-positions of these compound are flanked by only one directed metalation group, thereby rendering compounds **28** and **31** more stable than **27** and **30**, respectively.

III. Retrosynthetic analysis

With these considerations in mind, we chose to study the use of the halogen-dance reaction in the synthesis of the two pyridine-derived natural products, caerulomycin C and WS75624 B (Scheme 7). These compounds are potent endothelin converting enzyme (ECE) inhibitors.[17] ECE catalyses the proteolysis of the 38 amino acid peptide big endothelin 1 (big ET-1) to the 21 amino acid peptide endothelin 1 (ET-1). Both big ET-1 and ET-1 are vasoconstrictors, however ET-1 is about 100 times more active than big ET-1, and as such, inhibitors of ECE are potential antihypertensive agents.[18] Caerulomycin C and WS75624 B have similar architecture about the pyridine rings, and we wished to prepare them via a common synthetic strategy. A retrosynthetic analysis of these compounds that is consistent with the requirements of the halogen-dance reaction is provided in Scheme 10. We first disconnected the heterocyles at the 6-position of the pyridines, and planned to introduce them late in the syntheses via a metal-catalyzed cross-coupling reaction. We also substituted the carbonyl derivatives at the 2-position with a commonly employed DMG group, a diisopropyl amide, and arrived at the 6-halo derivative **34**. Disconnection of the halogen from compound **34** provides compound **35**, which contains sites of differing anion stability; the 3-position, which is stabilized by two DMG groups, and the 6-position which is only stabilized by one. As such, this compound is well suited for functionalization by the halogen-dance reaction. In the forward direction, incorporation of the halogen at the 3-position would be accomplished by lithiation and trapping, and would be followed by a 1,4-halogen-dance reaction to migrate the halogen to the 6-position. Synthesis of **35** was envisioned to occur via the corresponding diiodo derivative **36**, which would be subjected either to nucleophilic aromatic substitution or Ullmann coupling in order to install the methoxy groups. Compound **36**, in turn, could be prepared from **37** via a 1,3-halogen-dance reaction, as **37** also contains sites of differential anion stability; the 5-position is ortho to one DMG group, while the 3-position is flanked by two. Synthesis of **37** could be readily accomplished from compound **38** by metalation and trapping with iodine. Finally, compound **38** is known and has been prepared from picolinic acid as described in Scheme 11.

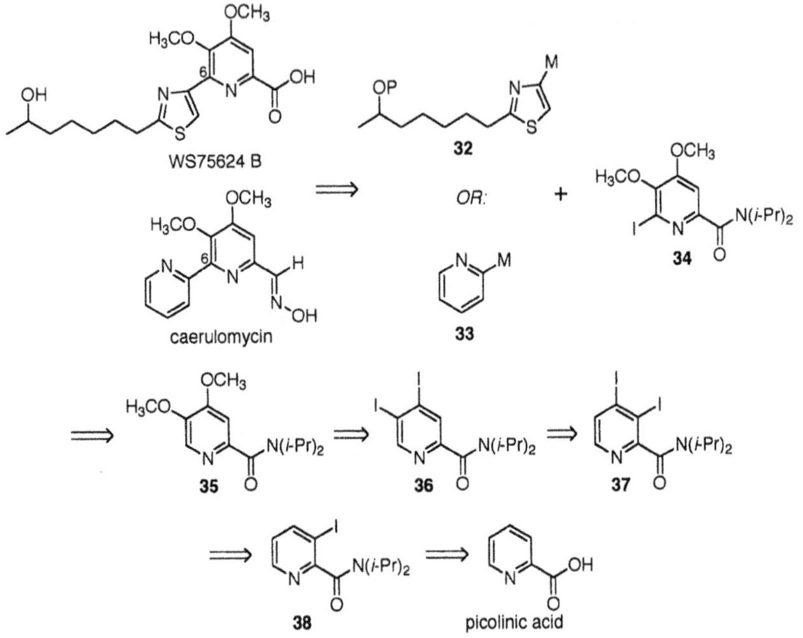

SCHEME 10

IV. Early Efforts, Limitations to the 1,3-Halogen-Dance Reaction

Our synthesis began with the diisopropyl amide of picolinic acid (**39**, Scheme 11). Introduction of an iodine at the 3-position was accomplished in 80% yield by lithiation with LDA and trapping with iodine. A second metalation with LDA and trapping with iodine provided the 3,4-diiodo species **41**, probably via the 3-lithio-4-iodo compound (**40**) that is produced by a 1,2-halogen-dance. Unfortunately, attempted preparation of **36** by treatment of **41** with LDA and quenching with water failed, and exclusively produced the 3,5-diiodo species **43**. We had hoped to induce either a 1,3-migration of the 3-iodo substitutent, or two sequential 1,2-migrations of the 4- and 3-iodo substituents. However, it appears that the 1,2-migration of the 4-iodo compound occurs to produce the 4-lithio species (**42**), which is stabilized by two ortho-iodides. This compound is either kinetically or thermodynamically unreactive, and does not undergo the second halogen-dance reaction.

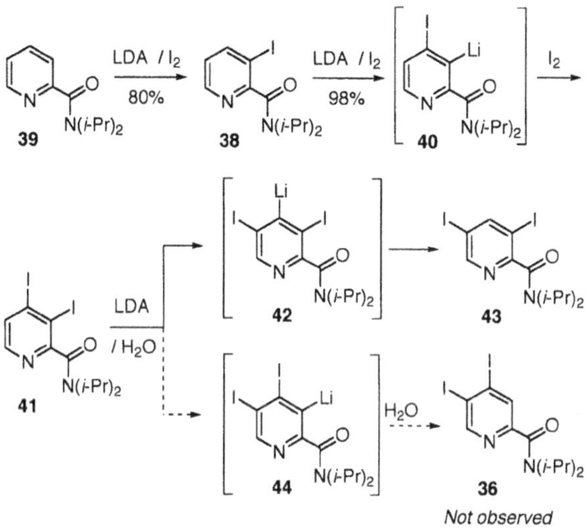

SCHEME 11

We therefore wished to avoid the formation of the 4-lithio species, and examined the use of a non-exchangeable group at the 4- position in place of the iodide. We chose to study the 4-chloropyridine derivative **45** based on the fact that chlorine is known to direct ortho-lithiations at −78 °C and does not undergo halogen-metal exchange. Chloropyridines are also susceptible to nucleophilic aromatic substitution with alkoxides, and as such this compound contains a handle for the introduction of the methoxy group at the 4-position as required later in the synthesis.

Our synthesis began with picolinic acid, which was treated with thionyl chloride in the presence of LiBr according to the procedure of Sundberg[19] to provide the 4-chloro acid chloride, which was then treated with diisopropylamine to provide diisopropyl amide **45** (Scheme 12). Ortho-lithiation of this species with LDA and trapping with iodine provided compound **46**. Unfortunately, this compound also proved to be a poor substrate for the halogen-dance reaction, and provided variable yields of products upon treatment with either LDA or lithium 2,2,6,6-tetramethylpiperidide (LTMP). While some of the desired product (47) was observed, the deiodinated product 45 was always the major product, and at times, it was the sole product. This result suggests that the iodide is undergoing a nucleophilic dehalogenation by LDA or LTMP, and while this is an unusual reaction, it is likely facilitated by the enhanced leaving group ability of the 3-pyridyl anion as this species is stabilized by the proximal amide and chloro groups.

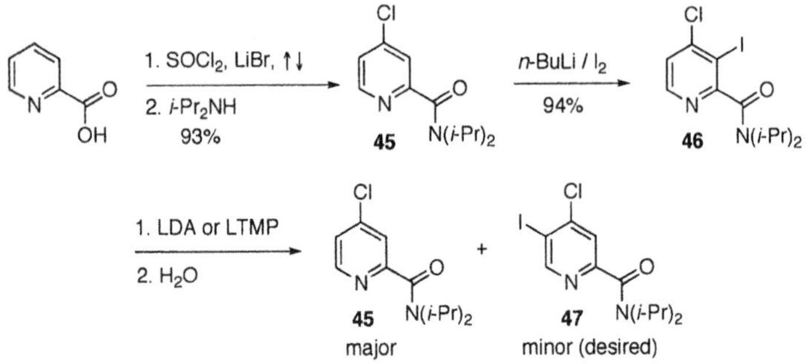

SCHEME 12

V. Successful 1,3- and 1,4-Halogen-Dance Reactions, Synthesis of the Core Pyridine Structure

In an effort to prevent this dehalogenation, we sought to decrease the leaving group ability of the pyridyl species by rendering the 3-pyridyl anion slightly less stable. We reasoned that methoxy groups are less stabilizing than chloro groups and, therefore, studied the 4-methoxy derivative. Our synthesis again proceeded through compound 45, which was subjected to nucleophilic aromatic substitution by treatment with sodium methoxide to provide 48 (Scheme 13). Metalation of this species with n-BuLi and trapping with iodine provided the 3-iodo compound 49 in 75% yield. *Treatment of 49 with LDA smoothly induced a 1,3-halogen-dance reaction to provide 50 in 88% yield.*

With the correct substitution pattern about the pyridine ring established, we wished to introduce the methoxy group at the 5-position. While the 4-methoxy substituent was smoothly introduced by nucleophilic aromatic substitution, this proved unsuccessful at the 5-position.[20] However, a modified Ullmann coupling procedure that

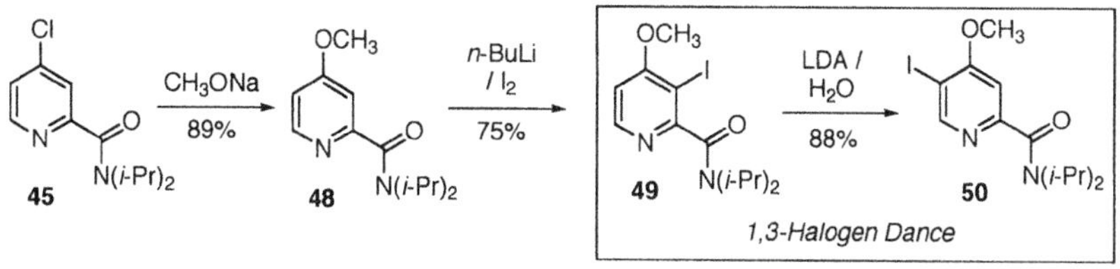

SCHEME 13

consists of treating **50** with sodium methoxide in dimethylformamide containing CuI at 80 °C provided **51** in 92% yield (Scheme 14).[21,22] In the absence of Cu(I) salts, or in less polar solvents, significant quantities of the reduced byproduct **48** are observed. With **51** in hand, we required the installation of our coupling substituent at the 6-position of the pyridine. As described above, our strategy was to take advantage of a 1,4-halogen-dance reaction and first install a halogen at the 3-position, then migrate it to the 6-position. In the event, treatment of **51** with *n*-BuLi and trapping with bromine provided the 3-bromopyridine **52** in 84% yield. *Subsequently, treatment of **52** with LDA smoothly induced a 1,4-halogen-dance, and after an aqueous workup provided the 6-bromo derivative **53** in 80% yield.*

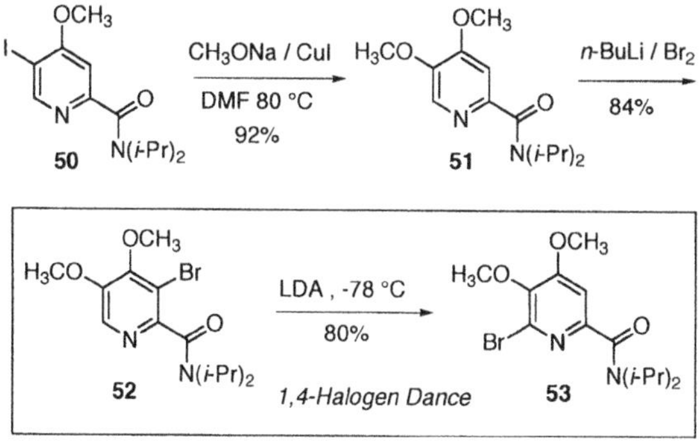

SCHEME 14

VI. Completion of the Synthesis of Caerulomycin C.

Completion of the synthesis of caerulomycin C required a cross-coupling with a suitably functionalized pyridine, followed by functional group manipulations to convert the amide to an oxime. We found the best procedure for this cross-coupling to be a Negishi coupling[23] using

Pd$_2$(dba)$_3$/Ph$_3$P as the catalyst, and obtained an 80% yield of bipyridyl **54** (Scheme 15). Conversion of the amide of **54** to the oxime was accomplished in two steps in 44% overall yield by reduction of the amide to the aldehyde with diisobutylaluminum hydride to provide **55** followed by condensation with hydroxylamine. Material prepared in this way displayed ^{1}H and ^{13}C NMR spectra identical to those reported for caerulomycin C.[1]

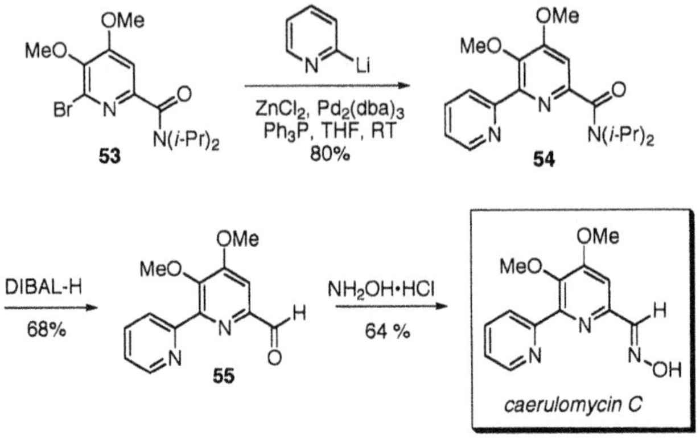

SCHEME 15

VII. Completion of the Synthesis of WS75624 B, Application of the Halogen-Dance Reaction to Thiazoles.

Completion of the synthesis of WS75624 B was more complicated than that of caerulomycin C due to the presence of the substituted thiazole. As previously mentioned, our retrosynthetic analysis involved the disconnection of the thiazole from the pyridine, and required that we synthesize an appropriately functionalized thiazole coupling partner bearing a metal at the 4-position. Our success with the halogen-dance reaction on pyridines prompted us to examine its use in the synthesis of substituted thiazoles.

As described above, the halogen-dance reaction takes advantage of the carbanion stabilizing properties of a directing group to produce a thermodynamically more stable carbanion from an initially formed less stable carbanion. However, the directing group is not a required component of this reaction, and if a substrate has sites that are intrinsically different in acidity, then a thermodynamic driving force exists for the rearrangement of a lithio species. For example, thiophene is more acidic at the 2-position than at the 3-position, and this difference in acidity has been utilized to drive halogen-dance reactions on this heterocycle. Thus, treatment of 2,5-dibromothiophene (56) with LDA followed by methyl iodide provides 2-methyl-3,5-dibromothiophene (59, Scheme 16). [24] This reaction likely proceeds via initial lithiation at the 3-position of the thiophene to provide 57, followed by a halogen-dance reaction to produce 58 wherein the anion resides at the 2-position.

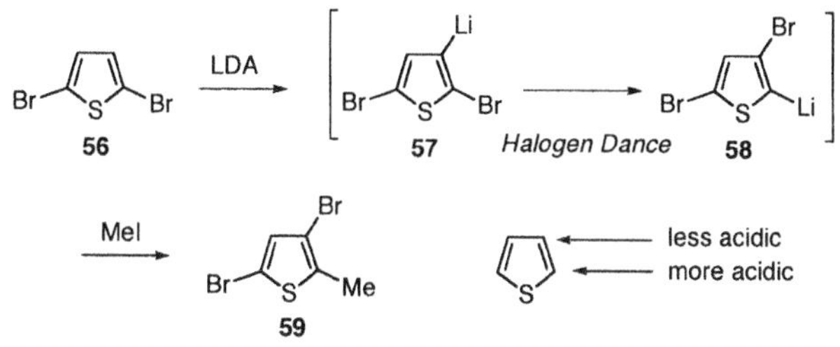

SCHEME 16

Similarly, thiazole contains sites of differential acidity, with the 2-position being the most acidic, the 5-position less acidic, and the 4-position the least acidic (Figure 2). [25] Because we required a substituent at the least acidic position of the thiazole (the 4-position), we wished to force a metalation at this position, and migrate a halogen from either the 2- or 5-position to this site.

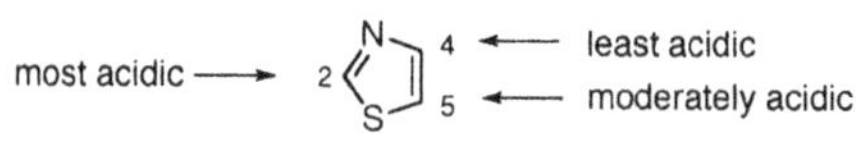

FIGURE 2. Differential acidity of thiazole.

In order to test the viability of the halogen-dance reaction on a simple thiazole substrate, we studied the use of 2-bromothiazole in this reaction (60, Scheme 9). This compound was treated with LDA at −78 °C in ethereal solvents, however, under all conditions examined a dark intractable material was produced. We suspect that this is a result of lithiation at the 5-position to provide 61, followed by migration of the bromine to provide 2-lithio-5-bromothiazole (62). This species can then undergo an α-elimination and open to the unstable isocyanide (63), which decomposes to a complex mixture. Although 2-lithiothiazole is known be stable at −78 °C, we postulate that the electron withdrawing nature of the bromide at the 5-position stabilizes the thiolate of **63**, and facilitates the ring opening.

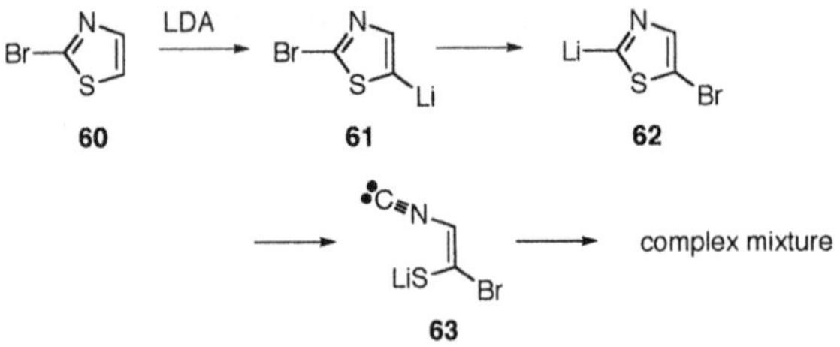

SCHEME 17

We therefore decided to block the 2-position of the thiazole, and examine the halogen-dance reaction between the 4- and 5-positions. We prepared 2-silyl-5-bromothiazole derivatives by treating 2-bromothiazole with *n*-BuLi and trapping with a variety of chlorotrialkylsilanes at –78 °C. The 2-silyl species were then metalated with LDA and quenched with bromine (Scheme 18). Of all the silanes we examined (trimethylsilyl, triethylsilyl, *tert*-butyldimethylsilyl, and triisopropylsilyl), we found that only the triisopropylsilyl (TIPS) derivate **64** underwent the bromination reaction without desilylation. *In the event, 2-triisopropylsilyl-5-bromothiazole (65) was subjected to the halogen-dance conditions by treatment with LDA at –78 °C, and it smoothly rearranged to the 4-bromo derivative 66 in 86% yield.* Confident that the halogen-dance reaction is viable on thaizole derivatives, we returned to examining the use of 2-bromothiazole as a starting material.

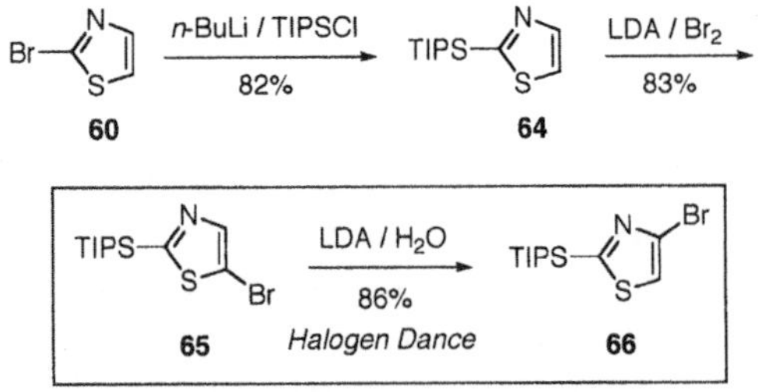

SCHEME 18

Our previous attempts at the use of 2-bromothiazole as a partner in this reaction suffered from the lability of the 2-lithiothiazole produced by the halogen-dance. As previously mentioned, it is likely that this lability is enhanced by the presence of a halogen at the 5-position of the thiazole, and we, therefore, wished to prevent the migration of the bromide to that site. In order to block this position under the reaction conditions, we studied a reaction wherein 2-bromothiazole was treated with 2.2 equivalents of LDA at –78 °C in the presence of one equivalent of TIPSCl. After stirring for 2 hours, methyl iodide was added to the reaction mixture, and after workup, 2-methyl-5-triisopropylsilyl-4-bromothiazole (**67**) was isolated in 82% yield (Scheme 19).

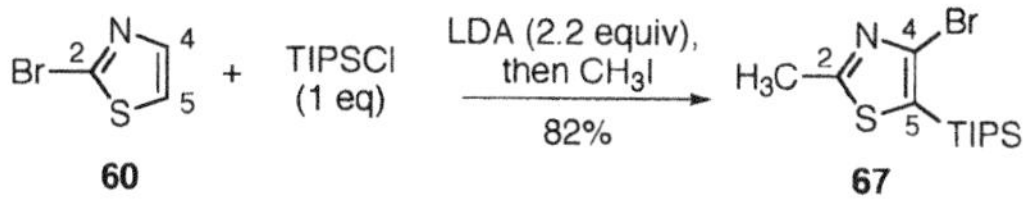

SCHEME 19

This reaction likely proceeds by a mechanism in which LDA first lithiates compound **60** at the 5-position to provide the 5-lithiothiazole **68**. In order for the observed product to be formed, trapping of this species with TIPSCl must be faster than migration of the bromide to the 5-position. Because typical silyl chlorides do not react with LDA at –78°C, we can run this reaction using an internal quench procedure and trap the 5-lithio species with TIPSCl as it is produced to provide the 5-TIPS thiazole intermediate **69**.[26] Lithiation of **69** with the remaining equivalent of LDA then provides the 4-lithio thiazole, **70**. This compound can only undergo the halogen-dance reaction since all the TIPSCl has been

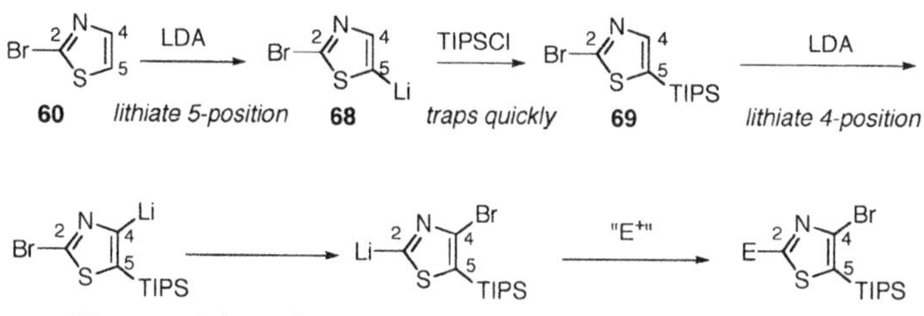

SCHEME 20

consumed in the first lithiation/trapping step, and provides the 2-lithiothiazole derivative **71**. This compound is stable at –78 °C due to the presence of the electron donating TIPS group at the 5-position. Trapping of this species with an electrophile, in this case, methyl iodide, then provides the product (**72**).

This reaction is remarkably clean considering the lability of the intermediates and the complexity of its mechanism, which requires that the initial lithiation and trapping with TIPSCl at the 5-position be faster than other processes, including the second lithiation at the 4-position.[27] However, it suffers from the lack of reactivity of the 2-lithiothiazole. At −78°C, this material is unreactive towards hindered lactones and primary alkyl halides, though it will react with more reactive alkyl halides and aldehydes, and at temperatures above about -70 °C, it is unstable and undergoes decomposition at a rate competitive with alkylation. A survey of various electrophiles is shown in Table 1. In cases that display diminished yields, the balance of the material consists of the protonated product **72** (E = H).

TABLE 1

THIAZOLE SILYLATION / HALOGEN-DANCE / TRAPPING SEQUENCE

"E+"	H₂O	CH₃I	⌇Br	CH₃CH₂I
Yield	85%	82%	20%	<10%

"E+"	OTMS ⌇⌇⌇I	O lactone	O lactone (Me)	OTMS ⌇⌇⌇CHO
Yield	0%	57%	0%	60%

In order to complete the synthesis of WS75624 B, we were faced with the problem of appending the sidechain at the 2-position of the thiazole. Since alkylation of the 2-lithio thiazole intermediate **71** with simple 1° alkyl halides proceeded in low yields, we sought to increase the reactivity of the alkylating agent and rendered the leaving group allylic. The synthesis of our coupling partner proceeded as shown in Scheme 21. Noyori hydrogenation of ethyl levulinate using the (*S*)-BINAP-derived catalyst provided lactone **73** in 91% yield and >98% ee.[28] This compound was converted to enoate **74** in a one pot procedure in which the lactone was reduced with diisobutylaluminum hydride at -78 °C, then subjected to a Horner-Emmons olefination in 63% yield.[29] The secondary alcohol was protected as the TIPS ether, and the ester was reduced with diisobutylaluminum hydride to allylic alcohol **76**. This compound was then activated as the chloride (**77**), bromide (**78**), or tosylate (**79**). Activation as the tosylate using tosyl chloride was problematic due to competing displacement of the tosylate by chloride ion. This could be suppressed using p-toluenesulfonic anhydride, and deprotonating the alcohol with n-BuLi.[30]

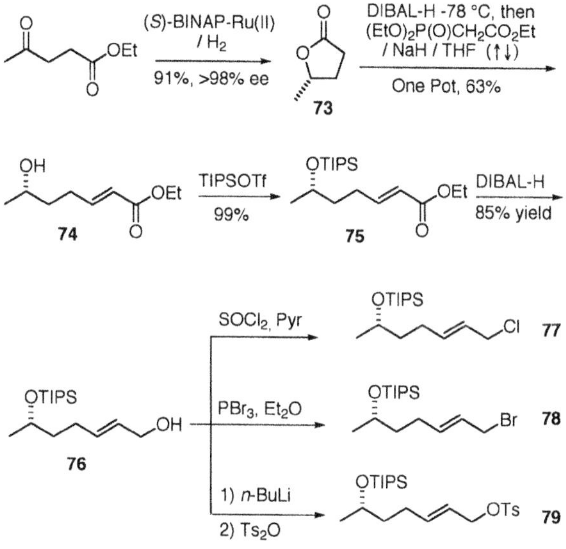

SCHEME 21

With the activated electrophiles in hand, we studied their use in the thiazole halogen dance/alkylation reaction. As expected, the alkylation was sluggish and provided low quantities of the desired product. However, in the presence of 10% CuI, a 40% yield of the coupled product was obtained with allylic tosylate **79** (Scheme 22). The corresponding bromide (**78**) and chloride (**77**) provided diminished yields and significant amounts of the S_N2' coupling product. Diimide reduction of the alkene occurred smoothly, and the TES blocking group on the thiazole was cleanly removed upon stirring in basic methanol at room temperature. In our prior studies, we used a TIPS blocking group on the thiazole, however, we found that removal of the TIPS group was difficult, and was accompanied by decomposition. This is not an issue with the more labile TES protecting group.

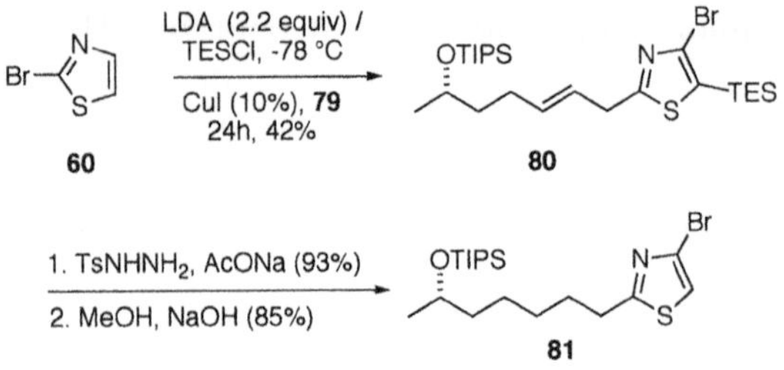

SCHEME 22

Completion of the synthesis required that we install a metal at the 4-position of the thiazole, and towards this end, compound **81** was treated with *t*-BuLi with an in situ quench of tributyltin chloride to provide **82** (Scheme 23). Due to the steric hindrance of the *t*-BuLi, halogen-metal exchange is more facile than attack on the tributyltin, allowing the use of the tin electrophile as an internal quenching agent. If the tributyltin chloride is not present during the halogen-metal exchange, diminished yields of product are observed, presumably due to proton exchange between the 4-lithiothiazole and the more acidic α-methylene. Coupling of the 4-stannyl thiazole **82** with iodopyridine **83** was then studied, and we were pleased to find that this reaction provided the coupled product **84** in 62% yield using Pd(PPh₃)₄ in dimethylacetamide at 60 °C for 36 hours. The completion of the synthesis requires hydrolysis of the amide and removal of the TIPS ether, however, we were unable to accomplish the amide hydrolysis under all conditions examined due to extensive decomposition. This result was not surprising given the steric hindrance of the amide, and we instead resorted to a two-step procedure wherein we reduced the amide with DIBAL to the aldehyde, then subjected the aldehyde to a Lingrin oxidation[31] to the acid. Finally TBAF deprotection of the TIPS ether occurred smoothly to provide (+)-(*S*)-WS75624 B in 67% yield for the three-step sequence. The optical rotation of our material was determined to be $[\alpha]_D$=+3.1 (c=8.0 mg/mL, CH₃OH), while that of the natural product was reported as being $[\alpha]_D$=+3 (c=8.0 mg/mL, CH₃OH). This allows us to assign the stereochemistry of the natural product as the (*S*) configuration at the carbon bearing the secondary alcohol.

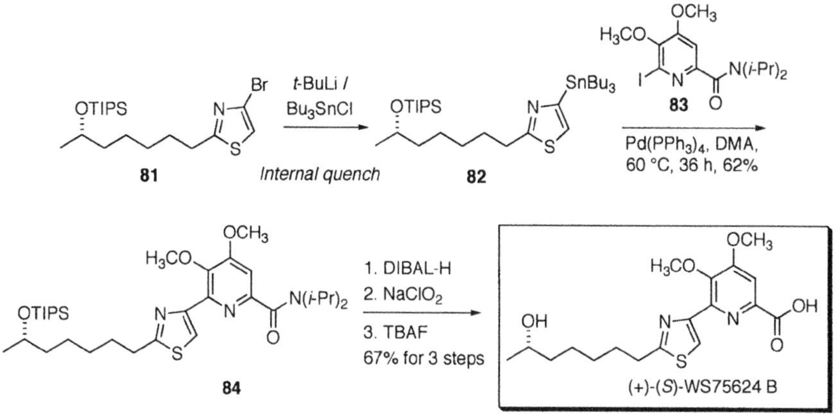

SCHEME 23

VIII. Conclusion.

This work provides another illustration of the use of the halogen-dance reaction in synthesis, and further evidence that it can be a valuable tool for the functionalization of heterocycles. Perhaps the most useful feature of this reaction is that it allows for the introduction of halogens at remote sites in a that are difficult to access by other means. However, this work also illustrates that this reaction can exhibit certain limitations, some of which are rather unexpected and surprising. Ultimately, as the reaction sees more use in synthesis, its scope will be better defined and it will be easier to gauge the success of the reaction in a particular system.

Acknowledgements

We thank the National Institutes of Health (GM48498), and Array Biopharma for financial support of this research. We also wish to thank Dr. T. Brian Hurley for helpful discussions and for initial work in our labs which prompted us to study the halogen-dance reaction. Eric Stangeland is a recipient of a graduate fellowship from Pharmacia.

References and Footnotes

1. Caerulomycin C: (a) Isolation: Funk, A., Divekar, P., *Can. J. Microbiol.* **1959**, *5*, 317. (b) structure determination and biological activity: McInnes, A. G., Smith, D. G., Wright, J. L. C., Vining, L. C., *Can. J. Chem.* **1977**, *55*, 4159.
2. WS75624 B: (a) Isolation: Yoshimura, S., Tsuruni, T., Takase, S., Okuhara, M. J., *Antibiotics* **1995**, *48*, 1073.

3. Others in the Schreiber group had also observed the hydrolysis of esters as a side reaction in sodium borohydride reductions.

4. Okuyama, T., Nagamatsu, H., Fueno, T., *J. Org. Chem.* **1981**, *46*, 1336.

5. Wulff, G., Lauer, M., Bihnke, H., *Angew. Chem. Int. Ed. Engl.* **1984**, *23*, 741

6. (a) Sammakia, T., Hurley, T. B., *J. Am. Chem. Soc.* **1996**, *118*, 8967. (b) Sammakia, T., Hurley, T. B., *J. Org. Chem.* **1999**, *64*, 4652. (c) Sammakia, T., Hurley, T. B., *J. Org. Chem.* **2000**, *65*, 974. (d) Sammakia, T., Wayman, K. A., *Org Lett.* **2003**, *5*, 4105.

7. For detailed mechanistic studies, see reference 6b.

8. For reviews of the halogen-dance reaction, see: (a) Frohlich, J. In *Progress in Heterocyclic Chemistry;* Suschitzky, H., Scriven, E.F.V. Eds. Oxford: New York, 1994; Vol 6, pp 1-35. (b) Frohlich, J., *Bull. Soc. Chim. Belg.* **1996**, *105*, 615. (c) Bunnett, J. F., *Acc. Chem. Res.* **1972**, *5*, 139. For leading references to more recent work, see: (a) Stanetty, P., Schnurch, M., Mereiter, K., Mihovilovic, M. D., *J. Org. Chem.* **2005**, *70*, 567. (b) Mongin, F., Rebstock, A.-S., Trecourt, F., Queguiner, G., Marsais, F., *J. Org. Chem.* **2004**, *69*, 6766. (c) Marzi, E., Bigi, A., Schlosser, M., *Eur. J. Org. Chem.* **2001**, 1371. (d) Comins, D. L., Saha, J. K., *Tetrahedron Lett.* **1995**, *36*, 7995.

9. Prior syntheses: (a) Trécourt, F., Gervais, B., Mallet, M., Quéguiner, G., *J. Org. Chem.* **1996**, *61*, 1673. (b) Sammakia, T., Stangeland, E. L., Whitcomb, M. C., *Org. Lett.* **2002**, *4*, 2385.

10. Prior syntheses: (a) Patt, W. C., Massa, M. A., *Tetrahedron Lett.* **1997**, *38*, 1297. (b) Massa, M. A., Patt, W. C., Ahn, K., Sisneros, A. M., Herman, S. B., Doherty, A., *Bioorg. Med. Chem. Lett.* **1998**, *8*, 2117. (c) Huang, S-T., Gordon, D. M., *Tetrahedron Lett.* **1998**, *39*, 9335. (d) Stangeland, E. L., Sammakia, T., *J. Org. Chem.* **2004**, *69*, 2381

11. For the synthesis of the related natural products caerulomycin B, WS75624 A, and Karnamicin B₁, see: (a) Mongin, F., Trécourt, F., Gervais, B., Mongin, O., Quéguiner, G., *J. Org. Chem.* **2002**, *67*, 3272. (b) Bach, T., Heuser, S. *Synlett* **2002**, 2089. (c) Umemura, K., Watanabe, K., Ono, K., Yamaura, M., Yoshimura, J., *Tetrahedron Lett.* **1997**, *38*, 4811.

12. For a brief discussion of substitution reactions on pyridines, see Eicher, T., Hauptmann, S. *The Chemistry of Heterocycles: Structures, Reactions, Synthesis, and Applications*, 2nd ed.; Wiley-VCH: Weinheim, Germany, 2003 pp 269–310. For a more comprehensive discussion, see: Comins, D. L., Joseph, S. P. Pyridines and their Benzo Derivatives: Reactivity at the Ring. In *Comprehensive heterocyclic Chemistry II* Katrizky, A. R.; Rees, C. W.; Scriven, E. F.; McKillop, A. Eds. Pergamon: Oxford, 1996; Vol. 5, p 37.

13. For a review of metalations of azaaromatics, see: Quéguiner, G., Marsais, F., Sniekus, V., Epsztajn, J. In *Advances in Heterocyclic Chemistry;* Katrizky, A. R. Ed. Academic Press: San Diego, 1991; Vol 52, pp 187-304.

14. Wotiz, J. H., Huba, F., *J. Org. Chem.* **1959**, *24*, 595.

15. (a) Bunnett, J. F., Moyer, C. E. Jr., *J. Am. Chem. Soc.* **1971**, *93*, 1183. (b) Bunnett, J. F., Scorrano, G., *J. Am. Chem. Soc.* **1971**, *93*, 1190. (c) Bunnett, J. F.,

McLennan, D. J., *J. Am. Chem. Soc.* **1971**, *93*, 1198. (d) Mallet, M., Quéguiner, G., *Tetrahedron* **1982**, *38*, 3035. (e) Mallet, M., Quéguiner, G., *Tetrahedron* **1985**, *41*, 3433. (f) Mallet, M., Quéguiner, G., *Tetrahedron* **1986**, *42*, 2253.

16. For examples, see: (a) Rocca, P., Cochennec, C., Marsais, F., Thomas-dit-Dumont, L., Mallet, M., Godard, A., Quéguiner, G., *J. Org. Chem.* **1993**, *58*, 7832. (b) Arzel, E., Rocca, P., Marsais, F., Godard, A., Quéguiner, G., *Tetrahedron* **1999**, *55*, 12149. (c) Arzel, E., Rocca, P., Marsais, F., Godard, A., Quéguiner, G., *Tetrahedron Lett.* **1998**, *39*, 6465. (d) Bury, P., Hareau, G., Kocienski, P., Dhanak, D., *Tetrahedron* **1994**, *50*, 8793. (e) Arzel, E., Rocca, P., Grellier, P., Labaeïd, M., Frappier, F., Guéritte, F., Gaspard, C., Marsais, F., Godard, A, Quéguiner, G., *J. Med. Chem.* **2001**, *44*, 949.

17. Tsuruni, Y., Ueda, H., Hayashi, K., Takase, S., Nishikawa, M., Kiyoto, S., Okuhara, M., *J. Antibiotics* **1995**, *48*, 1066.

18. Yanagisawa, M., Kurihara, H., Kimura, S., Tomobe, Y., Mitsui, Y., Yazaki, K., Goto, K., Masaki, T., *Nature* **1988**, *332*, 411.

19. Sundberg, R. J., Jiang, S., *Org. Prep. Proced. Int.* **1997**, *29*, 117.

20. Nucleophilic aromatic substitution at the 3- or 5-position of pyridines is known to be more difficult than at the 2- or 4-position. See: Schofield, K. *Hetero-Aromatic Nitrogen Compounds Pyrroles and Pyridines*; Plenum: New York, 1967, p 244.

21. Keegstra, M. A.; Peters, T. H. A.; Brandsma, L., *Tetrahedron* **1992**, *48*, 3633.

22. This product is contaminated by about 5% of the reduced compound **27**, which is difficult to separate at this stage, but is readily removed in the next step.

23. For a review, see: Negishi, E.-i. In *Metal-catalyzed Cross-coupling Reactions*, Diederich, F., Stang, P.J., Eds.; Wiley-VCH: New York, 1998; Chapter 1. The corresponding Stille reaction using 2-tributlystannyl pyridine was less efficient and provided the desired product in 50% yield under optimized conditions ($Pd_2(dba)_3$, $P(2\text{-furyl})_3$, CuI, DMF, 80°C). See: Farina, V., Krishnan, B., *J. Am. Chem. Soc.* **1991**, *113*, 9585, and Liebeskind, L. S., Fengl, R. W., *J. Org. Chem.* **1990**, *55*, 5359.

24. Kano, S., Yuasa, Y., Yokomatsu, T., Shibuya, S., *Heterocycles*, **1983**, *20*, 2035. For other examples of halogen dance reactions on thiophenes and furans, see: (a) Sauter, F., Frohlich, H., Kalt, W., *Synthesis* **1989**, 771. (b) Bury, P., Hareau, G., Kocienski, P., Dhanak, D., *Tetrahedron* **1994**, *50*, 8793.

25. For a review of the chemistry of lithiothiazole, see: Iddon, B., *Heterocycles* **1995**, *41*, 533.

26. For other examples where an *in situ* trapping with a silyl chloride proceeds faster than a halogen dance, see: (a) Kano, S., Yuasa, Y., Yokomatsu, T.; Shibuya, S., *Heterocycles*, **1983**, *20*, 2035. (b) Bury, P., Hareau, G., Kocienski, P., Dhanak, D., *Tetrahedron* **1994**, *50*, 8793. (c) Cochennec, C., Rocca, P., Marsais, F., Godard, A., Quéguiner, G., *Synthesis* **1995**, 321. (d) Arzel, E., Rocca, P., Marsais, F., Godard, A., Quéguiner, G., *Heterocycles*, **1999**, *50*, 215.

27. Alternatively, the lithiation at the 4-position could occur at a comparable rate to that at the 5-position, but the trapping with TIPSCl could be slow due to steric hindrance.

28. Ohkuma, T., Kitamura, M., Noyori, R., *Tetrahedron Lett.* **1990**, *31*, 5509.

29. For a related transformation, see: Burke, S. D., Deaton, D. N., Olsen, R. J., Armistead, D. M., Blough, B. E., *Tetrahedron Lett.* **1987**, *28*, 3905

30. This reaction required prior purification of the anhydride by recrystallization from ethylacetate in order to minimize the formation of several by-products.

31. (a) Lindgren, B. O., Nilsson, T. *Acta Chem. Scand.* **1973**, *27*, 888.   (b) Dalcanale, E., Montanari, F., *J. Org. Chem.* **1986**, *51*, 567.

STRATEGIES AND TACTICS IN ORGANIC SYNTHESIS, VOL. 6

Chapter 14

DIASTEREOSELECTIVE INTRAMOLECULAR 4+3 CYCLOADDITION AND AN ENANTIOSELECTIVE TOTAL SYNTHESIS OF (+)-DACTYLOL

Paitoon Rashatasakhon and Michael Harmata
Department of Chemistry
University of Missouri-Columbia
Columbia, Missouri 65211

"No passion in the world is equal to the passion to alter someone else's draft."

H.G. Wells

I. Introduction

The total syntheses of natural products allow practitioners to expose themselves to a series of obstacles and serendipities. These usually happen during the search for the best reaction conditions that can provide a desired transformation in the most regio-, diastereo-, or enantioselective manner. Rewards for these demanding efforts are the development of excellence in laboratory skills, the acquisition of intellectual and physical strengths, and the excitement when key reactions work or when the final targets are made. Moreover, most of the unexpected outcomes can lead to the discoveries or developments of novel methodologies that are the essence of organic synthesis.[1]

In our research program involving a total synthesis of (+)-dactylol, the above-mentioned remark could not be more accurate. Despite possessing a rather small structure with only a few stereocenters, dactylol provided us a lot of valuable lessons. We begin this report with a brief summary of dactylol syntheses and 4+3 cycloaddition reactions, followed by several stages in our synthesis that proceeded in accordance with our retrosynthetic plan. These include the preparation of the diene side chain and cycloaddition precursor, the 4+3 cycloaddition chemistry, and the conversion of the cycloadduct into (+)-dactylol. It was not our intention to develop this synthesis to merely show the utility of methodology that we had developed. Instead, the most important goal of this work was to prove our hypothesis regarding a diastereoselective 4+3 cycloaddition reaction. As the story continues, we hope the readers will experience our frustration with certain untoward outcomes, as well as our pleasure with all of our interesting discoveries and accomplishments.

II. Dactylol

Dactylol (**1**) is a sesquiterpene possessing an unusual *trans*-5,8-fused ring system. It was first isolated from a Caribbean sea hare, *Aplysia dactylomela* ,[2] and was later found in the red seaweed *Laurencia poitei*,

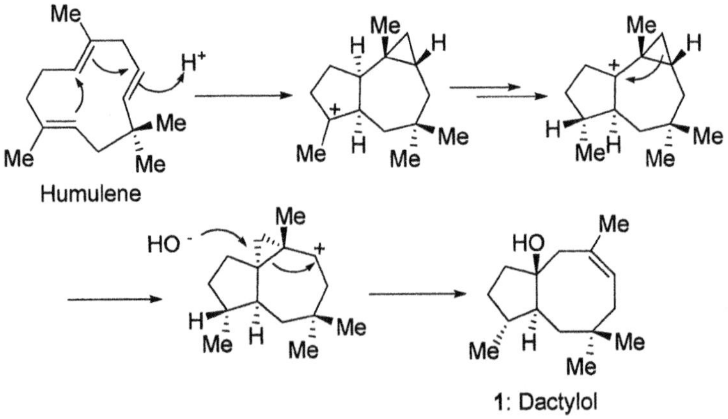

SCHEME 1

which is a food source of the Caribbean sea hare.[3]   The biosynthesis of this compound involves humulene undergoing an acid-catalyzed cyclization, a series of hydride shifts, and a *"cyclopropane sliding"* mechanism[4] (Scheme 1).

A biomimetic synthesis of dactylol was studied by a group of Japanese chemists (Scheme 2).[5]   The synthesis began with africanol (**2**), which could be prepared from the natural product humulene or via a synthesis from 4,4-dimethylcyclohexanone.[6]   Treatment of **2** with $POCl_3$ in pyridine followed by *m*CPBA oxidation afforded epoxide **3a** and **3b** in a ratio of 1:1.   Epoxide **3a** was isolated and treated with $BF_3 \cdot Et_2O$ to give a mixture of two dienes (**4a** and **4b**) along with some unidentified hydrocarbons. The major product (**4a**) was then partially hydrogenated at the five-membered ring olefin to afford dactylol in 16% yield from epoxide **3a**. This synthesis supported the existence of the *"cyclopropane sliding"* mechanism in the actual biosynthesis of dactylol.

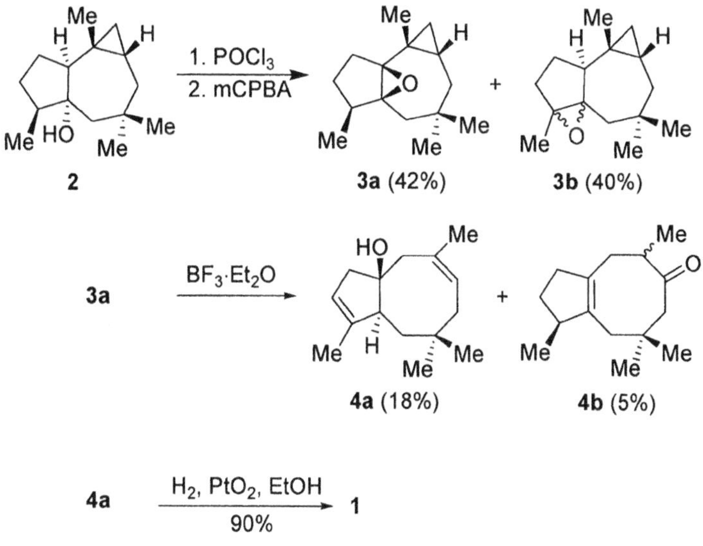

SCHEME 2

A number of nonbiogenetic total syntheses of dactylol in both racemic and optically pure forms have appeared in literature. The first example is the work reported in 1986 by Gadwood and co-workers. (Scheme 3).[7] This synthesis began with the conversion of enone **5** into cyclobutanone **6**. The cyclooctanoid framework was constructed using an oxy-Cope rearrangement. This key transformation involved treatment of **6** with lithium acetylide and stirring the resultant alcohol (**7**) at 50 °C. Addition of MeLi to the ketone **8**, PCC oxidation, and addition of Me₂CuLi produced ketone **9** with the required geminal dimethyl group. The ketone **9** was then transformed to ketone **10** by a series of reduction, protection, and oxidation reactions. Reduction and removal of the –SEM group provided the natural product (±)-poitediol (**11**), which could be converted into (±)-dactylol upon treatment with sodium in liquid ammonia.

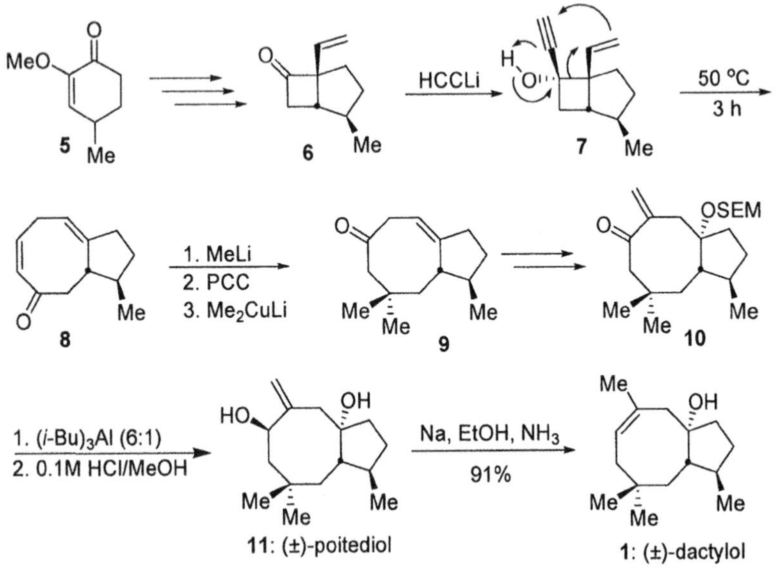

SCHEME 3

During the study of a stereoselective 6+2 photocyclization, Feldman and associates reported a total synthesis of racemic dactylol (Scheme 4).[8] The photocyclization of **12** produced 6+2 cycloadduct **13** in 41% yield along with 8+2 cycloadduct **14** in 12% yield. The Baeyer-Villiger oxidation of **13** using Ph(CH₃)₂COOLi and subsequent reduction led to diol **15** in 73% yield. The diene moiety was partially hydrogenated and the primary alcohol functional group was treated with acetyl chloride. Irradiation of ester **16** in HMPA-H₂O mixture cleanly removed the acetate group and produced dactylol in 50% yield.

The first non-racemic synthesis of dactylol was demonstrated by Molander and Eastwood using a 5+3 annulation strategy (Scheme 5).[9] In the presence of a Lewis acid ($TrSbCl_6$), 1,5-dicarbonyl **17** reacted with bis-trimethylsilylketene acetal **18** to give a 5+3 cycloadduct **19** in 77% yield. Decarbomethoxylation and Tebbe olefination led to cyclic ether **20**. A double bond isomerization was accomplished using $RhCl_3$. Two regioisomers of alkene **21** were obtained in 96% yield in a ratio of 12:1. Finally, the mixture of **21** was treated with Li metal in dimethoxyethane and ethylene diamine to afford (+)-dactylol in 25% yield along with 36% yield of undesired hydrogenation product (**22**).

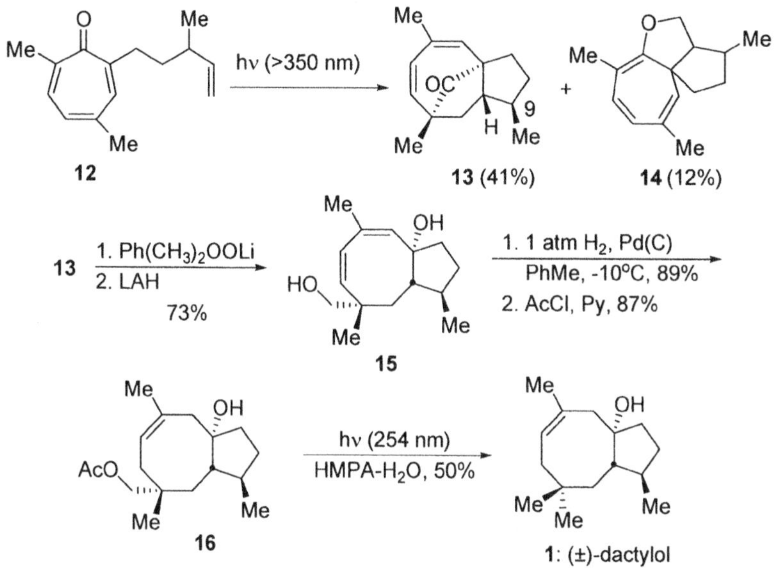

SCHEME 4

One of the most recent total syntheses of dactylol was reported by Fürstner and Langemann (Scheme 6).[10] The synthesis began with a conversion of 2-cyclopentenone (**23**) into ketone **24**. The addition of a Grignard reagent led to two diastereomeric products in 80% yield in a ratio of 1:1.2. The minor product (**25**) was converted to dactylol by protection of the hydroxy group, ring-closing metathesis using the Schrock catalyst (**26**), and deprotection.

III. 4+3 Cycloaddition Reactions of Cyclic Oxyallyl Zwitterions

For the construction of medium-sized ring systems, cycloaddition reactions are known to be among the most synthetically efficient methods.

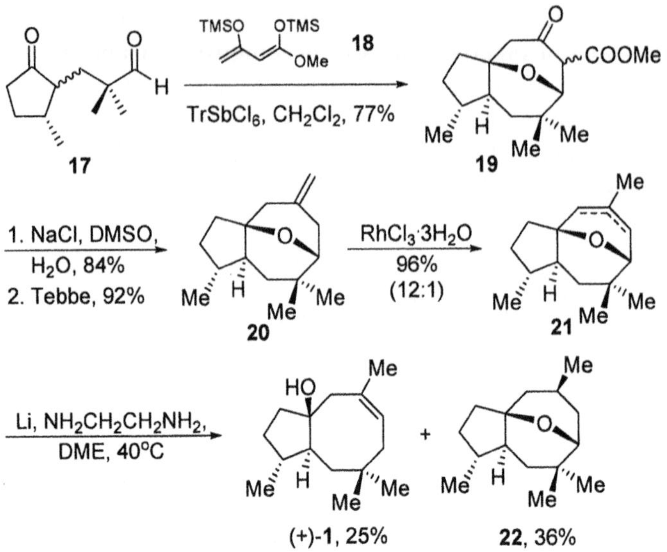

SCHEME 5

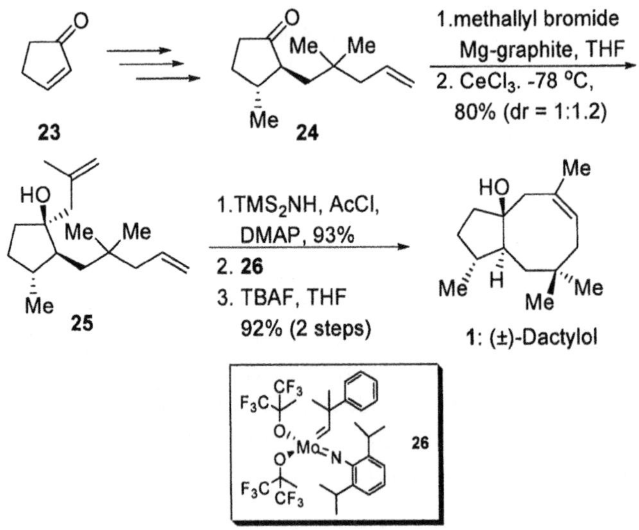

SCHEME 6

Not only can they form two bonds in a single step, the attainment of architecturally complex molecules from relatively simple substrates is also a distinctive advantage of these types of transformations. This is particularly true for the formation of seven-membered carbocycles using the most direct and efficient route, the 4+3 cycloaddition reaction.

In a general 4+3 cycloaddition reaction, a diene reacts with an allylic cation possessing a carbocation terminating group (X) that will lead to a corresponding functional group in the cycloadduct (Scheme 7). The allylic cation is generally an oxyallylic (X = O⁻), an alkoxyallylic (X = OR), or a trimethylsilylmethyl allylic (X = TMSCH$_2$) cation.

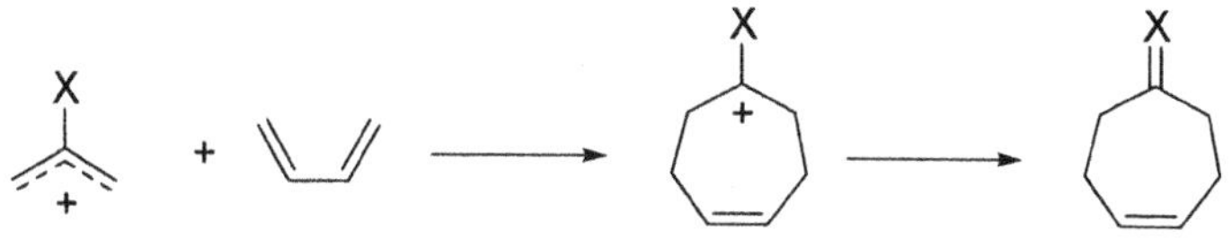

SCHEME 7

The cycloaddition reaction of oxyallylic cations was first reported by Fort during an apparent study of the Favorskii rearrangement.[11] Since then, the research in this area has received increasing attention and several reviews have appeared in literature.[12] One of the most exciting modifications is the use of cyclic oxyallyls in the 4+3 cycloaddition reaction, which allows an opportunity for creating more complicated and

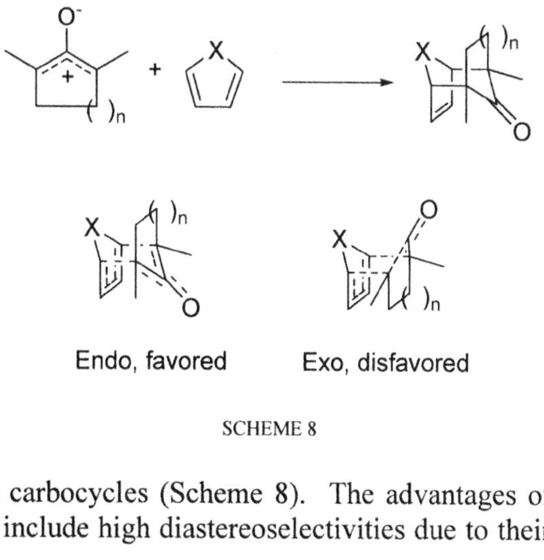

Endo, favored   Exo, disfavored

SCHEME 8

functionalized carbocycles (Scheme 8). The advantages of using cyclic oxyallyls also include high diastereoselectivities due to their well-defined structures and a bias that favors the *endo* (compact) transition state over the *exo* (extended) one.[12b] In addition, the ketone group in the cycloadducts could serve as functional group for further elaboration of cycloadducts.

IV. Synthesis of Cyclooctanoids via Intramolecular 4+3 Cycloaddition Reactions

The intramolecular mode of 4+3 cycloaddition reactions between allylic cations and dienes has been comprehensively summarized in several reviews.[13] Nevertheless, the intramolecular 4+3 cycloaddition reactions are not as well established as the intermolecular ones. One of the very first examples of this reaction was reported in 1979 by Noyori and co-workers, who had previously studied the intermolecular reaction between polyhalogenated ketones and dienes. For instance, treatment of dibromoketones **27** or **28** with diiron nonacarbonyl in refluxing benzene provided cycloadducts **29** or **30** in 41% and 38% yields, respectively (Scheme 9).[14] Although the stereoselectivities of these reactions were high, the lachrymatory nature of dibromoketones and the difficulty associated with the synthesis of dibromoketones limited the development of this process. Thus, no other examples of the intramolecular 4+3 cycloaddition using Noyori's methodology have been reported.

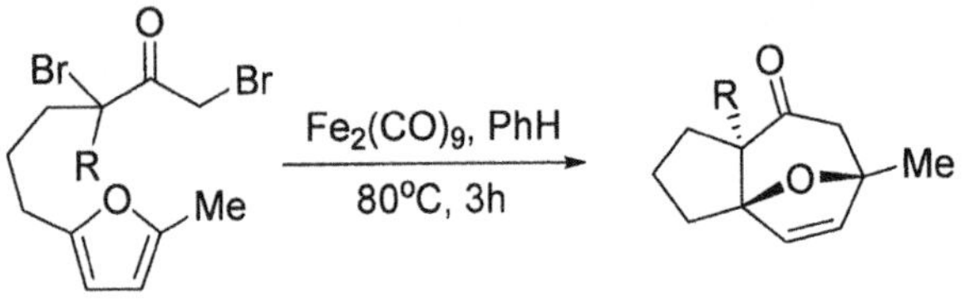

27: R = H
28: R = Me

Fe$_2$(CO)$_9$, PhH
80°C, 3h

29: R = H, 41%
30: R = Me, 38%

SCHEME 9

The intramolecular 4+3 cycloaddition reactions of cyclopentenyl oxyallylic cations have been studied and developed in our laboratory for quite some time. Our group appreciates the power of cyclic oxyallylic cations such as **31** that allow us to perform a 4+3 cycloaddition reaction to create a product that is a formal 4+(3+m) cycloadduct (**32**) (Scheme 10). So does Fred West's group.[15]

The intramolecular reaction of cyclic oxyallylic cations with tethered dienes (**33**) can lead to structurally complex molecules (**34**) that are

Printed and bound by CPI Group (UK) Ltd, Croydon, CR0 4YY

03/10/2024

01040420-0008

otherwise difficult to obtain using other methods. Hence, the scope of 4+3 cycloaddition reactions has been extended beyond the synthesis of seven-membered ring carbocycles.

During the early phase of our program, we examined the compatibility between oxyallyl cations and various tethered dienes. We observed that

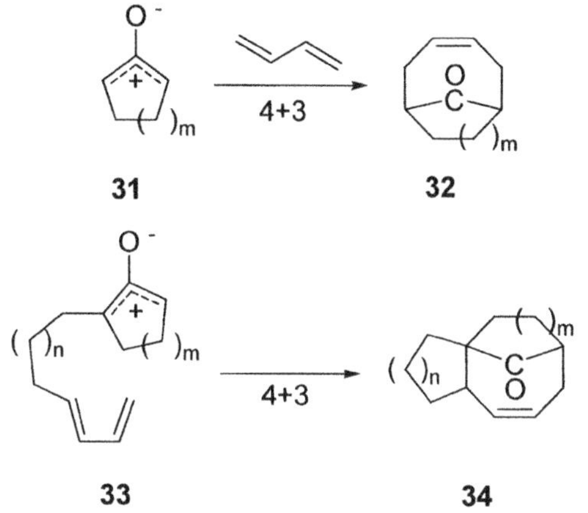

SCHEME 10

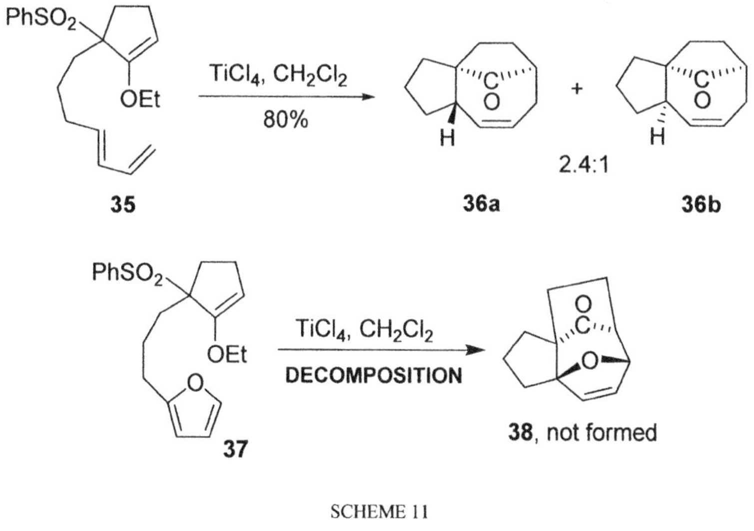

SCHEME 11

the more electrophilic cations react better with the less nucleophilic dienes, and *vice-versa*. For instance, the ethoxyallylic cation generated from an alkoxy allylic sulfone was more likely to give a better result when it reacted with a non-activated tethered diene such as butadiene. In the examples shown in Scheme 11, sulfone **35** reacted with $TiCl_4$ to give cycloadducts **36a** and **36b** in 80% yield in a ratio of 2.4:1. However, under similar conditions the reaction of **37**, which possesses a more nucleophilic diene (furan ring), led to a total decomposition of the starting material.

Fortunately, less electrophilic cations such as oxyallyl zwitterions were able to react with more nucleophilic dienes (Scheme 12), while reaction with less reactive dienes was more difficult.. In one of our studies, the reaction of chloroketone **39** with NEt_3 resulted in only the elimination product **40**, while chloroketone **41** afforded two diastereomers of cycloadduct (**42a** and **42b**) in 17:1 ratio, along with chloroketone cycloadduct **43** and elimination product **44**.

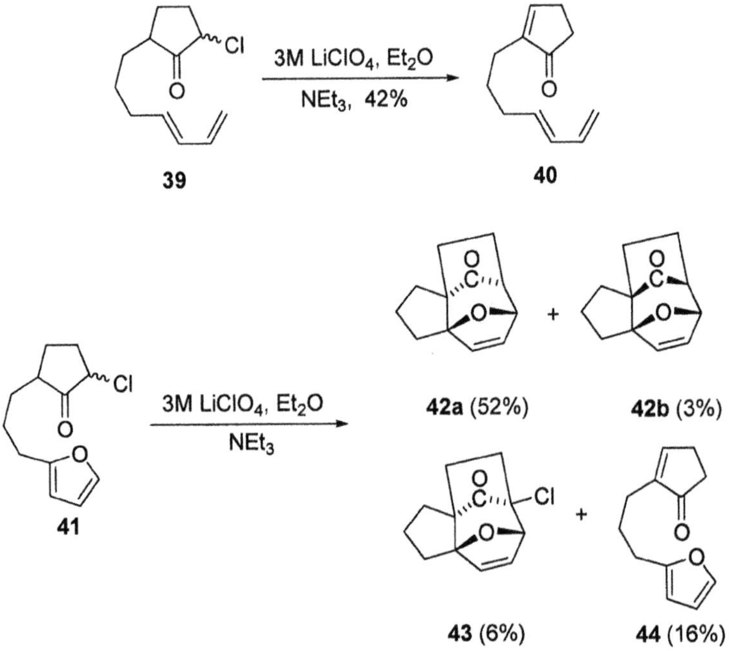

SCHEME 12

After developing some aspects of this chemistry, it became important to us to synthesize one or more of a number of natural products that possess a five-eight fused ring system. However, we also recognized the need to improve the diastereoselectivity of the intramolecular 4+3 cycloaddition reactions.   This is particularly true in the reaction of substrates with tethered butadienes.   In many cases, such reactions produced two isomers with no substantial diastereoselectivity.   For example, the reaction of **45** with TiCl$_4$ gave rise to the cycloadducts **46a** and **46b** in only 20% yield and almost 1:1 diastereoselectivity.[16]   The hemiacetal **47** derived from a competing 3+2 cycloaddition reaction was the major product in this case.

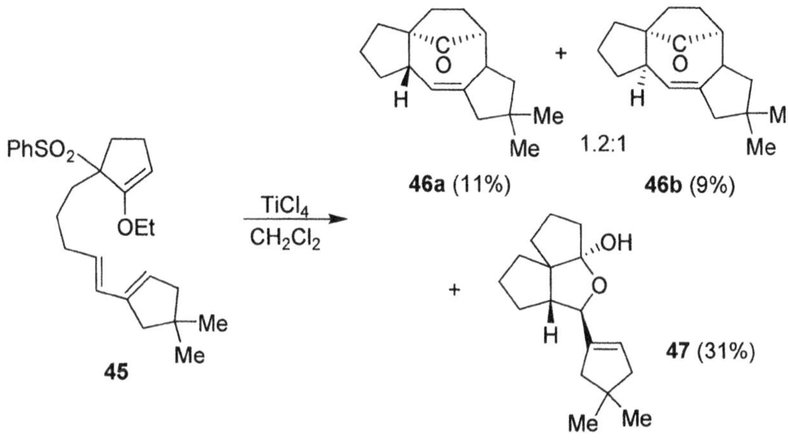

SCHEME 13

We also investigated the cycloaddition of a cyclopentenyl oxyallylic zwitterion derived from a chloroketone that contained the same tethered diene. The chlorination of ketone **48** at the less substituted α-position was accomplished by treatment with LDA and trifluoromethanesulfonyl chloride.   The crude product was then treated with sodium trifluoro-

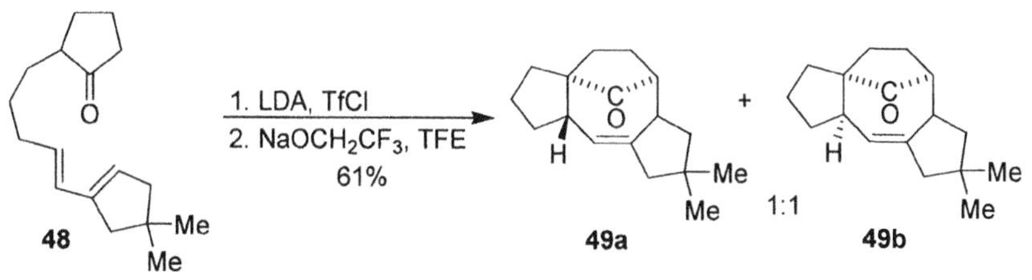

SCHEME 14

ethoxide to afford cycloadducts **49a** and **49b** in 61% yield. While the yield was satisfactory and could presumably be optimized, no degree of diastereoselectivity was observed. Nobody ever said life was fair.

The lack of diastereoselectivity in the intramolecular 4+3 cycloaddition reactions of cyclopentenyl oxyallylic cations with tethered butadienes has prohibited the application of this methodology to the synthesis of natural products. Therefore, the development of stereocontrol in cycloadditions of this type is of great significance. Our study in this area was conducted in the context of the synthesis of (+)-dactylol.

V. Diastereocontrol in 4+3 Cycloaddition Reactions

There have been several examples of good stereocontrol in 4+3 cycloaddition reactions. For instance, it is known that cyclopentenyl and cyclohexenyl oxyallylic zwitterions bearing a stereogenic center react preferentially with dienes from the face opposite to the substituent.[17] One of the precedents from our laboratory is the intramolecular reaction of ketones **50** and **52** (Scheme 15). Treatment of these ketones with LDA/TfCl afforded the corresponding α-chloroketones, which reacted intramolecularly with the tethered furans to yield cycloadducts **51a,b** and **53a,b** upon treatment with base.[18] The *endo* isomers (**51a, 53a**) dominated in both cases. Only those products derived from the transition states having furans approach the least hindered faces of the oxyallylic cations were produced.

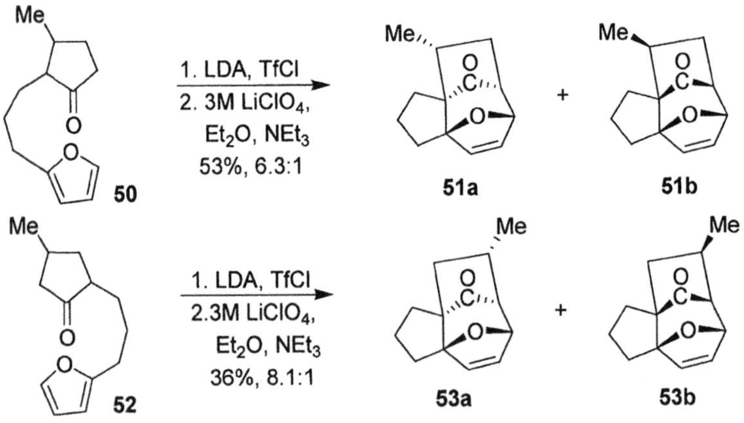

SCHEME 15

It has been reported that substituents on tethers also play an important role in the stereoselectivity of the reaction. Giguere and co-workers have reported an example of a highly diastereoselective 4+3 cycloaddition of trienol **54** upon treatment with triflic anhydride at low temperature (Scheme 16). The product **56** was formed in 82% yield in a ratio of 92:5:3:0 (only the major isomer shown).[19] The dramatic stereochemical outcome could be explained using a preferred transition state **55**. The methyl group on the tether occupies a pseudo-equatorial orientation on the puckered, incipient five-membered ring with the diene oriented so as to minimize gauche interactions.

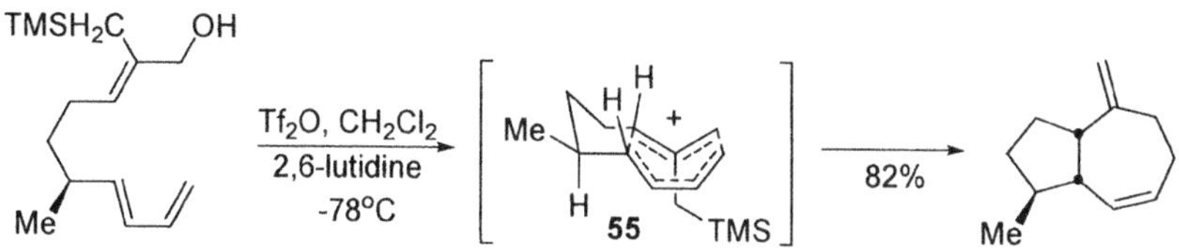

SCHEME 16

VI. Retrosynthetic Analysis of (+)-Dactylol

The two pieces of information mentioned above allowed us to envisage a natural product synthesis using a novel strategy for achieving a high diastereoselection. One of the particularly suitable targets for this study was the sesquiterpene (+)-dactylol (**1**). The retrosynthetic analysis of this compound is depicted in Scheme 17. We expected the synthesis of optically pure (+)-dactylol to be completed by functional group manipulation of 4+3-cycloadduct **57**. The key transformation would be the highly diastereoselective intramolecular 4+3 cycloaddition of ketone **58**. The cycloaddition was anticipated to proceed via a transition state resembling **63**, which involved the combination of the two stereocontrol elements discussed above. The methyl group on the cyclopentanone should control the facial selectivity of the oxyallylic zwitterion, while the dienylic methyl group should control the relative stereochemistry of the developing stereogenic centers. The required ketone **58** could be prepared by alkylation of the easily accessible ketoester **59** with the iododiene **60**. Ketoester **59** is a readily accessible compound which could be prepared from a commercially available (R)-pulegone (**61**) according to a literature procedure.[20] The iododiene **60** could ultimately come from optically pure hydroxyester **62**.

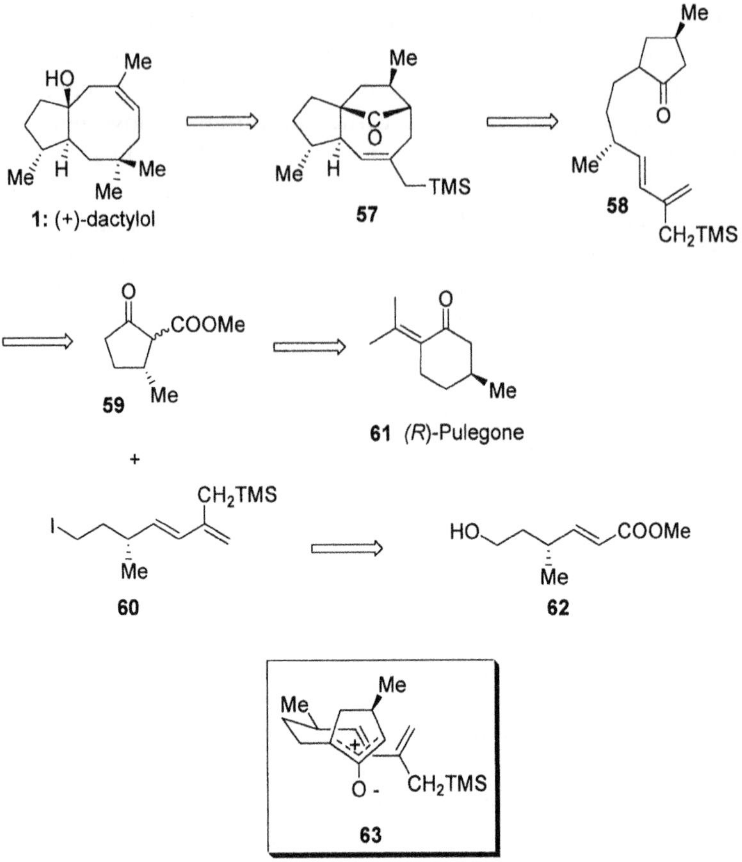

SCHEME 17

VII. Preparation of the Diene Side Chain

Perhaps the shortest synthesis of ester **62** could be carried out from optically pure lactone **65** (Scheme 18). Asymmetric hydrogenation of α-methylene butyrolactone (**64**) appeared to be the most straightforward way of making these compounds.[21] Since this was the very first reaction in the synthesis, it would require a large amount of the lactone **64**. However, the high cost of this compound, its volatility and the cost of the catalyst for the asymmetric hydrogenation made this idea impractical, at least at the time we started the work. It is not clear whether we were counting pennies at the time or just being cheap, but we did not pursue this process.

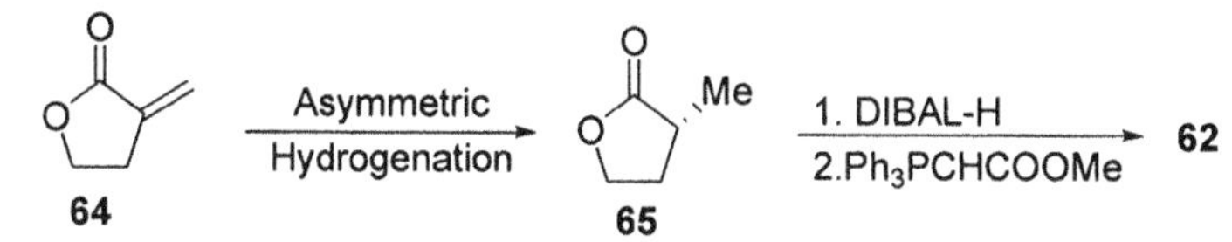

SCHEME 18

Instead, we turned our attention to the asymmetric alkylation of γ-butyrolactone or its ω-hydroxy esters derivatives. The work by Enders and co-workers using SAMP or RAMP lactone hydrazones has shown great success with six- and seven-membered ring lactones.[22] We decided to examine the use of this protocol in a five-membered ring analogue. (S)-Amino-2-methoxymethyl pyrrolidine (SAMP) was prepared from L-proline according to a literature procedure.[23] The synthesis began with the reaction of SAMP and the commercially available 4-chlorobutyric acid chloride (66) (Scheme 19). Treatment of the ω-chlorobutyro-hydrazide (67) with AgBF$_4$ resulted in butyrolactone-SAMP hydrazone (68). Deprotonation with LDA followed by trapping with MeI led to α-methyl butyrolactone-SAMP hydrazone 69. Unfortunately, the ^{1}H-NMR of the crude product mixture suggested a poor diastereoselectivity ($\sim$1:1 ratio). So much for being frugal.

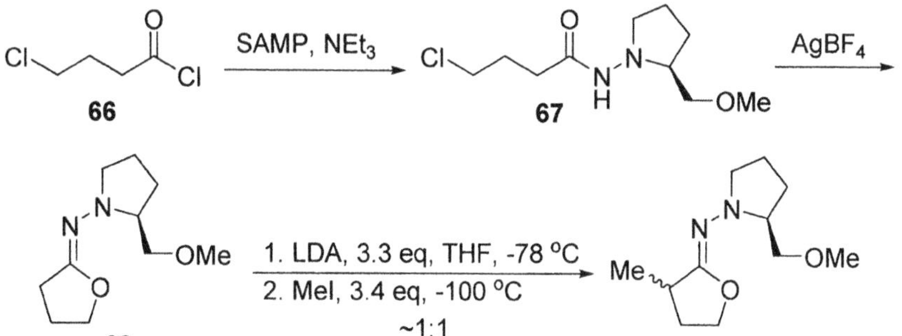

SCHEME 19

To circumvent this problem, a different route to access optically pure α-methyl butyrolactone (65) was attempted using Evans' chiral auxiliary chemistry.[24] N-Acylation of oxazolidinone with 4-pentenoyl chloride afforded 70, which was treated with NaHMDS and MeI to give 71

(Scheme 20). Ozonolysis of alkene **71** and NaBH$_4$ reduction of the resultant aldehyde **72** gave rise to lactone **65**. Reduction of lactone **65** with DIBAL-H provided the corresponding lactol, which was treated with a Wittig reagent to give ester **62**. However, the yield of the last two transformations was very low. This could be the result of over-reduction of the lactone or the high water-solubility of the lactol. In any case, attempts to find products failed. As a result, we needed to re-examine this route using a protection-deprotection protocol. On the bright side, these results partially vindicated our decision to not pursue the synthesis of **65** as described in Scheme 18.

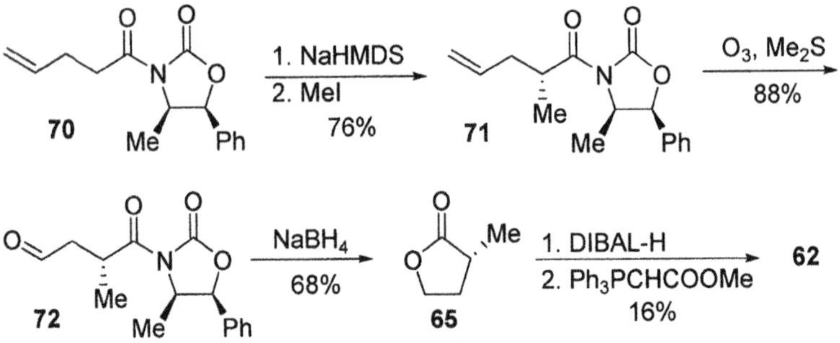

SCHEME 20

In the new strategy, the use of a protecting group would avoid the lactone formation during the removal of the chiral auxiliary. Not many choices of protecting group were compatible with the synthetic route. For example, silicon-containing protecting groups would not be stable under Jones oxidation conditions and the deprotection of a benzyl group in the presence of another double bond might not be suitable. We decided to protect the hydroxy group with an allyl group, which could be removed using a mild acid and a palladium catalyst.

Hence, monoprotection of 1,4-butanediol with NaH and trapping with allyl bromide led to 4-allyloxybutan-1-ol (Scheme 21). The crude product was treated with Jones reagent to furnish 4-allyloxybutyric acid (**73**). Treatment with oxalyl chloride in hexane at room temperature gave the corresponding acid chloride, which reacted with *N*-oxazolidinone enolate to give a substrate for asymmetric alkylation upon treatment with NaHMDS and MeI. Reduction of **74** with lithium aluminum hydride led to a primary alcohol and recovered chiral auxiliary. Swern oxidation and treatment with Ph$_3$PCHCOOMe afforded the α,β-unsaturated ester **75**.

When treated with Pd/C and TsOH in MeOH, the allyl protecting group was removed and ester **62** was obtained in good yield.[25] The mechanism of this deprotection may involve a regioselective double bond isomerization followed by a hydrolysis of the enol ether.

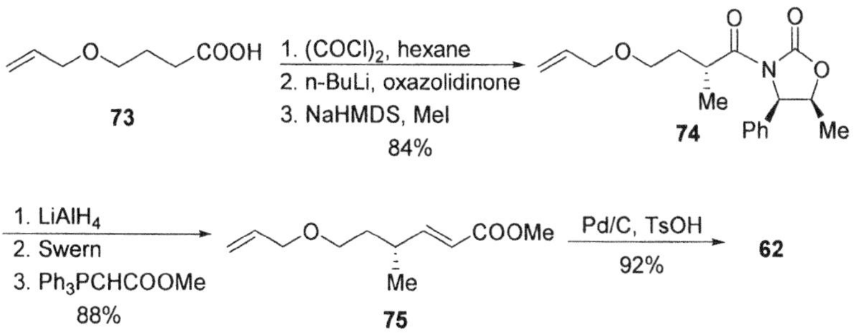

SCHEME 21

With ester **62** in hand, we carried out a Peterson olefination, which gave rise to diene **76** after acidic work-up. This compound was then treated with a salt of DCC and MeI[26] to afford the iododiene **60**.

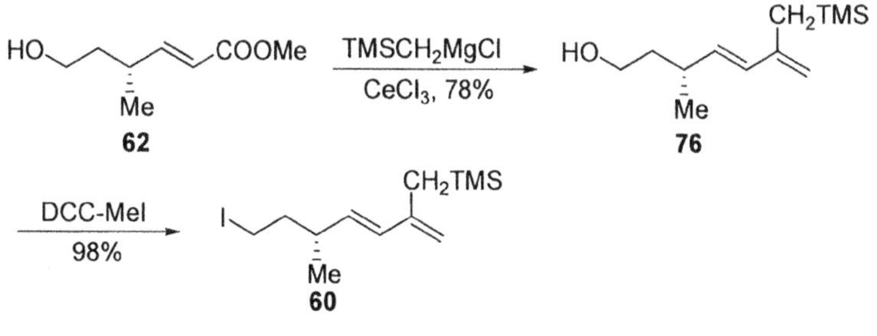

SCHEME 22

VIII. Preparation of Cycloaddition Substrates

Ketoester **59** was prepared from (R)-pulegone (**61**) according to a literature procedure.[20] Generation of the dianion of **59** using one equivalent of NaH and one equivalent of BuLi followed by the addition of iododiene **60** resulted in ketoester **77** (Scheme 23). The use of two equivalents of LDA also gave the same result in slightly lower yields. Optimum results were achieved when the starting materials were azeotropically dried with benzene. Compound **77** was converted to ketone **58** by treating with potassium cyanide in refluxing DMSO.[27] The right choice of reagent in the Krapcho decarbomethoxylation was essential, as halide salts (e.g., NaCl, NaBr, KCl) that are generally used for this reaction could destroy the sensitive allylsilane group.

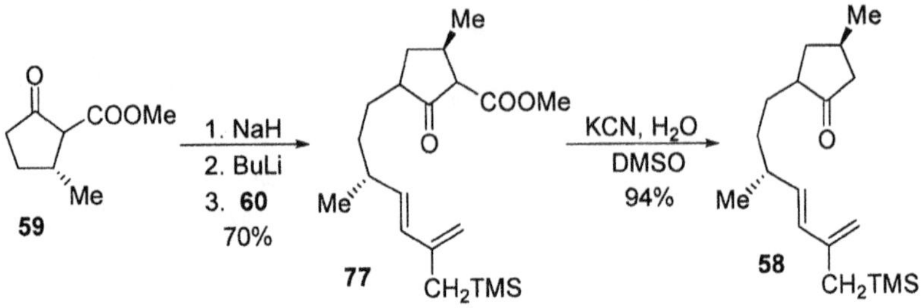

SCHEME 23

IX. The Key 4+3 Cycloaddition Reaction

In the 4+3 cycloaddition step, ketone **58** was treated with LDA and the resultant enolate was quenched with trifluoromethanesulfonyl chloride at low temperature to give an α-chloroketone.[28] This was not characterized

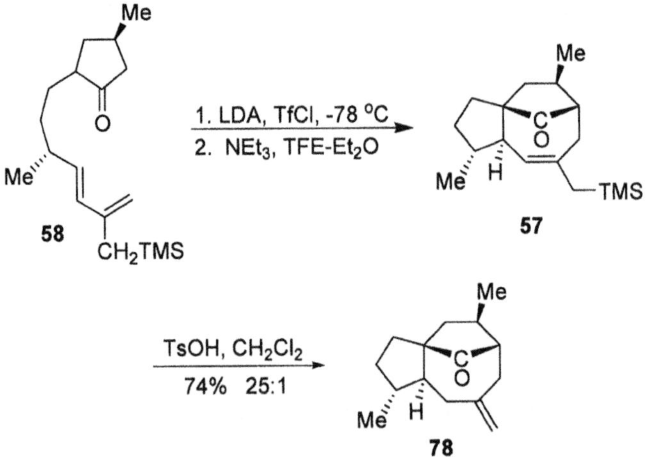

SCHEME 24

but was immediately subjected to typical cycloaddition conditions: stirring at -78 °C to room temperature in a 1:1 mixture of ether and trifluoroethanol in the presence of three equivalents of triethylamine (Scheme 24). Even though the purification of the cycloadduct **57** could be carried out at this stage, we found that some desilylation and double bond migration occurred during chromatography on silica gel. Therefore, the crude product was subsequently treated with tosic acid to give rise to the exocyclic alkene **78** as a 25:1 mixture of isomers in 74% yield for three steps. The diastereomeric ratio was determined by GC and ^{1}H NMR integration of the olefinic region of a crude product mixture. This result supports our hypothesis regarding the transition state **63**. Making a prediction based on a rigorous and logical analysis and seeing it bourne out by experiment is a real thrill.

X. From the Cycloadduct to (+)-Dactylol

An initial plan to transform cycloadduct **78** into (+)-dactylol (**1**) is shown in Scheme **25**. The Simmons-Smith reaction should convert the exocyclic double bond into a cyclopropane ring. Based on a standard mechanism, we anticipated obtaining lactone **79b**, which results from the migration of the tertiary carbon during the Baeyer-Villiger oxidation. Subsequently, the cyclopropane ring would be cleaved regioselectively by hydrogenolysis to install a geminal dimethyl moiety. Hydrolysis of lactone and decarboxylative elimination should lead to dactylol (**1**).

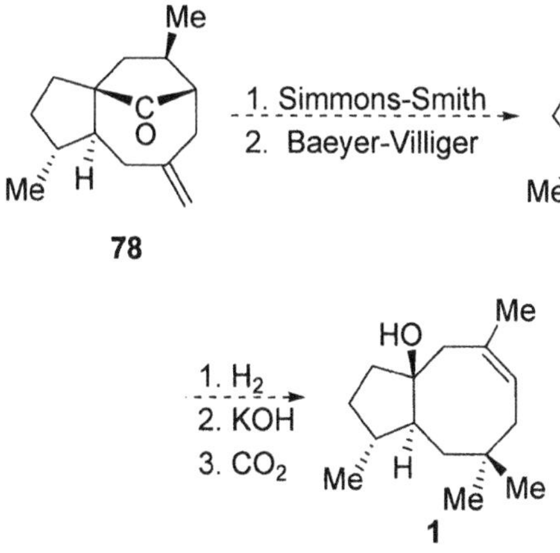

SCHEME 25

Hence, the final stage of the synthesis began with a Simmons-Smith cyclopropanation of **78**, which produced ketone **80** in 95% yield (Scheme 26). We chose to use Et$_2$Zn and CH$_2$I$_2$, since these conditions gave higher yields and the reaction was much cleaner than the use of Zn-Cu metals and CH$_2$I$_2$. The Baeyer-Villiger reaction of **80** using MMPP in DMF afforded lactone **79** as a 4:1 mixture of two regioisomers (**a** and **b**) after purification. Surprisingly, the major product (**79a**) was that which resulted from a migration of the less substituted carbon. Attempts to alter the regioselectivity and optimize the reaction using a variety of reaction conditions were unsuccessful as some reagents gave higher regioselectivity, but lower yields of the products (Table 1). The major isomer **79a** was separated and the cyclopropane ring was cleaved under a hydrogen pressure of 100 psi in the presence of PtO$_2$ to afford **81** in 98% yield. The structure and relative stereochemistry were confirmed by X-ray crystallography of **81**.

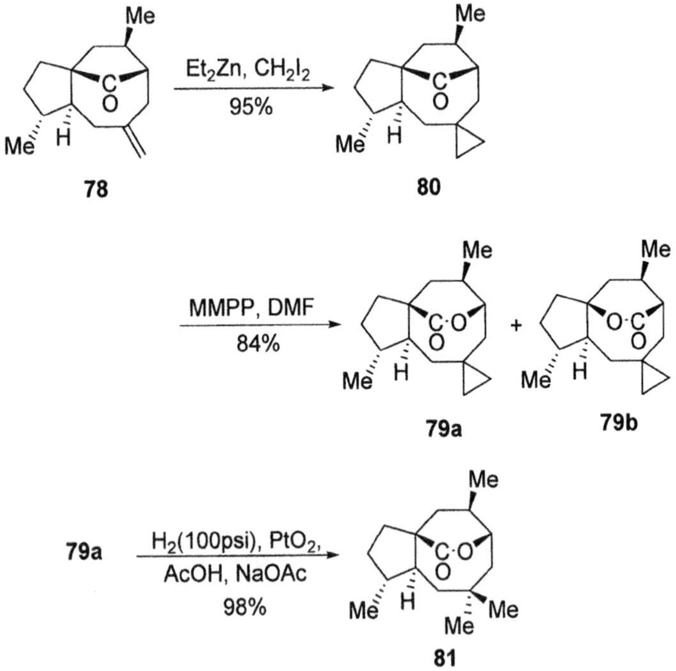

SCHEME 26

Me   Me   Me

Baeyer-Villiger →

80   79a   +   79b

TABLE 1
Baeyer-Villiger Reaction of Ketone **80**

Entry	Conditions	79a:79b*	Yield (%)
1	mCPBA/NaHCO$_3$, CH$_2$Cl$_2$, rt, 5 days	6:1	67
2	TFAA/H$_2$O$_2$, CH$_2$Cl$_2$, 0°C-rt, 3 days	5:1	36
3	mCPBA/TFA, CH$_2$Cl$_2$, 0°C-rt, 2 days	4:1	53
4	CH$_3$CO$_3$H/NaOAc, AcOH, rt, 2 days	9:1	31
5	MMPP, DMF, rt, 2 days	4:1	84

*Product ratio after chromatographic purification.

Since these regiochemical outcomes were unexpected based on simple predictions, we opted to investigate the Baeyer-Villiger oxidation of this type of polycyclic ketones in greater detail.[29] Our study suggested that such reactions are controlled by steric, electronic, and substituent effects and that at least some control over regiochemistry can be exerted by appropriate structural changes in the molecule and presumably the reagent as well. However, based on our studies with **80**, the direction of the regiochemical outcome with this substrate did not appear to be affected by the reagents or reaction conditions, except in a direction we considered undesirable.

At this point, the next synthetic goal was the installation of a hydroxy group on the C-1 dactylol with a correct stereochemistry. Our first attempt involved the treatment of lactone **79a** with Tebbe's reagent and hydrolysis of the resultant enol ether to ketone **82** (Scheme 27). To our great surprise, methyl ketone **82** reacted with MMPP in DMF to produce lactone **79a** in 86% yield. A similar result was obtained, although in lower yield (36%), in the reaction with mCPBA. We suspected that compound **82** might exist as a hemiketal that could be oxidized to lactone **79a**. Therefore, the protection of the hydroxy group with a benzyl and a *t*-butyldimethylsilyl (TBS) group was carried out in order to prevent this surprising and untoward result. Unfortunately, there was no reaction between the hydroxy-protected methyl ketones **83a,b** with MMPP, mCPBA, or TFAA-H$_2$O$_2$ at room temperature after 24 hours.

It appeared to us that the hydroxy group in ketone **82** had caused us several problems. We then chose to solve this issue by first installing a double bond between C3-C4 and converting the carboxylic acid group

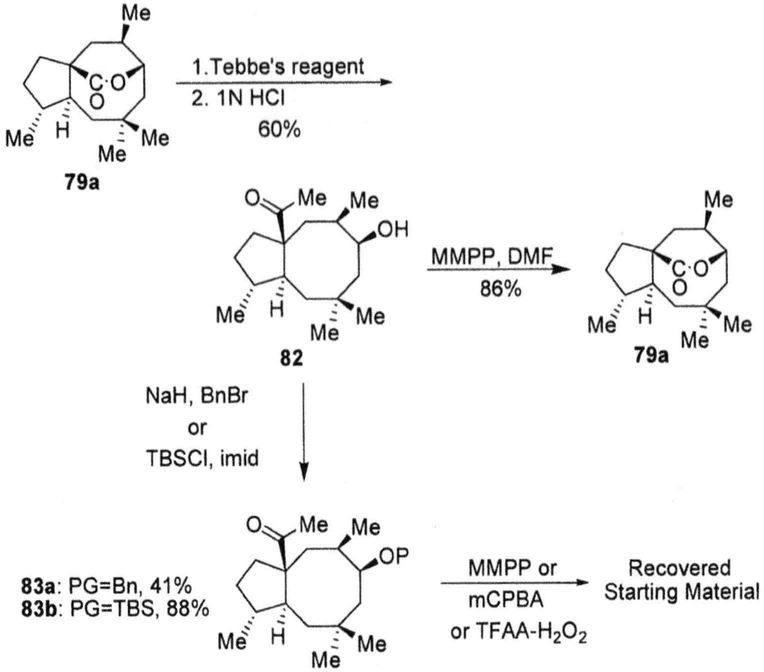

SCHEME 27

into a hydroxy group in the final stage of the synthesis. In an attempt to cleave the lactone ring in **79a**, we first undertook an S_N2-type reaction using a soft nucleophile such as sodium phenylselenide and expected to obtain a phenylselenyl ether (**84**) suitable for C3-C4 double bond formation. However, this experiment afforded the hydroxy acid **85** as the sole product in 51% yield, apparently via either the hydrolysis of lactone by contaminating hydroxide ions or attack of selenide ions at the carbonyl followed by the hydrolysis of selenyl ester (Scheme 28).

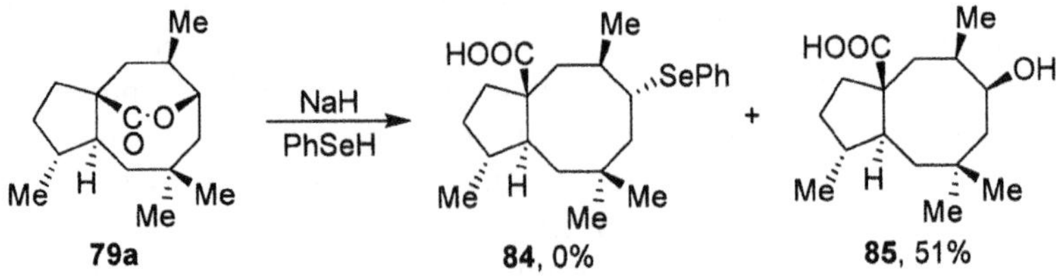

SCHEME 28

Even though this result was not exactly what we expected, it established the existence and stability of the 5-hydroxy carboxylic acid **85**. Therefore, the final sequence to (+)-dactylol began with a hydrolysis of lactone **79a**. Under reflux with 10% KOH in MeOH-H$_2$O for 12 hours, lactone **79a** was completely converted to the hydroxy acid **85**, which was treated with diazomethane to give hydroxy ester **86**. Another challenging step was the introduction of the double bond by dehydration. Regioselectivity was a problem, as were side reactions. For example, treatment of **86** with POCl$_3$ in pyridine, Tf$_2$O in pyridine, or Martin's sulfurane[30] resulted in the formation of a compound identified as **87**. Presumably, these reagents induced carbocation formation that was followed by a 1,5-transannular hydride shift and subsequent elimination to give tetrasubstituted alkene **87**. The structure of **87** was assigned on the basis of spectroscopic and analytical data.

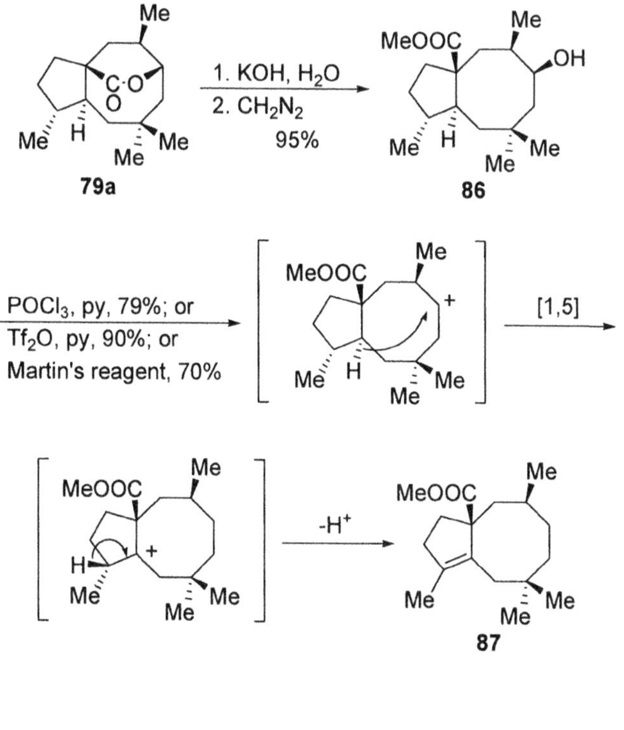

SCHEME 29

After an intensive literature search, we were able to overcome this regiochemical problem by taking advantage of the dehydration procedure developed by Trost.[31] A solution of **86** in HMPA was first added to POCl₃ with slow heating from room temperature to 50 °C. After the white precipitate that formed redissolved, pyridine was added and the heating continued to 100 °C for 45 minutes, at which point the dehydration was complete. Based on NMR data of the crude product, only alkene **88** was formed with no trace of **87** (Scheme 30). This was a great result! But now we had to figure out how to do an oxidative decarboxylation without disturbing that alkene.

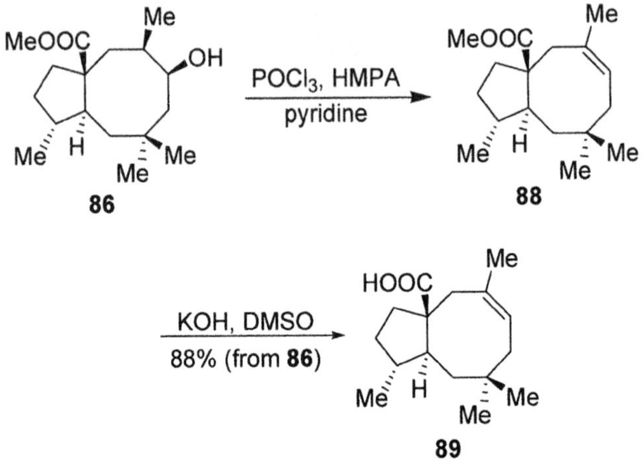

SCHEME 30

Due to steric hindrance, the hydrolysis of **88** under standard ester hydrolysis conditions failed to deliver acid **89**. This hydrolysis was effectively accomplished using KOH in refluxing DMSO to provide acid **89** in 88% yield.

At this stage, and not too surprisingly, we found that all typical procedures for acid chloride formation failed to give the acid chloride of the sterically hindered acid **89**. For example, stirring a solution of this acid in hexane with three equivalents of oxalyl chloride or heating a hexane solution of the acid with excess thionyl chloride resulted in no change after 24 hours. However, upon treatment with Vilsmeier's reagent[32] (DMF and phosgene) in refluxing toluene, acid **89** was cleanly converted to the acid chloride **90** (Scheme 31).

In a benzene solution containing pyridine and DMAP, this acid chloride reacted with mCPBA to form a mixed anhydride **91**, which

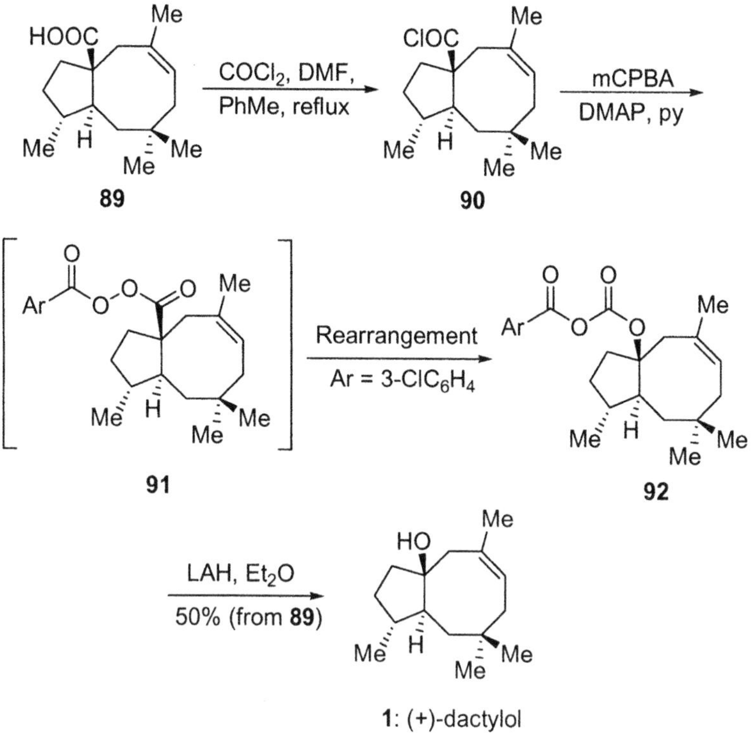

SCHEME 31

rearranged upon stirring for 24 hours at room temperature to a mixed carbonate **92** with retention of configuration.[33] A reasonable mechanism for this transformation is shown in Scheme 32. Finally, a reduction of the mixed carbonate then afforded (+)-dactylol in 50% yield overall yield from **89**. The spectral data, analytical data, and optical rotation of the sample were consistent with those published in literature.[7-10]

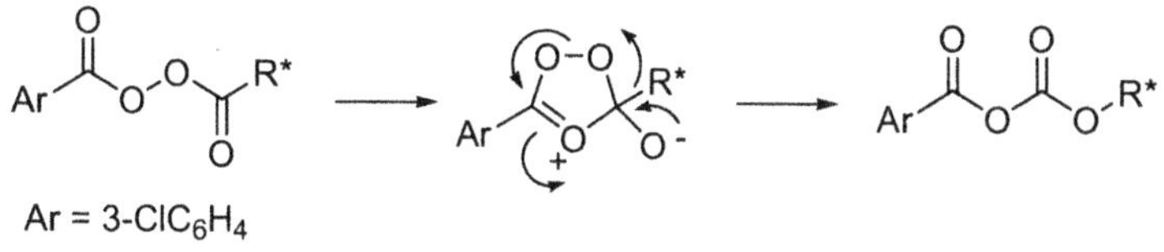

SCHEME 32

XI. Conclusion

The intramolecular 4+3 cycloaddition of cyclopentenyl cations can be used as a way to assemble the core framework of cyclooctanoid natural products. However, the diastereoselectivity in this type of reaction is generally low, especially in the systems with tethered butadienes, rather than furans. By combining some stereocontrol elements in the substrate, we attained a highly diastereoselective intramolecular 4+3 cycloaddition reaction between a cyclopentenyl oxyallylic zwitterion and a tethered butadiene. This result was achieved in the course of an enantioselective total synthesis of (+)-dactylol. We encountered several unforeseen problems along the way, for instance, the unexpected regioselectivity in the Baeyer-Villiger oxidation of a polycyclic ketone. With our flexible strategy and vigorous mindset (!), the total synthesis of (+)-dactylol was successfully executed.

Acknowledgments

This work was supported by the National Science Foundation to whom we are grateful. Special thanks go to Dr. Charles L. Barnes for acquisition of X-ray data.

References and Footnotes

1. Nicolaou, K. C., Snyder, S. A, *Proc. Nalt. Acad. Sci. USA* **2004**, *101*, 11929.
2. Schmitz, F. J., Hollenbeak, K. H., Vanderah, D. J., *Tetrahedron* **1978**, *34*, 2719.
3. Fenical, W., Schulte, G. R., Finer, J., Clardy, J., *J. Org. Chem.* **1978**, *43*, 3628.
4. Hayasaka, K., Ohtsuka, T., Shirahama, H., Matsumoto, T., *Tetrahedron Lett.* **1985**, *26*, 873.
5. Shirahama, H., Hayano, K., Kanemoto, Y., Misumi, S., Ohtsuka, T., Hashiba, N., Furusaki, A., Murata, S., Noyori, R., Matsumoto, T., *Tetrahedron Lett.* **1980**, *21*, 4835.
6. Paquette, L. A., Ham, W. H., Dime, D. S., *Tetrahedron Lett.* **1985**, *26*, 4983.
7. Gadwood, R. C., Lett, R. M., Wissinger, J. E., *J. Am. Chem. Soc.* **1986**, *108*, 6343.
8. (a) Feldman, K. S., Wu, M. J., Rotella, D. P., *J. Am. Chem. Soc.* **1989**, *111*, 6457. (b) Feldman, K. S., Wu, M. J., Rotella, D. P., *J. Am. Chem. Soc.* **1990**, *112*, 8490.
9. Molander, G. A., Eastwood, P. R., *J. Org. Chem.* **1995**, *60*, 4559.
10. Fürstner, A., Langemann, K., *J. Org. Chem.* **1996**, *61*, 8746.
11. Fort, A. W. *J. Am. Chem. Soc.* **1962**, *84*, 4979.
12. (a) Rigby, J. H., Pigge, F. C., *Org. React.* **1997**, *51*, 351. (b) Cha, J. K., Oh, J. *Curr. Org. Chem.* **1998**, *2*, 217. (c) Mann, J., *Tetrahedron* **1986**, *42*, 4611.
13. (a) Harmata, M. *Tetrahedron* **1997**, *53*, 6235. (b) Harmata, M. *Adv. Cycloaddit.* **1997**, *4*, 41. (c) Harmata, M. *Acc. Chem. Res.* **2001**, *34*, 595.
14. Noyori, R., Nishizawa, M., Shimizu, F., Hayakawa, Y., Maruoka, K., Hashimoto, S., Yamamoto, H., *J. Am. Chem. Soc.* **1979**, *101*, 220.

15. (a) Wang, Y., Arif, A. M., West, F. G., *J. Am. Chem. Soc.* **1999**, *121*, 876. (b) West, F. G., Hartke-Karger, C., Koch, D. J., Kuehn, C. E., Arif, A. M., *J. Org. Chem.* **1993**, *58*, 6795.

16. Harmata, M., Elomari, S., Barnes, C. L., *J. Am. Chem. Soc.* **1996**, *118*, 2860.

17. (a) Hoffmann, H. M. R., Wagner, D., Wartchow, R., *Chem. Ber.* **1990**, *123*, 2131. (b) Samuel, C. J., *J. Chem. Soc., Perkin Trans. 2* **1981**, *4*, 736. (c) Hirano, T., Kumagai, T., Miyashi, T., Akiyama, K., Ikegami, Y., *J. Org. Chem.* **1991**, *56*, 1907.

18. Harmata, M., Carter, K. W., Unpublished results from this laboratory

19. Giguere, R. J., Tassely, S. M., Tose, M. I., *Tetrahedron Lett.* **1990**, *31*, 4577.

20. Marx, J. N., Norman, L. R., *J. Org. Chem.* **1975**, *40*, 1602.

21. (a) Ohta, T., Miyake, T., Seido, N., Kumobayashi, H., Akutagawa, S., Takaya, H, *Tetrahedron Lett.* **1992**, *33*, 635.   (b)   Ohta, T., Miyake, T., Seido, N., Kumobayashi, H., Takaya, H., *J. Org. Chem.* **1995**, *60*, 357.

22. Enders, D., Gröbner, R., Runsink, J., *Synthesis* **1995**, 947.

23. Enders, D., Fey, P., Kipphardt, H., *Org. Synth.* **1987**, *65*, 173.

24. Evans, D. A., Ennis, M. D., Mathre, D. J., *J. Am. Chem. Soc.* **1982**, *104*, 1737.

25   Boss, R., Scheffold, R., *Angew. Chem., Int. Ed. Engl.* **1976**, *15*, 5582.

26 . Scheffold, R., Saladin, E., *Angew. Chem., Int. Ed. Engl.* **1972**, *11*, 229.

27. (a) Krapcho, A. P., *Synthesis* **1982**, 805. (b) Krapcho, A. P., *Synthesis* **1982**, 843.

28. Wender, P. A., Holt, D. A., *J. Am. Chem. Soc.* **1985**, *107*, 7771.

29   Harmata, M., Rashatasakhon, P., *Tetrahedron Lett.* **2002**, *43*, 3641.

30. Martin, J. C., Arhart, R. J., *J. Am. Chem. Soc.* **1971**, *93*, 4327.

31. Trost, B. M., Jungheim, L. N. *J. Am. Chem. Soc.* **1980**, *102*, 7910.

32. (a) Eilingsfeld, H., Seefelder, M., Weidiniger, H. *Angew. Chem.* **1960**, *72*, 836. (b) Marson, C. M., Giles, P. R. *Synthesis Using Vilsmeier Reagents*, CRC: Boca Raton, 1994.

33. (a) Denney, D. B., Sherman, N., *J. Org. Chem.* **1965**, *30*, 3760. (b) Kienzle, F., Holland, G. W., Jernow, J. L., Kwoh, S., Rosen, P., *J. Org. Chem.* **1973**, *38*, 3440. (c) Danishefsky, S., Tsuzuki, K., *J. Am. Chem. Soc.* **1980**, *102*, 6893. (d) Majetich, G., Hull, K., *Tetrahedron* **1987**, *43*, 5621.

INDEX

465

Q

quinoxapeptins, 3, 4

R

radical cyclization chemistry, 141
Raney nickel, 255
recognition properties, 95
reductive amination, 408
remediation of textile waste streams, 72
retro-aldol, 10, 291
retro-Michael reaction, 150
retro-Michael, 221
retrosynthetic analysis, 74, 162, 214, 225, 231, 232, 421, 449
reverse "5-endo-trig" process, 19
RhCl₃, 441
rifamycins, streptovaricins, 39
ring closing metathesis, 369
ring opening of A 1,3-dioxolanone, 110
ring-closing metathesis (RCM) reaction, 363
ring-closing metathesis, 102, 106, 441
RNA-dependant DNA polymerases, 42
RNAi, 354
Roche esters, 179
Ru(II)-(S)-BINAP, 184

S

salcomine, 64
salt of DCC and MeI, 453
samarium diiodide, 167
samarium(II) iodide (SmI₂), 137
samarium(II) iodide, 182
SAMP or RAMP lactone hydrazones, 451
sandramycin, 3
scale-up, 274
Schotten-Baumann conditions, 18
Schreiber ozonolysis, 15
Schrock catalyst, 441
Schrock's molybdenum complex, 104, 122
Schwartz reagent, 293
secondary orbital interactions, 139
second-generation Grubbs catalyst, 371
(+)-secosyrins 1 and 2, 212
Schizandra chinensis, 155
selective iodination, 364

selective tosylation, 375
self-assembly, 74, 75, 83, 85
self-association constant, 86
self-association, 83, 85
self-sorting, 83, 87, 97
serine inversion, 8
serine, 7
serine/threonine type 1 (PP1) and type 2A (PP2A) protein phosphatases, 354
serinylation of piperazic acids, 17
SES chloride, 378
SES deprotection protocol, 383
Sharpless AE (asymmetric epoxidation), 53
Sharpless asymmetric epoxiation, 250, 257
Sharpless catalytic RuO₄ oxidation, 11
Sharpless catalytic RuO₄ procedure, 9
Sharpless asymmetric dihydroxylation, 47
Sharpless-Kresze allylic amination, 360
sialic acid, 410
[2,3]-sigmatropic rearrangement, 359
signalling pathway, 390
silicon-assisted intramolecular cross-coupling reaction, 102
siloxane, 106
silver cyanide, 378
silver nitrate, 116
silver or manganese oxide, 59
2-silyl-5-bromothiazole, 428
silylation, 104
Simmons-Smith cyclopropanation, 456
Simmons-Smith reaction, 455
six-centered transition state, 264
Smiles-type rearrangement, 7
SOCl₂, 19, 159
sodium azide, 365, 375
sodium borohydride, 416
sodium hexamethyldisilazide, 290
sodium hydride, 372, 375, 383
sodium in liquid ammonia, 440
sodium iodide, 379
sodium phenylselenide, 458
sodium trifluoroethoxide, 447
solid phase synthesis, 25
Sonogashira conditions, 409
Sonogashira coupling, 411
spartadienedione, 144
spontaneous cyclodimerization, 23
S-shaped and C-shaped diastereomers, 83
stainless steel reactor, 283